CITRUS FRUIT

Dedicated

To

All scientists and technologists
Who have made today's
Advancements possible
In citrus industry
Around the World

CITRUS FRUIT

BIOLOGY, TECHNOLOGY AND EVALUATION

Milind S. Ladaniya
Principal Scientist (Horticulture)
ICAR Research Complex for Goa
Ela, Old Goa 403 402
Goa, India

ELSEVIER

Amsterdam • Boston • Heidelberg • London • New York • Oxford
Paris • San Diego • San Francisco • Singapore • Sydney • Tokyo

Academic Press is an imprint of Elsevier

Academic Press is an imprint of Elsevier
525 B Street, Suite 1900, San Diego, CA 92101-4495, USA
30 Corporate Drive, Suite 400, Burlington, MA 01803, USA
84 Theobald's Road, London WC1X 8RR, UK

First edition 2008

Library of Congress Cataloging-in-Publication Data
A catalog record for this book is available from the Library of Congress

British Library Cataloguing in Publication Data
A catalogue record for this book is available from the British Library

ISBN: 978-0-12-374130-1

For information on all Academic Press publications
visit our web site at books.elsevier.com

Printed and bound by CPI Group (UK) Ltd, Croydon, CR0 4YY

Transferred to Digital Print 2011

CONTENTS

6 Fruit Biochemistry 125

7 Growth, Maturity, Grade Standards, and Physico-Mechanical Characteristics of Fruit 191

10 Packaging 287

11 Precooling and Refrigeration 319

12 Storage Systems and Response of Citrus Fruits 333

Note: The trade names are used in this publication solely for the purpose of providing specific information. It is neither a guarantee or warranty of the products named nor the author and publisher want to endorse them. Inclusion of specific names in this publication does not signify that they are approved to the exclusion of others of suitable composition.

PREFACE

Citrus fruits rank first in the world with respect to production among fruits. They are grown commercially in more than 50 countries around the world. In addition to oranges, mandarins, limes, lemons, pummelos, and grapefruits, other citrus fruits such as kumquats, Calamondins, citrons, Natsudaidais, Hassakus, and many other hybrids are also commercially important. The contribution of the citrus industry to the world economy is enormous (estimated at more than 10 billion US$ annually) and it provides jobs to millions of people around the world in harvesting, handling, transportation, storage, and marketing operations.

Citrus fruit production recorded a handsome increase during the 1990s, and recently reached 100 million tons. Considering the therapeutic value of these fruits and the general health awareness among the public, citrus fruit are gaining importance worldwide, and fresh fruit consumption is likely to increase.

Postharvest biology and technology has evolved into a branch of science that combines biology and engineering. It has evolved rapidly over the past four or five decades, although scattered research efforts in various aspects of this field have been made previously all over the world. Increased citrus production combined with concern about growing population accelerated research and stimulated the development of new technologies in basic and applied areas.

With more and more emphasis by researchers on extending shelf-life and preserving the natural qualities of fresh produce to reduce losses and cater to the needs of ever-demanding, quality-conscious consumers, great advances have been possible. Now state-of-the-art postharvest handling technologies are routinely used in packinghouses. Efforts are continuing to develop and integrate technologies to ensure that all the fruit that leaves the packinghouse meets 100 percent quality standards, without any defects. Non-invasive or non-destructive evaluation techniques are in vogue now.

A comprehensive, single source on the subject that presents all related aspects such as postharvest biology, handling technology/machinery, organic and inorganic techniques of disease and pest management, marketing, quality evaluation, and biotechnology has been lacking. In fact, comprehensive books on the subject are very scarce. Available books are either very old or dwell mostly on mechanical handling aspects and the related issues concerning few developed countries. Some literature focuses exclusively on chemical constituents of citrus.

There have been considerable advances in the research on fruit physiology, biochemistry, handling, coatings, packaging, storage systems, and therapeutic value during last two or three decades. These advances call for an overall review of the subject. In this book, I have endeavored to present biology, technology, and all related aspects of fresh fruit. The information on biochemical compounds and how their reactions result in qualitative changes in citrus fruit is vital for students of postharvest science. The physical attributes of fruit determine their response to handling machines, and therefore machines have to be developed keeping physical and chemical attributes in mind. Postharvest management requires the proper blending of knowledge of biology and engineering/technology. In this book, with the help of about 1400 references, numerous tables, diagrams, and color plates, the subject matter is presented in 22 chapters. In addition to earlier findings, the book covers many new developments that have taken place between 1985 and 2007. The basic and applied scientific information is interwoven to make the reading interesting and comprehensible. Efforts have been made to provide information in totality. A vast group of readers, including students, teachers, nutritionists, trainers, quality inspectors, fruit handlers, transporters, exporters, growers, and postharvest technologists will find it useful.

Though this book contains some of my work at National Research Centre for Citrus, Nagpur, India, which I did during the past 20 years, I have tried to present existing scenario of citrus in the developing as well as the developed world. The subject matter is quite vast; page limitations do not allow me to cover all the fundamental and applied aspects in each topic. Nevertheless, it has been a great experience and a renewed education for me to write this book. I can only hope that I have made this subject easier to understand and interesting to read for others.

The introduction presents an overview of postharvest biology, technology, and quality evaluation. Knowledge of postharvest handling is incomplete unless one gains insight into postharvest losses – the extent and types of losses, their causes, and the stages of handling at which they occur. Pre-harvest factors have bearing on the postharvest quality and shelf life of fruit. Knowledge of fruit morphology, anatomy, physiology, and biochemistry is a prerequisite for thorough understanding of postharvest science. Proper fruit growth and physiological maturity determine fruit characteristics and also influence grade and internal quality. Therefore, separate chapters are devoted to these various aspects.

Exhaustive information is provided on fresh citrus varieties grown in different countries of the world with different climates, regions, and harvesting seasons.

The chapter on preparation for the fresh fruit market is quite extensive, since tremendous developments have taken place during past few decades in degreening, mechanical handling operations, and the types of coatings applied to fruit. Packaging containers and pre-packaging for the retail market (shrink-film packing and MAP) are discussed in detail. Precooling, storage systems, and transportation with cool-chains for fresh fruit are all vital aspects that are covered in separate chapters. Different citrus-growing countries have their own domestic systems of fruit distribution and hence, domestic market chains and the role of intermediaries/functionaries are presented separately.

Researchers have a considerable interest in physical and biological methods to control postharvest diseases. Organic management is a burning topic today. Organic methods (also called green, non-hazardous, and eco-friendly) to manage postharvest citrus spoilage are also discussed in detail along with chemical methods. Postharvest insect-pest problems, especially those caused by fruit flies are important from a quarantine point of view. Irradiation and its effects on citrus fruits are covered and the potential of irradiation in fruit-fly control is outlined.

There are great technological advances with respect to on-line quality evaluation for sorting and grading in packinghouses utilizing non-destructive, high-speed techniques; this information also finds its due place in the book. The quality assurance and control systems being used for fresh citrus fruit, as well as food-safety issues, are also discussed. Analytical methods and instruments required in postharvest management of fresh citrus are given. With the establishment of the WTO, the fresh fruit trade has become more competitive and SPS requirements and quarantine treatments have gained importance. The trade potentials, varietal preferences, and future prospects are outlined in the last chapter.

Any discussion of fresh citrus fruit remains incomplete without mentioning nutritive and medicinal value and the role of fibers, flavonoids, limonoids, pectins, minerals, and vitamins in the light of clinical findings. Information on the medicinal value of Bael (*Aegle marmelos*), which is a family relative of citrus, is included. Citrus and Bael find a place in the system of Ayurvedic medicine practiced in Asia, particularly in India since, ancient times.

There is a wide scope for biotechnological applications to citrus-fruit quality improvement. This is an emerging field of research that promises to improve postharvest fruit quality. There could be separate publication on this aspect and hence covered briefly in this book.

Practical tips for postharvest management, commercial products, and their sources particularly of coatings and plastic films, as well as an exhaustive glossary of terms are also given. The unit conversion tables are given in annexures.

Thermal properties of citrus fruits and insulating and construction materials will be an instant source of information for users of this book.

This book is under continuous review and I would greatly appreciate constructive comments to improve it further.

Milind S. Ladaniya
ICAR Research Complex
Ela, Old Goa
June 2007

ACKNOWLEDGMENTS

I am thankful to Dr. Autar K. Mattoo, Professor, Sustainable Agricultural Systems Laboratory, The Henry A. Wallace Beltsville Agric. Research Centre, Beltsville, MD, USA, for his critical comments and suggestions on chapters of Postharvest diseases, Medicinal and nutritive value of citrus, Quality evaluation and Biotechnological applications. I greatly appreciate valuable suggestions of Dr. Zora Singh, Associate Professor (Horticulture), Curtin Horticulture Research Laboratory, Curtin University of Technology, Perth, Western Australia, on the topic of Biotechnological applications. He has been always ready to extend his co-operation on scientific matters. I am also highly grateful to Dr. Mario Schirra, Professor, C.N.R. Instituto di Scienzedelle Produzioni Alimentari – Sassari, Localita Palloni, Oristano, Italy, for sharing his valuable data on storage aspects, suggestions on fruit fly management and also for reprints of his publications. I take this opportunity to thank Dr. M.S. Chinnan, Professor of food engineering, University of Georgia, Griffin, GA and Dr. Randolph M. Beaudry, Professor and Acting Chair, Department of Horticulture, MSU, MI, USA, who have spared their valuable time and shared views on modified atmosphere packaging. I am grateful to Dr. Mrs. P. N. Shastri, Head, Department of Food Science and Technology, Laxminarayan Institute of Technology, Nagpur, for going through chapter on Fruit Biochemistry and offering valuable suggestions. I am also thankful to Helen Ramsay,

Development officer, Department of Agriculture and Food, Western Australia, Australia, for sharing views about Australian citrus industry. I am greatly indebted to the technical editors of this book for their valuable and constructive comments on the text.

My teachers Dr. B.S. Dhillon, Professor of Horticulture and former Dean, Post-graduate studies, PAU, Ludhiana and Dr. Daryl Richardson, Professor of Horticulture (Postharvest Technology), OSU, Oregon, USA, and also Dr. K.L. Chadha, former DDG, Horticulture, ICAR, India, have been the source of guidance and inspiration for me in the profession of Horticulture.

The arduous task of completing this comprehensive book, that too single handedly, would not have been achieved without constant moral support from my family – my wife Mamta, my sons Makrand and Mayur and especially my mother – Mrs. (Sau) Sunder Ladaniya who is also my ideal for perseverance, sacrifice, and high values in life.

I express my gratitude for the editing advice and assistance in publication given by Christine A. Minihane, Acquisition Editor, Elsevier. She was associated with this project since beginning and played major role in shaping this book. I also thank Carrie Bolger, Development editor and Ramesh Munuswamy, Project manager, and all those concerned with this project in Elsevier, Academic Press for producing this book with care and high standards.

Milind S. Ladaniya
ICAR Research Complex for Goa
India

1

INTRODUCTION

I. CITRUS FRUIT PRODUCTION AND PROSPECTS

The role of citrus fruits in providing nutrients and medicinal value has been recognized since ancient times. Citrus fruits, belonging to the genus *Citrus* of the family Rutaceae, are well known for their refreshing fragrance, thirst-quenching ability, and providing adequate vitamin C as per recommended dietary allowance (RDA). In addition to ascorbic acid, these fruits contain several phytochemicals, which play the role of neutraceuticals, such as carotenoids (lycopene and β-carotene), limonoids, flavanones (naringins and rutinoside), and vitamin-B complex and related nutrients (thiamine, riboflavin, nicotinic acid/niacin, pantothenic acid, pyridoxine, folic acid, biotin, choline, and inositol). The flavonoids from citrus juices, particularly those from oranges and grapefruit, are effective in improving blood circulation and possess antiallergic, anticarcinogenic, and antiviral properties (Filatova and Kolesnova, 1999). Fresh grapefruits, pummelos, and oranges also provide fiber and pectin, which are known to reduce the risk of heart attacks if taken daily in the diet. Fresh citrus fruit consumption is important because the nutrients and health-promoting factors (especially antioxidants) from these sources are immediately available to the body and the loss of nutrients is negligible compared with processed juices.

Citrus is the most widely produced fruit, as a group of several species, and it is grown in more than 80 countries (Chang, 1992). Citrus cultivation not only is remunerative, but it also generates employment, and as stated previously, fruits have nutritive and therapeutic value. Citrus stands first among fruit crops in the world with respect to production leaving behind grapes, apples, and banana. World citrus production increased at the rate of 4.5 percent every year during the 1990s, which resulted in an output of 98.35 million tons during 2001–02, and that figure crossed the 100 million tons mark during 2003–04 (FAO, 2006). Almost half of this production was in the Americas (North and South), and about 10–12 percent was in Europe (Mediterranean). Production trends indicate that oranges constitute about 60 percent of the total citrus output, followed by mandarins, clementines, satsumas, and tangerines, which comprise about 20 percent of the output. The group of lemons and limes constitute 11–12 percent, and grapefruit and pummelos comprise roughly 5–6 percent. Since the 1970s production increased and doubled by 2004–05, and this increase was mainly due to increased orange production (Tables 1.1 and 1.2). Nearly 68.16 million tons of citrus fruits have been consumed as fresh produce (either domestically or by exports), while 26.63 million tons were utilized for processing out of 94.79 million total tons produced worldwide during 2004–05 (FAO, 2006). This trend clearly indicates importance of preserving the natural qualities of fresh citrus fruits after harvest for either domestic markets or export. World population is slated to be nearly 10 billion by 2050, therefore, projected citrus production has to increase at a suitable growth rate to meet this increasing demand.

Brazil leads in citrus production, with more than 18.90 million metric tons of fruit produced during 2004–05, followed by the United States and China. Brazilian citrus production is oriented toward processing, while U.S. citrus production is focused toward processing and the fresh fruit market. China's citrus production is growing steadily, with 1.3 million ha under citrus in 2001 (Xin, 2001) and new plantations yet to bear fruit. Total citrus production is expected to exceed 18.2 million tons in coming years in this country. China leads in tangerine/mandarin production, with more than 8.6 million-ton output. Production of high-quality navels and Ponkan will increase, with pummelos maintaining its

TABLE 1.1 Citrus Output of Some Major Producing Countries and the World (million tons)

Country	1970s (average)	1980s (average)	1990s (average)	2000–01	2001–02	2002–03	2003–04	2004–05
Brazil	7.39	11.67	16.90	16.49	19.91	17.73	21.39	18.90
USA	11.59	10.29	12.65	14.04	14.09	13.73	14.78	10.49
China	0.65	1.70	7.36	9.20	12.03	12.46	13.88	15.22
Mexico	1.90	2.48	4.42	6.14	6.35	6.08	6.58	6.91
Spain	2.68	3.47	5.03	5.40	5.75	5.94	6.23	6.18
India	1.72	1.89	3.30	4.31	4.39	4.63	4.66	4.66
Italy	2.68	3.10	3.15	3.01	3.01	2.76	2.75	3.32
Egypt	0.90	1.40	2.20	2.50	2.90	2.48	2.31	2.70
Total world production	47.13	57.77	81.12	89.70	98.35	94.08	100.85	94.79

Source: FAO (2003), FAO (2006).

proportion of production. Mandarins include satsumas (early) and Ponkan, and account for 55–60 percent, followed by oranges at 30 percent. Seedless pummelos, kumquats, and other citrus constitute 10 percent. About 95 percent of fruit is consumed fresh in China.

In the United States, Florida is the major orange-producing state (Table 1.3), and most of that produce is used for juice processing. Citrus fruits from California, Arizona, and Texas are sold for fresh consumption due to their excellent quality. Florida is a major grapefruit producing state, followed by California and Texas, and a considerable quantity of the produce is exported. Florida also leads in the production of tangerines, tangelos, Temples, and acid limes.

TABLE 1.2 Sweet Orange Production *Vis-à-Vis* Other Citrus Produced in Major Citrus Growing Countries of the World (million tons)

	1970s (average)		1990s (average)		2004–2005	
Country	Orange	Other citrus	Orange	Other citrus	Orange	Other citrus
Brazil	6.93	0.46	15.71	1.15	16.56	2.34
USA	8.75	2.84	9.88	2.77	8.41	2.08
China	0.349	0.26	1.79	5.60	4.46	10.76
Mexico	1.32	0.58	3.1	1.32	4.30	2.61
Spain	1.86	0.8	2.65	2.38	2.83	3.35
India	–	1.22	0.75	2.20	1.3	3.30
Italy	1.58	1.1	1.98	1.17	2.10	1.22
Egypt	0.77	0.13	1.54	0.66	1.75	0.95
Total world production	32.35	14.78	54.33	26.79	59.04	35.75

Source: FAO (2003), FAO (2006).

TABLE 1.3 Production of Citrus in the U.S. during 2001–02 ('000 boxes)

Citrus fruit	Arizona	California	Florida	Texas
Tangerines	620	2200	6600	–
Tangelo	–	–	2150	–
Temples	–	–	1550	–
Limes	–	–	150	–
Lemons	2800	19 000	–	–
Grapefruit	160	6000	46 700	5900
Oranges	270	3400	128 000	1530

Source: USDA National Agricultural Statistical Service (2003).

Citrus production in India is 4.575 million tons from 0.488 million ha under this crop (Anonymous, 2004). The production of citrus fruits is 10.39 percent of the total fruit production in the country. About 31 000 tons of fresh citrus were exported (APEDA, 2002–03), worth Rs. 34.07 crore (340 million rupees). The most important citrus fruits grown on a commercial scale in India are mandarins such as 'Nagpur,' 'Coorg,' 'Khasi,' 'Darjeeling' (*Citrus reticulata* Blanco), and 'Kinnow' (King × Willow leaf). Sweet oranges, such as 'Sathgudi,' 'Mosambi,' 'Jaffa,' 'Blood-Red,' and 'Valencia Late' (*Citrus sinensis* Osbeck), are the second most important citrus fruit in the country. Acid limes, such as 'Kagzi' (*Citrus aurantifolia* Swingle), and lemons, such as 'Assam lemon' and 'Galgal,' or Hill, lemon (*Citrus limon [L]*) Burm F., are important acid fruits. Grapefruits and pummelos are grown in small plantations and homestead gardens only. Fruits are mostly hand-picked, as more than 95 percent produce is sent to fresh fruit market.

Mediterranean countries produce huge amounts of citrus fruits destined for the fresh fruit market. Clementine mandarin and related fruit from Morocco and Spain dominate the easy-peelers category. The Mediterranean region exported 2.24 million tons of tangerines/mandarins/satsumas out of total exports of 3.13 million tons of these types of citrus in the world during 2004–05 (FAO, 2006).

Satsumas of Japan and red-fleshed grapefruits from South Africa, the United States, and Israel also have a strong presence in world trade. The Washington Navel with its clones, and Valencia are the major sweet-orange varieties worldwide. Navels from California and Spain, blood-red oranges from Italy, 'Shamouti' from Israel, 'Pera,' and 'Natal' from Brazil, and 'Jincheng' from China are popular with consumers.

Due to economic developments and the changing lifestyle of people in many cultures, fresh fruit consumption is increasing in the category of easy-peelers, such as tangerines/mandarins. These fruits are preferred for convenience in eating as they are seedless and small. Consumption of fresh oranges grew at the rate of 2.9 percent from 1986–88 to 1996–98. China is fast developing economy and with increasing income of the people, fruit consumption is also increasing. In populous areas, such as China, India, African regions, and most other Asian countries, citrus fruits are consumed fresh and are supplied by local citrus industries. In Central and Latin America, large segments of the population still prefer and can afford to consume only fresh citrus. Most of the increases in the consumption of domestic production are in India, Pakistan, China, Mexico, and Brazil. Although production of citrus fruits is increasing at the rate of 2–5 percent annually, the per capita annual consumption varies from 40 kg in Europe to 4 kg in Asian and African countries.

The 2010 orange production is projected at 64 million metric tons, and 35.7 million metric tons of that is expected to be consumed fresh. Production in Spain and South Africa would continue to grow. World tangerine/mandarin production is projected to be more than 20 million metric tons by 2010. Major producers of tangerines are China, Spain, India, Japan, Italy, Turkey, Egypt, Morocco, Argentina, and Pakistan. Spain alone accounts for more than 50 percent of the world's export of fresh tangerines. Tangerine production is expected to expand in Spain, China, Morocco, Brazil, and Argentina. Efforts are being made to produce clementines in California. Israel, the United States, South Africa, Spain, Australia, and Cuba are important suppliers of grapefruit. As grapefruit production is better suited to tropical climate, its production is scattered in countries near the equator in Latin America, Africa, and Asia. Projected grapefruit production in 2010 is more than 5.5 million metric tons (Spreen, 2001), and it is expected to increase in Cuba, Mexico, Argentina, and South Africa.

Lemons are widely produced in the cooler climate of southern California, which is free of frosts, and also in Spain, Italy, and Argentina. Lemons are adapted to the dryer climates of Egypt and Iran, while limes are grown mainly in Mexico, India, Brazil, the United States (Florida), and other countries with a tropical climate. With the present world production of 11.68 million metric tons of lemons and limes, the future production of these fruits is projected at 12 million metric tons by 2010. Lemon and limes do not face the same competition from other fruits (grapes, apples, bananas, etc.), because these are used primarily for salad garnishing and processing purposes and not as dessert fruits, like oranges and tangerines.

Citrus farming is being initiated in new areas as production in the old areas is decreasing due to rapid urbanization, diseases, and the abandonment of plantations injudiciously farmed (i.e., plantations on which water and soil resources were not conserved). Frost damage is a major threat for fruit production in cooler subtropical areas. As the costs of land, labor, and fuel rise, new technologies will have to be developed to increase the productivity of land for fruit to be competitive in the world market. Even in domestic markets, citrus fruit quality and price have to be competitive with other fruits. Increasing labor problems will lead to the development of better abscission chemicals and of robotic harvesting machinery, particularly in America and Mediterranean countries.

The prime concern in increasing production of all citrus fruits is the obstacles posed by biotic and non-biotic (environmental) stresses. The second concern is reducing dependence on agrochemicals during production through eco-friendly *good agricultural practices* (GAP). Nonhazardous postharvest management practices would result in safe food, and efforts are on to develop these methods.

New tools of biotechnology are being used extensively to develop varieties of fruit that can tolerate biotic and abiotic stresses in the field and still produce higher yields of better-quality fruit. Considering consumers' preferences worldwide, important objectives in fresh citrus fruit improvement programs are the production of seedless fruits with optimal size and shape, an easily removable peel, new organoleptic characteristics (flavor), and early or late ripening. Modern biotechnologies based on *in-vitro* cell, protoplast, tissue and organ culture, and gene introduction, in addition to conventional breeding methods, such as selecting clones from natural or induced mutations, nucellar selection, and hybridization, are being tried to improve the quality of existing cultivars and to produce new varieties. The modern tools of molecular biology should throw more light on the functions of enzymes, their pathways, and the genes controlling them. Using genetic engineering techniques, metabolic pathways can be modified with certain enzymes, forming desired metabolites and eliminating problems related to the flavor and eating quality of fruits. Studies to understand the molecular mechanisms and changes underlying resistance to chilling injury in some citrus fruits after certain treatments may lead to developing genetically engineered, low temperature-resistant cultivars (Sanchez-Bellesta et al., 2001). Similarly, the study of molecular changes and the mechanisms of maturation and senescence may lead to developing genetically modified cultivars that age slowly.

Citrus production faces varying problems in different regions of the world, and fruit quality varies with agro-climatic conditions. In subtropical regions, under arid conditions with low humidity, fruit quality is excellent, with very few blemishes on the fruit's surface, and pack-out can be as high as 95 percent if fruit meets the size requirements. In tropical climates with high humidity, however, pack-out can be less (50 percent of the produce harvested) because of blemishes on the fruit's surfaces. This is evident from the differences in the produce of Florida and California, two well-known citrus-growing regions, and of some areas of Israel, Spain, Italy, South Africa, and Australia. In tropical areas of India and many other Southeast Asian countries and Brazil, the incidence of fruit surface blemishes is high, and fruit rind color remains green even when the fruit is internally mature. In the cool climates (subtropical) and arid conditions of northwestern India, fruit quality is excellent with respect to color, size, and taste. Due to variation in fruit quality, fruit grades and internal standards differ among the domestic markets of many countries. The draft codex standards of Food and Agricultural Organization (FAO) are being evolved through discussions and consensus for world trade. Similarly, rules and regulations under sanitary and phytosanitary (SPS) treaties have been finalized and are being discussed under the new World Trade Organization (WTO) regime.

II. POSTHARVEST OVERVIEW

With increase in production, the increase in availability of the fruit and its supply involves preservation in terms of quality (external and internal including nutritional) and quantity. Breeding programs are leading to better color, unique flavors, and other enhancements, such as juiciness and TSS:acid ratio, but the real challenge for fresh fruit handlers and postharvest technologists is to retain or even enhance the 'garden-fresh' characteristics of the fruit.

Postharvest management starts in the orchard. Cultural practices must be carefully monitored to produce optimum results (i.e., required quality fruit). Careful harvesting and handling hold the key to getting desired results from postharvest treatments. Fruit's postharvest quality and shelf life are becoming increasingly important aspects, as consumers expect quality fruit to be available throughout the year. The need for proper postharvest management increases further in case the supply is short in some seasons and produce has to be utilized and distributed prudently, without losses. Losses during fresh fruit handling range from

5 to 10 percent in most developed countries, but from 25 to 30 percent or more in developing and underdeveloped countries. Although citrus fruits have a relatively long postharvest life compared with other tropical and subtropical fruits, the losses of these fruits are higher in most developing countries because the fruits are handled, marketed, and stored under ambient conditions with insufficient refrigeration. In developed countries, after harvesting and postharvest treatments, fruits enter into cool-chain and remain at lower temperature until they are consumed.

Handling methods, storage facilities, and marketing structures vary greatly among the citrus producing countries. Since the early twentieth century, packinghouse operations in many developed countries have included degreening, mechanized sorting, washing, waxing, and size grading. In many developing countries, however, conventional handling practices are still in effect, with only gradual changes toward mechanization and modernization. The concept of packinghouse methodologies, including degreening using ethylene gas and packing wax-coated fruit on a packing line, was introduced in Indian citriculture during late 1980s and early 1990s (Ladaniya et al., 1994). Many packinghouses in the country now process 'Kinnow' and 'Nagpur' mandarins. Nonetheless, the bulk of produce is handled with conventional procedures: Fruit is transported unpacked (loose) in trucks to distant markets for economic reasons. Acid limes are packed in gunny bags, each holding 500–600 fruits, and sent to distant markets by rail and road transport. The road transport is predominant in the postharvest handling of citrus. Refrigerated cool storage of mandarins and oranges is used for commercial storage in India's metropolitan cities. The evaporative cool-storage structures are being popularized for small scale, short storage at the farm level. The required infrastructure of mechanical handling and refrigerated transport is being developed for domestic and export marketing. Similar developments have recently been seen in many developing countries, including China.

Research and development in the field of postharvest management requires knowledge of the anatomical, physiological, and biochemical aspects of citrus fruits. This information helps researchers predict fruit's postharvest behavior, acceptability, and palatability, and is critical when fruit is harvested, packed, and transported for consumers in a large-scale commercial ventures that involve huge sums of money. Storage is the most important part of the marketing process, and all production operations, including harvesting, pre- or postharvest treatments, packaging, transportation, and temperature and humidity management during handling, influence the storage life of fruits.

Fruit senescence, physiological disorders, and diseases are the major causes of postharvest losses. The biochemical bases of physiological disorders and senescence are continuously studied. Similarly, the biochemical basis of chilling injury (CI) also is examined in depth. Recent findings indicate that membrane lipids and oxidative stress are related to chilling injury. These findings may lead to the development of suitable treatments to minimize this disorder in the coming years.

Although citrus fruits are non-climacteric and have a relatively long shelf life compared to mangos, bananas, and other tropical fruits, the dry, hot climate in most parts of the tropical world render fruit unsuitable for marketing if not properly handled and stored. Ethylene has a role in the ripening and abscission of citrus and needs to be used wisely to benefit consumers and handlers. Removing ethylene from the fruit storage environment is challenging, however, its removal extends shelf life. Other plant-growth regulating substances, such as cytokinins, auxins, gibberellins, and polyamines, also profoundly affect fruit quality. Understanding the mechanism of action and key processes regulated by the plant regulators and bio-regulators may lead to generate strategies for modifying fruit characteristics to improve quality and delay senescence. An ethylene inhibitor, 1-methyl cyclopropane (1-MCP), that apparently binds to the cellular ethylene receptor has been found to reduce chilling injury, stem-end rot, and volatile off-flavor in 'Shamouti' sweet oranges. This nontoxic, odorless compound also delays color development, so it has been recommended for use in green citrus fruits (Porat et al., 1999).

Wax coating, in different forms, has been practiced for centuries to extend the storage life of citrus fruits. Aqueous waxes, solvent waxes, wax solutions of different compositions reduce water loss and give fruit sheen. Fruit flavor is the critical parameter to be monitored after coating application. Alternative coatings known as edible coatings, which utilize sucrose esters, lipids, proteins, and cellulose, have been developed to enable packers to avoid the use of hazardous chemicals (Krochta et al., 1994; Khanrui, 1999).

Modified atmosphere packaging (MAP)/wrapping and air-tight sealing have shown great promise in extending fruit's shelf life and can serve as an alternative to refrigeration. When fruit is sealed in plastic film, a modified atmosphere is created by respiration (evolution of CO_2, utilization of O_2) and transpiration (water vapor causes saturated atmosphere). If the fruit is in equilibrium with its environment, the rate of gas exchange is the same in both directions. MAP also changes the concentration of respiratory gases inside the fruit. The thickness and type of film, the type of fruit, the fruit's weight, the fruit's maturity, the surrounding temperature, and microbial infection affect the modified atmosphere. When the optimization of various influencing factors is achieved for a particular citrus fruit, the shelf life can be reasonably extended without quality loss. Most plastic films are made from petroleum byproducts, but biodegradable films are being developed and hold great promise (Anonymous, 1998).

Pathogenic infection is a serious problem for fruit packers, because it results in off flavors and decay, rendering fresh fruit unsuitable for consumption and causing heavy economic losses. Unfortunately, the fungicides, insecticides, and other chemicals used to control various rots and insect-pests are toxic to human beings and pollute the environment. Many of these chemicals are in the process of being withdrawn by their manufacturers. Methyl bromide and some fungicides have already been withdrawn from postharvest use. Understanding the biological mechanism regulating plant–microbe interaction and interaction between the microbes is likely to provide means to develop control measures. The development of biocontrol agents and effectiveness of ozone has opened new avenues for eco-friendly, nontoxic, postharvest disease-control measures. High-temperature and ultraviolet-light therapy have great potential in inducing natural resistance in harvested citrus fruit. The future of eco-friendly treatments against biotic stresses lies in understanding the mechanisms of microbes, exploring new fungistatic compounds from plants, understanding the biochemical bases of changes in host tissue, and understanding host–pathogen interaction.

The prevention and cure of fruit-fly infestation is the major challenge in expanding the export of citrus fruits. Low-temperature treatment and ionizing radiation are effective in the disinfestation of crops. Irradiation at the dose of 300 Gy is lethal to all stages of *Ceratitis capitata* but does not alter the chemical composition of oranges (Adamo et al., 1996). The advantage of this treatment is the deep and uniform penetration of energy into the tissue, which guarantees disinfestation even in the center of the fruit without producing any undesirable effects. The treatment is effective in wrapped fruit also thereby preventing recontamination. In Florida, an irradiation dose of 1.0 kGy has been approved by the U.S. Food and Drug Administration (FDA) and found effective for controlling the Caribbean fruit fly in grapefruit shipments to Japan and Europe. The vapor heat treatment (2 h at 38°C) reduces the severity and incidence of peel injury by 50 percent in fruit irradiated by 0.5–1.0 kGy (Miller and McDonald, 1998). Thermotherapy (the use of very low and/or high temperatures) is useful in conditioning fruit for irradiation. Thermotherapy also is lethal to microbes and insect-pests, so it can be used to reduce decay and insect infestation.

Technological innovations in computer applications and robotics, and the development of various kinds of sensors for fruit handling, analysis, and quality control revolutionized postharvest management, making it a much more precise process than previously. Although most of the world's citrus fruit destined for fresh fruit market is harvested manually, continuing research on mechanical harvesting is leading to development of systems that cause the least damage to fruit at a lower picking cost. Various mechanical methods are being tried, including the use of shakers and robotic arms with machine vision.

Fresh citrus fruits, whether used as desserts or snack food, must compete with other products, such as dairy products and industrial baked goods, that exhibit a high consistency of quality. Thus, the high quality of fresh fruit must be made more consistent. The first step in the effort to increase quality assurance is developing sensors that are able to control product quality. These sensors could monitor external and internal quality. External quality includes fruit's size, shape, color, and surface defects. Cameras are being used more and more commonly to evaluate these characteristics. The aspects of internal quality include maturity and the absence of defects/disorder, rots, insects, and seeds, and these are being studied using X-ray transmission and tomography, sound waves, and nuclear magnetic resonance (NMR) imaging (Chen et al., 1989). The detection of degree of maturity as determined by sugar and acidity contents can be achieved with the use of near-infra-red spectroscopy (NIRS). All of these techniques involve measuring changes in transmitted electromagnetic waves according to the effect. Internal defects are by far the most damaging to a brand's reputation. In order to ensure internal quality with respect to desired total soluble solids (sugars) and acidity in satsuma mandarins, for example, a nondestructive online measurement technique utilizing near-infra-red spectroscopy in the transmittance mode is already being widely used in packing houses in Japan (Miyamoto et al., 1996).

Sensors also are being developed to monitor the quality of citrus fruits by measuring the relationship between the emission of volatile substances and the loss of quality during transportation (Conesa et al., 1994). The most frequent causes of quality deterioration during transportation are decay caused by pathogens and damage from handling. The volatiles emitted by disease pathogens (*Penicillium digitatum, Botrytis cinerea*, and *Geotrichum candidum*) are being studied in order to develop sensors capable of detecting them. Characteristic emissions from decay caused by *Penicillium digitatum*, for example, are ethanol and ethyl acetate. Research also is being done on sensors for limonene, alcohols, aldehydes (caused by aging), and ethyl acetate (Conesa et al., 1994). Two kinds of chemical sensors (tin dioxide gas sensors and polypirrol) and a biochemical sensor (enzymes) have been developed. The prototypes of these devices, called "Fruit Box" and "Fruit Sniffers," have been reported by Bellon (1994). The "Fruit Box" is made up of sensors (for temperature, humidity, and volatiles), signal circuits, and a microcontroller to record and store the data output from the sensors, thus enabling continuous monitoring of fruit condition during a shipment. Any physical damage or quality degradation in the fruit caused during transport can be identified with this device. "Fruit Sniffers" are reported to be able to check individual pallets and boxes for damaged and moldy fruit.

III. CONCLUSION

Efforts are continuing worldwide to develop methods, technologies, and gadgets to preserve the qualities of freshly harvested citrus fruit while it is taken from orchard to market shelves at distant places. For postharvest technologies to be effective, however, trials and evaluations are necessary on a commercial scale. Successful fruit-quality management requires a combined effort from producers, marketing and handling personnel, transporters, and researchers. Successful, sustainable, and increased citrus production is possible with sound postharvest management and marketing. Grade standards, international quality criteria, packaging needs, food safety norms, and sanitary and phytosanitary requirements have acquired greater significance in the post-WTO scenario, and these challenges need to be met for the healthy growth of the international fresh citrus fruit trade.

REFERENCES

Adamo, M., D'llio, V., and Gionfriddo, F. (1996). The technique of ionization of orange fruits infested with *Ceratitis capitata. Informat. Agrar.* (Italy) 52(4), 73–75.

Anonymous (1998). Biodegradable plastic films. *Indian Fd. Industry* May–June, 17(4), 35–38.

Anonymous (2004). Survey of Indian agriculture. *Hindu Publ.*, 122–123.

APEDA. (2003). Export statistics for agro and food products – India, 2002–03. Agricultural and Processed Food Products Export Development Authority, New Delhi, pp. 250–256.

Bellon, V. (1994). Tools for fruit and vegetable quality control: A review of current trends and perspectives. In *Post-harvesting operations and quality sensing. Proc. IV Int. Symp. Fruit and Vegetable Product Engineering* (F. Juste, Ed.), Vol. 2, 1–12, Valencia, Spain.

Chang, K. (1992). The evaluation of citrus demand and supply. *Proc. Int. Soc. Citric.*, Italy, Vol. 3, 1153–1155.

Chen, P., McCarthy, M.J., and Kauten, R. (1989). NMR for internal evaluation of fruits and vegetables. *Trans. ASAE*, 32(5), 1747–1753.

Concsa, E., Hoddells, P.J., and Emmanopoules, G. (1994). Citrus fruit life tracking systems. In *Post-harvesting operations and quality sensing. Proc. IV Int. Symp. Fruit and Vegetable Product Engineering* (F. Juste, Ed.), Vol. 2, 165–217, Valencia, Spain.

FAO (2003). Citrus fruit – fresh and processed, annual statistics, 2003. Commodities and Trade Division, FAO of the UN, Rome.

FAO (2006). Citrus fruit – fresh and processed, annual statistics, 2006. Commodities and Trade Division, FAO of the UN, Rome.

Filatova, I.A., and Kolesnova, Y. (1999). The significance of flavonoids from citrus juices in disease prevention. *Pishchevaya Promyshlennost* 8, 62–63.

Khanrui, K. (1999). Edible packaging: an eco-friendly alternative to the plastics. *Indi. Food Indust.* 18(1), 34–38.

Krochta, J.M., Baldwin, E.A., and Nisperos-Carrido, M.O. (1994). Edible coatings and films to improve food quality. Technomic Publication Co., Lancaster, PA, USA.

Ladaniya, M.S., Naqvi, S.A.M.H., and Dass, H.C. (1994). Packing line operations and storage of 'Nagpur' mandarin. *Indian J. Hort.* 51(3), 215–221.

Miller, W. R., and McDonald, R. E. (1998). Amelioration of irradiation injury to Florida grapefruit by pre-treatment with vapour heat or fungicides. *Hort Sci.* 33(1), 100–102.

Miyamoto, K., Kitano, Y., Yamashita, S., Honda, H., and Nakarishi, Y. (1996). Fruit quality control of Satsuma mandarin in packing houses with non-destructive measurement by Near Infra-Red Spectroscopy (NIRS). In *Proc. Int. Soc. Citric.* International Socety of Citriculture *VIII Int. Citrus Congress*, S. Africa, May 12–17, 1996, Vol. 2, 1126–1128.

Porat, R., Weiss, B., Cohen, L., Daus, A., Goren, R., and Droby, S. (1999). Effects of ethylene and 1-methyl cyclopropane on the postharvest qualities of Shamouti oranges. *Postharvest Biol. Technol.* 15, 155–163.

Sanchez-Ballesta, M.T., Lafuente, M.T., Granell, A., and Zacarias, I. (2001). Isolation and expression of a citrus cDNA related to peel damage caused by postharvest stress conditions. *Acta. Hort.* 553, 293–295.

Spreen, T.H. (2001). Projection of world production and consumption of citrus up to 2010. *Proc. China – FAO Citrus Symp.*, May 2001, 137–142.

USDA (2003). Agricultural statistics. National Agricultural Statistical Service, Utilized Production. http://www.usda.gov/nass.

Xin Lu, L. (2001). China's citrus production: retrospect, present situation and future prospects. *Proc. China – FAO Citrus Symp.*, May 2001, 151–165.

2

COMMERCIAL FRESH CITRUS CULTIVARS AND PRODUCING COUNTRIES

I. CITRUS CULTIVARS FOR FRESH FRUIT MARKET

A. Sweet Orange

1. Non-blood or Non-pigmented Orange

Fruits in this group are spherical, globose, or ovate. The important commercial cultivars in this group are as follows: • Hamlin: Fruits are almost spherical, commercially seedless (fewer than 9 seeds), ovule sterility, smooth and thin peel, juice content low (about 36–40 percent); TSS about 8–9 percent, flesh do not develop orange color, early (September–October) and can be harvested up to December, marketed fresh in Florida, Brazil, South Africa and many other countries, prolific yielder (60–80 metric tons per ha), fruits hold on the tree well. • Ambersweet: Hybrid between mandarin and orange, early in maturity and preferred by the people in place of early Hamlin in Florida. It has already taken some of the share of Hamlin. • Valencia: Bears its name from where it has spread, that is Valencia, Spain (it is thought to be of Chinese origin), most widely grown sweet orange in the world but most of the produce is processed, late maturing (February–October in Northern hemisphere), prolific bearer (up to 50–60 metric tons per ha). In Southern hemisphere it matures in summer, that is, September to March, holds well on the tree but regreening problem occurs, fruit with excellent quality orange color of rind and yellow-orange color of flesh, juicy, seedless (Fig. 2.1, see also Plate 2.1).

Late maturing selections/bud sports of Valencia orange such as Olinda, Campbell, Frost, Midknight, and Delta are widely grown due to suitability of harvesting time and quality for fresh fruit markets around the world. Problems of creasing and splitting occur due to holding of fruit on the tree. In Argentina a variation, 'Lue Gim Gong,' is grown, which is quite similar to Valencia Late. Pera: Sweet orange grown in Brazil and some other Latin American countries, export cultivar mid to late season orange for fresh fruit market but also processed; fruit medium-sized with eight or fewer seeds, TSS content lower than Valencia, oval to ellipsoid fruit, medium thin skin, smooth surface finely pitted, light orange peel at maturity, holds well on the tree (Romano and Donadio,

FIGURE 2.1 Valencia Orange – Widely Grown Late Maturing Orange World Over.

1981). • Shamouti: Excellent fresh fruit market cultivar from Israel, exported by many Mediterranean countries to EC, similar to Jaffa orange, fruit with very good orange color on peel and inside, ovate in shape, seedless, flat area at stem-end, rind can be removed by hand, maturity mid-season. • Pineapple: Mid-season orange grown in USA, Mexico, South Africa, Brazil, and India. Excellent quality due to high juice content, but seedy, develops orange color with relatively high TSS. • Mosambi: Widely grown cultivar in India, medium to large fruit, thin peeled, juicy nucellar selections available, acidity about 0.35–0.5 percent when fully mature and hence sweet to flat taste when stored, slightly acidic fruit with higher acidity and greenish-yellow color rind are better, early maturity (Fig. 2.2, see also Plate 2.2).

• Sathgudi: Globose fruit, medium to large-sized, juicy, orange colored flesh; moderately seedy, pale orange at maturity, finely pitted smooth peel, mid-season, good flavored, medium thick rind, extensively grown in South India, especially Andhra Pradesh. • Malta: Seeded, mid-season variety, juicy, well adapted to arid hot climate with irrigation facility. • Hongjiang: Seedless (seeds 1–2 per fruit) very popular sweet orange in Guangdong province of China, fruit color is dark orange to reddish, flesh red, juicy, flavor pleasant. • Qianyang seedless Dahong: Medium-sized fruit, weight 162 g, bright orange to red peel when fully mature, juicy, seedless, good eating quality, good storage ability. • Beipei seedless Jincheng: Commercially seedless (5–6 seeds per fruit), Mutant 2-2 is self incompatible, seedless, round-shaped, thin and bright orange colored-peel, juicy selection from Jincheng orange, pleasant sweet-acidic flavor, weight 180–190 g, storage ability good, one of the main orange cv of China. • Jiaogan Seedless 85-2: It is obtained from Tankan, matures in mid-January, medium large fruit, skin orange red, flesh soft, juicy, fruit stores well. • Marrs: This cultivar is popular in Texas (USA), early and heavy bearing, medium to large fruit size, bears fruits in clusters, round to oblate fruit without navel; juicy pale orange flesh, moderately seedy depending on cross pollination, medium thick finely pitted smooth peel, holds well on tree. • Bonanza: Early ripening, round-shaped fruit, deep orange

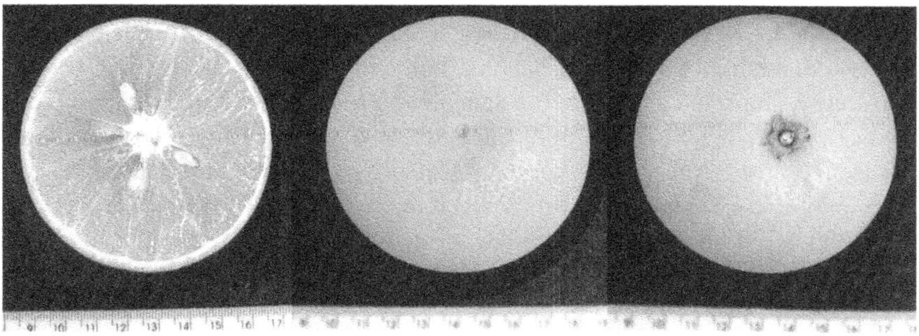

FIGURE 2.2 Mosambi Orange-Widely Grown Early Maturing Orange in India.

colored thin rind, juicy, rich flavored; seedless, peel separates easily, sweet eating quality (Singh et al., 2003). • Salustiana: Early, orange-fleshed, seedless, commercial cultivar in Spain, harvested for 6 months, similar to Cadenera, medium large fruit, sub-globose to spherical, juicy, sweet, fruits hold on the tree. • Cadenera: Grown mainly in Spain, fruit seedless, mid-season, with excellent flavor. • Belladonna: Widely grown in Italy, fruit oblong, medium-sized, pleasantly flavored, seeds few to none, deep orange colored, juicy flesh, early to mid-season variety. • Beledi: Seeded mid-season variety, grown in Egypt and Mediterranean countries. • Jaffa: Seedless mid-season variety, adapted to dry, irrigated areas, fruit juicy. • Natal: Brazilian orange variety of late maturity, resembles Valencia orange, medium-sized fruit, globose in shape, seeds few, flesh pale yellow colored, rind finely pebbled and yellow-orange in color. • Lamb's Summer: Medium small ellipsoidal to ovoid fruit, seeds few, late in maturity, holds on the tree; grown mainly in Florida. • Diller: Sweet orange cultivar grown in Arizona (USA), also known as Arizona Sweet due to very sweet taste, seeded, early to mid-season, round shaped, smooth rind, yellow-orange colored at peak maturity. • Comune: Comune or Criolla is medium to large-sized seeded orange, spherical to elliptical, pale orange colored, slightly thick skin, juicy.

2. Blood or Pigmented Oranges

These are so called because of light to deep blood-red colored anthocyanin pigments in rind and flesh of the fruit. Red color develops with warm days and cool nights that is the main feature of the Mediterranean climate. Even in other subtropical regions, this type of climate during fruit maturation result in red flesh color development although not as deep as in the Mediterranean region. In blood oranges, Doblefina group fruit are grown mainly in Spain, medium small, oval to oblong, seedless, smooth surface with yellowish-orange peel, flesh firm, red blood flecks scattered in flesh, moderately juicy, distinct flavor, mid-season. It has given rise to Entrefina and Ameliorее.

The common blood-oranges are • Malta Blood Red: Seedy, light blood but good flavor, rind thin, adherent, smooth. • Moro: Sub-globose to round fruit, medium to large size, juicy, deep red flesh, excellent flavor, peel orange colored medium thick and smooth, seedless, mid-season to late maturing. • Tarocco: Large fruited, deep red fleshed, seeds few to none, obovate to globose shape, collar at base, medium thick finely pebbled rind, red blush at maturity, mid-season in maturity, ships and stores well. • Sanguinello Moschato: Deep blood colored, seedless, juicy, holds on the tree, mid-season. • Sanguinelli: Nearly seedless, intense red flesh, oval fruit shape, thick leathery peel, late, harvested February–June in Spain. • Sanguinello: Fruits are round to slightly oblong with slightly loose rind, red fleshed, seedless, mid to late season.

3. Navel Oranges

Navel orange fruits are ovate to oblong in shape. This is an important group of fresh fruit orange varieties due to excellent quality. Fruits differ from round shaped oranges either blood or non-blood in that these bear a small navel at the stylar-end, which is in fact a small fruit within the fruit, and hence the fruit is somewhat oblong (Fig. 2.3). This very small fruit, which is somewhat protruded at stylar-end, can be seen without cutting the fruit, and hence it is called the navel of the fruit.

Fruits are seedless due to pollen sterility and ovule sterility. Fruits are large, weighing up to 300 g or more. • Washington navel: So called because it has spread from Washington DC (USA) to other areas after having been received in Washington DC from Brazil. It is the most widely grown navel cultivar in the world. Fruits are juicy, aromatic flesh, seedless, orange colored, excellent for fresh fruit market but do not keep well on the tree, not adapted to hot arid climate. Limb sports of Washington navel orange reported, tree smaller than Washington navel, fruit excellent quality, holds on the tree. • Navelate: Late maturing bud-sport of Washington navel selected in Spain, good yielder. • Newhall, Atwood, Fisher: These cvs are grouped in one since they originated from Washington navel at California. These are quite early in maturity. • Baianinha: Most widely grown navel orange in Brazil, fruit is smaller than Washington navel, can be grown in hot and arid conditions also. • Lanelate and Leng: These originated in Australia from the Washington navel. Lanelate is a selection, whereas Leng is a limb-sport. Leng fruit is smaller, whereas Lanelate is similar to the Washington navel orange. Lanelate is a late cultivar and is widely grown in Australia. Fruit has thin skin (5 mm), juicy (45 percent), high TSS (16.5 percent) and 1.23 percent acid, giving a palatable TSS/acid ratio of 14. • Palmer: Originated as a nucellar seedling of the Washington navel in South Africa. It holds on the tree well and is a heavy yielder.

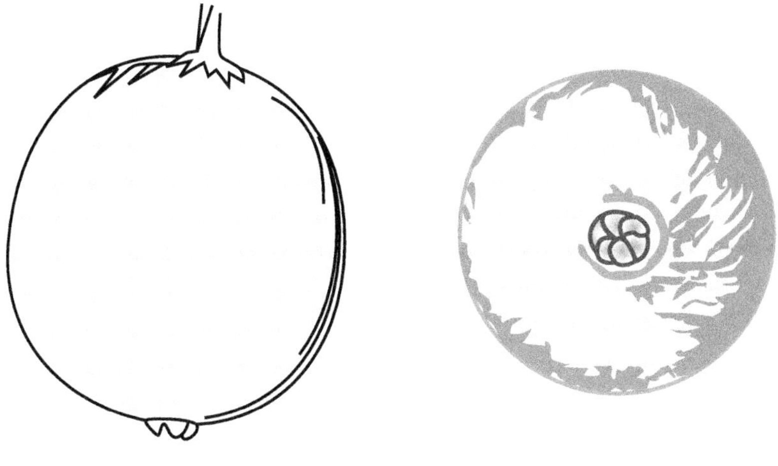

FIGURE 2.3 Navel Orange-One of the Fine Orange Fruit for Table Purpose.

B. Mandarin

Commercial citrus fruits included in this group belong to separate species, hybrids (either human made or natural), selections, and mutants. It is the most important group in fresh citrus marketing because premiums are paid on these fruits due to attractive appearance, pleasant taste, and convenience due to seedlessness and easy peeling characteristics. The terms *tangerine* and *mandarin* are used interchangeably and indicate easy-peelers only. The difference is that tangerines have deep orange to reddish-orange color and are smaller in size than standard mandarins (*Citrus reticulata*, Blanco).

1. Common Mandarin (*Citrus reticulata*, Blanco)

• Ponkan: This is the most important mandarin (*Citrus reticulata*, Blanco) cv grown in China, India, and many other south Asian and southeast Asian countries with different names, Very good eating quality, seeded. In a tropical climate, fruits do not develop orange color and puffs when over-mature. Fruit also develops collars under certain conditions. Fruit quality improves and rind gets thinner as the trees grow old as observed in 'Nagpur' mandarins. 'Nagpur' mandarins or Nagpur *santra* is similar to Ponkan grown in China; acidity decreases as fruit maturity advances on the tree or in storage. At harvest fruits are prone to plugging or buttonhole injury due to brittleness of peel, hence they need to be clipped. Most of the time the fruit can be snap harvested by twisting and pulling as the abscission zone develops, and fruits separates easily. Quality varies considerably with soil type and climate. In the light type of sandy loam soil, excellent quality fruit can be produced in sub-tropical climates. In humid tropical climates and under clay soils, fruits do not develop color and may get puffy early. Fruit matures within 8–9 months in Central India. Fruit is obovate with a long neck or collar in humid and warm conditions while ovate to globose in cool dry climates (Fig. 2.4, see also Plate 2.3).

TSS is 9–11 percent, and acidity is 0.7–0.9 percent in crops that mature in November–December. Quality is far better in crops that mature in February–March with fruit having 12–14 percent TSS and 0.6–0.8 percent acidity. Trees need support to avoid breaking of branches. It is a heavy bearer (1500–2000 fruit/tree) with alternate bearing as a common feature. • Khasi: The name of this mandarin is derived from Khasi hills of Meghalaya, India, where it is grown extensively; quality is excellent in that region. Fruit is deep orange colored, juicy, 13–14° Brix, 0.6–0.8 percent acidity, weighs 90–100 g, seeded. Fruits grown at higher ranges (1000–1500 m elevation) develop deep orange color and are more acidic, whereas at lower elevations, they remain yellowish-green (Fig. 2.5, see also Plate 2.4).

• Darjeeling and Sikkim mandarin: These are similar to the Khasi mandarin, but the TSS is lower. Most other fruit characteristics are similar. • Coorg mandarin: Grown in the Coorg district of Karnataka state. Fruit is similar to that of 'Nagpur' mandarins but smaller in size. • Emperor: Early to mid-season, grown mostly in Australia, seeded, orange colored smooth skin, fruit weight 90–100 g. • Clementine: It is of Chinese origin and resembles the Canton mandarin.

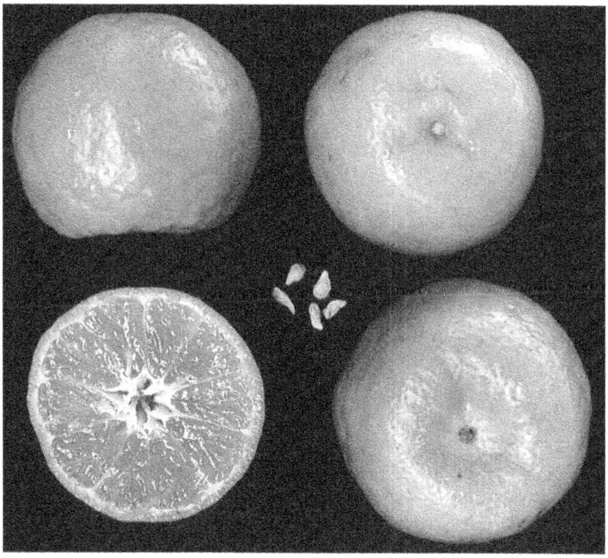

FIGURE 2.4 'Nagpur' Mandarin – the Leading Mandarin Cultivar of Very Fine Taste.

FIGURE 2.5 'Khasi' Mandarin – Widely Grown Mandarin in North-Eastern India.

It derived its name from Father Clement Rodier who worked in Oran, Algeria (Hodgson, 1967). Various clones of Clementine are commercially very important because the export trade of Spain and many other Mediterranean and North African countries is based mainly on clementines. Fruits are seedless except Monreals, but size is small. Clementines have not performed well in humid tropical climates. Trees are different than common mandarin trees in that they are densely foliated and cover the fruit, whereas Ponkan or Nagpur santra trees are upright with an open canopy, and fruit on the top of the canopy may get sunburned. Selections such as Marisol and Oroval are early maturing, whereas Nules (de Nules) or Clemenules are late in maturity. Nules hold on the tree well. Monreal is self-fruitful and produces none or very few seeds. Seedy when pollinizers used. The Fina group of Clementines are mid-season in maturity with excellent quality and very important in Spain. • Dancy: This is an important cultivar in Florida and a true tangerine with adequate orange-red peel color even in tropical humid areas. Best color develops in full sunlight, shaded fruit does not develop color, and fruit has a higher heat-summation requirement for maturity. It derived its name from Colonel Dancy, who discovered the tree near Orange Mill, Florida in 1857 (Davies and Albrigo, 1998). Like many other mandarins, it does not hold on the tree – juice sacs dry out, and the fruit must be spot-picked. Fruit size is small (80–90 g) due to heavy bearing, and branches may break if not supported. • Tardivo di Ciaculli: Italian commercially important variety, bud mutant of Avana mandarin, matures in February–March, juicy with fragrant flavor. • Murcott Afourer: Popular in California, late maturing, holds well on the tree, seeded (few seeds), easy-peeler with excellent flavor, attractive orange rind.

2. Mediterranean Mandarin (*Citrus deliciosa* Tenore)

It actually originated in China but derived its name as Mediterranean because it is widely grown in that region. It is also called *Willow leaf mandarin* in the USA due to lanceolate leaves like a willow tree. Trees are densely foliated and fairly tolerant to freezes. Fruits are seeded, moderate size, thin rind, loosely adherent, oblate in shape, mid-season maturity, very pleasant flavor. Trees have alternate bearing habit. Clementine mandarin has replaced this mandarin in most places. • Comun de Concordia: Grown in Argentina. Fruit medium sized, flattened at base (stylar end) with a smooth or slightly pitted fine skin, juicy, sweet taste, seedy, yellow-orange color, mainly grown in Entre Rios region.

3. King Mandarin (*Citrus nobilis* Loureiro)

It is considered to have originated in the Indo-China region (Malaya, Vietnam, Cambodia). Fruit is large, seeded, with thick orange colored peel that is rough to sometimes warty (Hodgson, 1967). Fruit is moderately juicy with rich flavor, late in maturity, stores well on the tree.

4. Satsuma Mandarin (*Citrus unshu* Marc.)

The most important and widely cultivated mandarin of Japan; later on it spread to China, the Mediterranean region, and other parts of the world where it has adapted. There is a very wide range of mutants and selections of Satsumas those are early, mid-season, and late and varying in quality attributes. Satsuma is considered to have originated in Japan from material that came from China.

It grows well in cool sub-tropical to temperate regions of Japan, central China, Spain, and many other countries. It is very cold-hardy and withstands temperature as low as –9°C. It is also grown in South Africa. Trees have drooping and spreading habit with dense foliage, and fruit requires lower heat-summation to mature than other mandarins. Fruits are seedless, medium to large, and oblate to obovate in shape, without neck, firmly fleshed, with characteristic of peel puffing. • Miyagawa and Okitsu Wase: *Wase* in Japanese means *early maturing*; matures in September when field grown, and in polyhouses with controlled conditions, fruit matures June and July. Miyagawa Wase is popular in Japan, Italy, and Spain. Miyagawa fruit is larger with smooth and thin bright colored peel, juicy, excellent quality. • Owari Satsuma: Widely planted in Japan. In Spain it matures mid-season (October–November). Owari fruits are medium in size, flattened surface, orange colored, and sweet (11–12 percent TSS and 0.7–0.8 percent acidity). Clausellina is an early maturing bud-mutant of Owari and commercially important in Spain. • Sugiyama Unshu: It is mid to late maturing (November), large fruited, oblate-shaped, thin-skinned, excellent quality Satsuma. • Hayashi unshu: Very good keeping quality.

5. Natural Mandarin Hybrids

• Temple: It is the most widely grown natural mandarin hybrid (tangerine x tangor) in Florida and South Africa. Fruits have many seeds but high TSS and very good flavor and deep orange rind color because it is a hybrid with sweet orange blood in it. Unlike mandarins, fruits are not easy to peel. It matures from January to March in Florida along with Murcott, which is also a natural mandarin hybrid. • Murcott: It is also known as *Honey* or *Honey Murcott* or *Smith*. Fruits are medium in size, seeded (seeds few), oblate in shape, peel deep orange to reddish in color, easy to peel, flesh orange colored, juicy, excellent flavor, fruit ships well. Murcott bears heavy crop and is prone to alternate bearing. Crop is held for late marketing (February). • Imperial: Chance hybrid of Mediterranean mandarin, medium sized (fruit weight 80–90 g) oblate fruit with short collar or neck, few seeds, rind thin, orange color, matures early (April) in Australia. • Ellendale: Considered as natural tangor and grown mainly in Australia; fruits are medium to large, slight neck or rounded, apex slightly depressed, seeded, juicy (fruit weight 100–135 g), bright orange to red peel, smooth, hollow axis, do not store well on tree. • Tankan: Natural hybrid of orange and mandarin. It looks and tastes like a sweet orange but is easily peelable. • Ortanique: Natural tangor with smooth adherant orange colored peel, speciality fruit, sweet, juicy but seeded, ships well. • Topaz: Tangor, large fruit, seeded, adherant skin, good flavor.

6. Hybrids Developed in Cross Breeding Programs

Orlando and Minneola (*C. reticulata* × *C. paradisi*) Cross of Dancy × Duncan grapefruit. These are seedless when not pollinized. Robinson, Lee, Osceola and Nova: These are the hybrids of Orlando tangelos and Clementine mandarins developed in Orlando, Florida. • Robinson: Fruits are small, oblate, seedless when set parthenocarpically, easy-peeler, deep orange color, separating segments, hollow core, matures in October–December in Florida. • Nova: Matures late November to late February in Israel, medium large fruit, oblate to subglobose, apex flat and sub-globose, reddish orange peel color. • Page: Minneola × Clementine hybrid developed in Florida, resembles sweet orange. • Orah: Temple × Dancy, developed in Israel, seeded, easy peeler, medium size, sweet, bright orange red, matures December–January, holds well on the tree. • Kinnow: Hybrid of King × Willowleaf mandarin, oblate shaped fruit flattened at base and apex, adherent and smooth peel, develops deep orange color, seeded, juicy, highly flavored, mid-season variety, holds well on the tree, very good shipping and storage quality Sibling of Wilking, which is also a hybrid of King × Willowleaf mandarin. • Sunburst: Tangerine developed through cross of Robinson and Osceola, fruits are deep orange to reddish colored, thin peeled, and medium-sized, oblate, flat at stem-end, mature during October and November, ships better than Robinson or Dancy, develop seeds when cross pollinated with Temple, Orlando, or Nova. • Fallglo: Cross between Bower × Temple, slightly earlier than Sunburst and Robinson, oblate shaped, seedy depending on cross pollination, deep orange colored, easily peelable. • Ambersweet: Cross between Clementine × Orlando, mandarin hybrid × mid-season sweet orange (15–3), early season maturity (early October), medium in size, similar to temple in shape, dark orange in color, seedless to seedy depending on degree of cross pollination, taste and aroma like sweet orange (characteristic of one of its parents). • Mapo and Cami: Hybrids are also grown in Italy. Cami [(*C. clementina* Comun × *C. deliciosa* Avana) × Mapo tangelo], ripens in late December to mid-March, seedless, size 6–7 cm diameter and about 120–150 g weight. • Fortune: Clementine × Dancy, medium to large and oblate fruit, reddish orange color, peelable rind, fruit seeded, late maturing, chilling sensitive, unsuitable for long-term storage.

C. Grapefruit (*Citrus paradisi* Macfadyen)

Grapefruit derived its name as fruits are borne in clusters like grapes. Among citrus fruits grown in the tropics, grapefruits have the highest quality and are the best with the commercial possibilities as fresh fruits. This may be attributed to origin of this fruit in the tropics. This is one of the large citrus fruits (although smaller than pummelos) with diameters of 10–15 cm. Grapefruit is considered to have originated in the West Indies and has many characters similar to Pummelos (*C. maxima*). This is a typically tropical fruit with a high heat-summation requirement. Fruits grown in tropical areas develop high juice content, thinner peel, and lower acidity. People in North America, Europe, and Japan prefer fresh grapefruit. Grapefruit can hold on the tree very long from September to the following July in the northern hemisphere. However, holding too long on the tree is not desirable because granulation may take place and seeds may germinate within the fruit, although acidity decreases. There are red-fleshed and white-fleshed cvs. Red flesh is due to lycopene and carotenoid pigments.

1. Red-Fleshed Grapefruit

Most of today's commercial cultivars are selections or mutations from earlier ones, which were replaced gradually with newer ones. Consumers prefer red-fleshed seedless fruits, so white grapefruits are slowly being replaced. • Red Blush and Ruby Red: The rind develops red blush on fruits. These are mutations (limb sports) of Thompson pink, which is a limb-sport of Marsh (seedless grapefruit). Red Blush and Ruby Red are red-fleshed grapefruit with almost identical fruit and tree characters and selected in Texas. • Ruby Jaguey: Originated from Ruby Red, these are grown widely in Cuba. It is high yielding, has fewer seeds, a thinner rind, and a higher TSS. • Henderson, Ray Ruby, and Rio Red: Henderson has deeper red flesh and peel color than Ruby Red. Ray Ruby has even deeper red flesh than Henderson. Ray Ruby is a mutation of Ruby Red. Rio Red is a seedling selection from Ruby Red. Flesh color of Rio Red is deeper than Ray Ruby with higher yields. In South Africa, Rio Red matures when most other grapefruits are harvested. • Flame: Originated from Henderson as a seedling. Peel color is similar to Ray Ruby and internal color is similar to Star Ruby. In Ray Ruby, Flame, and Rio Red, deep red color of flesh persists longer, hence marketing can be extended. Fruits of these cvs develop red color on rind and flesh even in tropical warm areas. • Star Ruby: Fruits very deep red fleshed with reddish blush on peel. Smaller fruit size, seedless, matures 6 weeks before Marsh seedless. Fruit has good flavor, but this cv. is sensitive to cold. Fruits are borne inside canopies, so there is less damage due to frost.

2. White-Fleshed Grapefruit

• Marsh: Most popular grapefruit due to seedlessness and good holding characteristic on the tree thus extending the marketing season from September to March in the northern hemisphere. • Duncan: Fruit large, oblate, seeded, pale yellow colored, peel medium thick, smooth, juicy, rich flavored, early maturity. • Foster: Seeded, globose to oblate fruit, pale yellow colored smooth peel, flesh juicy, flavor good. • Saharanpur Special: It is grown in Uttar Pradesh and Uttaranchal states of northern India. Fruits are white fleshed, matures from November to February, seedless, light yellow pulp, vesicles are juicy.

D. Pummelo or Shaddock (*C. grandis* or *C. maxima*)

Originated in southern China. It is found growing wildly in China, many northeastern states of India, and also in the Malay region of southeast Asia. Fruits and flowers of pummelo are the largest among commercially grown citrus in the world. Fruits that are 20–25 cm in diameter are common; even up to 30 cm in diameter can be seen. Fruits are liked in China and southeast Asia very much. The major difference between grapefruit and pummelo is that pummelo fruits are much bigger with thick peels and are less juicy. Flesh of the pummelo is firm with crisp carpellary membranes and juice sacs. Commercial pummelos do not exhibit traces of bitterness as exhibited in grapefruit on eating. Fruits are obovoid to pyriform in shape and seeded. Chandler, a Thai variety, and an Indonesian variety, Djeroek Deleema Kopjar, are pink-fleshed. Chinese varieties Goliath, Mato, and Shatinyu are white fleshed. Israel produces Goliath, Chandler, and Tahiti varieties for domestic and export markets. • Chandler: Hybrid of Siamese pink and Siamese sweet, pink fleshed, early ripening fresh fruit, maturing in December–April, globose, seeded, medium oblate to globose, smooth peel, stores well (Singh et al., 2003). • Kao Ponne and Kao Phueng: These are white fleshed, seeded Thai cultivars. Kao Phueng fruits are pyriform with distinct necks, whereas Kao Ponne fruit is globose. Kao Ponne matures earlier than Kao Phueng, good flavored juicy pulp. • Banpeiyu: It is a Malaysian cultivar widely grown in that country, high-quality fruit, large size sub-globose in shape, seeded, light-yellow fleshed. • Wendan and Chumen Wendan: Early, flesh white, soft, juicy, peel thin, widely grown in Fujian province of China. • Pingshan: Major variety of pummelo in Fujian, China. Large oblate fruit weighing 1 kg or more, matures mid-September, seedless, pale red albedo, superior eating quality (TSS 11–12 percent, acids 6.3–6.4 percent), edible part 55–60 percent, good storage ability. • Anjiangxiang: It is a pummelo of the Hunan province of China, seeded, tolerant to cold and drought, matures in late September to early October, flesh is juicy and fragrant. • Diangjiang: Pummelo grown in Sichuan province of China, fruit oval, large, juicy, few seeds, sweet, fragrant, matures early November. • Shatinyu: Grown in Guangxi province of China, 0.6–1.5 kg fruit weight, pyriform fruit, light-yellow flesh, crisp, sweet, late maturing (October–December), fruit can be stored until the following April (Chen and Lai, 1992).

E. Hybrids of Pummelo and Grapefruit

Hybrid Oroblanco (acidless pummelo × grapefruit tetraploid) is a triploid, matures about 6 weeks before Marsh seedless grapefruit. This fruit has many of the characteristics of grapefruit and is much liked in Japan. 'Melogold' is also a triploid of sweet pummelo × tetraploid grapefruit.

F. Lemon (*Citrus limon*)

Lemon fruit quality is excellent in semi-arid irrigated areas and coastal areas. In humid tropics, lemon trees produce fruit with coarser peels. A Mediterranean-type climate is better suited for lemons. Lemons are also suited to the sub-Himalayan region and foothills in India where they are believed to have originated. In coastal climates lemons continue to flower throughout the year, whereas in Mediterranean climates, flowering mainly takes place twice a year. Like grapefruit and other citrus fruits, flowers are also borne in clusters but are purple. The fruit size and shape of lemon varies greatly from spherical in Baramasi lemon or Eureka to oblong in Assam lemons. Fruits develop nipples or mammila at the stylar end, which is almost absent in some cvs and prominent in others. Composition and seed content of the fruit also vary greatly. • Verna or Berna: Also known as Vernia or Bernia, major cultivar of Spain and similar to Lisbon lemon. There are two crops of this cv maturing mainly from February through August, and a second crop (also called *Verdelli*) in September to October in that country and in the northern hemisphere in general. Fruits of Verna are seedless and elongated with prominent mamilla. Verna fruits hold well on the tree. • Fino: Fruits are smaller than Verna with smoother rinds and high juice content. Fruits are spherical and have no neck. Seeds are 5–7 per fruit, Fino is also known as Mesero, Primofiori, and Blanco. • Feminello: Italian cvs mainly of the Feminello group produce fruit throughout the year. There are four crops in a year. Primofiore (harvested from September to November), Limoni (harvested from December to May), Bianachetti (April–June), and Verdelli (due to withholding of water flowering in August–September, Verdelli are harvested the following May–July). Feminello commune, Feminello stracusano, and Feminello St. Teresa are well known in Italy. Fruits are seeded, elliptic, have blunt nipples, are medium sized, juice content not high but fairly acidic, ships and stores well. • Eureka: The most important cultivar of lemon in the world as far as its cultivation is concerned. It is grown in most parts of the world as commercial crops including Australia, California, South Africa, and North Africa besides Israel and other Mediterranean countries. Fruits are ovate to round with rounded apical mamilla. Fruits are commercially seedless (4–6 – usually fewer than 10) with high juice and acid content. Clonal selections with more vigor such as Allen, Cascade. Cook and Ross are available (Morton, 1987). • Lisbon: This is heavy yielding cv grown in California, Australia, Argentina, and Portugal from where its name comes. Fruits are seeded and otherwise similar to Eureka but with pronounced mamilla. Unlike Eureka trees, Lisbon trees have a dense canopy; hence frost damage is usually less in Lisbon than in Eureka under California conditions and elsewhere in the regions that have frost. • Villa franca: Similar to Eureka, smooth but thick rind, medium-large fruit, ovate-oblong, bright lemon yellow when mature, seedy, juicy, grown in Australia. • Assam lemon (Nepali oblong): Fruit matures in December and January, almost seedless, blooms throughout year, oblong to obovate fruit, 10–15 cm long and 3–4 cm wide, apex nippled (Fig. 2.6, see also Plate 2.5), fruit rind medium thick, central axis partially filled or hollow.

• Baramasi lemon: Seedless, medium-sized, medium-thick rind, round fruit, juicy, TSS 5.5–6 percent, acid 2–3 percent, good flavored, trees produce fruit year-round. • Galgal (Hill lemon): It is also growing in wild conditions along the foothills of the Himalayas. Small-scale plantations exist in Punjab and Himachal Pradesh (India). Medium-large to large fruit (250–400 g), 7–8 cm diameter, 9–11 cm length, smooth skin with slight furrows, oblong to ellipsoid shape with nipple at apex, slight collar at stem-end, medium thick (6–7 mm) peel, juice 24–25 percent, seeded (40–55), 11–12 segments, white coarse textured fleshed, hollow axis, TSS 7 percent, acidity 5.5 percent (Fig. 2.7, see also Plate 2.6).

FIGURE 2.6 Nepali Oblong (Assam Lemon) – Common Lemon Cultivar in North-Eastern India.

FIGURE 2.7 Galgal or Hill Lemon – Popular Lemon Cultivar in North and North-Western India (Left), Possible Lemon x Citron Hybrid (Right).

• Seedless lemon: Fruit is seedless, ellipsoid to ovate with nipple at apex, base round, peel smooth, thin, axis hollow, segments 10–13, pulp light yellow and juicy, matures in November and January (Fig. 2.8, see also Plate 2.7).

• Pant lemon-1: Variety grown widely in the Nainital district of Uttaranchal state, India and surrounding areas in northern India. Fruit weight 90–100 g, fruit

FIGURE 2.8 Seedless Lemon – A Promising Seedless Lemon Variety of India.

length 6.7–7.0 cm, breadth (diameter) 5.6 cm, peel thickness 4 mm, TSS 7.8 percent, acidity 5.82 percent, juice 48–50 percent. • Interdonato: Considered a lemon × citron hybrid, commercially grown in Italy, thornless and vigorous tree, fruit oblong to cylindrical with nipple, few seeds, juicy, acidic. • Meyer: Hybrid of lemon × mandarin, introduced from China to US by Frank N. Meyer, fruit obovate, medium-sized (6–7 cm diameter), round at base with short nipple, juicy flesh, moderately acidic, seeded (small seeds).

G. Acid Lime

Fruits *C. aurantifolia* Swingle are small whereas that of *C. latifolia* Tanaka are large. Fruits of *C. aurantifolia* Swingle are commercially important in India and neighboring countries, Mexico, and Florida. Fruits are highly acidic with titratable acidity up to 8 percent, mainly seeded but seedless selections are also commercially available. Trees are susceptible to frost damage and hence not grown in frost-prone areas. • Key: Also known as Mexican or West Indian lime with small fruit size. In India it is known as Kagzi or Kagdi nimbu due to very thin paper-like peel almost 2 mm or even less (Fig. 2.9, see also Plate 2.8). Fruits are spherical to ovate in shape, 35–40 mm in diameter, and harvested green. Some markets prefer yellow fruit. Mature harvested fruit turn yellow without any treatment within few days. Acid content (7–8 percent) and juice contents (50–55 percent) are high. Used mostly for garnishing salads, curries, and vegetable preparations in south Asian countries. Processed products such as cordials and squashes are very common in India. Makes refreshing drinks in summer.

FIGURE 2.9 Key or 'Kagzi' Lime – Common Small Fruited Acid Lime of the World.

• Persian or Tahiti: Large fruited (*C. latifolia* Tanaka). This is also known as a Bearss lime. Fruit shape is prolate spherical, seedless due to ovule sterility, and much bigger in size. Maturity period is longer than that of the key lime.

H. Sweet Lime (*Citrus limettioides* Tanaka)

It is grown in India as Mitha nimbu mostly in Punjab and northern India and in the state of Tamil Nadu. It is also grown in Egypt, Palestine, Israel, and neighboring countries. Fruits are sweet, medium sized, with few seeds. Mature fruit are of flat taste due to lack of acid. Peel is yellow to yellow-orange in color, quite smooth, thin and adherent. Flesh is straw-yellow colored.

I. Citron (*Citrus medica* Lin.)

Citron fruits are oblong in shape like some of the lemons (long-oval or ellipsoidal to obovate in shape), but fruits are very seeded, quite large with yellow colored very thick, rough and bumpy peel at maturity. These are grown since ancient times in India and considered to be native to India. In Assam state it is widely available and known as *Bira-Jira* or *Bakel-Khowa-Tenga*. The fruits of Mitha-Jara are sweet (juice and peel both), weighing 1.5–2 kg whereas Bira-Jara fruit weighs 3–4 kg and only peel is sweet, which is eaten fresh or used for making candied peels.

The Etrog or Ethrog citron is widely cultivated in Israel due to its importance in Jewish religious ceremonies. Fruit should be relatively small and symmetrical with persistent style (style part of the flower at apex of the fruit) for this

FIGURE 2.10 Calamondin-Small Mandarin – like Fruit with Edible Peel.

purpose. Corsican is the French variety, whereas 'Diamante' is grown in Italy. 'Diamante' and 'Etrog' are acid citrons, whereas 'Corsican' is sweet.

J. Calamondin (*Citrus madurensis* Loureiro)

It is a mandarin-like fruit but quite small (3–3.25 cm in diameter) weighing 20–30 g, orange colored peel that is smooth and very thin (1–2 mm). Fruit seeded with orange colored flesh, which is acidic in taste, but the peel is sweet and edible (Fig. 2.10, see also Plate 2.9). It is used for culinary purposes and for marmalade making. Calamondin is commonly grown in southern China, Taiwan, Japan, Philippines, and northern India.

K. Natsudaidai (*Citrus natsudaidai* Hayata)

This is grown widely in Japan (*natsu* means summer) and China. Considered to be a hybrid (tangelo) of pummelo and mandarin. It is known as *Natsumikan* in Japan. Fruits are grapefruit-sized, oblate to obovate in shape, with few seeds, sweet, yellow to yellow flesh, and quite refreshing. Peel can be removed by hand when mature, peel surface pebbled to slightly rough. It is late and hence preferred. Fruits remain on the tree and become sweet and juicy in summer so are preferred in southern and coastal areas of Japan where winter is not that severe. Fruit quality is known to improve in storage.

L. Hassaku (*Citrus hassaku* Tanaka)

It is a hybrid of pummelo and mandarin. Fruit looks like mandarin, medium in size (7–8 cm diameter), oblate in shape, stem and stylar ends depressed, yellow-orange colored, coarsely pebbled and moderately adherent peel, early to mid-season in maturity, sour in taste, not very juicy, seeded, stores well.

M. Kumquat (*Fortunella* spp.)

These are the fruits of the family Rutaceae (Genus *Fortunella*) to which citrus belongs, but due to morphological differences are placed in the genus *Fortunella*. There are several similarities with citrus fruits as well. Fruits are grown mainly in China, Japan, and Philippines. Uniqueness of these fruits is that they are eaten without peeling as the peel is also edible and not bitter. Meiwa kumquat (*Fortunella crassifolia*) fruit are spherical whereas Nagami kumquat fruit (*Fortunella margarita* (Lour.) Swingle) are elliptical or oval in shape. These fruits weigh 10–11 g, 15–16 percent juice content, 15–17 percent Brix, 4–5 percent acid and 50–55 mg ascorbic acid/100 ml. The composition may vary under different growing conditions. Marumi kumquats (*Fortunella japonica*) are grown in Japan; the species grows wildly in that country. Fruits are as small as Key limes with 2–3 cm diameter and are seedy. Fruits develop deep orange to reddish color when fully ripe. Kumquat trees could also be used as ornamental citrus trees in home gardens.

N. Bael (*Aegle marmelos* (L.) Correa.)

It belongs to the family Rutaceae (same as that of citrus) but genus is *Aegle*. The tree can be seen growing in tropical to subtropical regions throughout India and adjoining countries. It is indigenous to India and is also known as *Bengal quince* (John and Stevenson, 1979). The tree has great mythological and religious significance in the country. The leaves (the trifoliate with larger terminal leaf than the pair) are offered during prayer to Lord Shiva, the Hindu God. Fruit has considerable medicinal properties. Fruit is a hard shelled, seeded berry with smooth greenish-yellow pericarp, globose to ovoid in shape, peel nearly 4–5 mm thick, pulp is yellow-orange colored, fragrant and pleasant when ripe. The weight of fruit varies from 0.5 to 1 kg. Fruit takes 10–11 months to mature and like citrus it is non-climacteric. Varieties Mirzapuri, Rampuri Azamati and Khamaria produce good quality fruit (Singh, 1961). Kagzi Gonda cv produces fruit with thin rind and soft yellow pulp with excellent flavor (Teaotia et al., 1963). Narendra Bael-5 is a promising selection with round smooth fruit weighing about 1 kg with pleasantly flavored pulp, fewer seeds and mucilages. Flesh is soft with less fiber (Pathak and Pathak, 1993). Fruit stores for 10–15 days under ambient conditions and for 2.5–3 months at 9–10°C with 85–90 percent RH.

II. COUNTRIES, VARIETIES GROWN, AND HARVESTING SEASONS

Citrus fruits are grown in most of the countries of the world, but roughly 20 of them contribute to 90 percent or more of the total world production and trade. Important citrus-producing regions of the world and some major countries are covered here. Two major regions of the world differentiated by climatic conditions, northern and southern hemispheres, are further divided into geographic regions and countries to describe varieties grown, climate, and harvesting seasons keeping in view many similarities in countries that are in geographical proximity to each other.

A. Northern Hemisphere

In the northern hemisphere, from the equator towards the north pole, winter sets in from October to November onwards and continues to March–May depending on latitude. With onset of spring, in March or April, citrus trees flower and fruit harvesting starts in October–November. In late varieties, harvesting continues up to May or June the following year or even later, thus flowers of the next season and fruits of the previous year can be seen on the same tree. In areas near the equator (tropics), trees flower twice or three times a year and produce as many crops each time.

1. South Asia

This is one of the important regions of citrus growing countries (Fig. 2.11), and many species are native to this region, particularly the Indo-Chinese sub-region. This region comprising India, Pakistan, Bangladesh, Nepal, Bhutan, and Sri Lanka contribute one-third of citrus produced in the Asia-Pacific region and grow mostly easy-peeler mandarins. Oranges, limes, and lemons are other important citrus. Grapefruit and pummelos are grown in homestead gardens in Punjab and the North-Eastern Hill (NEH) region and in Nepal and Bhutan.

(a). India: Very wide range of climatic conditions are available in this country (Fig. 2.12) for citrus cultivation, and total citrus production passed the 4 million metric ton mark in the year 2000. Sweet orange cultivars popularly grown are Sathgudi and Mosambi, which have commercial importance in Andhra Pradesh and Maharashtra states, respectively. Sathgudi production is 427 635 metric tons from a 35 513 ha area in Andhra Pradesh (Anonymous, 1999). Mosambi sweet orange growing districts are Jalna, Aurangabad, Beed, Nanded, and Parbhani in the Marathwada region of Maharashtra. The total area in Maharashtra where this crop is grown is 41 018 ha with 369 000 metric tons of annual production. The Marathwada region alone has 26 220 ha area under this crop with 230 000 metric tons of annual production (Shinde and Kulkarni, 2000) thus contributing to 63 percent of Mosambi production in the state. Malta, Jaffa, and Blood Red are grown in Punjab, Himachal Pradesh, and Rajasthan.

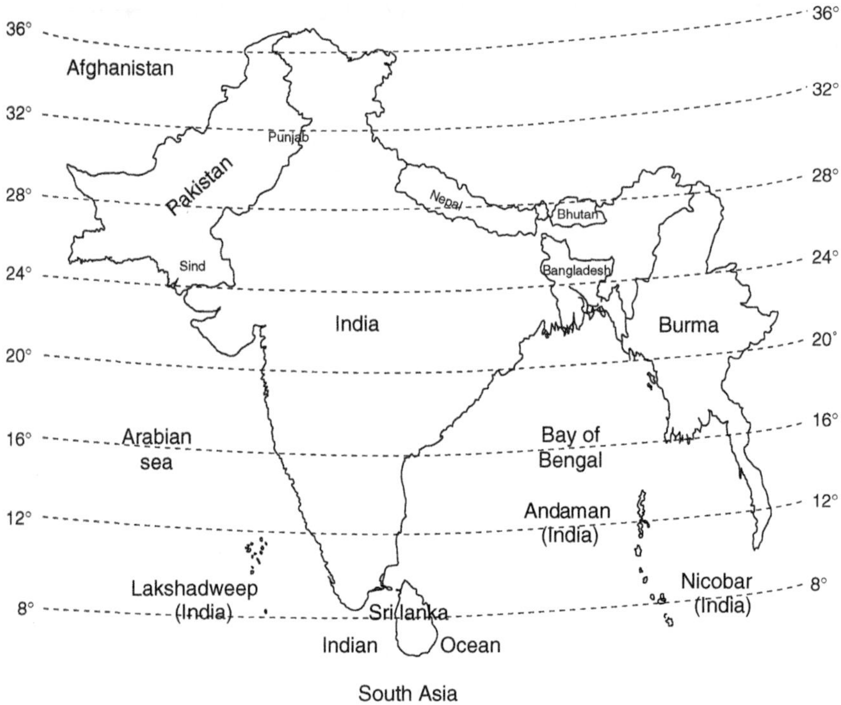

FIGURE 2.11 South Asia – Major Citrus Growing Countries of Indian Sub-continent.

Mandarins constitute more than 50 percent of the total citrus production in the country. Well known cvs are 'Nagpur', Khasi (derived its name from Khasi hills and a tribe from Meghalaya state), and Kinnow (Hybrid). Kinnow is commercially grown in Punjab and neighboring areas of Rajasthan, Himachal Pradesh, and Haryana. Coorg mandarins are grown at limited scale in Karnataka state on hilly areas of the Coorg district in elevations of up to 1000 m. This region receives 150 cm rainfall annually with temperatures ranging from 10°C to 30°C. 'Nagpur', Khasi, and Coorg mandarins are similar to Ponkan of China. The region around Nagpur (central India) is warm and dry with 40–46°C maximum temperatures in summer and 8–10°C minimum temperatures in winter. Rainfall is 100 cm from June to September. Khasi is well adapted to Brahmaputra plains of Assam (at 90 m elevation) and up to 1500 m elevation of Sikkim and also Meghalaya where annual rainfall is 400–500 cm.

'Nagpur' mandarin trees produce excellent quality fruit on sandy and medium (10–20 percent clay) type of loamy soil. The quality of crop that matures in February and March is better than that matures in November and December. Seedless clones of Nagpur mandarin are available. In the subtropical climate

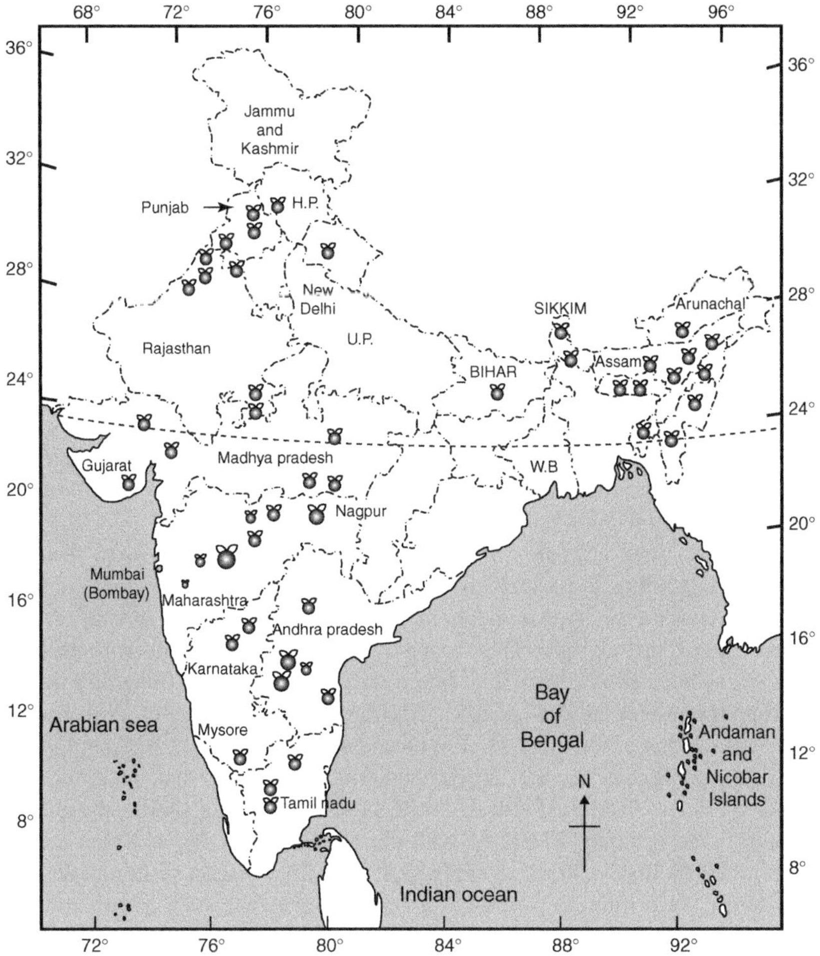

FIGURE 2.12 Major Citrus Growing States of India.

of the north, Nagpur cv is being tested. Khasi mandarin is grown in northeastern states on terraces on hill slopes. The NEH region of India and parts of Bangladesh experience high rainfall (300–400 cm). The states of Assam, Meghalaya, Mizoram, Nagaland, Tripura, Manipur, and Arunachal Pradesh together have about a 50 000 ha area in which Khasi mandarin trees are growing. It is a mostly warm humid area at foothills with a cooler climate at 1000–1500 m elevation. Sikkim and Darjeeling mandarins are grown at the elevation of 500–1500 m. Almost all mandarin orange orchards in the region are of seedling origin (Ghosh, 1985). Trees of 40–60 years of age are common. All commercial mandarin cvs of the region are seeded and exported to neighboring countries of south Asia through land routes. In Assam, mandarin (locally known as *Kamala*) is concentrated in Kamrup, Tinsukia, and North Cachar districts, whereas in Meghalaya, it is grown mostly on the southern slope of Khasi hills bordering Bangladesh and in Jayantia and Garo Hills. In Sikkim, the mandarin (locally called Suntala) growing area comprises East, South, and West districts.

There are two crops of sweet oranges and mandarins in central and south India. The spring blossom crop is harvested during November and December, whereas the monsoon blossom crop is harvested during February and March. In north and northeast India, only the spring-blossom crop is harvested, from October onwards up to January or February depending on elevation and cultivar (early or late). Valencia late is recommended for harvest up to the first week of March in some parts of Punjab; thereafter, granulation takes place with reduction in juice. Hamlin and Mosambi, which are early, are harvested in the months of October and November in northern India. The monsoon rains occur in the north and northwest part from July to September. Because of the subtropical climate, good quality Kinnow mandarin and sweet oranges are grown in Punjab, Rajasthan, and Himachal Pradesh.

The acid lime (*C. aurantifolia* Swingle) is a major citrus fruit of Maharashtra, Andhra Pradesh, Gujarat, Tamil Nadu, Karnataka, and Madhya Pradesh. In Gujarat it is grown mainly in the Kheda, Baroda, and Surendra Nagar districts. Approximately 9000 ha is covered by this crop in Gujarat with nearly 6 metric ton/ha average productivity. Maximum yields are up to 20 metric tons per ha, and production in the state is about 70 000 metric tons annually. With an area of 38 878 ha and production of 583 170 metric metric tons per year (Anonymous, 1999), Andhra Pradesh is the largest acid lime producing state in India. The Nellore district of Andhra Pradesh is the largest acid lime producing area in the country and occupies 21 017 ha with the production of 315 255 metric tons. Gudur, in the Nellore district, is the biggest acid lime market in India (Wanjari et al., 2003). Acid lime crops grown in Maharashtra covers an area of roughly 9358 ha. In northern and western Maharashtra, acid limes are grown mostly in Ahmednagar, Solapur, Dhule, Jalgaon, Pune and Nasik districts. In Vidarbha, Akola, Nagpur, Amravati, Buldana, and Wardha districts are major areas of acid limes. Cultivation of acid limes is negligible in the coastal Konkan region of the state. The total production of the acid lime in the state is roughly 1 50 000 metric tons considering productivity level of 12–15 metric tons per ha. The area has increased in the state after 1994–1995 since the introduction of the employment guarantee scheme by the state government in which planting material is subsidized to certain extent. There are acid lime processing units in the state that utilize fruit for peel oil and pectin production. Fruits are processed into pickles, and juice is utilized for squashes and cordials (drinks with syrup and juice).

Lemons cv 'Galgal' and 'Baramasi' are common at foothills and submountainous regions of Himachal Pradesh, Punjab, and Uttar Pradesh. Assam lemon (Nepali oblong) is widely grown in Assam and surrounding areas because it is used in black tea flavoring. It is harvested from December to January in that region. Grapefruits grown in India are Marsh Seedless, Saharanpur Special, and Red Blush. The plantations are small mainly in Punjab and adjoining areas of Himachal Pradesh and Uttar Pradesh.

(b). Sri Lanka: This country, with its tropical climate, grows citrus on a 9000-ha area with production of 15 000 metric tons. Major citrus fruits are acid limes and sweet oranges. Sweet orange, cv. Bibile, is seeded but juicy with greenish-yellow to yellow-orange rind color. Citrus is grown as a mixed crop with tea and banana. There is no monoculture of citrus. Almost all citrus is grown as seedlings. Due to rains for 8 months of the year, stress for an induction of flowering is a problem. Mandarins mature in May and June whereas oranges mature in February and April. Most of the fruit is consumed locally, and there is no export whatsoever. A mandarin variety is a clonal selection from Clementine. Acid lime cvs Monargala and Thimbutan are also grown.

(c). Pakistan: Citrus is an important fruit crop of Pakistan occupying more than one-third of the area of fruit crops. Sweet oranges and lemons are grown for domestic consumption, whereas Kinnow is grown for export to Gulf countries (Saudi Arabia, UAE, Dubai, Bahrain). Malta sweet orange, Eureka lemon, and Marsh grapefruit are other important citrus fruits. Citrus is widely cultivated in Lyallpur, Montogomery, and Sargodha districts and Peshawar valleys. The citrus areas reached 147 000 ha by 1988 in Pakistan. Kinnow mandarin and Feutrell's Early are the major citrus fruit grown in the Punjab province of Pakistan. The drawbacks of Kinnow fruit are high seed content, late maturity, strong alternate bearing, and firmly adherent peel. It has good keeping quality. Blood Red and 'Mosambi' sweet oranges, and limes are also grown.

(d). Nepal: Citrus is grown mainly in the eastern part of the country in the Dhankuta district and in the Pokhara valley in the western part (northwest of Kathmandu at 800 m elevation). Suntala or Ponkan type of seeded mandarin is the main cultivar, and quality is equally good to that of mandarins grown in Bhutan. The Junar sweet orange, Nepali oblong lemon, and acid limes are grown in some parts. Pummelos are grown in backyards of homes.

(e). Bangladesh: Bangladesh produces mainly mandarins (Khasi mandarin). Pummelos and acid limes (*C. aurantifolia*) for domestic consumption. The Chittagong Hill tract of the eastern part and the Sylhet district in the northeast part of the country produce mandarins. These are of seedling origin. The annual production is about 50 000 metric tons.

(f). Bhutan: Excellent quality mandarins are produced in this country. Fruits are similar to those produced in Sikkim state of India. Fruit is deep orange colored and their size is medium to large. In the cooler climate of higher hills (1500 m), fruit is compact and thin-skinned with a good sugar-acid blend, whereas in the foothills (hot, humid climate) the same variety produces a thicker rind and a loosely adherent peel. Fruit is exported to India and Bangladesh. Mandarin is a cash crop with export value up to $10 million. Fully grown trees bear up to 400 kg of fruit equivalent to 40 metric tons per ha (Van Schoubroeck, 1999).

2. Southeast Asia, China, and Japan

(a). China: China is a native home of mandarins and has a cultivation history of over 4000 years. There were three main cvs. Huangpi ju, Huang gan, and Zhu ju before 618–907 A.D., and present cultivars of mandarins have evolved through the centuries (Li, 1992). China (Fig. 2.13) is also a home of sweet oranges. According to Candolle (1866) as mentioned in the book *Origin of Cultivated Plants*, sweet oranges were introduced by Arabs and Portuguese from South China to Lisbon and Valencia around 1488.

Citrus is mostly grown between 20–30° north latitude on hilly land at 700–1000 m altitude. Between 30°N and 35°N latitude, citriculture is limited due to severe winters. Acreage is scattered along and south of the Yangtze River valley. As many as 19 provinces grow citrus commercially, and citrus has attained the status of a cash crop in south China. The major provinces growing citrus are Sichuan, Chongqing, Hubei, Fujian, Guangdong, Guangxi, Zhejiang, Hunan, and Yunnan. China projected 1 million ha of citrus up to the year 2000 with a target of per capita consumption of 10–13 kg year. About 95 percent of the harvest goes to the fresh fruit market (Aubert, 1991). Mandarins, sweet oranges, and Pummelos are the three most important citrus fruits of commercial importance in that order in Chinese citriculture. Citrus production in China has been reallocated since the 1990s with acreage increasing in advantageous areas such as Jiangxi, Hunan, and Zhejiang. Zhejiang province, mainly the Taizhou prefecture, has a history of 1700 years of citrus cultivation. The average annual temperature in this province is 14–18°C with minimum in January of 5–7°C and a maximum in July of 33–41°C. Annual rainfall ranges from 1100 to 2000 mm. Another important province is Guangdong, which is in the southern part of the country. It has

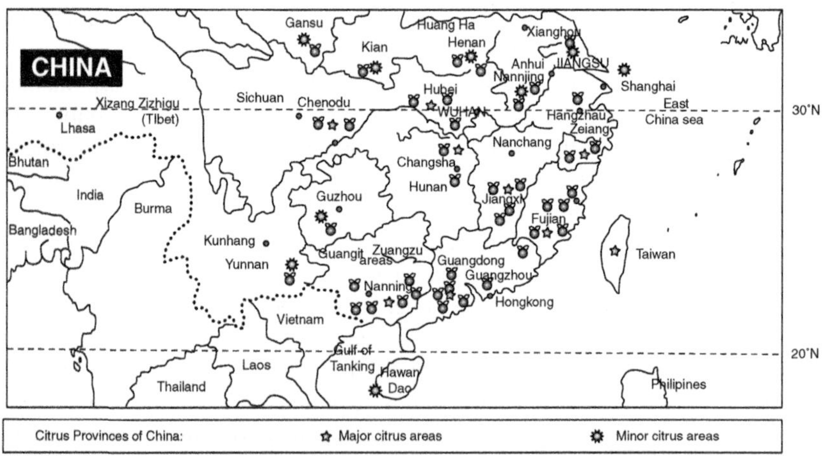

FIGURE 2.13 Citrus Growing Regions of China.

a subtropical climate, the rains occur from April to September, and annual rainfall is 1400 mm. Every city and county grows citrus in this province. Family-type farms with all manual operations also exist. Big state-owned citrus farms are also there. Yields of 22–60 metric tons/ha are obtained in high-density plantings. The total citrus production of this province in the early 1990s was 1.1 million metric tons (Liansheng, 1991). Fruits are transported to north China markets. Jincheng is the high-quality sweet orange grown in China (Table 2.1). Fruit is bright orange colored with strong flavor, high TSS : acid ratio, and good storage life. The seeds are up to six per fruit. A spontaneous seedless mutant of this cultivar, Seedless mutant 2 2, is now being grown (Li, 1991). The Hunan province has a subtropical humid climate, and there exists good scope to grow navel oranges. China is replacing old varieties by top working (grafting), and navel and other high-quality fruits are emphasized. There is a major thrust to commercialize the industry. Tien Cheng (seedless), Shan-sui-cheng, Luicheng, and Kwang Kwo are important oranges grown on the southern coast of the country. In pummelos, Buntan, Sha-Tein, Pei-Yu, Ma-Tao, and Sui-Chi-Pao are importent. Liuyue You (*You* meaning *pummelo* in Chinese) ripens in August, and Wanbai You ripens the following January. Fourseason You pummelos flower and produce fruit almost year-round (Chen and Lai, 1992). Chinese people like pummelos more than grapefruit. Pummelos are sweet, and grapefruit are not liked probably because they are more acidic. North of Canton, Ponkan and Tankan are grown. The Huangkan tangerine/mandarin, which are similar to Dancy, are grown in Fujian province.

China exported about 200 000 metric tons of citrus (2000–2001), which was 2 percent of total production in that country. Major export destinations are Canada, Russia, and neighboring countries (Xiuxin, 2001).

A very wide variety of citrus fruits can be seen in the markets of this country. Communes or community areas produce and market the citrus fruit. Harvesting season in China is from October to December, with late harvesting up to February indicating more winter-ripening varieties. Efforts are on to plant late types of oranges and Tankans. Valencia Late, Olinda, and late ripening Shikan is increasing in the southern and western parts of the country where harvesting is extended up to April. Pummelos are very widely grown in southern China.

Citrus fruits from other parts of the world are being introduced in China and evaluated. The mandarin hybrid Nova was introduced from Spain in 1987 and is performing well in the Chongqing area of Sichuan. The fruits mature in mid-November, are medium-sized (average weight 130 g), oblate in shape, and have an orange-red and glossy rind. The flesh contains 10.2 percent soluble solids, 0.76 percent acids, and 1–3 seeds/fruit. Storage quality of fruit is good (Peng, 1999).

(b). Taiwan: It is a small island with production of good quality pummelos, Ponkans, and Tankans. Taiwan produced 313 067 metric tons of citrus from 31 301 ha during 1989 and also exported 5 percent of it (En-Usiung, 1991). Taichung, Chianan, and Hsinchu prefectures in the central-west part are important for Ponkan mandarin cultivation. Taipei and Tungtai grow Tankan. Oranges are grown mainly in Chianan prefecture. Pummelos are grown mainly in Tungtai. The Ponkan and

TABLE 2.1 Chinese Provinces, Climate, Citrus Grown, and Harvesting Seasons

Region and provinces	Climate	Citrus fruits grown	Harvesting season
North: Beijing, Tianjin, Hebei, Shanxi, Inner Mongolia	Too cold; citrus cultivation nil	–	–
Northeast: Lioning, Jilin, Heilonjiang	-do-	–	–
East: Zhejiang	Suitable for citrus	Bendizao: mandarin Zaoju: early mandarin	October–November onwards
Shanghai	-do-	Jincheng orange, kumquats, Ponkan	-do-
Fujian	-do	Jincheng orange, kumquats, Ponkan, Tankan, Wendan pummelo	Early September and onwards
Jiangxi	Suitable for citrus	Satsuma and Ponkan	September–November
Central and South China: Henan	Sub-tropical	Kumquats, Ponkan, Tankan Jiaogan seedless	
Hubei	-do-	Jincheng orange, Ponkan, Tankan, satsuma	Oranges: October–April
Hunan	-do-	Qiangyang seedless Dahong orange, kumquats, Ponkan, Tankan, Satsuma	Oranges: October–April
Guangdong	-do-	Hongjiang seedless, Jincheng orange, kumquats, Ponkan, Tankan, pummelo	Oranges: October–April
Guangxi	-do-	Jincheng orange, kumquats, Ponkan, Tankan, Satsuma, pummelo	Oranges: October–April
Hainan	-do-	Jincheng orange, kumquats, Ponkan, Tankan	-do-
South and southwest: Sichuan	Tropical to sub-tropical	Phoenix pummelo, kumquats, Ponkan, Tankan, Jincheng orange	Pummelos: mid September–November; Oranges: October–April; Tangerines: October onwards
Chongquing	-do-	Nanchang, kumquats, Ponkan, Tankan	
Yunnan		Lemon, kumquats, Ponkan	
Guizhou		Sweet oranges: Huazhou Cheng, Hong Jiang Cheng, Feng Cai, Liu Cheng	

Satsuma harvest season is from October to December, whereas pummelos mature from September to February. Tankans mature from January to March. The sweet orange varieties Lue Gim Gong and Pineapple are grown to some extent.

(c). Japan: Citrus is confined mainly to the southwestern region of this country below 36°N latitude (Fig. 2.14). Citrus is grown near the coast line because there's less of a frost problem. Rainfall in citrus growing region ranges from 1200 to 3000 mm. Citrus is produced in some 20 different prefectures, the leading being Ehime, Wakayama, Saga, Shizuoka, Kumamoto, Nagasaki, and Hiroshima. Japan is a leading producer of Satsuma mandarins (*C. unshu*). A wide range of Satsuma fruits (mutants and selections) are grown, from early Miyagawa Wase (early September) to mid-season (Nakate unshu), to late Owari cvs. Okitsu Satsuma matures one week earlier than Miyagawa Wase. Miyamoto, Yamakawa, and Tokimori are other early Satsumas. Owari is harvested from November to January in different parts of Japan. Bud mutants such as Juman and Aoshima are late ripening (December–January) with Brix as high as 13 (Iwahori, 1991). Satsuma is called *Unshu mikan*. In Japanese, *mikan* is a common word for *orange*. But a satsuma is a loose-skin mandarin. There are over 100 bud variants reported in *C. unshu* in Japan. The unshus are grown in Kanagawa prefecture to Kagoshima prefecture with main areas including Shizuoka, Ehime, Wakayama, Hiroshima, Kumamoto, Kanagawa, and Saga. Plantations are on hill terraces on slopes with 3 by 3 m spacing. Due to steep slopes, harvesting and other cultural practices are assisted by trolley cables suspended on poles. Due to severe winter, paddy straw is used to cover the trees. Satsuma on Trifoliate is very cold-hardy and can tolerate temperatures several degrees below zero for many days. Most of the fruit packing and marketing is done through cooperatives. Satsumas are available from October to April.

Natsudaidai (*Citrus natsudaidai* Hayata), Hassaku (*C. hassaku* Hort. Ex. Tanaka), and Iyo (*C. iyo* Hort. Tanaka) are other important citrus fruits grown in Japan. Ponkan mandarin, hybrid Tankan, Kumquats, Fukuhara oranges (*C. sinensis* Osbeck), and navel oranges are also grown to some extent. Sweet oranges, lemons, and grapefruit are not very successful in Japan because temperatures are too low for late-maturing citrus such as grapefruit and Valencias, and there is heavy rainfall in early summer. Winter in most parts of Japan is very severe from December through February. Satsumas are successful due to superior quality, cold-hardiness, ripening in wide range of season, and seedlessness. Through storage techniques and various strains of Satsuma grown in the open field, fresh fruits are made available from October to April. Japanese citriculture is unique in producing commercial citrus indoors under heated plastic (vinyl) houses, thus Satsumas are harvested from June onwards. These are called house mikans and are of high sugar and excellent quality. The control of water application and other inputs is possible for a high degree of uniformity in fruit quality. These fruits are sold at 4 times higher prices than normal field grown ones (Iwahori, 1991). The fruit is relatively green and needs degreening. Natsudaidai is usually marketed from December through July, Hassaku from December

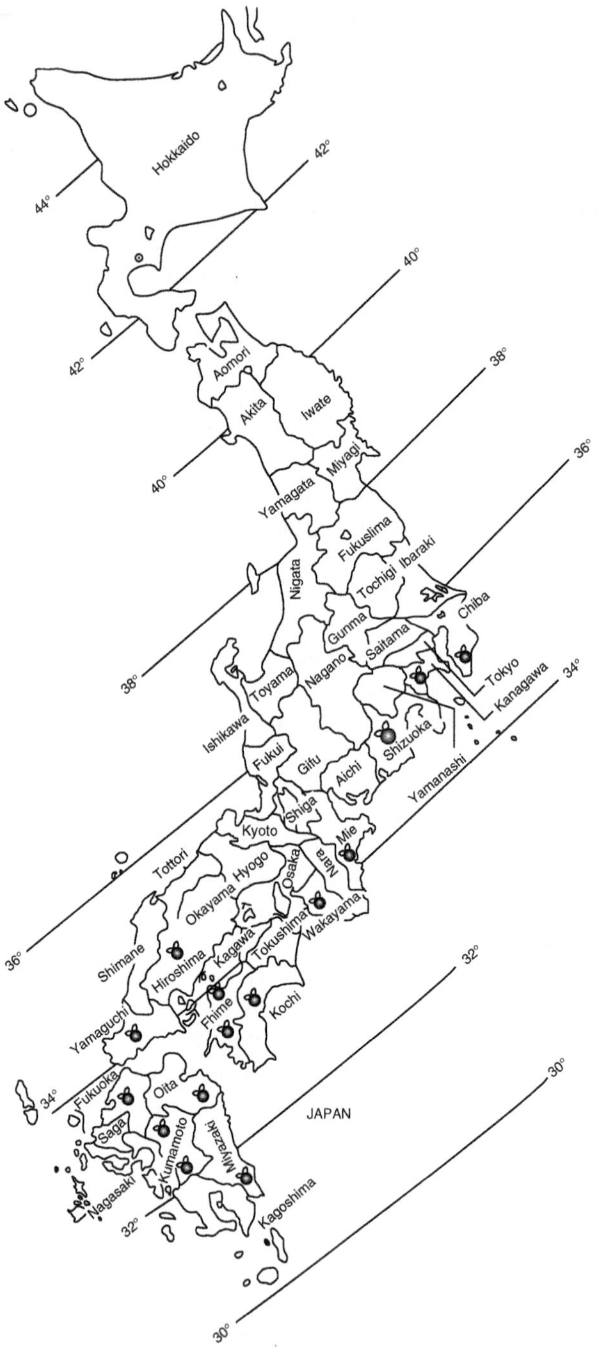

FIGURE 2.14 Citrus Growing Prefectures of Japan.

through June, and Iyokan from January through May. The navel oranges and Ponkan mandarins are grown in non-heated vinyl houses under the roof. Fruits are larger and smoother without damage on the rind, and color develops earlier (Iwahori, 1991).

(d). The Philippines: The Batangas region of the Luzon and Davao region of Mindanao are growing citrus. King mandarin, Ladu and Szinkom mandarin, plus Kao Puang pummelos and Valencia oranges are the main types grown.

(e). Thailand: The central coastal region and mid-northern inland areas are well-known for citrus cultivation. Mandarins (cv. *Somkeowan*), pummelos (*som-o*), orange (*somkliang*), and limes (*manao*) are grown in dry northern and northeastern provinces. In tidal coastal areas of Nakorn Patom, white- and pink-fleshed pumelos are grown. The mandarin Somkeae Wan known as *Siam* in Indonesia or *Limau lang kat* in Malaysia is providing 80 percent of the national production. This mandarin belongs to the group of *Citrus suhuiensis* Hort. ex. Tanaka. Fruit is spherical in shape, with bright yellow color in northern provinces (Nan, Chiang mai, Chiang Rai) but remains green in central and south. A chance hybrid of Somkeae Wan and Ponkan is also grown. Acid limes are another important fruit. Sweet orange is called Somkliang and Somtra is the cultivar that is grown. The seedless pummelo cvs Kao Phuang and Khao Pan are superior in quality.

(f). Vietnam: This country grows sweet oranges, mandarins, limes, and pummelos mostly for internal consumption. The Campbell and Olinda Valencias and Hamlins are important cultivars of oranges.

3. Mediterranean Region

This is a very important region for citrus cultivation (Fig. 2.15). This region produces some of the finest mandarins, oranges, and lemons in the world. Important citrus-producing countries are Spain, Portugal, Italy, Morocco, Greece, Israel, Turkey, Tunisia, Algeria, Cyprus, and Lebanon. Climate in general is dry in summer and moist in winter. Coastal areas are moist in summer also. This region is famous for excellent quality blood-red sweet oranges. Non-blood (yellow-fleshed) oranges are also grown and called *Blanca* in Spain or *Blond* in Italy. Harvesting of early varieties start in October and November. In some of the blood oranges, rinds, flesh and juice are all red (Sanguinella or Sanguigna of Italy).

(a). Spain: This is one of the leading fresh citrus growing countries of the world, and most of the produce is exported to European Union members. Commercial citriculture is more than a century old. Eastern and Southern regions facing the Mediterranean Sea grow citrus (Fig. 2.16). Interior river valleys (closer to the sea coast) also grow citrus, but most of the northern and central parts of the country are too cold to cultivate citrus. Most of the mandarins are grown in Castellon and Valencia provinces, which are on the east coast. Alicante, Tarragona, and Murcia provinces of this region also grow mandarins. Most of the sweet oranges come from Alicante, Valencia, and Castellon. Murcia,

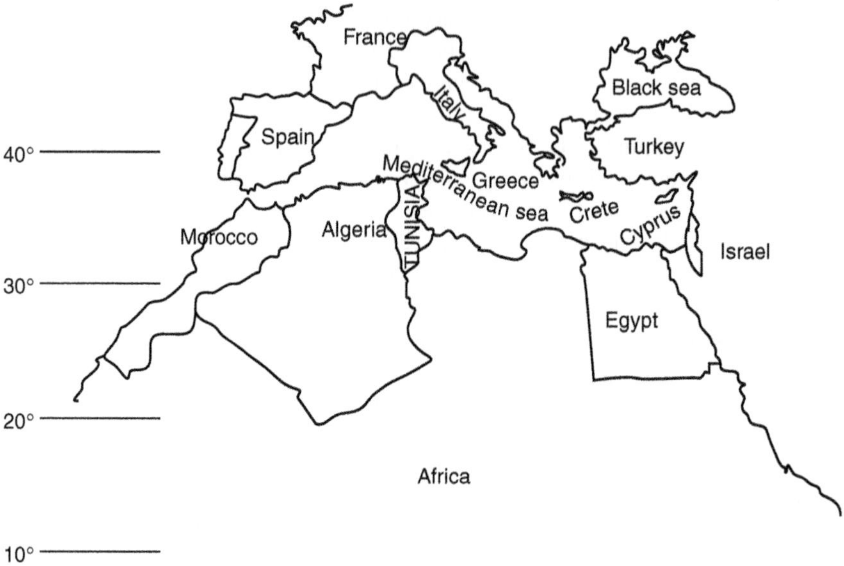

FIGURE 2.15 Important Mediterranean Region Countries Growing Citrus.

Valencia, and Alicante are leading in lemon cultivation. Grapefruit is also grown in Valencia and Alicante to some extent. The east of Andalucia, comprising provinces of Almeria, Granada, and Malaga also grow lemons, sweet oranges, and mandarins. The western Andalucia region with provinces of Cadiz, Sevilla, Cordoba, and Huelva grow sweet oranges and mandarins. Sour oranges are grown in Sevilla (Del Rivero, 1981).

Temperatures in the Valencia region start dropping in September, and in October, daytime temperatures are 12–15°C and about 10°C in mornings and evenings. Rains are received in September and October, Winter starts in November and it is quite severe in December–February in some parts of Spain. Oranges are available during October–July. Satsumas and early navels start appearing in the markets from September onwards. Important citrus cultivars and harvesting seasons are given in Table 2.2.

Navelina, Newhall, Washington navel, and Navelate are four main navel group oranges grown in Spain. Newhall is the earliest to mature. Lanelate is introduced for late harvesting. Valencia Late is available for the longest duration. Ricalate is a spontaneous bud mutant of the Washington navel orange and is late in harvesting (June–July). Fruits retain their firmness and do not drop until June. Fruit of this variety are oval in shape with smooth rinds and are uniformly orange in color. Salustiana is non-blood orange that is widely grown in Spain.

Among mandarins, Marisol, Esbal, Tomatera, Clemenules, Hernandina, Satsuma, and Clausellina, and the hybrids Nova and Fortune are grown. Two new spontaneous mutations, Clemenpons and Loretina, in the Clementine group, are early, seedless, and promising (Bono et al., 1997).

FIGURE 2.16 Citrus Growing Regions of Spain.

Grapefruit is grown in southern Valencia and northern Alicante provinces. Marsh, Red Blush, Star Ruby, and Shamber are the main varieties with very good quality grown on sour orange and Troyer. Harvesting starts in November and continues up to March.

Lemons are grown mostly in Murcia and Alicante (southeast) along the Mediterranean coast. The region has a semi-arid climate with annual rainfall less than 300 mm, long dry summers, and mild short winters. Fino-49 (early), Eureka, and Verna-62 are extensively grown. Fino is also known as *Mesero*, *Primofiori*, or *Blanco* in Spain.

(b). Portugal: Algarve in southern Portugal is an important orange (Valencia Late) growing province. Rains up to 45 cm occur in summer. Mediterranean mandarins and Washington navel oranges are also grown. Lisbon lemons are widely grown in this country.

(c). Italy: Citrus is grown mainly in three climatic regions of this country – Ionian, South Tyrrhenean, and Sardinia (Russo, 1981). The regions have varied climate and rainfall conditions because of mountain ranges and sea coast (Fig. 2.17). (1). The Ionian region lies along the Ionian arch (Gulf of Taranto) and farther south and has a Mediterranean climate. The areas of Puglia, Basilicata, Calabria, and Messina and the provinces of Syracuse, Ragusa, and Sicily are in this region. The provinces of Puglia, Basilicata, Calabria, and Ragusa produce the best quality oranges, clementines, and mandarins. The Catania areas, with mild summers and cold winters (southwest side of Mt. Etna) and the Catania plain, with diurnal variation, produce Italy's famous blood oranges because the climate is best for anthocyanin pigment production. The interior areas of Catania and Ragusa also produce the best quality Avana mandarins and Clemetines because the pigment development and color is best. The Ionian coast of the

TABLE 2.2 Citrus Fruits Grown and Harvesting Season in Spain

Citrus cultivar	Harvesting season	Remarks
Sweet orange: Doblefina group	December–January	Mid-season
Salustiana	November–May	
Sanguanelli, Washington sanguine	February–May	Blood group oranges
Washington navel	November–May	Good color, 11–12% TSS, 1–1.2% acidity
Navelina	Harvest starts in November–March	Earliest orange, harvesting continues up to February
Navelate	January–June	Very late cultivar
Cadenera	January–April	Holds well on the tree
Valencia	April–July	Retained on tree for longest time
Clementine: Fina	November–December	Fina fruit: small, seedless, excel flavor, holds on the tree
Marisol, de nules, Oroval, Monreal	September–October or November	Marisol and Oroval mature earlier than Fina, and de nules matures later. Monreal is self fruitful and produces few seeds
Arrufatina and Esbal	Early November	
Hernandina	December–January	Marketed up to February
Satsuma group: Clauselina, Other range of Satsumas (Owari)	September–February October–November	Orange colored, seedless, sweet
Grapefruit	October–June	Mostly red-fleshed cvs
Lemons: Fino, Masero	October–February	
Verna (Berna), Verna-62, Lisbon, Eureka	February–August, September–October	Very good quality lemons, almost throughout the year

Messina area produces the best quality lemons. (2). The South Tyrrhenean region facing the Tyrrhenean sea, with its mild winters and summers receive winter rains of 1500–2000 mm. The coastal Calabria, plain of Gioia Tauro, and coastal Campania produce oranges (Biondo Comune), mandarins, and lemons. (3). In the Sardinia region, summers are dry, and rain occurs in fall and spring. This region produces a small quantity of citrus, mostly Washington navels, Avana mandarins, and Biondo Comune oranges. Fresh citrus of one or the other type is harvested year-round in this country (Table 2.3).

Italy is one of the leading exporters of fresh citrus fruits. Moro Italian sweet oranges (deep blood variety) are harvested in January near Catania, Sicily and contain about 44 percent juice 10 percent TSS, and 1.77 percent acids. Valencia oranges harvested in this region in June contain nearly 41 percent juice, 12.3 percent

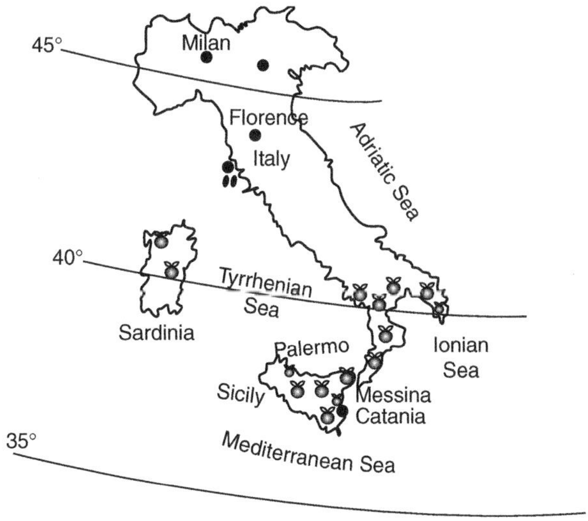

FIGURE 2.17 Important Citrus Regions of Italy.

TSS, and 1.27 percent acids (Cooper and Chapot, 1977). Another important blood variety of Italy is Tarocco. Among the navel group, Navelina is important due to globe-shaped-to-spherical fruit, very good quality, earliness, and suitability to different environments. A new selection is Navelina ISA 315, which produces uniform-sized fruit with pronounced color. The temperate climate in some parts of Italy pose a threat of freezing damage to fruit. Bionde oranges (non-blood) and navel and Valencia oranges are widespread in Calabria, Basilicata, Puglia, and Sardinia.

Major mandarin cvs are clementines and Satsuma (Table 2.3). The red-skinned (like the Dancy tangerine) variety is Mandarino sanguigno. Fremont and Malvasio mandarins are late ripening and are grown along with clementines (Oroval, Nules, Comune). In southern Italy (Apulia), clementines and mandarins are important. Mapo tangelos (commune mandrin × Duncan grapefruit) are widely grown in Sicily, Calabria, Sardinia, Apulia, and Basilicata.

Italy is a large producer of lemons. Most of the Italian lemons are grown in the Messina, Palermo, (38°N), Catania, and Syracuse areas. One of the important lemon varities is Feminello Ovale, which produces medium sized-fruit with a nipple year-round. Two other varieties are Monachello and Interdonato. Interdonato is juicier, early, and large fruited. Malsecco (*Phoma tracheiphila*) is a fungal disease of importance in Italy and Monachello, and Interdonato are resistant to some extent. Summer lemons are called *Verdelli crop* (July–August), the spring fruit is called *bianchetti crop* (April–June harvest), and the winter crop of lemoni is harvested from November to March.

(d). Morocco: In Morocco, citrus is grown on a 76310 ha area with production of about 1.4 million metric tons (El-Otmani, 2003). Climate in general

TABLE 2.3 Citrus Cultivars and Harvesting Seasons in Italy

Citrus group	Cultivar	Harvesting season	Remarks
Oranges			
Pigmented (blood)	Tarocco	December–April	Deep blood red
	Moro	December–April	Deep blood red
	Sanguinello	December–April	
	Sanguinello Moscato	December–April	
	Sanguinello Comune	December–April	
	Sanguigno		
Blond (non-pigmented)	Valencia	May–June	
	Navels	November	Early, seedless
	Navelina	October – November or December	Early, seedless
	Biondo Comune	November–December	Mid-season, yellow-orange flesh, good color in January
	Ovale		
	Belladona	November–December	Holds on the tree
	Calabrese	January–February	Few seeds, holds on the tree, late
Mandarins/tangerines	Satsuma, (Wase, Miyagawa, Okitsu, Owari)	October or November–February	Seedless
	Clementine Comune	November–February	Seedless
	De Nules		Seedless
	Oroval		Seedless
	Monreal		Seedy
	Tardivo-di-Ciaculli	February–March or April	Fewer seeds than Avana
	Mapo tangelo	October	
	Avana	November–December	Seedy
Lemon	Feminello group	Available year round	Average resistance to Malseco
	Monachello group	Available year round	High resistance to Malsecco
	Interdonato	Available year round	Very few seeds, juicy, acidic, early maturing
Citron	Diamante		Sour in taste

is Mediterranean type (cool autumn and winter nights while springs and summers are warm). Citrus is grown in the ever-humid climate of the Atlantic coast near Rabat to the arid hot climate in the interior locations of Mechra Bel Ksiri, Sidi Shimane, Fes, and Marrakesh. Clementines are grown mainly in the Souss (200 mm rainfall) region in the south and Oriental regions of the country where an arid dry climate prevails and excellent quality fruit is grown under the irrigation facility on sour orange rootstock (El-Otmani, 1992). Valencias are grown in the Souss and Gharb regions (northwest). Navels and blood red oranges are predominantly grown in the Gharb region (warm and humid). Among sweet orange varieties, Valencia, Washington navel, Washington Sanguine (Washington blood orange), and Salustiana are important. Washington navels grown in the interior arid areas of Haouze (Marakkesh) and Tadla produce fine quality fruit, but the rind is thick. Fruit grown in coastal areas also compares well with fine quality citrus of Spain. Navelate and Lane late mature in February and March, whereas Maroc late (as Valencia is known in Morocco) is available up to June (El-Otmani, 2003). Washington Sanguine (blood red) performs very well in Morocco. Among mandarins, clementines are grown in Loukkos in northern areas (about 500–600 mm rainfall). Satsumas of a wide range of maturity are also grown. Miyagawa wase are early (September), and Owari are mid to late (December–January). In Clementine, the selections Bekria (early September) to Muska and Nour (late) are grown.

The Moroccan Clementine is early, has excellent rind color (deep red), is seedless and juicy with a thin rind and excellent eating quality. Quality even surpasses clementines grown in Spain. This fruit is also grown in Algeria, Spain, Corsica (France), and Tunisia. That is why Morocco exports more than 50 percent of the citrus produced in the country, and Europe is the major market (El-Otmani 1992). Murcott, Nova, Fortune, and Ortanique are widely grown, but areas of these cvs are decreasing due to certain problems. Afourer or white Murcott or Delite is a selection from Murcott. It is seedless (in solid blocks), widely cultivated, easy peeling, intense orange red in color, and late maturing. TSS is about 12–14 percent with 1 percent acidity.

Grapefruits are grown in the Souss Valley east of Agadir in a humid climate. Very little tonnage of grapefruit and lemons are produced compared to mandarins.

(e). Algeria: Varieties of oranges grown in Algeria are Hamlin, Cadenera, Maltaise Sanguine, Thompson navel, and Washington navel. Hamlin oranges grown in Algeria develop reddish orange color with unblemished fruit; however, skin is rough, and juice amount is relatively small as compared to fruit grown in the humid climate of Florida. The Algerian Clementine matures late (November–February), fruits are small to medium, deep reddish color with few seeds, sweet and juicy.

(f). Tunisia: Major sweet oranges grown in this country are: Valencia and Washington navel. Maltaise Sanguine (Blood Red) is distinctly a table fruit and grows in fine quality in the Mediterranean climate.

(g). Egypt: Citrus is grown mainly in the delta area of the Nile river near Tanta, Benha, and Cairo. Navel and Beledi oranges and clementine mandarins are major citrus fruits. The Beladi orange, a seeded variety with 13–14 percent TSS and light blood-red flesh color is also grown. Valencias are exported to Middle-East countries and Russia. Local acid limes and Eureka lemons are harvested from August to November, whereas Valencias remain on the trees up to August.

(h). Israel: Citrus is grown on the Mediterranean coast of the country with its dry, warm and sub-tropical climate and also in dry and hot climate (semi-arid) of the inland valleys and Negev desert. During 1990 and 1991, the citrus area was around 43000 ha with 1.5 million metric ton production (Cohen, 1991). Presently, the area is around 62000 acres with 1 million metric tons of production, 85 percent of which is exported. The citrus area is shrinking due to a water shortage, reduced profitability, and urbanization. The climate difference is dramatic from the coastal Mediterranean one to interior semi-arid valleys as one travels from the coast to the interior. In summer there is no rain, whereas in winter, rainfall is about 700 mm in the north near the border of Lebanon to 150 mm in the south. The southern region of the country is better suited for lemons and Washington navel oranges, and hot and dry inland areas are useful for grapefruit. Export season is from November to May (Cohen, 1991). The main orange variety is Shamouti, whereas Washington navels and Valencias are also grown. In grapefruits, Marsh, Red Blush, Star Ruby, and Flame are important. The Eureka lemon is another fruit of export importance. Shamouti of Israel is medium to large, oval shaped, seedless, mid-season, orange colored at maturity, with much less rag, full of juice, and equally as good as Spanish fruit. This variety is believed to be a limb sport of the Beladi orange grown near Jaffa. The Shamouti orange grown in dry areas of Israel has very good quality with a much lower rot percentage. Washington navels also produce fruits of excellent quality. Among mandarins/tangerines, Nova (clementine × Orlando tangelo) and Niva (Valencia × Wilking) are grown in addition to clementines, Satsumas, and Dancy tangeries. The Nova matures from late November to late February, and the Orah matures in December–January. Israeli citrus is of very good quality and it is exported to the EU and other countries of Europe. The Minneola tangelo is an easy peeling cv harvested from mid-December to mid-January. Towards the end of the harvest season, the fruit suffers from creasing, loses its firmness, and deteriorates. 'Ortanique' and 'Murcott' are also grown to some extent. The important cvs and harvesting seasons of citrus in Israel are given in Table 2.4.

Israel's marketing season starts in September and continues until the following April. More than half of the Shamouti crop is exported. Israel is also a main exporter of grapefruit, lemons, and mandarins (specialty fruits) to the EU.

(i). Lebanon: The Shamouti orange is the main cv, whereas Beladis (Beledi) and Washington navels are also grown. Beledi is a non-blood orange that is seeded and thin-skinned. Blood Red Shamouti is moderately colored with reddish blush on the rind when mature.

(j). Syria: Citrus fruits of the same varieties as in Lebanon are grown in Syria.

(k). Cyprus: Cyprus grows Shamouti oranges and Eureka lemons of high quality, which are also exported to European countries. The main areas are Famagusta on the eastern end of the island and Limassol on Cape Gata.

(l). Turkey: The Mediterrranean coastal areas of Antalya, Adana, Mersin, Ivcel, and Hatay are well known for citrus cultivation, mainly of oranges and lemons. Coastal parts of the Aegean and eastern Black sea regions also grow citrus. The Aegean region grows Satsuma mandarins. Lemons have spread in the eastern Mediterranean region, especially in Mersin and Adana provinces. The lemon

TABLE 2.4 Citrus Cultivars and Harvesting Season in Israel

Citrus cv	Harvesting season
Orange: Shamouti	November–March
Washington navel	October–December
Valencia	March–May
Grapefruit: Red Blush, Flame, Sweetie	September–April
Mandarin: Clementine, Satsuma, Nova, Orah, Minneola	October–March
Lemon: Eureka	August–April

var Kutdiken (belonging to the Eureka group) provides high yields, very good fruit quality, and storage viability. Fruits of this variety are long, cylindrical, and typical of a lemon shape. Orange varieties of Shamouti, Washington navel, Valencia, and Parson Brown are grown. Akcay Sekerlisi, an early, non-blood, seedless variety of orange is also grown. Satsuma mandarins are grown in the Rize and Izmir areas.

(m). Greece: Greece is situated in the eastern part of the Mediterranean basin. It has a Mediterranean climate with annual rainfall of 1000 mm in most parts of the country. In the east and south, rainfall is 400–800 mm. Citrus is grown mostly in the central and southern parts of the country in Crete province (Chania and Heraklion districts), Peloponnesus (Corinth, Achaia, Laconia, and Argelis districts), Central Greece (Magnissia and Etolokarnania districts) and Ipiros (Arta and Preveza districts). There is a coastal climate in many parts of the country, and thus the mean minimum temperature is not very low. It is 7°C in the north in the month of January and 18°C in July. Mean maximum temperature is 13°C in January and 31–32°C in July, whereas in the south, the mean maximum temperatures are 16°C in January and 28°C in July. There is not much variation in minimum and maximum temperatures in the southern part.

Lemons are grown between Corinth and Patrai. Important varieties are Maglene and Eureka. Maglene trees flower in the months of April and May, and fruits are harvested from September to January. Lemons are exported from October to April. Oranges are harvested from November to March. Fruits grown in Preveza, Arta, and Magnesia suffer frost damage. Important mandarin cultivars Encore (Willow leaf × King) and Fortune (Clementine × Dancy) are grown in the Crete area of Greece. The Malvasio mandarin is also grown. Late mandarins are harvested from February to April.

4. Eastern Mediterranean Region and Central Asia

Iraq and Iran grow citrus mainly for domestic consumption. In Iraq, citrus growing has potential near the areas of the Diyali River, Baghdad, Karbala, Babylon, and Kut. Beladi sweet oranges, Washington navels, Shamoutis, Moros, and Taroccos are grown. Mandarins known locally as Indian mandarin (similar to mandarins grown in India) are grown. It was probably introduced from India. It is a seeded mandarin with fairly good quality, but it puffs when ripe. Among grapefruits, Marsh and Red Blush are grown. Clementine mandarins of various types are grown. Climate in this region is arid with temperature about 35°C or higher with 20–30 percent RH and 12–15 cm rainfall.

In Iran, Beledi, Maltaise Sanguine, Shamouti, Thompson navel, Hamlin, and Salustiana oranges are grown. Mazandaran province is one of the important citrus producers. In southern Iran, Mexican limes, locally called Shirazi, are grown. Among grapefruits, Marsh and Red Blush are important.

5. North and Central America and Caribbean Region

(a). United States of America: This is one of the leading countries in citrus cultivation with Florida, California, Arizona, Texas, and Louisiana being citrus producing states (Fig. 2.18). As per the information of the U.S. National Agricultural Statistics Service, the value of the 2004–2005 U.S. citrus crop has been estimated to be around $2.39 billion as packinghouse door equivalent. A fairly good quantity of fresh citrus is exported from the United States California and Arizona export mainly oranges and lemons, Texas exports good quality grapefruits, and Florida exports tangerines, their hybrids, and grapefruits. California produces navel oranges and its variants, whereas Florida produces Hamlin, Pineapple, and Valencia. Sweet orange industries of California, Texas, and Arizona are based on fresh fruit marketing, whereas Florida oranges are destined mainly for processing. Florida, California, and Texas produce fairly large quantities of grapefruit for the fresh fruit market. The citrus crop is vulnerable to freezes and hurricanes in the states of Florida, Texas, and California. The micro-sprinklers are the preferred form of irrigation to reduce the damage caused by freezes. The canker has been reduced to an acceptable level from Florida. Fruit flies are a problem in some areas of California, Florida, and Texas (Coggins, 1991).

(i) Florida: Most of the fruit grown in Florida is processed, and due to Florida's orange production, the U.S. stands second in the world in citrus processing after Brazil. The Valencia orange is the most common fruit processed to juice including frozen concentrated orange juice (FCOJ). Florida citrus acreage dropped from 931 249 in 1968 to 748 555 acres in 2004. Oranges occupy 622 821 acres, grapefruit are grown on 89 048 acres, and specialty fruits (tangerines/mandarins and their hybrids) grow on 36 686 acres (Anonymons, 2004). Indian River, Polk, St. Lucie, Hendry, and De Soto counties are major producers of citrus (Fig. 2.19).

Major orange varieties of Florida are Hamlin, Parson Brown, Valencia, and Pineapple. Ambersweet (*C. reticulata* × [*C. paradisi* × *C. reticulata*] × Mid season orange), Sunstar, Midsweet, and Gardner are some of the new mid-season varieties, which are also planted on a considerable area. Persian limes and lemons are also grown, and harvesting is short in the summer season. Because of freezes in north Florida in the past, citrus cultivation has shifted to southern coastal areas and the central ridge. The climate is hot and humid, so problems

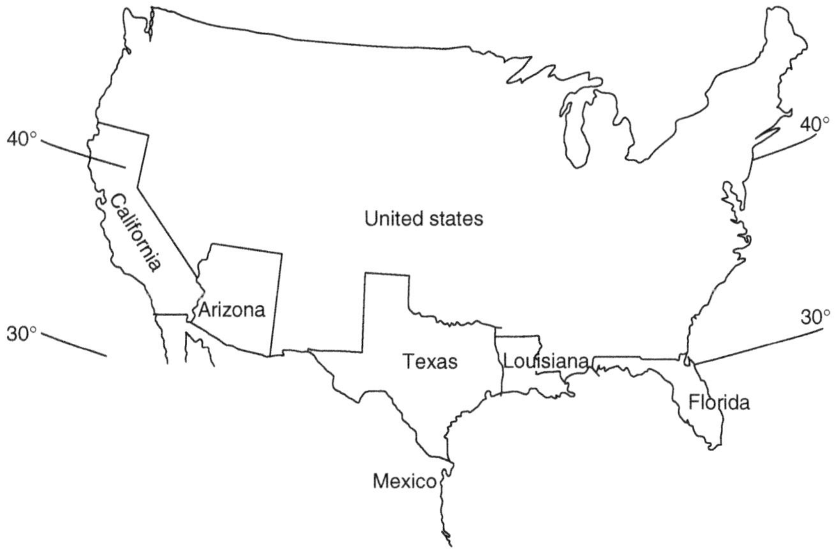

FIGURE 2.18 Citrus Producing States of the United States.

such as insect-pests affecting rind quality and causing poor color development (although internal quality is achieved) render fruit relatively inferior in quality as compared to fruit from California. Neverthless, Florida produces very good quality grapefruits, which are exported to Japan and EC countries. Red Blush and Flame (deep red flesh) are important seedless grapefruit cultivars besides Marsh.

The tangerine cultivars, namely Dancy, Robinson (*C. reticulata* × [*C. paradisi* × *C. reticulata*]), Fallglow, and Sunburst (Robinson × Osceola) are important for the domestic and export markets. Temples and Murcott (Honey) are major cvs of specialty fruit in Florida. Afourer, a new seedless selection from Murcott, was introduced from Morocco and is harvested from November to January. Similarly, Nules and Marisole (bud mutations from clementines) are also being tried in Florida. Orlando, Minneola, and Seminole tangelos (Duncan grapefruit × Dancy tangerine) and Nova hybrids are important. Nova and Robinson are early. Of the total citrus produced in Florida, not more than 6–7 percent of oranges, about 1/3 of grapefruits, and about 5 percent of the specialty (tangerines) fruit are sent to the fresh fruit market, and the rest is processed. Harvesting starts with Hamlin for the fresh fruit market in October–November and continues up to May as the Valencias and grapefruits are harvested until that time. June–September is a gap, and efforts are on to develop new cvs of grapefruit and specialty fruit to fill this gap.

(ii) California: It is a major fresh citrus producing region in the United States besides Arizona and Texas. Citrus is grown mainly in southern and central California. The San Joaquin Valley region is the largest orange and tangerine growing belt with more than 65 percent of the state's citrus acreage. This

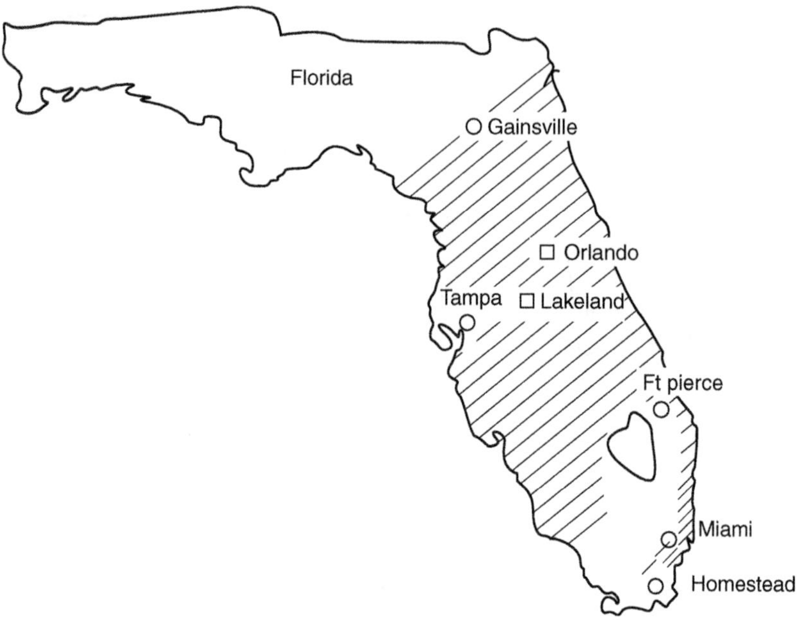

FIGURE 2.19 Important Citrus Growing Parts (Hatched Areas) of the Florida.

region has hot and dry summers and cold and wet winters. This region grows mainly Washington Navel and Valencia oranges. The coastal areas have a mild Mediterranean-like climate. The desert areas are in Coachella and the Imperial valley, where the climate is similar to that in Israel. Rainfall is received mainly during winters (November–March or April); summers are dry. The interior regions such as western Riverside and San Bernardino counties, inland parts of San Diego, Orange, and Los Angeles counties are dry and have hot climates. These areas are colder than coastal parts in the winter (Fig. 2.20). Washington navel oranges grown in the interior areas of California have outstanding eating quality. Coastal areas are relatively cool and humid. In cool humid areas and hot deserts, navel quality is not so good. Navels are well adapted to the Mediterranean type climate. Valencias are well adapted to hot deserts (Coachella valley) and also to heat-deficient coastal climate. Washington navels (mature within 9–10 months) are available from November to February, whereas Valencias (mature within 12–15 months) are available from March to September, thus the industry is based on these two orange varieties.

In southern California, the Indio, Santa Paula, and Riverside areas have reasonable acreage under citrus. Lemons are mainly grown in southern California, specifically Ventura and San Diego counties. Lemons are picked when silvery green and then cured. Olinda and Cutter Valencias (late April–September), Washington navel, and Lanelate are leading sweet orange varieties of California. Bonanza sweet oranges are becoming popular. Oroblanco and Malogold are the triploid grapefruits successful in inland citrus valleys of California. Oroblanco is also exported to Japan. Among lemons, Eureka and Lisbon are major varieties.

(iii) Arizona: In Arizona, major areas growing citrus are desert valleys where Washington navels, Valencias, Marsh grapefruit, and lemons are grown. Citrus areas include along the Salt river (Salt River valley), Tempe near Phoenix, Welton-Mehawk, and desert Yuma areas. In Arizona, summer temperatures above 38–41°C are common around Yuma. Rainfall is 18–20 cm annually in this area.

(iv) Texas: The lower Rio Grande valley delta is well known for high-quality grapefruits. Ruby Red, Star Ruby, Flame. and Rio Red are major cvs for the domestic and export markets. Marsh seedless is white fleshed grapefruit still preferred by growers. Marrs is a popular sweet orange in Weslaco, Texas, and it matures from October to November.

(b). Mexico: Mexico is an important producer of oranges, grapefruits, tangerines, and acid citrus. Marsh and Red Blush grapefruit and oranges are grown mainly in Veracruz, Nuevo Leon, San Luis Potosi, Tamaulipas, and Sonora states. The Mexican lime (*C. auratifolia* Swingle), a small fruited acid lime also known as Limon Mexicano, is grown. The Tahiti lime (*C. latifolia*) is also grown, and fruits are harvested from April to November due to several flushes of flower. From December to March, there is less fruit in the market, and prices are high. Limes are produced on a large scale on the west coast in Colima, Michoacan, Tamaulipas, Guerrero, and Veracruz states. Among oranges, Valencia, Corriente, Hamlin, Pineapple, Washington navel, and Malta, are harvested from November to March. May–September is the off-season, and prices are high. Oranges are grown in the Montemorelos district in a sub-tropical climate (at 900 m elevation) and at Valles-Tamasunchale, Veracruz, Jalapa, and Cardoba districts in a tropical climate. Dancy, Fremont, Murcott Honey, and Minneola are important mandarin fruits. Among lemons, Genoa, Eureka, Lisbon, and Villa-Franca are important cvs.

(c). Cuba: This country has a tropical climate, and leading citrus producing provinces are Matanzas, Pinar del Rio, Isla dela Juventad (Isle of Youth), and Santiago. It is a major exporter of fresh citrus. Isle of Youth exports high quality grapefruit. This area has very favorable climate and soil. Marsh Jiarito is a local selection and a commercial variety harvested in August. Other grapefruits grown are Thompson, Frost, Marsh, Duncan, and Ruby Red. Sour orange and Cleopatra are the important rootstocks. Integrated pest management (IPM) has produced very good results in the area thus reducing pesticide use and consequently its residue hazard in fresh fruit. Use of bioregulators and rational use of insecticides has produced exportable quality fruit (Pardo et al., 1992).

Olinda, Frost, Campbell Valencia, and Criolla are grown among sweet oranges, and Honey, Kinnow, and Dancy are the important mandarins cultivated. Eureka lemons, acid limes (Tahiti and Key) are the major acid fruits. In the warm humid climate of the Jaguey Grande valley, Valencia oranges are widely grown.

Recently, two hybrids, Valentina (Clementine × Valencia Early) and Clementina (Clementine × Hamlin) have been found to be performing well at

FIGURE 2.20 Citrus Growing Regions of California.

Jaguey Grande (Bello, 1997). These hybrids have a deep orange color of peel and juice, are easy peeling, and have high juice content and nice flavor.

(d). Honduras, Costa Rica, and Nicaragua: These are the citrus producers in Central America. Honduras grows oranges, limes, and lemons mostly for domestic consumption.

(e). Caribbean Region: Jamaica, Puerto Rico, and Haiti are the important citrus producers in the region. Valencia oranges and Marsh seedless grapefruit of good quality are grown. Grapefruit in fact originated on Caribbean islands, and large juicy fruit of very good quality are grown here.

B. Southern Hemisphere

The southern hemisphere is characterized by winter during March–October and summer during November–February. Hence countries of the southern hemisphere that produce citrus can market fresh citrus fruits during March–October.

1. Australia and New Zealand

(a). Australia: Australia lies between 10°S and 40°S latitude and is a relatively dry country if overall climate is considered. Coastal areas receive good rainfall. On the eastern coast of Australia (Queensland), annual rainfall up to 250 cm/year occurs in the summer, whereas on the south coast (Melbourne, Victoria, and Perth) rainfall occurs in winter. On the east coast of New South Wales, rainfall is about 125 cm.

In Australia, citrus is grown mainly in New South Wales, Victoria, South Australia, and Queensland provinces (Fig. 2.21). The Riverland area of South Australia province (Murray River) comprises the large area of the citrus output (25%) of Australia. South Australia grows Valencia, navels, Marsh grapefruit, and mandarins. Sunraysia and Mid-Murray areas of Victoria and New South Wales (Riverina) contribute more than forty percent of the citrus produced. The Murray basin is most suited to citrus because it is closer to big markets such as Melbourne, Adelaide, and Sydney. Fast growing urbanization is leading to development of production centers near markets. The Murrumbidgee irrigated area and other irrigated areas of New South Wales contribute about 10 percent of the citrus produced in the country. About 5 percent comes from the central coast of New South Wales. Queensland contributes 15 percent of the production, and 4–5 percent comes from areas around Perth in Western Australia. Dry inland areas of New South Wales receive annual rainfall averaging 300 mm. Good quality Washington navels and Valencias are grown on irrigated land with a dry climate because the pest problems are lower, so use of chemicals is also lower. Recent selections from navel orange are similar to Washington navels in taste but have good on-tree storage capacity like Valencias. Orange production is about 80 percent of Australian citrus. The central coast of New South Wales on the eastern side of the country is a major lemon producing region. Eureka lemons are favored because they constitute a more profitable summer crop. Lisbon is grown less due to thorniness. Valencia, Washington navel oranges, mandarins, and grapefruits are also grown in the eastern coastal area of Gosford. Queensland province grows high-quality mandarins – Emperor, Ellendale, and Beauty. Seedless easy-peelers are high in demand. Fresh fruit maturity varies due to varied climatic conditions, and harvest starts from April in some coastal areas and northern Queensland until late October in southern inland areas.

Australian citrus fruits are available during most of the year (Table 2.5). Mandarins are available from April to November. Lemons and grapefruits are in short supply during January and February and are imported for supply to domestic market. Navels and Valencias from southern Australia are exported to

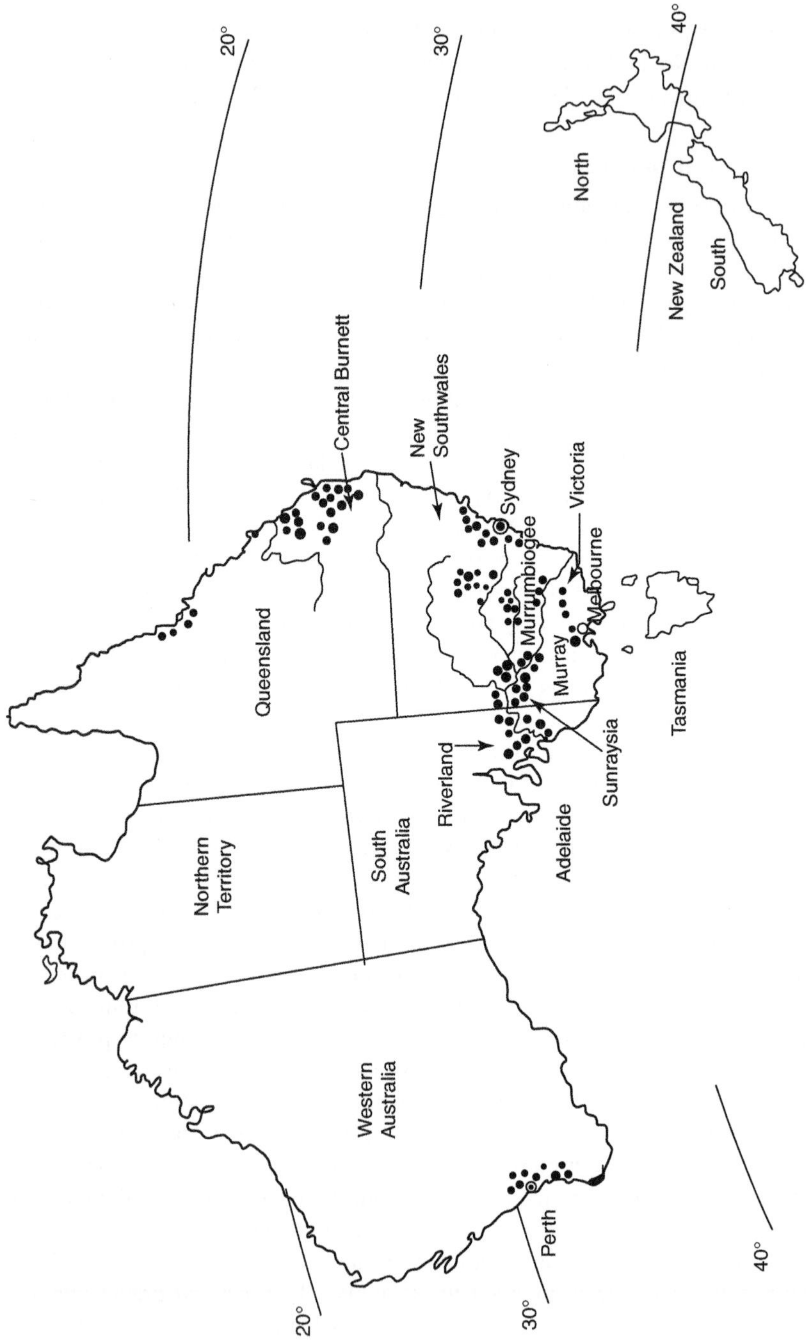

FIGURE 2.21 Major Citrus Growing Areas of Australia.

New Zealand and southeast Asia. Australian grapefruits are finding markets in Hong Kong and the rest of China also.

(b). New Zealand: New Zealand is a small island country (North and South Islands) with a moderate climate affected by the ocean (Fig. 2.21). Citrus is grown in the northern part of the country, which lies at about 35–45°S latitude and has a warm temperate climate. It receives rainfall of 1700 mm. In summer (January), temperatures around 24°C maximum and 14°C minimum are common. In July (winter), maximum temperature is 14°C and minimum is 5–6°C. This country has modest citrus production and very good scope for export of citrus to Pacific rim countries, mainly Japan. Citrus cultivation is possible in frost-free areas and eastern areas of the North Island. Among sweet oranges, navels and Valencias are grown. Miyagawa and Okitsu Wase (*C. unshu*) are important Satsumas. Clementines have also been tried, and Caffin is the early selection (Harty et al., 1997). The Seminole tangelo is also cultivated with some lemons.

2. South America

(a). Brazil: This country is situated between 4°N latitude and 33°S latitude, and equator passes through it (Fig. 2.22). In general, the climate is tropical in the north where the Amazon rain forests are and sub-tropical in southern and southeastern regions. Brazil leads citrus production in the world. Brazilian citrus depends heavily on rainfall – only about 10–15 percent of the citrus plantation is irrigated. The State of Sao Paulo (24°S latitude) in the southeast is the major producer of citrus and is more advanced technologically than Sergipe and Bahia states in the northeastern region. Rainfall averages 1400 mm with a mean maximum temperature of 31°C and mean minimum temperature of 11°C. The highest temperatures are 36–37°C. Rio Grande do Sul state is the leading producer of citrus, and it rains there almost every month in this state, so irrigation is not required. Rio de Janeiro, Menas Gerais, Santa Caterina, and Parana are the other citrus producing states. Winter season begins from March and lasts up to July or August. Summer starts in September and lasts up to February or March. Bahia, Baianinha, Hamlin, Natal, Navel, Pera, and Valencia (Liu Gim Gong) are important orange cvs. Hamlin is early, Pera and Shamouti are mid-season, and Valencia and Natal are late varieties. Most of the sweet orange crop is processed, and in fact, Brazil is the Number One citrus processor in the world. Among mandarins, Ponkan, Cravo (early), Mediterranean, Murcott (high quality and late ripening), and Dancy are important. Siciliano lemons and Tahiti and Mexican limes are grown in Southern states and Sao Paulo.

(a). Argentina: This country lies between 24°S and 56°S latitude. Citrus is grown mostly from 24 to 40°S latitude (Fig. 2.23). Winter season starts in March and lasts up to September. Summer is from October to February. A wide variety of citrus fruits are grown and harvested most of the year (Table 2.6).

The main citrus regions are as follows: (a) Mesopotamia: It lies from 26°S to 34°S latitude. The climate is semi-tropical to subtropical with a temperature range of 1–45°C in winter and summer. Rainfall received in the region is

TABLE 2.5 Important Commercial Cultivars and Harvesting Seasons of Citrus in Australia

Citrus cultivar	Harvesting season	Remarks
Oranges		
Hamlin	May–June	Early
Washington navels, Navelina	April–September	Late hanging selections of navels from Lanelate named Hutton and Wiffen can extend marketing up to December (Gallasch, 1997)
Leng navel	Early (August–September)	Thin-peeled, seedless bright colored thin rind, juicy fruit
Lanelate	November–December	Late var, remains on the tree up to May, peak harvest September–December
Valencia late	September–March	
Mandarins/tangerines		
Satsuma (Wase, Owari, Okitsu)	April–July	Seedless
Clementine Comune	May–July	Seedless
De Nules		Seedless
Oroval	April–May	Seedless
Monreal		Seedy when pollinizer used
Ellendale	Mid-July–August	
Dancy tangerine		Mid-season
Emperor	June–August	
Imperial	April–July	Early
Murcott		
Grapefruit		
Marsh seedless	July–October	
Red Blush	July–December	
Tahiti		
Lime	June–August	
Lemon		
Eureka	Mid-June–October	
Lisbon		
Villafranca		

1000–1500mm. The provinces of Missions, Corrientes, and Entre Rios are in this region. Citrus fruits grown are mandarins, oranges, grapefruits, and lemons. The climate resembles that of Florida. (b) The northwest region lies from 21°S to 28°S latitude with temperatures ranging from 2–45°C and rainfall of 500–900mm. The provinces are Jujuy, Salta, and Tucuman. This region is well known for lemons and grapefruit due to the warm humid climate of summer and mild winter. Marsh, Thompson, and Red Blush are important grapefruits (Table 2.6). Valencia and Pineapple oranges are also grown in this region. (c) The Littoral

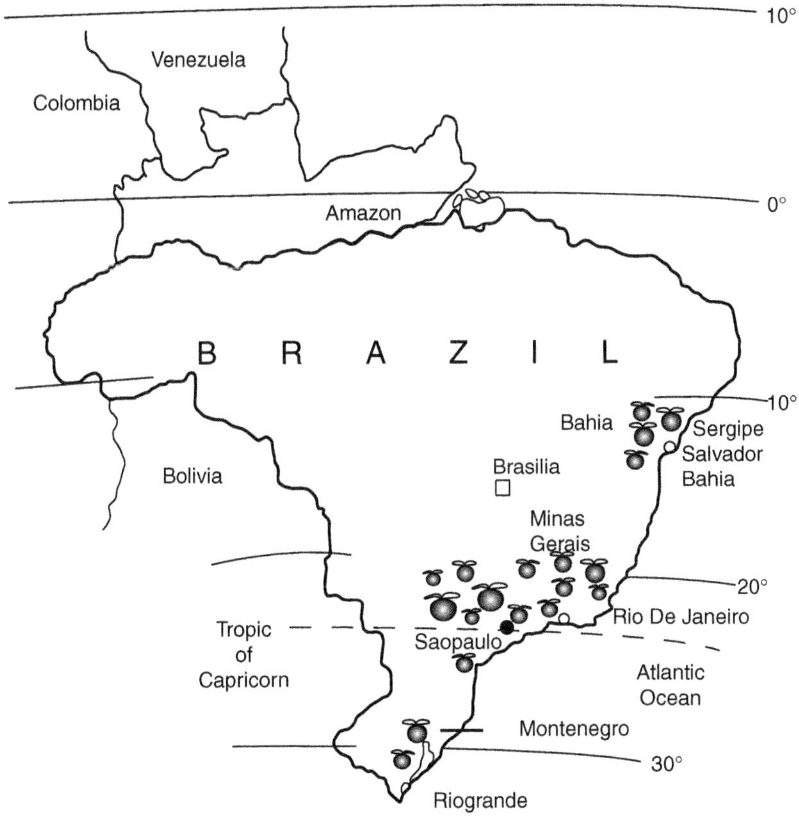

FIGURE 2.22 Brazil's Citrus Growing Regions.

region lies at 31–34°S latitude. It has a coastal climate. The interior climate is subtropical to temperate with temperatures ranging from 1°C to 24°C. This covers the east coast of Santa Fe province, northeast Buenos Aires province, and all of Entre Rios and Buenos Aires provinces. Mandarins are mainly grown in this region. (d) The central region covers Cordoba and Sao del Estero. This region grows oranges, mandarins, and grapefruit. (e) In the northeast region, in Formosa province, oranges, mandarins, grapefruits, and lemons are grown. (f) In the Andean region, covering the Catamarca area, mainly oranges, mandarins, and lemons are grown (Sartori, 1981). Two crops of lemons are importamt in Argentina. The winter harvest is from June to September, and summer harvest is from November to February. Eureka is harvested year-round in some areas, and fruit receives good price in summer. The main season of Argentine exports is from June to November. Valencias and lemons are major export fruits. Nagami kumquats (oval shape) with 3–4 cm length and 2–2.5 cm breadth with smooth

FIGURE 2.23 Citrus Regions of Argentina.

and well colored peel and 2–4 seeds are also grown in Entre Rios province. Citrus is imported from Brazil and Paraguy, the neighboring countries.

(c). Colombia: This country grows excellent quality sweet oranges (Washington navel and Valencias) and tangerines. The climate is tropical to sub-tropical.

(d). Venezuela: This country has a tropical climate as it lies between 2°S and 11°S latitude. In the central part of the country in the states of Yaracuy and Carabobo, citrus is grown. The states of Miranda, Aragua, Portuguesa, Guarico, and Monagas also produce citrus. The high altitude areas up to 500 m produce very good quality citrus due to the cool climate. Valencias, Washington navels, and Pineapple are important oranges. Dancy tangerines, Marsh grapefruits, and Mexican limes are also important crops.

TABLE 2.6 Citrus Fruit Grown and Their Harvesting Seasons in Argentina

Citrus fruit	cv	Harvesting season	Remarks
Oranges	Valencia	September–February	
	Washington navels, Navelina	March–July	
	Hamlin	March–April	Early
	Jaffa	June–September	
	Comun	April–September	Early to mid-season
Mandarins/ tangerines	Satsuma (Wase, Owari, Okitsu)	February–May	Seedless
	Clementine Comune	February–May	Seedless
	De Nules		Seedless
	Oroval		Seedless
	Monreal		Seedy
	Comun de-Concordia (Mediterranean type mandarin)	May–September	June–August peak
	Dancy tangerine	August–September	Mid-season
	Malvasio	July–October	Late ripening
	Campeona	July–October	Late ripening, large fruited, good eating quality but poor shipping quality
	Murcott	July–October	
Grapefruit	Duncan	April–November	
	Marsh seedless	April–November	
	Thompson, Red Blush, Ruby Red	April–November	
Lemon	Genova	March–October	Genova fruits are similar to Eureka
	Eureka	September–February	
	Lisbon		
	Villafranca		

(e). Uruguay: In the north of the country, Artigas, Salto, Paysandu, Rio Negro, Tacuarembo and Rivera, and in the south, Montevideo, Canelones, San Jose, and Calonia are important areas growing citrus. Uruguay produces excellent quality tangerines and oranges, and more than one-third of the produce is exported to European and North American markets. Mandarins such as Satsuma, Clementine, Lee, Ellendale, Malvasio, Ortanique, and Murcott are grown extensively.

(f). Peru: Maximum citrus cultivation is found around Lima, the capital. In the dry coastal valleys, citrus is grown. The areas of Ica and Piura (Morropon), the interior Amazon region, and areas of the state of Junin are suitable for citrus. Important vars. of orange are Washington navel, Valencia, Hamlin, and Criolla. Mexican lime is also a major crop. Dancy, Satsuma, and 'Ponkan' mandarins are grown. 'Marsh' is an important cultivar among grapefruits.

(g). Chile: Citrus cultivation is quite old in this country, but dynamic citrus cultivation is known only since the 1980s. The area of citrus is about 14000 ha (Ortuzar et al., 1997). Fruits are harvested from May to August during winter. Genova is the major lemon variety, although Lisbon and Eureka are also grown. Grapefruits are available from September to April. Thompson and Washington navel are the main oranges. Clementines and local mandarins are also grown. All fruit for the domestic market is packed in packinghouses after brushing or washing.

3. Africa

The African continent south of Sahara comprises important well-developed commercial citrus producers such as Republic of South Africa, Zimbabwe, Swaziland, and Mozambique. Nigeria, Kenya, Tanzania, and Ivory Coast also grow citrus, which is utilized for local consumption. The most important and technologically advanced citrus is being produced in South Africa mostly by Outspan/Capespan.

(a). Republic of South Africa: South Africa lies in the southern hemisphere in the African continent at about 23–34°S latitude (Fig 2.24). The climate in general is temperate – Mediterranean to sub-tropical. The autumn season (12–18°C minimum and 27–31°C maximum temperature) starts from March to May followed by winter (6–12°C minimum and 24–26°C maximum temperature) from June to November and summer (17–21°C minimum and 29–32°C maximum temperature) from December to February. The harvesting of fruit is mostly from February onwards until October.

In the Republic of South Africa, Transvaal, Natal, and Eastern and Western Cape provinces are the main citrus growing regions. Northern, eastern, and central Transvaal areas grow mainly sweet oranges (Valencia), lemons, and grapefruit, which are of export quality. The Grabouw area grows mostly clementines, whereas in Wolseley, Satsumas are grown. In the Saron area, Midknight and Delta Valencias are grown. Eastern Transvaal and Swaziland are low-lying areas that are hot and humid at less than 600 m altitude. The grapefruit produced in these areas are of good quality although they have a thick rind. In hot dry areas, fruit develop a 'sheep nose' shape (Barry and Veldman, 1997). Red Blush, Rio Red, Flame, Ruby Red, and Star Ruby are major varieties, which develop deep red internal color and TSS acid ratio of 8:1. The yields are as high as 110–140 metric tons per ha. Grapefruits are harvested from March and April to August. The Sundays River valley is the major area in Eastern Cape. Eastern and Western Cape produce more than 60 percent of the lemons exported from South Africa. The cool inland areas with high altitudes (more than 900 m) in Western Cape are suitable for Satsumas, clementines, navels, and Valencias. Lemons and Eurekas are also grown in these areas.

Midknight Valencia (medium late maturity April–September) matures 2–6 weeks earlier than Valencia. This variety is popular due to excellent quality seedless fruit. Generally Valencias are harvested from June to November. The mandarin/tangerines are harvested from March to September.

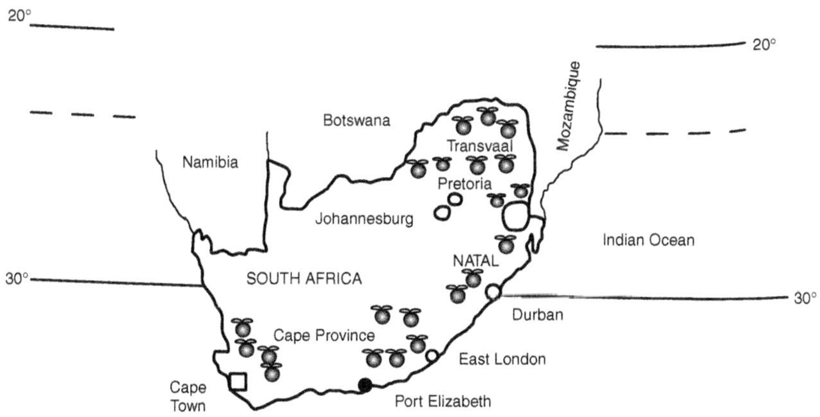

FIGURE 2.24 Citrus Growing Provinces of South Africa.

Fino lemons mature during April–August. Verna fruits are harvested in two seasons – from February–July and October–February. The summer harvesting (October–February) is of Verdelli or Rodregos crop. Lisbon and Eureka are also grown. Eureka sets multiple crops and can be harvested during three periods. Fruit is medium to small, and seeds are few to none. Lisbon's peels are medium thick and smoother than that of Eureka. In general, March–August is the peak season for lemons and limes.

(b). Zimbabwe: Large parts of Zimbabwe have a semi-tropical climate suited to citrus cultivation. Sweet oranges (Valencias, mid-season oranges, and navels), mandarins and tangerines, grapefruits (Marsh seedless), and lemons (Eureka) are grown.

(c). Swaziland: Citricultural practices are similar to those of summer rainfall areas of Eastern Transvaal in the citrus region (Nelspruit) of South Africa.

(d). Mozambique: Citrus culture in this country is similar to that prevailing in adjacent parts of Eastern Transvaal (South Africa) and Swaziland. Sweet oranges and grapefruits are widely grown. The South African cooperative citrus exchange handles the exports of Mozambiquan citrus fruit.

REFERENCES

Anonymous (1999). *State agric profile: Andhra Pradesh.* Commissionerate of Agriculture, Hyderabad, India.

Anonymous (2004). Florida citrus acreage drops to 748 555. *Citrus Ind.* 85(11), 23.

Aubert, B. (1991). What citriculture in South-East Asia for the year 2000. *Proc. Int. Citrus Symp.*, Guangzhou, China, pp. 37–58.

Barry, G.H., and Veldman, F.J. (1997). Performance of nine grapefruit cultivars under hot, subtropical Southern African conditions. *Proc. Int. Soc. Citric*, South Africa, Vol. 1, pp. 113–115.

Bello, L. (1997). Valentina and Clementina: two promising hybrids. *Proc. Int. Soc. Citric*, South Africa, pp. 219–220.

Bono, R., Soler, J., and Fernandez de Cordova, L. (1997). New spontaneous mutations of clementines. *Proc. Int. Soc. Citric*, South Africa, pp. 174–179.

Chen, Z., and Lai, Z. (1992). The introduction and research on pummelo germplasm of China. *Proc. Int. Soc. Citric*, Vol. 1, Italy, pp. 48–52.

Coggins Jr., C.W. (1991). Present research trends and the accomplishments in the USA. *Proc. Int. Citrus Symp.*, Guangzhou, China.

Cohen, E. (1991). Investigations on postharvest treatments of citrus fruits in Israel. *Proc. Int. Citrus Symp.*, Guangzou, China, pp. 732–735.

Cooper, W.C., and Chapot, H. (1977). Fruit production – with special emphasis on fruit for processing. In *Citrus science and technology* (S. Nagy and W. Veldhuise, eds.), Vol. 2. AVI Publication Co., CT, USA, pp. 1–127.

Davies, F.S., and Albrigo, L.G. (1994). *Citrus*. CAB International, UK, 254 pp.

Del Rivero (1981). Citriculture in Spain. *Proc. Int. Soc. Citric*, Vol. 2, pp. 939–949.

El-Otmani, M. (1992). Present situation and future outlook of Moroccan citriculture. *Proc. Int. Soc. Citric*, Acireale, Italy, Vol. 3, pp. 1164–1166.

El-Otmani, M. (2003). Citriculture in Morocco. *Citrus Ind.* 84(11), 22–23.

Gallasch, P.T. (1997). Evaluating new selections of late hanging navel orange. *Proc. Int. Soc. Citric*, South Africa, pp. 193–197.

Ghosh, S.P. (1985). Citrus in South–East Asia. FAO, Regional Office Bangkok, 70 pp.

Harty, A, Sutton, P., and Machin, T. (1997). Clementine mandarin evaluation in New Zealand. *Proc. Int. Soc. Citric*, South Africa, pp. 177–180.

Hodgson, R.W. (1967). The horticultural varieties of citrus. In *The citrus industry*, Vol. 1. Division Agricultural Science, University of California, USA, pp. 431–591.

Iwahori, S. (1991). Present research trends and accomplishments of citriculture in Japan. *Proc. Int. Citrus Symp.*, Guangzhou, China, pp. 14–24.

John, L., and Stevenson, V. (1979). *The complete book of fruit*. Anques and Robertson Publication.

Li, Y.T. (1991). A seedless mutant from cv. Jincheng. *Proc. Int. Citrus Symp.*, Guangzhou, China, pp. 116–120.

Li, W.B. (1992). Origin and distribution of mandarins in China before Song Dynasty (AD 960–1279). *Int. Citrus Congress Proc.*, Italy, Vol. 1, pp. 61–66.

Liansheng, G. (1991). The state of citrus production in Guangdong province. *Proc. Int. Citrus Symp.*, Guangzhou, China, pp. 174–175.

Morton, J. (1987). Lemon. In *Fruits of warm climate*, pp. 160–168.

Ortuzar, J.E., Gardiazabal, F., and Maghdal, C. (1997). The citrus industry in Chile. *Proc. Int. Soc. Citric*, South Africa, pp. 304–307.

Pardo, A., Martinez, C., Betancourt, M., Del Val, I., Esterez, I., Sanchez, R., Lopez, A., and Proenza, M. (1992). Grapefruits from the Isle of Youth. *Proc. Int. Citrus.*, Italy, Vol. 1, pp. 108–109.

Pathak, R.K., and Pathak, R.A. (1993). Improvement of minor fruit. In *Advances in horticulture* (K.L. Chadha, and O.P. Pareek, eds.), Vol. 1. Malhotra Publishers, New Delhi, pp. 88–94.

Peng, T.H. (1999). The performance of a mandarin and orange hybrid "Nova" in the Chongqing area. *China Fruits* 1, 59.

Romano, R., and Donadio, L.C. (1981). Characteristics of Brazilian sweet orange cvs Pera, Natal and Bahia orange. *Proc. Int. Soc. Citric*, Vol. 1, pp. 76–77.

Russo, F. (1981). Present situation and future prospect of the citrus industry in Italy. *Proc. Int. Soc. Citric*, Vol. 2, pp. 969–973.

Sartori, E. (1981). Argentina citriculture. *Proc. Int. Soc. Citric*, Vol. 2, pp. 950–964.

Schoubroeck, Van F. (1999). Learning to fight a fly: developing citrus IPM in Bhutan. Thesis, Wageningen University and Research Centre, CIP-DATA Koninklijke Biblioheck, Den Haag.

Shinde, N.N. and Kulkarni, R.M (2000). Citrus industry in Maharashtra: Marathwada region. *Souvenir. XIth group discussion of AICRP on tropical fruits*. MPKV, Rahuri, 5–8 January 2000, pp. 32–38.

Singh, A., Naqvi, S.A.M.H., and Singh, S. (2003). *Citrus germplasm rootstocks and varieties*. Kalyani Publishers, India.

Singh, L.B. (1961). Some promising selections of Bael. *Hort. Res. Inst. Ann. Rep. 1960–61*, Saharanpur, pp. 111–119.

Teaotia, S.S., Maurya, V.N., and Agnihotri, B.N. (1963). Some promising varieties of Bael (*A. marmelos*) of eastern district of Uttar Pradesh. *Indian J. Hort.* 20, 210–214.

Wanjari, V., Ladaniya, M.S. and Singh, S. (2003). Marketing and assessment of post-harvest losses acid lime in Andhra Pradesh. *Indian J. Agric. Marketing* 16(2), 32–39.

Xiuxin, D. (2001). China's import export of citrus fruits and its products. *Proc. China/FAO Citrus Symp.*, China, pp. 41–45.

3

POSTHARVEST LOSSES

Every sound postharvest management program aims to minimize losses and to preserve nutritional quality and freshness of fruit. It also ensures better economic returns to growers and higher availability of fruits to consumers in off-season. Before planning for any postharvest management program, it is imperative to ascertain the cause and nature of loss and stages at which it occurs so that suitable technologies are developed. Losses are physical (weight loss and decay), nutritional, cosmetic (loss of appearance as a result of shrinkage), and economic in nature. Causes can be many, including infection, infestation, bruises, insect attacks, and disorders.

Losses in citrus are influenced by pre- and postharvest factors. Preharvest factors include climatic conditions, especially relative humidity, rain, temperature, cultivation practices, tree health, stage of fruit maturity, and fruit type. Postharvest practices such as harvesting, handling, treatments, packaging, and marketing greatly influence fruit losses. Losses take place at various stages of handling, from harvesting until fruit reaches consumers. Usually higher losses are encountered in mechanically harvested citrus (Recham and Grierson, 1971). Losses are quantitative as well as qualitative. Appearance is the criterion used by most consumers while purchasing fruit. Consumers dislike fruit that is shriveled, lusterless, and soft. These conditions arise from water loss and lead to a reduction in price. Decay or rotting makes the fruit unfit for consumption and buyers will reject it outright, leading to economic loss that is directly proportional to the extent of the decay. Various workers have offered different estimates of losses

Citrus Fruit: Biology, Technology and Evaluation

in different countries depending on the commodity, the way it is handled, and the conditions of packing, transport, season, and duration of marketing. Losses occur from field level until storage at consuming markets. In California-grown citrus fruit, losses resulting from decay during the year 1979–1980 were estimated to be 10.48 million cartons with value of 84.15 million US$ (Anonymous, 1980). The stakes are quite high during exports; losses may occur from physiological disorders and sometimes decay, despite effective control measures. In citrus fruits imported in Japan during 1983, losses of US$5–10 million were sustained (Kitagawa and Kawada, 1984). In Israel, as a result of excellent postharvest management practices, losses resulting from decay are only 0.1–1 percent (Cohen, 1991).

Losses resulting from decay in South African citrus fruits exported to England (UK) were estimated at $3.2 million in 1979 and $1.5 million in 1980 (Pelser and Grange, 1981). Postharvest losses in oranges at retail and consumer level in New York markets were 0.8 percent from mechanical damage and 3.4 percent from parasitic and nonparasitic diseases (Harvey, 1978). The average losses for the years 1988–91 in citrus fruit handled by producers' cooperatives in Italy were 6 percent for oranges, 2 percent for lemons, and 6.6 percent for mandarins (Zarba, 1992). Postharvest losses of fruits in Thailand in general are higher both in terms of quality and quantity. Legal standards for fruits and vegetables are largely nonexistent. Losses are caused by both lack of storage and shelf-life extension practices and lack of regular inspection. With some postharvest treatments and storage facilities, relatively lower losses are encountered in citrus fruit (Siriphanich, 1999).

In developing and underdeveloped countries, mostly in the tropics, high losses result from inadequate storage facilities and improper transport and handling. In economic terms, losses are high since repacking of fruits (as a result of the decay of some fruits) increases costs and reduces profit margins. In developed countries losses are relatively low because of advanced technologies, available facilities, and awareness among growers and marketing personnel. Estimated losses in less-developed countries were pegged at an average of 28 percent of the total citrus fruit production in these countries according to a survey by the U.S. National Science Foundation (Anonymous, 1978). The range of losses was 20–95 percent in these countries.

In India and most other developing countries, the causes of loss include: (1) harvesting of under- or over-mature produce; (2) rough/careless harvesting; (3) containers such as bamboo baskets and wooden boxes, which have rough surfaces; (4) high temperatures during harvesting and handling season; (5) very rough handling during loading and unloading of loose (unpacked) produce; (6) transportation on poor roads and careless driving; (7) wholesaling/auctioning under hot, dry conditions or wet weather because of a lack of sheds; (8) lack of temperature-control and relative humidity-control equipment during retailing; (9) lack of knowledge about produce physiology. Various estimates of losses are available. Wastage in different types of citrus fruits in India – including unripe, culled, and bruised fruit – has been reported to be 10 percent (DMI, 1965).

According to Jain (1981) postharvest losses are estimated at 25–30 percent in fruits (including citrus) and vegetables. Studies by the author reveal that on an average, losses are 15–20 percent in citrus fruits. Losses are relatively more (20–25 percent) in mandarins/easy-peelers. Higher perishability is also attributed to the puffiness of 'Nagpur' mandarin fruit. In India, handling and marketing of citrus is done mostly by conventional methods without postharvest treatments. Because they are loose (without packaging), fruits are handled 20–25 times from harvest until retailing.

Loss figures in citrus fruits vary depending on methodology adopted for the study, time of sampling, size of the sample, crop season, and climatic conditions at harvest. The supply and demand conditions at the time of harvest determine the price of the commodity; any drop in price can discourage harvest and thus increase losses.

I. MANDARIN

The rind of mandarin fruit is very fragile, thus these fruits are very prone to bruises, cuts, and infections. In 'Coorg' mandarins, nearly 3.8 percent of market arrivals were discarded, while another 38 percent were sorted and sold at 80 percent reduced price (ICAR, 1991). During marketing and storage at Delhi, 17.1–18.36 percent fruits were lost (ICAR, 1991). Losses in mandarin were 14–23.3 percent at Jabalpur (Ratnam and Nema, 1967) and nearly 20 percent at Malda in West Bengal (DAWB, 1964). In 'Nagpur' mandarin fruit that were transported loose, the losses were 26 percent although 35 percent more fruit could be carried as opposed to packed fruit (DMI, 1982). Losses in mandarins at wholesale and retail markets of North India were roughly 5 and 15 percent, respectively (Chauhan et al., 1987; Anonymous, 1991).

A. 'Nagpur' Mandarin

In Central India, 'Nagpur' is a major mandarin variety, and two crops – monsoon blossom ('*Mrig*') and spring blossom ('*Ambia*') – are grown, with harvesting in February–April and October–December, respectively. Fruits are harvested by snap method (twisting and pulling) using ladders. After the harvest, fruits are collected at one place in the orchard, sorted to remove culled fruit, transported loose by bullock-carts, tempo/truck to wholesale market. Paddy straw is used as a cushioning material. At local wholesale markets fruit is auctioned. The buyers send it to distant markets after packing it in wooden boxes or loose by road or rail transport. Some growers sort and grade the fruit in the orchard itself before transporting it to distant markets. The injuries to 'Nagpur' mandarin fruit are considerably less in clipping and hence the lower decay losses during storage. In the snap method of harvesting, 8–9 percent fruits are injured (Sonkar et al., 1999). Seasonal variation in losses has also been observed. Fruit that is infected in the field or injured during harvesting is sure to rot within 2–3 days while marketing.

During sorting at the farm, the losses vary from 4 to 5 percent. The losses are mostly due to culls, harvesting injury, and insect damage. The loss in terms of culled fruit is not due to improper harvesting or handling but it is a loss resulting from improper cultivation practices. The losses resulting from culls are realized only after harvesting since it is difficult to estimate the loss when the fruit is hanging on the tree. At the wholesale market, sorting is done twice, first according to size and then blemishes. The first sorting is done by the grower before auction at the time of arranging the fruits in a circular heap (each containing 1000–1500 fruits) and the second sorting is done by the buyer/trader before packing in wooden boxes. During sorting at the wholesale market, the losses vary from 4 to 8 percent depending on the quality of produce in monsoon blossom crop fruit. It included culled, that is very small and deformed (49 percent), insect-damaged (23.5 percent), pressed (20 percent), rotted (5.5 percent), and with sunburn injury (2 percent). At the retail level, losses were up to 10 percent mostly as a result of rotting, pressing, and insect damage. Therefore total losses during marketing in the production area were up to 24 percent (5 percent at farm, 8 percent at wholesale, and 10 percent at retail level) (Ladaniya, 2004). The economic loss is estimated to be nearly Rs. 50 crore (500 million Indian Rupees) every year in this mandarin crop alone, considering 25 percent loss of the total produce from farm to retail level (at prices about Rs. 5000/ton during 2004–05).

'Nagpur' mandarins are stored in cold storage meant for apples and potatoes; hence chilling injury occurs at low temperatures. Separate storage chambers exclusively for citrus fruits do not exist at markets. The normal recommended storage temperature for 'Nagpur' mandarin is 6–7°C, whereas temperatures are maintained at 1–4°C in commercial potato and apple storage rooms. Because of the prolonged storage of mature fruit, losses from *Alternaria* stem-end rot are also observed to the tune of 20 percent during post-storage sorting. *Penicillium* rots also develop at low temperatures, causing severe losses. The losses further increase depending on ambient conditions when the stored produce is sent for retailing.

B. 'Coorg' Mandarin

Crop season and time of harvesting influence the extent of losses in 'Coorg' mandarins. There are two crop seasons: 'Monsoon crop' harvested in June and July and 'Main crop' harvested in October and November. Fruits are harvested 32–36 weeks after fruit set. The harvesting of 'Coorg' mandarin is not to be delayed beyond 36–38 weeks in any case as it leads to shriveling and heavy fruit drop (CFTRI, 1989). Improper harvesting methods are another cause of losses. The common method of picking is by twisting the fruits angularly, which leads to injury and postharvest losses. Around 30 percent losses have been reported by Ramana et al. (1973). According to Subramanyam et al. (1970) 16 percent spoilage occurred in 'Monsoon crop' of 'Coorg' mandarin stored for 35 days. Surveys reveal that nearly 8.5 percent of fruit was sorted out in the orchard itself while another 3.8 percent was removed after transportation in the wholesale market (Table 3.1).

TABLE 3.1 Postharvest Loss in 'Coorg' Mandarin During 1990

Handling events	Number of fruits
Harvested fruit	100.0
Discarded fruit at farm	8.5
Fruits brought to wholesale market	91.5
Fruits discarded in wholesale market	3.8
Fruits sold at wholesale level	87.7
Fruits marketed in retail	46.5
Fruits discarded at retail level	3.0
Fruits marketed at reduced price in retail	38.2
Economic loss per 100 fruits (Indian Rupees)	13.18

ICAR (1991).

At the retail level, 3–5 percent of fruit was discarded in the Bangalore market and 2.5 percent in the Hassan and Chickamagalur markets of Karnataka, India.

C. 'Khasi' Mandarin

The 'Khasi' mandarin is the most important commercial citrus crop grown in the northeastern Hill region of India and covers the largest area in Assam, Meghalaya, and Tripura. Other states of the region, such as Mizoram, Arunachal Pradesh, and Nagaland, also grow this mandarin. Winters are quite cold and trees flower from February until April, with fruit maturing between November and January depending on elevation. Siliguri is the major wholesale market for mandarins grown in Sikkim and the Darjeeling area of West Bengal. From there, the produce is sent to different markets in the country and also exported to Bangladesh. Shillong, Guwahati, Dibrugarh, and Tinsukia are the other major markets in the region.

During sorting at the farm in production areas of Sikkim, losses were reported to be mostly due to harvesting injury (1 percent) and insect damage (0.5 percent). In production areas of Assam, losses from insect damage and culled fruits were 0.5 percent each (Ladaniya and Wanjari, 2002). At the retail level, losses are much less at Gangtok market in Sikkim.

In Guwahati market, during sorting at wholesale level, the losses were mostly due to rupturing (2 percent) and bruising (0.5 percent). These injuries occurred during handling and transportation. At the retail level, losses from bruising and pressing were 0.5 percent at the Gangtok and Guwahati markets. The losses from rotting were 1.5 percent at Guwahati and 1 percent at the Gangtok market. A very low level of rotting during marketing was attributed to low temperatures (10–15°C) during harvesting season (Ladaniya and Wanjari, 2002). Transportation time is about 2–8 h as the fruit is brought to these markets from nearby production areas. The total losses from farm to retail level in Sikkim and Assam

TABLE 3.2 Effect of Crop Season and Transport Mode and Time on Postharvest Losses in 'Nagpur' Mandarin

Transport mode and crop season	Time taken for transportation (days)	Percent losses
Rail, spring blossom crop	4	10–11.5
	5	12–15
	6	18–20.5
Rail, monsoon crop	4	15–16.7
	5	18–20.2
	6	28–31.7
Road, spring blossom crop	2 and ½	6–7.5
	3	6–8
	4	8.4–10
Road, monsoon blossom crop	2 and ½	7.5–10
	3	10–11
	4	10.6–12

were 3 and 6.15 percent, respectively. Total losses were considerably lower at Siliguri (3.2 percent), Shillong (2 percent), and Kalimpong (2.5 percent) as recorded in November–December.

In the postharvest handling chain, maximum losses can occur during transportation if it is not done properly. The mode of transport, the time allowed for transport, and the season are important factors that determine the extent of losses. Losses increase in rail transport (Table 3.2) and in summer season compared to road transport in winter. Postharvest losses of mandarins varied from 5 to 31.7 percent depending upon the mode of transport, transit time, and season. According to the Directorate of Marketing and Inspection of the Indian government, transit losses in mandarins vary from 8 to 28 percent depending upon the mode of transport (DMI, 1982).

Mandarins packed in bamboo baskets and loosely loaded into trucks for transport to distant markets showed up to 13 percent wastage after 5 days of transportation (Table 3.3). Transport in rail wagons took 6 to 8 days and upon reaching the destination resulted in 10 percent loss from decay (Dalal, 1988). Packing in CFB boxes with postharvest treatment resulted in negligible losses.

In market-packed 'Nagpur' mandarin fruit, total losses were observed to be 18.5–23 percent when transported by road (truck) and 22.5–28 percent when transported by rail up to the Delhi market. The higher losses in rail transport were attributed to multiple rough handling of fruit and a time lapse of 6–7 days after harvest. High temperatures during transportation (particularly during March and April) also leads to higher losses as the rail cars are not refrigerated. The losses in the farm-packed fruit are relatively low: 15–20 percent when transported by road and 20–21 percent when transported by rail. The lower losses in farm-packed fruit are due to less handling and a shorter time to reach market. An average of 5 percent of fruit is lost from physical injuries during marketing

TABLE 3.3 Wastage in Mandarins Packed in Various Containers During Transport

Origin and destination	Container	Mode of transport	Time taken (days)	Loss (%)
Assam–Calcutta	Basket	Road	3–5	12–13
Darjeeling–Calcutta	Basket	Road	2	2–3
Darjeeling–Calcutta	Basket	Rail	4	9–10
Nagpur–Delhi	Basket	Rail	6–8	9–10
Nagpur–Delhi	Wooden box	Rail	6–8	5–8
Nagpur–Delhi	Basket	Road	3–4	7–8
Nagpur–Delhi	Wooden box	Road	3–4	3–4
Coorg–Madras*	CFB box	Road	1–2	Nil
Coorg–Singapore*	CFB box	Sea	4–5	0.05

Dalal (1988), * Fruit with postharvest treatment.

TABLE 3.4 Postharvest Losses in 'Nagpur' Mandarin Transported Through Different Modes from Central India to New Delhi During Monsoon Blossom Crop Harvest Season in March (fruit packed in wholesale market at Nagpur)

	Loss percent	
Causes of losses	Road	Rail
Diplodia, Alternaria, Phomopsis rots	5–6	5–6
Anthracnose	0.5–1	0.5–1
Sour rot	2–3	5–7
Penicillium rots	<1	1–1.5
Aspergillus rot	<1	1
Injuries	5	5–6
Sunburn	1.0	1.0
Culled fruit	5	5
Total	18.5–22.0	22.5–28

to distant wholesale markets. Of the total loss in road and rail transport, nearly 50 percent was due to various types of rots. Infection by pathogens at different stages of fruit development in the orchard and during handling cause these losses. The breakdown of losses by different factors in the 'Nagpur' mandarin is given in Table 3.4.

Delayed transport of citrus fruits without any pre- and postharvest treatment in unventilated trains, multiple rough handling, and conventional packing were primarily responsible for huge losses. Out of every 400 wooden boxes carried in the truck, 15–20 are extensively damaged, crushing 15–25 percent of 'Nagpur' mandarins in those boxes (Ladaniya and Sonkar, 1996).

II. ACID LIME

Small-fruited acid lime, or 'Kagzi,' (*Citrus aurantifolia* Swingle), is an important commercial crop in India and grown in several states. Important producing states are Andhra Pradesh, Maharashtra, Gujarat, Tamil Nadu, and Bihar.

The Nellore district of Andhra Pradesh is the largest acid lime-producing center in the country, with Gudur being the biggest market. Surveys of this area revealed that total losses were 3.89–4.08 percent from farm to retail level. These relatively low losses were attributed to green and hard fruit, which can withstand rough handling. While harvesting, some small-sized immature fruits also fall down; such fruits are included in culled fruits. Thorn injury and splitting/cracking are other important causes of losses. Splitting occurs after an abrupt increase in soil moisture. Surveys indicated that fruits damaged by mite attack were negligible: 0.08–0.1 percent (Table 3.5). At the wholesale level, the losses were observed to

TABLE 3.5 Losses in Acid Lime Fruit at Various Stages of Marketing in Andhra Pradesh (India)

Stage of loss	Types/causes of losses	Loss (%)	
Farm level		Hyderabad	Gudur (Nellore)
	1. Insect/mite damaged	0.08	0.10
	2. Very small-sized	0.21	0.10
	3. Thorn injury	0.17	0.19
	4. Bruises	0.11	0.09
	5. Splitting	0.12	0.16
	Total	0.69	0.64
Wholesale level			
	1. Bruises	0.14	0.14
	2. Rotting	0.23	0.19
	3. Rupture	0.18	0.11
	4. Very small-sized	0.14	0.12
	5. Insect damaged	0.11	0.14
	6. Over-mature	0.07	0.12
	Total	0.87	0.82
Retail level			
	1. Rotting	1.20	1.40
	2. Bruises, crushing, splitting	1.20	0.85
	3. Insect-damaged	0.12	0.18
	Total	2.52	2.43
	Grand total	4.08	3.89

Ladaniya and Wanjari (2002).

be due to bruising, rotting, and rupture. These injuries take place during handling and transport. The losses at wholesale level were less (0.82 percent) at Gudur because the produce was brought by the growers from nearby areas. In Hyderabad market, fruit was brought from a distance of 200 km and thus the losses were slightly higher (Ladaniya and Wanjari et al., 2002). The dry season of summer in the month of May helped in reducing decay, which usually occurs more often in yellow and soft fruit during rainy season (Wanjari et al., 2002). Taking into account the total losses of 3.89 percent out of 315 255 tons of production in the Nellore district, the total monetary loss for 12 263 tons of fruit was estimated to be Rs. 9.81 crores (98.1 million Indian Rupees) annually at the prices of Rs. 8000/ton during 2001–02. Losses would be higher in the marketing channels, in which produce is sent through several intermediaries to distant places in North India.

Maharashtra is another important acid lime-growing state. Akola, Ahmednagar, and Nagpur are important acid lime-growing districts in the state. The losses were as high as 30 percent during the glut season of June–September and November–December in Vidarbha region (PKV, 1992). In surveys conducted by the author, losses were found to be 19–23 percent during rainy season (June–August) as a result of early color-break and aging of fruit and higher atmospheric humidity in comparison to winter and summer crops (6–14 percent losses). In rainy season crop, losses – attributed to diseases, bruises, and thorn injury – were 8.21 percent at farms near Nagpur. At the wholesale market (Nagpur) wastage was 8.58 percent as fruit was yellow-colored and ripe. Most of the losses in yellow limes were due to decay. At retail level losses were 20.5 percent mostly from decay and aging of fruit. In green fruits, losses were only 1–2 percent during April–May (summer season crop) as the keeping quality was better (Ladaniya and Wanjari, 2002).

At Akola, the total losses from farm to retail level were 22.34 percent in the yellow-colored fruit harvested during the rainy season (June–July). At the farm level, 5–6 percent of the fruit was sorted out. At wholesale market, losses were 7.4 percent. During March–May (summer season) losses were 3–4 percent only, as the fruit was green and hard with much less decay. In Ahmednagar district, the losses at the farm level were nearly 4 percent in crops harvested in November (winter) and 2–3 percent in summer. Total losses were 5 percent (farm to retail level) in the fruit marketed locally.

III. SWEET ORANGE AND GRAPEFRUIT

Market surveys in New York City revealed that losses in Valencia oranges shipped from Florida were 3.2 percent, while navel oranges from California recorded 4.2 percent loss. Losses were mainly due to fungal pathogens (Ceponis and Butterfield, 1973). In grapefruit losses were 1.4 percent at the wholesale level and 3.6 percent at the retail level (Ceponis and Capellini, 1985). Hamlin, Pineapple, Navels, and Valencias are the major fresh fruit orange varieties world over. External physical injury to Hamlin oranges grown in Florida as a result of picking comprised stem-end tears, torn buttons, bruising, scratching, and plugging (Burns and Echeverria, 1989). After sorting in the packinghouse, however, only stem-end tears remained, and no further external handling injury occurred. Decay was not detected until the retail- and consumer-handling stages (14 and 21 days after harvest, respectively), but it markedly increased at all handling stages after storage at 15–20°C for 4 or 8 weeks. Total loss from decay was greatest at the retail-handling stage.

'Sathgudi', the major sweet-orange variety grown in Andhra Pradesh, India, recorded less than 5 percent losses in the wholesale market. Losses included mostly very small, insect-damaged, and diseased fruit. A total loss during transportation from farm (A.P.) to Nagpur varied from 4 to 5 percent. Damaged fruit comprised 60 percent culled, 27 percent rotten, 11 percent bruised, and 2 percent insect-damaged. Besides this, around 5–6 percent fruit of very small size was sold at reduced price. At the retail level, losses were 2–3 percent, mostly from rotting.

'Mosambi' is the major sweet-orange variety grown in Maharashtra state, especially in the Marathwada region. Losses vary depending on the crop season. During '*Ambia*' crop season (spring blossom with harvesting in October–November), losses at farm level were 1.2 percent and this included insect-damaged, over-ripe, and very small-sized fruits. If fruit is held on the tree for better price, it is likely to over-ripen and losses increase from insect damage (fruit-sucking moths) and rotting. Losses from poor harvesting practice (snapping) were negligible. Fruits are transported loose with paddy straw as cushion. At distant wholesale markets such as Pune and Nagpur, the losses were 1 percent, which included mostly pressed, ruptured, and bruised fruit. Shriveling and rotting were higher at the retail level. Higher decay at the retail level was due to degreening with calcium carbide powder kept in paper pouches in the fruit. Rotting increases as slight softening takes place with degreening. Loss was 5 percent from decay at the retail level. Hence, total loss of 'Mosambi' fruit in spring blossom crop season was 7.2 percent (Ladaniya and Wanjari, 2003). Unlike decay, shriveling does not lead to complete loss. Shriveling causes loss of weight and softening, resulting in a reduction in price.

In the '*Hasta*' crop of 'Mosambi' (blossom during October–November and harvesting in July–August – the rainy season), the fruit quality is not so good. Nevertheless, growers attempt this crop by forcing the bloom (water stress followed by irrigation) in order to fetch off-season high prices. The losses in this crop were due to diseases and culls (very small or very large fruit with thick skin, insect-damaged, or deformed). During sorting at the wholesale and retail levels at Nagpur, the losses were 7.5 and 12.5 percent, respectively. Loss was comprised of 40 percent rotten, 40 percent culled/small, and 15 percent insect-damaged; 5 percent of the fruit had sunburn injuries. Rotting was caused by stem-end rot, bruises, and damage done by fruit-sucking moths.

These findings indicate that quantitative and monetary losses were greatly dependent on crop season/climatic conditions, time taken for transportation, marketing processes, number of handlings, and form of handling.

IV. MEASURES TO REDUCE LOSSES

1. Losses from splitting/cracking can be reduced if mature fruits are harvested before onset of rains. Heavy irrigation should be avoided after long dry spells.
2. Good tree health and general sanitary conditions in the orchard can reduce preharvest infection.
3. Reduction in transportation time and market chain/intermediaries can reduce losses. Losses can be minimized if produce is sent from production areas directly to consumption centers (towns and cities). Modifications in the existing marketing system are necessary.
4. In Asian, African, and some Latin American countries, fruits are marketed under ambient condition at roadside or open markets. Higher temperatures and very low or very high relative humidity contribute to losses. Less loss occurs in properly packed and displayed fruit in sheds, as observed in supermarkets. Freeze-damaged fruit is prone to decay and needs to be sorted out.
5. Careful handling of harvested fruit during marketing and transportation can substantially reduce losses.
6. Proper cushioning of fruit and air suspension in transport vehicles can reduce losses.
7. Storage facilities in marketing chain.

REFERENCES

Anonymous (1978). Report of the steering committee for study on postharvest food losses in developing countries. National Research Council, National Science Foundation, Washington DC., 206 pp.

Anonymous (1980). Imazalil – a new weapon in the fruit decay battle. *Citrograph* 65, 95–96.

Burns, J.K., and Echeverria, E. (1989). Assessment of quality loss during commercial harvesting and postharvest handling of 'Hamlin' oranges. *Proc. Fla. Sta. Hort. Soc.* 101, 76–79.

Ceponis, M.J., and Butterfield, J.E. (1973). The nature and extent of retail and consumer losses in apples, oranges, lettuce, peaches, strawberries and potatoes marketed in greater New York. USDA, Market Research Report, 996, 23 pp.

Ceponis, M.J., and Cappellini, R.A. (1985). Wholesale and retail losses in grapefruit marketed in Metropolitan New York. *HortScience.* 20, 93–95.

CFTRI, (1989). Mandarin orange in India – Production, preservation and processing. CFTRI, Mysore, Industrial Monograph Series, 40 pp.

Chauhan, K.S., Sandooja, J.K., Sharma, R.K., and Singhrot, R.S. (1987). A note on assessment of certain prevailing practices for marketing of commercial fruits. *Haryana J. Hort. Sci.* 16, 229–232.

Cohen, E. (1991). Investigations on postharvest treatments of citrus fruits in Israel. *Proc. Int. Citrus Symp.*, Guangzou, China, pp. 32–35.

Dalal, V. B. (1988). Losses in fresh fruits and vegetables and their remedial measures. *Indian Fd. Industry* 7(3), 1–13.

DAWB (1964). Darjeeling oranges – A study in quantitative loss. Market Research Section, Agricultural Marketing Board, Government of West Bengal, Calcutta.

DMI (1965). Marketing of citrus fruits in India. Directorate of Marketing and Inspection, Government of India, Marketing Series No. 155.

DMI (1982). Transportation trials on fruits and vegetables. MPDC Report No. 17. Directorate of Marketing and Inspection, Nagpur.

Harvey, J.M. (1978). Reduction of losses in fresh marketing fruits. *Ann. Rev. Phytopath.* 16, 321–341.

ICAR (1991). Final report of Indo-USAID subproject on post-harvest technology of horticultural crops. ICAR, New Delhi, 271 pp.

Jain, H.K. (1981). Postharvest losses. Report of Postharvest food conservation. *INSA-US NAS Joint Workshop*, 3–7 December 1979, INSA, New Delhi, pp. 41–45.

Kitagawa, H., and Kawada, K. (1984). Decay problems of imported citrus fruits in Japan. *Proc. Int. Soc. Citric.*, Vol. 1, pp. 500–503.

Ladaniya, M.S. (2004). Reduction in postharvest losses of fruits and vegetables. Final Report of NATP Project, Submitted to ICAR, New Delhi.

Ladaniya, M.S., and Sonkar, R.K. (1996). Harvesting, storage, maturity standards, packing and transport of 'Nagpur' mandarin. Annual Report for 1995–1996 NRCC, Nagpur,. pp. 54–60.

Ladaniya, M.S., and Wanjari, V. (2002). ICAR network project on marketing and post harvest loss assessment in fruits and vegetables. Final Report, Submitted to ICAR, New Delhi.

Ladaniya, M.S., and Wanjari,V. (2003). Marketing and post-harvest losses of 'Mosambi' sweet orange in some selected districts of Maharashtra. *Indian J. Agric. Marketing.* 17, 8–10.

Naqvi, S.A.M.H., and Dass, H.C. (1994). Assessment of post-harvest losses in 'Nagpur' mandarin – a pathological perspective. *Pl. Disease Res.* 9, 215–218.

Pelser, P. du, and La Grange, J.M. (1981). Latest development in the control of postharvest decay of citrus in South Africa. *Proc. Int. Citrus Cong.*, Japan. Vol. 2, pp. 812–814.

PKV (1992). Twenty-five years of research on citrus in Vidarbha (1967–1992). Directorate of Extension, Dr. Punjabrao Deshmukh Krishi Vidyapeeth, Akola, 54 pp.

Ramana, K.V.R., Saroja, S., Setty, G.R., Nanjundaswamy, A.M., and Moorthy, N. V. N. (1973). Preliminary studies to improve the quality of 'Coorg' monsoon mandarin. *Indian Fd. Packer.* 27, 5–9.

Ratnam, C.V., and Nema, K.G. (1967). Studies on market diseases of fruits and vegetables. *Andhra Agric. J.* 14, 60–65.

Reckham, R.L., and Grierson, W. (1971). Effect of mechanical harvesting on keeping quality of Florida citrus fruit for the fresh fruit market. *HortScience* 6, 163–165.

Siriphanich, J. (1999). Postharvest problems in Thailand: Priorities and constraints. In *Postharvest technology in Asia – A step forward to stable supply of food products* (Y. Nawa, H. Takagi, A. Noguchi, and K. Tsubota, Eds.). *5th JIRCAS International Symposium. Series*, No. 7, pp. 17–23.

Sonkar, R.K., Ladaniya, M.S., and Singh, S. (1999). Effect of harvesting methods and postharvest treatments on storage behaviour of 'Nagpur' mandarin (*Citrus reticulata* Blanco) fruit. *Indian J. Agric. Sci.* 69, 434–437.

Subramanyam, H., Lakshminarayan, S., Moorthy, N.V.N., and Subhadra, N.V. (1970). The effect of iso-propyl *N*-phenyl carbamate and fungicidal wax coatings on 'Coorg' mandarins to control spoilage. *Trop. Sci.* 12, 307–313.

Wanjari, V., Ladaniya, M.S., and Gajanana, T.M. (2002). Marketing and assessment of post-harvest losses of acid lime in Andhra Pradesh. *Indian J. Agric. Marketing* 16, (May–August), 32–39.

Zarba, A.S. (1992). Recent developments in the co-operative movement for marketing citrus produce in Italy. *Proc. Int. Soc. Citric.*, Aciriole, Italy Vol. 3, pp. 1185–1188.

4

PREHARVEST FACTORS AFFECTING FRUIT QUALITY AND POSTHARVEST LIFE

The developing fruit is subjected to a host of internal and external influences that may modify its inherent anatomical, chemical, and physical characteristics and physiological behavior to some extent. The particular rootstock-scion combination, type of fruit set (either parthenocarpic or with adequate pollination followed by seed abortion or pollination with fertilization leading to seed development), availability of essential nutrients, endogenous and exogenous (applied) plant growth regulators (PGRs), water supply, position on the tree (i.e. the direction that determines exposure to light, wind, and other environmental factors), pesticide sprays, and crop load are major factors that determine or modify fruit size and external and internal qualities and storage ability.

Cultivation practices significantly affect the shelf-life of citrus fruits. The orchard health, training and pruning, source of nutrients (organic or chemical), quantity and time of irrigation, preharvest sprays of fungicides and other chemicals, and tillage operations (weed control) affect the fruit physiology, biochemical composition, and the latent infection by pathogens thereby influencing the keeping quality. The management of losses in citrus fruits should take the integrated approach from the preharvest stage until it reaches the consumer thus involving major production practices and postharvest handling procedures (harvesting, handling, packaging, transportation, storage, and marketing).

I. ORCHARD HEALTH

A. Diseases That Affect Fruit Quality in the Field

Tree health and sanitary conditions of the orchard affect fruit quality and storability of citrus fruits. Postharvest decay by *Penicillium* rots and stem-end rots can be reduced by orchard sanitation. The infected and rotten fruits that have fallen on the ground must be removed promptly (Hough, 1970). Pruning of dead wood reduces the inoculum of *Diplodia* and *Alternaria* stem-end rots. Storage decay due to stem-end rots are lower (1–4 percent decay) in Nagpur mandarin (*Citrus reticulata* Blanco) fruits from healthy trees than fruits from neglected trees (8–69 percent decay) with 10–25 percent blighted twigs (Naqvi, 1993a). Mandarin fruits (Nagpur) with insect blemishes on the fruit surface are likely to rot earlier than fruit without blemishes.

1. Melanose

This disease is caused by *Diaporthe citri* Fawcett Wolf and is a prime disorder lowering the fruit grade in mandarins and sweet oranges. It is prevalent in high rainfall areas (moist conditions) when fruit is developing. The symptoms are pustules or continuous crust and wax-like lesions that are amber brown to dark brown or nearly black in color. The feel is rough like sandpaper on the surface of the fruit rendering it unfit for market. This pathogen also forms tear-stains like melanose symptoms if rain water drops down the surface thus spreading the infection. In arid areas, this disease is very rarely seen.

2. Scab

This is another disease affecting fruit appearance in the field and is caused by *Elsinoe* spp. It is more prevalent on Temple oranges and lemons and less severe on grapefruits, mandarins, and limes. In shaded parts and under moist conditions (rainy conditions), fungus grows faster. Sweet orange scab affects sweet oranges and mandarins. Symptoms are numerous round pustules that sometimes are merged and raised causing a corky appearance on the fruit surface. The fungus can be controlled with neutral copper sprays or Bordeaux mixture (3:4:100 ratio of copper sulphate:lime:water).

3. Anthracnose

This disease is attributed to the fungus *Colletotrichum gloesporioides* (Penz.) Sacc. It is a weak pathogen and survives on dead wood. The spore spreads to the fruit by rain, wind, and insects. Mandarin-type fruits and their hybrids are most susceptible mainly after ethylene treatments. Decay usually starts at the stem end or sides of the fruit as silvery gray to brown spots. The affected part initially is only the skin, which is sunken as compared to the healthy parts. The disease can be very well controlled with measures that are applied for other stem-end rots (field sprays of benomyl) with careful handling and regulated short-term degreening treatments.

4. Greasy Spot

This field disease caused by *Mycosphaerella citri* affects fruits and leaves. Affected fruit loses marketability. The fungus infects through stomata and kills guard cells. Yellow to dark brown and black lesions of circular shape are formed on the fruit surface. The lesions coalesce to form blotchy, slightly raised blemishes, and therefore it is also called *greasy spot rind blotch*. In severe cases it takes a pinkish cast. Occurrence is high during wet seasons and under high humidity conditions. It is a problem in the Caribbean basin (Mondal and Timmer, 2006). In Florida, this disease is observed more in early- and mid-season varieties such as Hamlin and Pineapple. Grapefruits are also affected. The disease is observed more from April to July and less in winter. The suggested control measures are sprayings with oil and oil with copper fungicide. Febuconazole and Strobilurin are recommended in Florida to protect new flush from March through June. The fungicide Gem (a synthetic strobilurin of Bayer Corporation) is reported to be effective against Greasy spot, *Alternaria* leaf spot, melanose, and Scab.

5. *Alternaria* Brown Spot

This is caused by *Alternaria citri* and renders fruit unmarketable by infecting the surface. It is more prevalent in rainy seasons. Copper fungicide and fungicides such as Abound (azoxystrobin, a product of Syngenta Inc.), Gem, and Headline are recommended in Florida.

6. Citrus Canker

This is one of the most important diseases of citrus (particularly acid limes), which affects fruit appearance and renders it completely unfit for marketing. Small circular-shaped (0.1–1 mm up to 3 mm in diameter) cankerous lesions in the form of raised pustules with a yellow halo develop on fruit surfaces, twigs, and leaves. It causes a rough texture on the fruit surface, but the internal part of fruit is not affected. These symptoms are quite unsightly from a fruit marketing point of view. Mandarins and lemons are resistant to it. It is caused by the bacterium *Xanthomonas compestris* pv. Citri. The disease gets dispersed by wind mostly during rains with splashed water because the Xanthomonads have a mucilaginous coat, and they easily suspend in water and disperse in droplets (Koizumi et al., 1997). Droplets are carried by wind up to 6 meters or more. This bacterium survives in wet weather, but fortunately this disease is not observed in drier citrus growing regions. The disease is common in most tropical regions.

The eradication program in Florida resulted in uprooting and burning of millions of trees, but citrus canker has nonetheless been reported recently in some parts of Florida (Collier county). A quarantine program has been undertaken.

B. Insect-Pests and Mites

There are over one thousand insect-pests and mites that attack citrus trees around the world, but few of them directly influence fruit quality and cause significant economic loss. Although the number of insects attacking fruits is relatively low compared with insects that attack leaves and other plant parts, severity of incidence determines the extent of losses, which could be 10–30 percent of the mature crop or even more in some cases if not attended to immediately.

Some of the important insect-pests and mites that attack citrus fruits are invariably reported in most growing countries around the world. If an attack occurs on grown-up fruit, it causes serious economic losses because intense labor and energy has already been expended to produce it.

1. Aphids, Psyllids, White Flies, and Black Flies

These insects affect fruit quality indirectly. These insects attack new growth, suck the sap from leaves, and excrete honeydew on leaves and growing fruits, thus conditions become favorable for development of a black sooty mold fungus that is difficult to wash off. In Nagpur mandarin growing areas in central India, black fly attacks were so severe during the late 1980s and early 1990s that orchards on thousands of hectares of land were blackened by sooty mold, leaves and fruits together. This also affected photosynthesis and overall production. Consequently, fruit size remained small, and black mold formed on fruit surfaces, which had to be washed in a detergent solution before marketing.

2. Fruit Piercing and Sucking Moths

There are many genera and species of these insects reported worldwide in different citrus-growing countries. These are relatively large insects with a specialized proboscis for piercing in the fruit to feed at night. These moths are active at night and fly during the day when disturbed. Punctures made by the mouth parts are distinct holes, and decay organisms soon enter the affected fruits, which drop within few days (Fig. 4.1, see also Plate 4.1).

If fruit is picked up, it may enter a packhouse where it may be packed and decay causing heavy economic loss. Fruit sucking moths cause heavy losses to Nagpur mandarin crops in central India in the Vidarbha region of Maharashtra state and the Chhindwara district of Madhya Pradesh state. Fruit sucking moths prefer fruits that are ripening or already ripened (Jeppson, 1989). Use of poison baits and light traps are quite common for control of these insects. The other recommended control measures are (a) malathion (2 ml per liter of water for high volume sprays); (b) smocking of orchards during evening and night; and (c) prompt disposal of affected fruits (Shivankar and Singh, 1998).

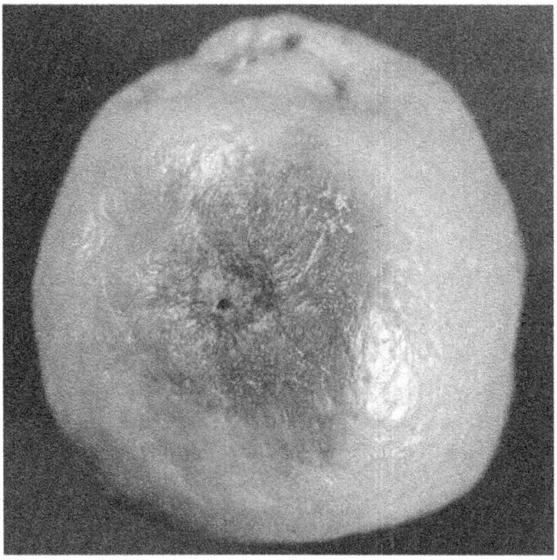

FIGURE 4.1 Damage to Nagpur Mandarin by Fruit Sucking Moth.

3. Lepidopterous Larvae that Attack Fruit

Larvae of a few lepidoptera attack orange fruit. These are commonly called *orange worms*. The orange tortrix (*Argyrotaenia citrana* (Fern) is an important pest in citrus growing areas of the USA and Spain (Jeppson, 1989). These larvae feed on fruits and new flush. Fruits are also scarred around sepals, and holes are eaten in the peel. Such fruit ultimately drops or if inadvertently packed, they are sure to rot and may spoil other fruits too. The fruit tree leaf roller also damages the mature fruit in a manner similar to that of orange worms. The pink scavenger caterpillar (*Pyroderces rileyi* Walsm) is reported to make holes in ripe fruit causing drop and some fruit decay during handling, storage, and marketing. The caterpillar has caused losses as high as 35 percent of oranges in California's San Diego and Orange counties (Jeppson, 1989).

4. Citrus Rind Borer (*Prays Endocarpa Meyr*), Orange Fruit Borer (*Isotenes Miserana* (Walk), and Citrus Peel Miner (*Marmera Salictella* Clem.)

These are insects of minor importance that cause blemishes and damage to fruit rendering them unacceptable for market (Jeppson, 1989).

5. Grasshoppers, Mantids, Katydids, and Cut Worms

These insects chew the fruit rind when fruits are small. As the fruit grows, the affected areas also grow proportionately thus causing blemishes, which lowers the fruit grade – sometimes the fruit becomes totally unacceptable. Katydids feed on fruit that are in the shady part of the canopy. The characteristic depressed grayish or golden-colored rough areas are seen on the orange fruit as katydid damage.

6. Fruit Flies

These insects, mainly of the family *Tephritidae*, are of economic importance. Fruit flies attack mostly mature or maturing fruit. In appearance, a fruit fly is about the same size as a common housefly and looks similar to a housefly to a layman. The pattern and color of wings and body parts vary in fruit flies of various genera and species. These flies are known to travel great distances, usually from 6 to 130 km. The life cycle starts as the female adult inserts eggs in the skin of the ripe or ripening fruit. The oil glands are not punctured because oil kills the eggs. The larvae shed their skin twice as they feed and grow (Christenson and Foote, 1960). After completion of the third instar, larvae emerge from the fruit and form puparia. After a few days, the adult emerges and mates, and a new cycle begins as females lays eggs under the fruit rind. There are over 4000 species of fruit flies described all over the world (Moreno et al., 2000). These are present in all citrus growing countries in the world. Flies cause fine invisible punctures to fruit for egg laying and larval development that result in fruit drop causing serious economic loss. In Sao Paulo state in Brazil, fruit fly infests orange fruit as small as 50 percent of development. Fruit flies of the families Tephritidae and Lonchaeidae were found to infest the citrus fruits in this state. Fruit flies of *Neosiiba* spp. were also observed. Cravo mandarins and sour oranges were most infested (Raga et al., 2000). In Argentina, *Anastrepha fraterculus* Weid. (South American fruit fly) is a major threat, while in South Africa, the Mediterranean fruit fly causes serious damage. In the Americas, the Mexican fruit fly (*Anastrepha ludens*) and *A. fraterculus* are most serious. The Chinese citrus fly (*Bactrocera minax*) is reported to be spreading in China since the 1940s; they are found in Bhutan also. The fly lays eggs up to September and development of eggs to maggots takes 1–4 months. Maggots start feeding in October and November, and the fruit drops (Van Schoubroeck, 1999).

Mediterranean fruit flies (*Ceratitis capitata* Wied), Oriental fruit flies (*Bactrocera dorsalis*), South American fruit flies, Mexican fruit flies, and West Indian fruit flies (*Anastrepha obliqua*) are most important as far as quarantine and trade is concerned. The Mediterranean fruit fly prefers mature fruit only, and in Israel, Jaffa (Shamouti) is the favored host. This fly generally does not prefer lemons; this may be due to acidity or some other factor of resistance. The fruit fly has not been seen infesting acid lime (*Citrus aurantifolia* Swingle) fruit, which are more acidic (usually 7–8 percent titratable acidity) than lemons. Mediterranean flies are not active when the temperature is below 15–16°C because the flies do not reproduce below this temperature. Therefore, fruit such as clementines, Washington navel oranges, and grapefruits that mature before the onset of severe winter suffer attacks of this fly. In Valencia, Spain, fruit maturing after December escape the attack due to cold weather in winter up to the following June because fruit flies are inactive. Thereafter, very little fruit is left in most orchards. In Egypt, because the winter is not that severe, the fruit fly is active all through the season.

Poison baits containing insecticides is the commonly used method to control fruit flies. In California, the eradication procedure for Mediterranean fruit flies includes aerial spraying with a bait containing malathion. The quarantine and interception of infested fruit at ports is feasible and could be an effective means of reducing this problem (Coggins, 1991). The 20 percent malathion with 80 percent Nulure mixture is standard in eradication programs (Moreno et al., 2000). New chemicals such as spinosad, abamectin, emamectin, thiamethoxam, imidacloprid, and Phloxine B have been tested as spot sprays with good efficacy.

In South Africa, SENSUS traps are used, which use the modified trimedlure attractant (Capilure) or the patented Ceratitislure to catch male flies, or they are used with a female attractant (Questlure). In an organic citrus culture, a novel trap M3 bait station is used with attract-and-kill technology. The station is active against female flies only and is fruit-fly specific. It is environmentally friendly (non-target insects are not affected) and is not affected by rain (Buitendag et al., 2000).

Fruit fly infestation requires quarantine treatment during trade between the states in the same country and between two countries. Considerable research has been done and is continuing to develop suitable eco-friendly, non-hazardous treatments for various citrus fruits (see Chapter 18).

7. Thrips

The citrus thrip (*Seirtothrips citri* Moult) is a pest of serious economic importance. In California and Arizona, this pest has been reported to cause huge losses to growers. It is almost endemic in southern coastal districts of California where lemons are grown. Thrips are also found in arid low-land valleys of southwestern California. Thrips are a serious pest on all citrus fruits, although they have some preferences. In acid limes, attacks of thrips are seen as negligible, whereas on mandarins (Nagpur), they are a serious pest with losses that can exceed 25 percent of fruit affected in unattended orchards. In California, thrips prefer Washington navels over Valencias as a host. The characteristic scarring in Nagpur mandarins are formation of rings around the fruit peduncle (Fig. 4.2, see also Plate 4.2). This type of damage is also reported in sweet oranges (Jeppson, 1989).

The scarring can be seen on other parts of the fruit surface also. Affected fruit loses its market value. The scarring is due to sucking of sap by the insect particularly on tender fruitlets. The other species (*Scirtothrips aurantii* Faure), which is reported in South Africa (Transval region), the injury caused is similar to that of citrus thrips and also includes tear-like stains during rainy seasons. Tear-stained areas on the orange as it approaches maturity are characterized by a smooth surface that gives fruit a varnished look. The greenhouse thrips (*Heliothrips haemorrhoidalis* Bouche) attack the fruit that are in bunches and in contact with leaves. As a result, damaged fruit becomes grayish or silver in color. Thrips are controlled with timely sprays usually just before or just after flowering using systemic insecticides with residual effects of up to 15–20 days. In California, chemical

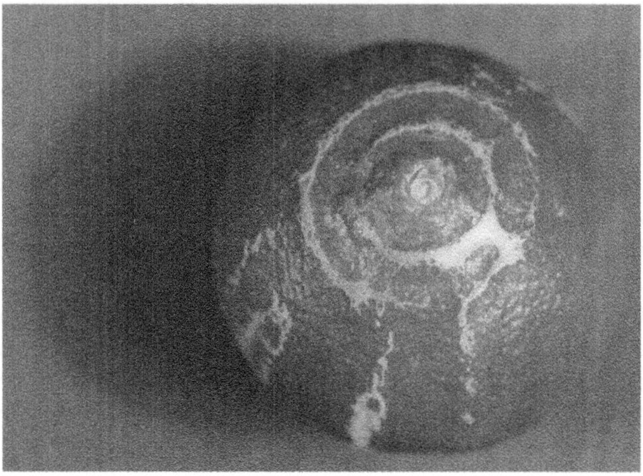

FIGURE 4.2 Very Common and Peculiar Damage by Citrus Thrips in Ring Form Around the Fruit Stem of Nagpur Mandarin.

control is initiated when 75 percent petal-fall is complete. The critical period is from petal-fall up to when fruit reached 4 cm in diameter. The pest causes maximum damage when fruit is developing from May to October. The organophosphorus and carbamates have offered the best control and are used considering spray schedules (without using the same insecticides again and again).

8. Mites

Mites are important economic pests because the damage to fruit rinds can be extensive. They emerge as a serious pest in a short span of time and require immediate chemical control. Mites penetrate their stylet in cells and cause rupturing of the fruit surface or oil gland. The mites also suck out cell content and inject a toxin in the tissue. Citrus rust mites cause golden russeting or black russeting on oranges in Florida. In lemons, this mite causes a silver effect, and in grapefruits, damage is known as *shark skin*. The citrus bud mite (*Aceria sheldoni* Keif) attacks fruit buds and results in deformation (Jeppson, 1989). Red mites are reported to be a problem in Japan.

9. Citrus Scales

High humidity and a decrease in light accompanied with dense vegetation favors growth of scale insects. These are present in most citrus growing areas in the world. Infestation of this pest on fruit results in elimination of those fruits from shipment to the market. California red scale is a occasional problem in southern California. There are reports of control of red scale with the parasite *Aphytis melinus* De Bach and with oil sprays (Luck, 1981). Citricola scale, wax scale, and brown soft scale extract juices from plants and fruits and excrete honeydew on which sooty mold fungus grows thus reducing marketability of the fruit.

10. Fuller Rose Beetle (*Pantomorus cervinus*)

The adult lays eggs in masses of 10–60 under the fruits' calyx and crevices of trunk and stem bark. It is an important pest in California and a phytosanitary issue for fruit exported to Japan. In cultural practices, skirt pruning of trees and Sevin and Kryocide sprays are recommended (CCQC, 2005).

11. Mealy Bugs

These can be a serious postharvest insect-pest from a quarantine point of view, especially in exports if insects remain in the underside of the calyx that is not easily washed in packing operations.

II. TREE NUTRITION AND CULTIVATION PRACTICES

Effects of tree nutrition (macro-and micronutrients) on physico-chemical quality and storage ability of citrus fruits are significant. It is a general belief that increasing doses of the chemical nitrogen adversely affects fruit-keeping quality. Increased nitrogen supply to Hamlin, Pineapple, and Valencia orange trees increases soluble solids, acid percentage, juice percentage in fruits, and green fruit percentage and reduces individual fruit weight and percentage of packable fruit. Reese and Koo (1975) reported that incidences of decay in storage were greatest with the lowest nitrogen rate whereas Grierson and Hatton (1977) found that effects of fertilizer treatments particularly on storage quality are inconsistent. Neverthless, it is generally recognized that a lot of nitrogen decreases storage potential of fruits. During ambient and refrigerated storage, decay and weight loss in Nagpur mandarin fruits grown on only farm yard manure (25 kg/plant/year) remained on par with fruits grown on chemical fertilizers (600 g nitrogen through urea + 200 g phosphorus through single super phosphate + 100 g potassium through muriate of potash/plant/year) combined with farm yard manure at 25 kg/plant/year (Huchche et al., 1998). Application of 800 g nitrogen/plant/year reduces fruit weight loss and increases retention of Vitamin C content in acid limes (*C. aurantifolia* Swingle) during storage at room temperature for 12 days (Prasad and Putcha, 1979). In the sweet orange cv. Mosambi, higher doses (800–1200 g/tree) of nitrogen reduced storage decay and fruit weight loss as compared with lower or no application of nitrogen (Govind and Prasad, 1981).

Phosphorus deficiency results in a thick peel with a hollow core particularly in sweet oranges. The fruit rind texture becomes rough, and fruit develops higher acid content with delayed maturity.

Potassium is considered indispensable in consistent production of good quality citrus fruits. Its deficiency not only reduces yields but results in smaller fruit size, soft and thin peels, and increased decay. Excess potassium results in larger and coarse-textured fruit with high acid content. Potassium application reduces splitting and increases fruit size and acidity in Valencia oranges if leaf potassium is low (Bar-Akiva, 1975). The source of potassium has limited influence on the storage life of citrus fruits. Fruit grown on potassium chloride retains more juice and has better organoleptic characteristics during storage than fruit grown on potassium nitrate, potassium-magnesium sulfate, and potassium sulfate (Berger et al., 1996). Foliar application of monopotassium phosphate, which contains 52 percent P_2O_5 and 32 percent K_2O affected fruit quality of Star Ruby grapefruit. Fruits from sprayed trees were larger than those of non-sprayed trees and had thinner peels and increased pulp diameter. The higher ratio of pulp to peel increased the juice percentage and quality, whereas juice acidity was lower (Lavon et al., 1996). The least storage losses were observed in mandarin fruits from trees fertilized with phosphorus and potassium + calcium + FYM (Tsanva et al., 1976). The Zinc (0.5 percent) + Boron (0.4 percent) micronutrient treatment and Karna Khatta rootstock were better than Troyer citrange for yield, fruit weight, fruit juice, and ascorbic acid content in Kinnow mandarins (Mishra et al., 2003).

Balanced nutrient management through chemical fertilizers and organic manures is the key for producing good quality citrus fruit with desired storage ability. Nutrient management needs to be done through soil and foliar applications as the need arises.

Rootstocks also influence fruit ripening and postharvest quality. Shelf life of acid lime fruits grown on Rangpur lime rootstock has been longer with lower fruit weight loss during storage as compared with fruits grown on Troyer and Trifoliate stocks (Jawaharlal et al., 1991). A higher color score (34.2) for juice of Hamlins on *Poncirus trifoliata* and Chu Kag mandarin (*C. reticulata*) was recorded compared with 32.9 on sour orange (*C. aurantium*) (Wutscher and Bistline, 1988).

Losses in fruits of Ruby Red grapefruit due to decay and chilling injury at 4°C or 12°C during storage for 6 weeks were generally lower in fruits from trees budded on rough lemon rootstock than on *Citrus ambylecarpa* (Reynaldo, 1999). In Marsh seedless grapefruit, the TSS percentage was more on Troyer than on rough lemon rootstock. Heavier fruits with less peel percentage were also recorded on Troyer and Carrizo citranges than on rough lemon rootstock (Mehrotra et al., 1999).

Irrigation practices may also influence citrus fruit quality. Monselise and Sasson (1977) reported effects of differential irrigation on postharvest behavior of Shamouti oranges. Fruits from frequently irrigated trees had a higher rot percentage but more firmness and better appearance after long storage. The electrical conductivity of fruit peel and malic acid content of juice were highest whereas total soluble solids were lowest. After 9 years of experiments, Kuriyama et al. (1981) established that soil moisture during fruit development and maturity has a considerable effect on Satsuma fruit quality. When the soil was moist during the fruit development period of July to September and kept dry in October and November when fruit matured, orange color developed quickly, there were less puffy fruits, juice was highly colored, and overall quality was considerably improved. To obtain concentrated juice, these researchers observed that it is necessary to dry the soil to the fruit's initial wilting point. When the soil was kept dry from July to September and wet in October and November, fruit puffiness was high, fruit contained less juice, and coloration was delayed. Mulching with vinyl film and use of drainage pipes is effective in controlling soil moisture in the orchard. Deficit irrigation in Mihowase Satsuma (*Citrus unshiu*) to a soil water tension of −60 kPa at 60 cm soil depth followed by initiation at 0–4 weeks in clay loam or at 0–2 weeks in sandy loam increases TSS content (Peng and Rabe, 1998).

Clean cultivation decreased yield and mean fruit weight but increased the fruit color index in Satsuma mandarins. There was no difference in yield, fruit degreening, and sugar and acidity contents between grassed and mulched plots (Daito, 1986). Control of weeds and grasses is a must for blemish-free fruit and high pack-out. In countries with high rainfall, it is very challenging, but it is necessary to keep fields clean in order to minimize insect–pest problems.

Bearing angle is also known to affect rind roughness, pigmentation, and sugar and organic acid concentrations in the juice of Satsuma mandarin fruits in Japan (Kubo and Hiratsuka, 1998). The rind surface of sideways fruits has been smoother as the fruits grew, whereas upward fruits developed rough skins until late September. Fruits from lightly cropped trees also remained rough until late October.

The Satsuma fruit quality and yield increased, and time of harvesting advanced by about 4 months by growing Wase (early) Satsumas in plastic houses in Japan (Inoue and Harada, 1981). Under the climatic conditions of southwestern Japan, Wase Satsuma flowers in April, and fruit matures in October, but when grown in plastic houses, flowering took place in late December, and fruit matured in June as the temperatures in the plastic houses maintained 20–25 °C temperatures. Fertilization efficiency is quite high in plastic houses. This practice resulted in availability of fruit right from June onwards up to February of the following year.

III. PREHARVEST SPRAYS

Preharvest applications of plant growth regulators (PGRs), fungicides, and several other chemicals significantly affect quality and storage life of citrus fruits. Plant bioregulators, or PGRs, are non-natural compounds produced by utilizing fungi in industrial units on a large scale as in the case of gibberellins (GAs) or by chemically synthesizing them in the factories. Naphthalene acetic acid (NAA) and 2,4-D (2,4-Dichloro phenoxy acetic acid) are the chemically synthesized compounds, but their actions are similar to natural auxins present in plants. Some compounds, such as paclobutrazole, are chemically synthesized, and they are utilized to retard growth because they interfere with actions of natural hormones. Some of the important uses of plant bio-regulators in citriculture concern: (1) Fruit size: Auxins are utilized extensively for this purpose. The action is in three ways: (a) by thinning the crop, NAA and IZAA or Ethychlozate (Figaron) increase the growth of remaining fruitlets; (b) by directly enhancing fruit growth when applied after June drop; and (c) by enhancement of growth of some fruitlets and increase in late fruitlet abscission when applied at flower openings (Guardiola, 1997). Abscission of fruitlets in the case (c) is not related to the auxin-induced ethylene synthesis (as in thinning). (2) Fruit maturity: The IZAA improves internal maturity and accelerates degreening in Satsumas. The similar action is seen in cases of the use of ethephon and ethylene. The GAs delay fruit pigmentation and harvest, thereby marketing of fruit can be manipulated depending on demand. GA can be applied with auxin to reduce preharvest fruit drop and delay carotenoid accumulation. The yield in the following year may be affected if harvesting is delayed beyond certain limits. If applied before color break, GA effectively reduces *creasing* (a physiological disorder in oranges) and *puffing* (as in Satsumas). (3) Fruit loosening: Ethephon and cyclohexamide have been tried for this purpose. (4) Delay in fruit senescence: Use of GA and 2,4-D in postharvest management of fresh fruit.

A. Auxins

Plant growth regulators such as 2,4-D, 2,4,5-T, and NAA are effective in improving fruit quality, particularly size, extending keeping quality, and retarding deteriorative changes. Best results for fruit size improvement are obtained when cell division ceases and stage II begins with cell enlargement. At this stage, vesicles completely fill the locule (fruit segment) and start the accumulation of juice.

There is a large premium paid by consumers for large size fruits – size is the main factor determining consumers' acceptance. Growers apply various growth promoters and nutrients, especially micronutrients. NAA and 2,4-D are used commercially for this purpose. 2,4-Dp (2,4-dichlorophenoxy propionic acid) at 50 ppm applied just at the end of June drop (when clementine fruit reaches 20 mm in diameter) resulted in satisfactory fruit growth (Agusti et al., 1992). Sprays of 2,4-D (20 ppm) and NAA (10–100 ppm) 1 month prior to harvest increased juice content in Coorg mandarins. In monsoon crop fruit, 2,4,5-T (25–50 ppm) and CLPA (Chlorophenoxy acetic acid at 25 ppm) sprays before harvest resulted in 28–35 percent less weight loss than the control during 35 days' storage at 5–6°C and 85–90 percent RH. Treated fruit had better marketability, although respiration was unchanged (Rodrigues et al., 1966). Cytozyme (containing chelated minor elements, enzymes, and growth regulators, mainly cytokinins) at concentration of 4:1000, applied 1 month after fruit-set, increased fruit weight and juice percentage in Campbell Valencia oranges. Fruit quality was generally better when treatments were applied at full bloom (Daulta and Beniwal, 1983). Fruit weight, diameter, juice percentage, TSS, and ascorbic acid contents in kagzi limes (*Citrus aurantifolia* Swingle) were improved with Zn (0.6 percent) + 2,4-D (20 ppm) when applied in January for the spring flush and in May for the summer flush (Babu et al., 1984).

In citrus, fruit size is inversely related to the number of fruits on the tree. Usually heavy bearing trees produce fruit of smaller size. The cycle of alternate bearing also develops in mandarins with heavy bearing. The thinning of fruit is also done by various means, and chemicals are most economical for this purpose as manual operation would be labor intensive, tedious, and uneconomical. In the case of Dancy tangerines, application of ethephon (500 ppm) for thinning was economical when fruitlets were 1 cm in diameter and fruitlet density was 143 fruit/cum foliage (Gallasch, 1988). Thinning of Murcott and Sunburst tangerines grown in Florida with NAA applied 6–8 weeks after full bloom (about ½ inch fruit diameter) increases fruit size, mean fruit weight, pack-out, and fruit appearance.

The effects of NAA and 2,4-D increase when combined with potassium. 2,4-D (18 ppm) + KNO_3 (4 percent) or NAA (300 ppm)+KNO_3 (4 percent) applied to Star Ruby grapefruit in the month of May when fruit size was 12–18 mm increased fruit size. NAA also increased fruit size when applied without KNO_3. Ethephon (300 ppm) applied at the same time also increased fruit size but reduced yield (Greenberg et al., 1992a). The mode of action of ethephon and NAA is by a thinning effect thus reducing competition among growing fruits. Ethephon treatment has been shown to increase ethylene production in young fruits and leaves (Wheaton, 1981) within 2 days. NAA application also resulted in a rise in ethylene but it is delayed (after 7 days). Ethylene enhances the activity of cellulytic and pectolytic enzymes in the abscission zone of citrus fruits, which controls abscission (Iwahori and Oohata, 1976). Foliar application of XGR-373 (auxin of M/S Dow-Elanco) at 20–40 ppm as an amino salt after the June drop (fruit size 26 mm) significantly increased fruit size in Fino lemons without altering internal quality parameters. Fruit thinning was also not observed (Garcia-Lidon et al., 1992).

B. Gibberellins

Among growth regulators, maximum work has been done with respect to the effect of preharvest applications of GA_3 on citrus fruit quality. Gibberellic acid is now recognized as a valuable tool in the control of citrus fruit maturation and senescence, both preharvest and postharvest. Pioneering work of Coggins and coworkers in California led to use of this PGR worldwide in citrus fruits. Although the increased peel viability is important along with delay in senescence, the interference of gibberellins in rind coloration is not acceptable. The timing of the treatment should be in such a way that next crop is not affected. In Israel, Minneolas are grown as easy peelers and are harvested in December and January. Toward the end of harvesting season, fruit suffers severe creasing, loses firmness, and deteriorates. The GA_3 summer treatment (August) at 10 ppm concentration is sufficiently effective in maintaining fruit firmness and control of creasing while not having the disadvantage of interfering with peel coloration at normal harvest time. The autumn treatments can be given in case harvesting is to be delayed, but a certain reduction in peel color development and in the following year's yield must be taken into account. Autumn application is also effective in preventing creasing. Acidifying of the treatment solution using 0.1 percent H_3PO_4

or acidifying surfactant (0.15 percent B-5) are equally effective (Greenberg et al., 1992b). The interference with subsequent seasons' fruiting is definitely due to the inhibitory effect of GA_3 on flower bud induction, which occurs in Israel during November and December (Monselis, 1985). The metabolism of applied GA_3 in citrus albedo may be slow, and this can explain the effectiveness of summer treatment in preventing creasing.

Preharvest applications of gibberellic acid in general maintains fruit peel quality on the tree and in storage and reduces postharvest decay in several citrus fruits (El-Otmani and Coggins, 1991). GA_3 (10 ppm) application to Washington navel fruits prior to color break reduced the increase in epicuticular wax and delayed the decrease of CO_2 conductance, resulting in less of an increase in internal CO_2 (El-Otmani et al., 1986). GA_3 (0–100 ppm) sprays applied before color break caused Marsh grapefruit rinds to be more resistant to puncture and delayed yellow color development (McDonald et al., 1987). In Kinnow mandarins, preharvest sprays of GA_3 (75 ppm) reduced weight loss during storage (Ahlawat et al., 1984). Color development of Kinnow fruits was delayed with four sprays of GA_3 at 300 ppm (Sandhu, 1992). In Nagpur mandarins, GA_3 (10–20 ppm) treatment at color break improves fruit firmness, delays color development and puffiness, and reduces fruit weight loss during storage (Fig. 4.3, see also Plate 4.3). Acidification of treatment solution (lowering of pH to 4) was compared without acidification with respect to delay in color development, but the difference was not significant (Ladaniya, 2004).

FIGURE 4.3 Gibberellic Acid Sprays Delay Fruit Color Development and Facilitate On-Tree Storage of Nagpur Mandarin.

Two sprays of GA_3 (10 ppm) also help in retaining fruit on the tree up to 4 weeks beyond normal harvest time thus facilitating on-tree storage without affecting fruit quality at the time of harvest and yield of subsequent crops (Ladaniya, 1997). The pigment changes in the peel of seedless fina clementines were retarded by GA_3 (10 ppm from late September to late November). Response was greatest when GA_3 was applied during the 10-day interval between the onset of chlorophyll degradation and the onset of carotenoid accumulation (Garcia-Luis et al., 1992). Exogenous application of GA_3 (100 ppm) did not significantly affect abscissic acid content, but it did inhibit the decrease of endogenous GA_3 and zeatin contents. The treatment also delayed chlorophyll degradation and inhibited carotenoid beta-cryptoxanthin biosynthesis and its accumulation, which inhibited the development of fruit color and luster (Tao et al., 2002).

Nodiya and Mikaberidze (1991) reported that Satsuma fruits treated with gibberellic acid (60 ppm) before the fruits showed signs of yellowing (late September to early October) took longer to ripen and had more vitamin C content but had thicker and relatively rougher skin. As a result of GA treatment, Shamouti oranges were firmer, evolved less ethylene and showed slower malonic acid accumulation in juice and albedo in prolonged storage at 17°C with no humidity control (Monselise and Sasson, 1977). A combination of GA (10 ppm) with Bezyladenine (BA) (10 ppm) enhanced and prolonged the effect of GA_3 in delaying degreening by 2 months in clementine mandarins. Multiple applications at 15-day intervals were effective. Treatment is effective when fruit is dark green before color break (about October 15th) for fruit harvested in January. Later application had a small effect on coloration but reduced the fruit cracking that is a problem in Clementine Nules (Baez-sanudo et al., 1992).

The GA_3 preharvest applications at color break are reported to increase juice percentage from 2–5 percent in Hamlin, Pineapple, and Valencia. The peel was thinner and firmer in treated fruit (Davies and Stover, 2003).

Gibberellic acid is commonly used for seedless crops of Robinson, Nova, Orlando, and Minneola tangelos and other self-incompatible mandarin hybrids. The application is made at full bloom or at $^2/_3$ petal-fall. This provides a required growth promoter to the seedless fruit. Clementines are a commercially important mandarin in Spain, but they are self-incompatible and have very low ability to set fruit. To increase fruit size and fruit-set, it is a common practice in Spain to spray GA at full bloom (5–15 ppm depending on cultivar). For fina, a 5-ppm concentration is used, whereas for de Nules it is 3–6 ppm (Fornes et al., 1992). If pollinators are used in clementine orchards, the result is seeded fruits, which have a low market value. GA_3 applications have also affected other fruit quality attributes. In the grapefruit cultivars Cecily and Hudson Foster, treatment of immature fruits with GA_3 at 1000 ppm mixed with lanolin paste resulted in a decrease in total acid concentration and an increase in fruit size. GA_3 also significantly reduced the naringin concentration in juice, juice sacs, and the peel. ABA and BA at 1000 ppm mixed with lanolin paste significantly decreased fruit size and increased the naringin concentration but had little effect on the soluble solids and acid contents (Berhow and Vandercook, 1992).

Several factors, including temperature, relative humidity, pH of the spray solution, and surfactant affect gibberellic acid uptake. Uptake of GA (14C-GA_3) increases with temperature in the range of 5–35°C (Greenberg and Goldschmidt, 1988), and light has no influence in absorption. Uptake increases with RH and acidification of treatment solution (pH 4). A combination of GA with petroleum oil (used to control red scale) reduces the efficacy of GA against creasing as reported in case of Palmer navel oranges grown in the Sunday River valley of the Eastern Cape of South Africa. Acidification of GA spray to a pH of 4 improved efficacy with and without presence of oil in the spray (Gilfillan and Cutting, 1992). The surfactant L-77 (concentration 0.025 percent) gave the best result with respect to uptake and biological activity of GA. Findings indicated that GA gets in to the fruit in lipophilic, non-dissociated form. Experiments with several wetting agents and surfactants indicated that the surfactant that had the best wetting or spreading capacity did not necessarily demonstrate greater GA efficacy. Triton- B 1956 was least effective, whereas L-77 was very effective (Henning and Coggins, 1988). GA_3 (23 ppm) with the surfactant Silwet L-77 (0.05 percent) preharvest treatments early in October reduced postharvest pitting of Fallglo tangerines grown in Florida (Loxahatchee) after 1 week of storage (10°C), but GA_3 effects did not last long as storage time and pitting severity progressed. Despite the greater peel strength, GA treatments did not significantly reduce postharvest decay or disorders, other than postharvest pitting. In Minneolas, early GA_3 with Silwet treatment significantly reduced TSS (10.50° Brix, compared to 11.10° Brix in the control) and the TSS:TA ratio (9.62, compared to 11.52 in the control) at harvest (Ritenour and Stover, 2000).

C. Other Chemicals

Besides PGRs, several other chemicals are also being tried to improve fruit quality and reduce postharvest disorders. Arsanilic acid sprays (600–1500 ppm) reduced acidity in grapefruits and were not affected by simulated rain showers. Spraying with Citrus-10 (a stabilized enzyme solution) (20 ml/100 gallon, 3–4 sprays 2–4 days apart) and triacontanol (1–5 ppb single spray) reduced acidity in oranges and increased soluble solids but did not reduce acidity in grapefruits (Wilson et al., 1988). Ezz (1999) reported that preharvest proline foliar sprays (0.5–1.5 percent, applied at full bloom, fruit set, and 4 weeks before harvest) significantly reduced chilling injuries during cold storage (5°C, 80 percent RH) of Marsh grapefruits and Washington navel oranges. Treatments eliminated fruit surface pitting, and this effect was clearer in Washington navel oranges than in Marsh grapefruits. Proline foliar sprays caused a pronounced increase in fruit juice ascorbic acid content and a decrease in total soluble solids and acidity. Calcium nitrate (1 percent) spray improved the keeping quality of Kinnow mandarins by reducing decay (Bhullar et al., 1981). The activity of pectin methyl esterase (PME) and peroxidase enzymes has been lower during the storage of Kinnow fruit treated with calcium compounds (Dhillon, 1986). Calcium nitrate (2 percent) applied 2 weeks before harvest was more effective than calcium chloride. Sprays of calcium nitrate + carbendazim (0.05 percent) extended the marketability of Kinnow fruits stored up to 56 days at ambient conditions (Surinder Kumar and Chauhan, 1989) and up to 98 days in refrigerated (4 ± 1°C) conditions (Surinder Kumar and Chauhan, 1990). Recently, Ozeker et al. (2001) reported that $CaCl_2$, $CaCl_2$+ 2,4-D and $CaCl_2$ + GA_3 applied before harvest can result in better wound healing by lignin formation in flavedo and albedo in Satsuma mandarins after harvest as compared to non-treated fruit.

D. Ethephon and Other Chemicals for Fruit Color Improvement

Ethephon has been extensively tried for improving color of citrus before harvest. Its preharvest applications give unpredictable responses, and the concentrations for coloring and defoliation greatly overlap (Stewart, 1977). Nevertheless it is still being used although on a limited scale.

Ethephon proved successful for degreening Ponkan (*C. reticulata*) mandarins in Japan. By spraying 100–200 ppm ethephon from late October through November (Iwahori, 1978), coloration is accelerated, and thus harvesting advanced by 1–3 weeks without causing appreciable leaf drop (less than 5 percent). There was no fruit drop. Fruit quality other than rind color, such as sugar and acid content, is not affected nor is the ethanol and acetaldehyde that may be related to citrus fruit ripening. Ponkans are usually harvested during the middle and end of December and stored until January for marketing because of poor color.

Ethephon is unsuccessful for degreening Satsumas on-tree because of leaf drop. Adding 1 percent calcium acetate to ethephon solution (200–300 ppm) almost completely prevents defoliation and fruit drop that otherwise might occur. Degreening may be somewhat delayed by an addition of calcium acetate, but sprays are still effective in advancing coloration (Iwahori and Oohata, 1980). Application of 200 ppm of ethephon to trees of Oroval and Marisol mandarins (early ripening clementine cvs) improves fruit color. The best results are obtained when sprays are applied 20–25 days before color break. Addition of calcium salts (both acetate and nitrate at 0.45 percent) reduces leaf drop by 50 percent but also reduces effectiveness of ethephon to some extent (Pons et al., 1992). Different clones of clementine have different responses to ethephon with respect to color development and fruit removal force (abscission) when applied at the rate of 480 ppm (Protopapadakis and Manseka, 1992). Application of ethephon (250 ppm) covering the entire tree canopy have been found to be effective to degreen Nagpur mandarins (Sonkar and Ladaniya, 1999). The addition of calcium acetate (1 percent) and carbendazim (500 ppm) minimized leaf and fruit drop and storage decay. Paclobutrazol applied once at 1000 ppm in November at the end of the fruit drop period, enhanced fruit color at harvest without making the peel puffy in Owari satsumas on rough lemon rootstocks in South Africa. Ethephon (250 ppm) applied 7 days before harvest had little effect on fruit color but made the peel puffy and resulted in excessive decay, mainly *Diplodia* and *Alternaria* stem-end rot and anthracnose caused by *Glomerella cingulata* (Gilfillan and Lowe, 1985). The application of Camposan (50 percent ethephon) at 0.3 ml/liter and Hydrel [bis-(2-chloroethyl-phosphonate) hydrazinium] at 0.3 ml/liter before yellowing of Satsuma fruit (late September to early October) advanced ripening by 15–20 days and also resulted in smoother and thinner peel and better chemical composition (Nodiya and Mikaberidze, 1991). Application of 0.01–0.05 percent Phenosan (B-(3,5-diter–butyl-4-hydroxyphenyl) – propionic acid advanced mandarin fruit ripening and affected kinetics of ethylene production (Serebryan et al., 1991). Ethychlozate, Ethyl 5-chloro-1 H-3-indazolylacetate, also known as IZAA, at 70–100 ppm concentration also hastens fruit degreening, increases Brix and sugar content in juice, and decreases acidity in Satsuma mandarins. Applications performed 65–105 days after full bloom twice at 15–20-day intervals is quite effective (Iwahori, 1991). This chemical is also useful for fruit thinning at 100–200 ppm applied 40–50 days after bloom. CPTA (2-(4-chlorophenylthio)-triethylamine hydrochloride (1250–5000 ppm) is also reported to stimulate lycopene biosynthesis and accumulation (Daito, 1986).

Ethephon (400 ppm) application enhanced development of orange-red fruit color in Meiwa kumquat fruits also. Color increased more rapidly in treated fruit than in untreated fruit stored at 10°C. Average fruit weight increased with ethephon application by about 5 percent and increased the ratio of peel to fruit and juice to fruit. Ethephon accelerated the reduction in fruit acidity, but sugar composition was not affected (Hashinaga and Itoo, 1985).

Leaf drop problems would not arise if fruits were dip treated on-tree, but that would be practically impossible. With preharvest applications of etherel (2000–4000 ppm) as a dip treatment for Washington navel fruits, degreening could be achieved within 6 days (Chauhan and Rana, 1974). Etherel (1000–5000 ppm) as a preharvest dip was effective in grapefruit color development also (Chundawat et al., 1973; Soni and Ameta, 1983). Ethrel (200 ppm a.i.) applied twice before color break to Beladi oranges enhanced the degradation of chlorophyll and at the same time increased carotene synthesis in the fruit rind. In contrast, 2,4-D treatments (200 ppm) retarded chlorophyll degradation and reduced rind carotene contents (Al-Mughrabi et al., 1989). In order to prevent regreening of Valencia oranges, preharvest treatments with carotogenic bioregulators such as [(*N,N*-diethylamino) ethoxy]benzophenone, *N,N*-diethyloctylamine or (*N,N*-diethylamino)ethyl p-bromobenzoate have been found to reduce chlorophyll biosynthesis in the flavedo of fruit thus reducing regreening. This treatment also increased total xanthophyll content and resulted in fruits having a better color (Hsu et al., 1989).

E. Fungicides

Preharvest applications of fungicides are very effective in minimizing field infection of citrus fruits thereby reducing decay losses. *Diplodia* and *Phomopsis* stem-end rots and *Penicillium* decay are controlled by a benomyl application

3 weeks before harvest (Brown and McCornack, 1969; Brown, 1974). Preharvest sprays of thiabendazole (500 ppm) have been effective in reducing postharvest rot in Kinnow (Gupta et al., 1981). Three preharvest applications of Benlate (0.1 percent) minimized more than 80 percent of postharvest decay in Nagpur mandarins stored up to 21 days at ambient condition (Naqvi, 1993b). Fungicides can be combined with ethephon or PGRs during spray application.

IV. CLIMATIC FACTORS

Hail storms, freezes, and untimely incessant rains adversely affect the fruit quality and storage life and cause heavy losses to the growers. Losses due to freezes can be reduced by applying irrigation and use of gas heaters in the orchard. In Japan, growers cover the tree canopy with dry paddy straw, which helps to reduce losses due to chilling temperatures when fruit is ready for harvest. Fruit injury is not severe if hailstones are of smaller size. Immediate benomyl sprays were found to be effective to control preharvest and postharvest decay. In case of severe injury due to large hailstones, fruit drop losses are quite high. Cultivars such as Kinnow that bear fruit inside the canopy suffer less loss due to light hailstorms. Fruits of Nagpur mandarins are born mostly in the outer part of the canopy, and the tree canopy is relatively open, so losses are greater.

REFERENCES

Agusti, M., Almela, V., Aznar, M., Pons, J., and El-Otmani, M. (1992). The use of 2-Dp to improve fruit size in citrus. *Proc. Int. Soc. Citric.*, Italy, Vol. 1, pp. 423–427.

Ahlawat, V.P., Daulta, B.S., and Singh, J.P. (1984). Effect of pre-harvest application of GA and Captan on storage behavior of Kinnow. *Haryana J. Hort. Sci.* 13, 4–8.

Al-Mughrabi, M.A., Bacha, M.A., and Abdelrahman, A.O. (1989). Influence of preharvest application of ethrel and 2,4-D on fruit quality *J. King Saud Univ. Agric. Sci.* 1, 95–102.

Babu, R.S.H., Rajput, C.B.S., and Rath, S. (1984). Effects of zinc, 2,4-D and GA_3 in kagzi lime (*Citrus aurantifolia* Swingle). IV. Fruit quality. *Haryana J. Hort. Sci.* 11, 59–65.

Baez-sanudo, R., Zacarias, L., and Primo Millo, E. (1992). Effect of gibberellic acid and Bezyladenine on tree-storage of Clementine mandarin fruits. *Proc. Int. Soc. Citri.*, Italy, Vol. 1, pp. 428–431.

Bar-Akiva, A. (1975). Effect of potassium nutrition on fruit splitting in Valencia orange. *J. Hort. Sci.* 50, 85–90.

Berger, H., Opazo J., Orellana, S., and Galletti, L. (1996). Potassium fertilizers and orange postharvest quality. *VIII. Int. Citrus Congress*, *Proc. Int. Soc. Citric.*, Vol. 1, pp. 759–762.

Berhow, M.A. and Vandercook, C.E. (1992). The reduction of naringin content of grapefruit by applications of gibberellic acid. *Pl. Growth Regulation* 11, 75–80.

Bhullar, J.S., Dhillon, B.S., and Josan, J.S. (1981). Preliminary studies on the effect of pre-harvest calcium nitrate treatment on the storage behavior of Kinnow fruits. Paper presented in the Symposium on Recent Advances in Fruit Development, PAU, Ludhiana, 14–16 December.

Brown, G.E. (1974). Postharvest citrus decay as affected by Benlate application in the grove. *Proc. Fla. State Hort. Soc.* 87, 237–240.

Brown, G.E., and McCornack, A.A.(1969). Benlate, an experimental preharvest fungicide for control of post-harvest citrus fruit decay. *Proc. Fla. State Hort. Soc.* 82, 39–40.

Buitendag, C.H., Naude, W., and Ware A.B. (2000). Recent developments in fruit fly monitoring and control in Southern Africa. ISC Congress presentation, 3–7 December. Florida, Abstract (P 71), 103.

CCQC (2005). California Citrus Quality Council document 2005 (http://www.citrusresearch.com/ccqc).

Chauhan, K.S. and Rana R.S. (1974). Effect of ethrel on degreening of Washington navel orange. *Indian J. Hort.* 31, 154–156.

Christenson, L.D., and Foote, R.H. (1960). *Biology of fruit flies. Ann. Rev. Entom.* 5, 171–192.

Chundawat, B.S., Gupta, O.P., and Singh, J.P. (1973). Effect of 2-chloroethyl phosphonic acid (Ethrel) on degreening and quality of Ganganagar Red grapefruit (*C. paradisi* Macf). *Haryana J. Hort. Sci.* 2(3–4), 45–49.

Coggins, Jr. C.W. (1991). Present research trends and the accomplishments in the USA. Bangyan H and QuianY. (eds), *Proc. Int Citrus Symp.*, Guangzhou, China, 1990.

Daito, H. (1986). Maturity and its regulation in Satsuma mandarin fruit. *Japan Agric. Res. Quarterly.* 20, 48–59

Daulta, S., and Beniwal, V.S. (1983). Effect of time of application, chemicals and concentrations of plant growth regulators on composition of fruit in sweet orange (*C. sinensis* Osbeck) cv. Campbell Valencia. *Haryana J. Hort. Sci.* 12, 168–172.

Davies, F. and Stover, E. (2003). New uses for old plant growth regulators. Citrus Industry, November, 21.

Dhillon, B.S. (1986). Bioregulation of developmental process and subsequent handling of Kinnow mandarins. *Acta Horticulturae.* 179 (1), 251–256.

El-Otmani, M., and Coggins, Jr. C.W. (1991). Growth regulators effect on retention of quality of stored citrus fruits. *Scientia Hort.* 45, 261–272.

El-Otmani, M., Coggins, Jr. C.W., and Eaks, I.L. (1986). Fruit age and gibberellic acid effect on epicuticular wax accumulation, respiration, and internal atmosphere of navel orange fruit. *J. Am. Soc. Hort. Sci.* 111, 228–232.

Ezz, T.M. (1999). Eliminating chilling injury of citrus fruits by pre-harvest proline foliar spray. *Alexandria J. Agric. Res.* 44, 213–225.

Fornes, F., Van Rensburg, P.J.J., Sanchez-Perales, M., and Guardiola, J.L. (1992). Fruit setting treatments' effect on two clementine mandarin cvs. *Proc. Int. Soc. Citric.*, Italy, Vol. 1, pp. 489–492.

Gallasch, P.T. (1988). Chemical thinning of heavy crops of mandarins to increase fruit size. *Proc. 6th.* Int. Citrus Symp., Israel, Vol. 1, pp. 495–500.

Garcia–Lidon, A. (1992). Early harvesting of Fino lemons using a new synthetic auxin. *Proc. Int. Soc. Citric.*, Italy, Vol. 1, pp. 473–474.

Garcia-Luis, A., Herrero, V.A., and Guardiola, J.L. (1992). Effects of applications of gibberellic acid on late growth, maturation, and pigmentation of the clementine mandarin. *Scientia Hort.* 49, 71–82.

Gilfillan, I.M., and Cutting, J.G.M. (1992). Creasing reduction in navel oranges: Lower efficacy of gibberellic acid in spray mixture containing petroleum oil. *Proc. Int. Soc. Citric.*, Italy. Vol. 1, pp. 527–529.

Gilfillan, I.M., and Lowe, S.J. (1985). Fruit color improvement in Satsumas with paclobutrazol and ethephon – preliminary studies. *Citrus Subtrop. Fruit J.* 621, 4–6, 8.

Govind, S. and Prasad, A. (1981). Effect of nitrogen nutrition on storage of sweet oranges cv. Mosambi (*C. sinensis* Osbeck). *Progressive Hort.* 13, 39–43.

Greenberg, J., and Goldschmidt, E.E. (1988). The effectiveness of GA_3 applied to citrus fruit. *Proc. 6th Int. Citrus Congress*, Israel, Vol. 3, 339–342.

Greenberg, J., Hertzano, Y., and Eshel G, (1992a). Effects of 2,4-D, ethephon and NAA on fruit size and yield of Star Ruby grapefruit. *Proc. Int. Citrus Congress*, Italy, Vol. 1, pp. 520–523.

Greenberg, J., Oren, Y., Eshel, G., and Goldschmidt, E.E. (1992b). Gibberellin A_3 (GA_3) on Minneola tangelo: Extension of the harvest season and improvement of fruit quality. *Proc. Int. Citrus Congress*, Italy, Vol. 1, pp. 456–458.

Grierson, W., and Hatton, T.T. (1977). Factors involved in storage of citrus fruits. A New Evaluation. *Proc. Int. Soc. Citric.*, Vol. 1, pp. 227–231.

Guardiola, J.L. (1997). Future use of plant bio-regulators. Proc. Int. Soc. Citric. Vol. 1. South Africa, 456–459.

Gupta, O.P, Singh, J.P., and Charia, A.S. (1981). Effect of pre-and post-harvest application of chemicals on the storage behavior of Kinnow mandarin. Paper presented in the Symposium on Recent Advances in Fruit Development, PAU, Ludhiana 14–16, December.

Hashinaga, F. and Itoo, S. (1985). Effect of Ethephon on the maturity of Meiwa Kumquat fruit. Bulletin, Faculty of Agric., Kagoshime Univ., 35, 43–47.

Henning, G.L., and Coggins, Jr. C.W. (1988). Bioassay used to determine the impact of surfactants on the biological effectiveness of exogenous gibberellic acid. *Proc. 6th Int. Citrus Congress*, Israel, Vol. 3, pp. 325–331.

Hough, A. (1970). Control of green mould by orchard sanitation. *S. African Citrus J.* 442, 11, 13, 15.

Hsu, W.J., DeBenedict, C., Lee, S.D., Poling, S.M., and Yokoyama, H. (1989). Preharvest prevention of regreening in Valencia oranges (*Citrus sinensis* (L.) Osbeck). *J. Agric. Fd. Chem.* 37, 12–14.

Huchche, A.D, Ladaniya, M.S., Lallan Ram, Kohli R.R., and Srivastava, A.K. (1998). Effect of nitrogenous fertilizers and farm yard manure on yield, quality and shelf life of Nagpur mandarin. *Indian J. Hort.* 55(2), 108–112.

Inoue, H. and Harada, Y. (1981). Nutritional problems of Satsuma mandarin in plastic house. *Proc. Int. Soc. Citric.*, Tokyo, Japan, Vol. 2, pp. 556–559.

Iwahori, S. (1978). Use of growth regulators in the control of cropping of mandarin varieties. *Proc. Int. Soc. Citric., Int. Citrus Congress*, Australia, pp. 263–270.

Iwahori, S. (1991). Present research trends and accomplishments of citriculture in Japan. Proc. Int. *Citrus Symp.*, Guangzhou, China., 14–24.

Iwahori, S., and Oohata, J.T. (1976). Chemical thinning of Satsuma fruit by NAA: role of ethylene and cellulase. *Scientia Hort.* 4, 167–174.

Iwahori, S., and Oohata, J.T. (1980). Alleviative effects of calcium acetate on defoliation and fruit drop induced by 2-chloroethyl phosphoric acid in citrus. *Scientia Hort.* 12, 265–271.

Jawaharlal, M., Thangaraj, T., and Irulappan, I. (1991). Influence of rootstock on the post-harvest qualities of acid lime fruit. *South Indian Hort.* 39, 151–152.

Jeppson, L.R. (1989). Biology of citrus insects, mites, and mollusks. *In* Citrus Industry. (W. Reuther, E.C. Calavan, G.E.Carman, Eds.), Vol. 5, pp. 2–81. Div. Agric. Natural Resources, Univ. California.

Koizumi, M., Kimijirna, E., Tsukamoto, T., Togawa, M., and Masui, S. (1997). Citrus bacterial canker. *Proc. Int. Soc. Citric.*, South Africa, pp. 340–344.

Kubo, T. and Hiratsuka, S. (1998). Effect of bearing angle of Satsuma mandarin fruit on rind roughness, pigmentation, and sugar and organic acid concentrations in the juice. *J. Japanese Soc. Hort. Sci.* 67, 51–58.

Kuriyama, T., Shimoosako, M., Yoshida, M., and Shiraishi, S. (1981). The effect of soil moisture on the fruit quality of Satsuma mandarin. *Proc. Int. Soc. Citric.*, Tokyo, Japan, pp. 524–527.

Ladaniya, M.S. (1997). Response of 'Nagpur' mandarin to pre-harvest sprays of gibberellic acid and carbendazim. *Indian J. Hort.* 54, 205–212.

Ladaniya, M.S. (2004). Reduction in post-harvest losses of fruits and vegetables. Final report of NATP Project, Submitted to ICAR, New Delhi.

Lavon, R., Shapchiski, S., Mohel, E., and Zur, N. (1996). Fruit size and fruit quality of 'Star-Ruby' grapefruit as affected by foliar spray of mon-potassium phosphate. *Proc. Int. Soc. Citri.* Vol. 2, pp. 730–736.

Luck, R.F. (1981). Integrated pest management in California Citrus. *Proc. Int. Soc. Citric.*, Tokyo, Japan, pp. 630–635.

McDonald, R.E., Shaw, P.E., Greany, P.D., Hatton, T.T., and Wilson, C.W. (1987). Effect of GA on certain physical and chemical properties of grapefruit. *Trop. Sci.* 27, 17–22.

Mehrotra, N.K., Vij, V.K., Harish Kumar, Aulakh, P.S., and Raghbir Singh (1999). Performance of Marsh Seedless cultivar of grapefruit (*Citrus paradisi* Macf.) on different rootstocks. *Indian J. Hort.* 56, 141–143.

Mishra, L.N., Singh, S.K., Sharma, H.C., Goswami, A.M., and Bhanu Pratap (2003). Effect of micronutrients and rootstocks on fruit yield and quality of Kinnow under high density planting. *Indian J. Hort.* 60, 131–134.

Mondal, S.N. and Timmer, L.W. (2006). Greasy spot – a serious endemic problem in Caribbean basin. *Pl. Disease*, 90, 532–538.

Monselis, S.P. (1985). Citrus. *In* Handbook of flowering (A.H. Halvey ed.) Vol. 2, pp. 275–294. CRC press, Boca Raton.

Monselise, S.P. and Sasson, A. (1977). Effects of orchard treatments on orange fruit quality and storage ability. *Proc. Int. Citrus Congr.*, Vol. 1, pp. 232–237.

Moreno, D.S., Harris, D.L., Burns, J., and Mangan, R.L. (2000). Novel chemical approaches for the control of fruit flies in citrus orchards. ISC Congress presentation, 3–7 December. Florida, Abstract (P 68), 103.

Naqvi, S.A.M.H. (1993a). Influence of pre- and post-harvest factors on export oriented production of Nagpur mandarin. Proc. Nat. Seminar on Export Oriented Hort. Production – Research and Strategies, 5–6 December, P.K.V. Akola, and V. Naik Pratisthan, Nagpur, 182–185.

Naqvi, S.A.M.H. (1993b). Pre-harvest application of fungicides in 'Nagpur' mandarin orchards to control post-harvest storage decay. *Indian Phytopath.* 46, 190–193.

Nodiya, M.G. and Mikaberidze, V.E. (1991). Effect of ethylene-producing agents and gibberellic acid on ripening and quality of Unshiu mandarins. *Subtropicheskie-Kul'tury.* 4, 85–89.

Ozeker, O., Sen, F.Z., Karakali, I., Yildiz, M., Kinay, P., and Yildiz, F. (2001). The effects of some preharvest treatments on wound healing in Satsuma mandarin after harvest. *Acta Hort.* 553, 73–75.

Peng, Y.H and Rabe, E. (1998). Effect of differing irrigation regimes on fruit quality, yield, fruit size and net CO_2 assimilation of 'Mihowase' satsuma. *J. Hort. Sci. Biotechnol.* 73, 229–234.

Pons, J., Almela, V., Juan, M., and Agusti, M. (1992). Use of ethephon to promote color development in early ripening Clementine cvs. *Proc. Int. Soc. Citric.*, Italy, Vol. 1, pp. 459–462.

Prasad, A. and Putcha, R.K.M. (1979). Studies on N nutrition in Kagzi Lime (*C. aurantifolia* Swingle). V. Studies on storage behavior of fruits. *Progressive Hort.* 10(4), 32–35.

Protopapadakis, E., and Manseka, V.S. (1992). Effect of ethylene-releasing compounds on color break and abscission in five clones of Clementines. *Proc. Int. Soc. Citric.*, Italy, pp. 463–464.

Raga, A., Presetes, D.A.O., Souza, Filho, M.F., Salo, M.E., Siloto, R.C., and Zucchi, R.A. (2000). Fruit fly infestation in citrus cultivated in the state of Sao Paulo, Brazil. ISC Congress presentation. 3–7 December. Florida, Abstract (P 70), 103.

Reese, R.L. and Koo, R.C.J. (1975). Response of Hamlin, Pineapple and Valencia orange trees to nitrogen and potash application. *Citrus Industry* 50(10), 6–8.

Reynaldo, I.M. (1999). The influence of rootstock on the post-harvest behaviour of Ruby Red grapefruit. *Cultivos-Tropicales* 20(2), 37–40.

Ritenour, M.A. and Stover, E. (2000). Effects of gibberellic acid on the harvest and storage quality of Florida citrus fruit. *Proc. Florida State Hort. Soc.* 112, 122–125.

Rodrigues, J., Dalal, V.B., and Subramanyam, H. (1966). Effect of pre-harvest sprays of plant growth regulators on size, composition and storage behavior of Coorg mandarin. *J. Sci. Fd. Agric.* 17, 425–427.

Sandhu, S.S. (1992). Effect of pre-harvest sprays of gibberellic acid, Vipul, calcium chloride and Bavistin on the tree storage on Kinnow fruits. *Acta Horticul.* 321, 366–371.

Schoubroeck, Van Frank (1999). Learning to fight a fly: developing citrus IPM in Bhutan. Thesis Wageningen Univ. and Res. Centre. CIP-DATA koninklijke Biblioheck, Den Haag.

Serebryanyi, A.M., Kacharava, M.M., Bulantseva, E.A., and Zoz, N.N. (1991). Effect of Phenosan and hydroquinone on mandarin fruit ripening and ethylene production. *Fiziologiya Biokhiniya Kulturnykh Rastenii* 23(6), 606–610.

Shivankar, V.J. and Singh, S. (1998). Management of insect-pests in citrus. National Research Centre for Citrus, Nagpur, Technical Bulletin No. 3, 99.

Soni, S.L., and Ameta, S.L. (1983). Effect of pre-harvest application of ethrel on coloration and physico-chemical composition of grapefruit. Proc. Int. Citrus Symp. Bangalore, Horticultural Society of India, pp. 309–315.

Sonkar, R.K., and Ladaniya, M.S. (1999). Effect of pre-harvest sprays of ethephon, calcium acetate, and carbendazim on rind color, abscission, and shelf life of 'Nagpur' mandarin. *Indian J. Agric. Sci.* 69, 130–135.

Stewart, I. (1977). Citrus color–A review. *Proc. Int. Citrus Cong.*, Florida, Vol. 1, pp. 308–311.

Surinder, K. and Chauhan, K.S. (1989). Effect of certain fungicides and calcium compounds on post-harvest behavior on Kinnow mandarin. *Haryana J. Hort. Sci.* 18, 167–176.

Surinder, K. and Chauhan, K.S. (1990). Effect of fungicides and calcium compounds on shelf life of Kinnow mandarin during low temperature storage. *Haryana J. Hort. Sci.* 19, 112–121.

Tao, J., Zhang, S.L., Chen, K.S., Zhao, Z.Z., and Chen, J.W. (2002). Effect of GA_3 treatment on changes of pigments in peel of citrus fruit. *Acta Hort. Sinica*. 29, 566–568.

Tsanva, N.G., Sardzhveladze, G.P., and Burchuladze, A.S. (1976). The effect of nutrition on mandarin storability. *Subtropicheskie Kultury* 3–4, pp. 61–65.

Wheaton, T.A. (1981). Fruit thinning of Florida mandarins using plant growth regulators. *Proc. Int. Soc. Citric.*, Italy, Vol. 1, pp. 263–268.

Wilson, W.C., Obreza, T.A., and Cooke, A.R. (1988). Methods for controlling acidity in citrus fruit. Proc. Plant Growth Reg Soc., America, pp. 208–212.

Wutscher, H.K. and Bistline, F.W. (1988). Rootstock influences juice color of 'Hamlin' orange. *HortScience* 23, 724–725.

5

FRUIT MORPHOLOGY, ANATOMY, AND PHYSIOLOGY

I. FRUIT MORPHOLOGY

Citrus fruit is a modified berry or a specialized form of berry (hesperidium) resulting from a single ovary. In addition to citrus, this type of fruit is observed in five more genera: *Poncirus* (*Trifoliate* orange), *Fortunella* (kumquat which is eaten as is, with peel), *Microcitrus*, *Eremocitrus*, and *Clymenia* in the subfamily Aurentioidae of family Rutacae. The usually five-pointed or five-lobed calyx and button-like receptacle are attached to the peduncle – as seen in the orange fruit – at the stem end. The calyx remains attached to the branch if the fruit is naturally separated by abscission. Usually the small, green-colored button and calyx at stem-end is preferred in market. The size of citrus fruits range from nearly 2.25 cm for kumquats (*Forlunella* spp.) to more than 20 cm in diameter for pummelo (*C. grandis*). The shape is also variable: oblate in grapefruit, mandarins, and tangerines; globose to oval (spherical or nearly so) in sweet oranges; oblong in lemons (*C. limon*), (*C. medica*); and spherical in limes (*C. aurentifolia*). The rind, or peel, is leathery – more so when it loses some moisture. It is fragile and breaks on folding when turgid. Fruits generally have 8–16 segments. Grapefruits and pummelos have 17 or 18 segments. Seeds vary in number from zero in a few cultivars to many, leading to quite seedy fruit. Tahiti lime (*C. latifolia*) and navel oranges can be called truly seedless and have almost no seeds, while grapefruit and pummelo have 40–50 seeds. Seed size and shape also varies greatly among species.

A. Fruit Characteristics

1. Mandarin

Most mandarins are easily peelable and deep orange to reddish-orange in color when fully mature. Fruit is small to large (5–8 cm in diameter at the equatorial axis); globose to oblate; base with a low to high collar, a deeply depressed apex and a medium-thick, loosely adherent rind; a relatively smooth surface sometimes pebbled with prominent, sunken oil glands; segments numbering about 10–17; a large and hollow axis; orange-colored flesh; tender, melting, and juicy; mild and strong flavor; seedless to seeded with nearly 3–7 or more seeds; small and plump; early to late season in maturity. Most mandarins lose quality and the rind 'puffs' if not picked when internally ripe.

Oblong and pyriform fruit has to be removed as it is not true-to-type and hence damages the impression of the whole lot of fruit. Fruit of small to medium size are preferred; very large fruits are associated with puffiness by the trade. Fruit with puffiness and over-maturity lack keeping quality. Some markets in the East prefer large mandarin fruits. Fruit with small neck or collar and near-flat base (at stem-end) are preferred for export as they suffer less damage. Most consumers like yellow-orange to deep-orange color, although fruit with a slight greenish tinge is acceptable. Fruit should be juicy. Flavor is another important attribute of fresh mandarins; this is the quality affected most in long-term storage and by various handling conditions. Most Indian consumers dislike excessive acidity, while a lack of acid renders fruit flat in taste. In European markets, seedless fruits with higher acids are preferred. Clementine and Satsuma mandarins are seedless. Seedless clones of the 'Nagpur' mandarin are being developed.

2. Sweet Orange

Unlike common sweet oranges, Navel oranges have navel-like structure at the stylar end, or apex. This difference is anatomical in nature and consists of a navel that is a rudimentary secondary fruit embedded in the primary fruit. Navel oranges, particularly the Washington Navel orange variety, are somewhat obovate or ellipsoid in shape. The fruit surface is usually smooth in all orange fruits of commercial importance; fruits with rough surfaces are removed while packing. In Navel oranges, the surface is moderately pitted and pebbled. Sweet oranges are mostly globose but oval to ellipsoid (Shamouti or Palestine Jaffa) are also common. Common Valencia is oblong to spherical. Fruits are flattened at the base in most varieties. Surface is finely pitted but smooth. Italian 'Moro' is deep blood-orange variety on the inside, while 'Sangunello Moscata,' an Italian variety, and 'Doblefina' of Spain are both light blood oranges. Fruits of these varieties are deep orange to reddish in color from outside at maturity. Most sweet oranges in general have rind that is yellow to yellow-orange, thin to medium-thick, and firm and leathery. Flesh color is yellow to yellow-orange. The presence of a green or yellowish-green calyx and button at the stem end of an orange indicates the freshness of the fruit.

3. Lemon

Lemon fruit vary greatly in shape, size, color, rind texture, and juice content. Fruits of Assam lemon and Verna (Berna or Vernia of Spain) are long, obovate to oblong in shape, and medium to large in size. Fruits of Eureka are medium-small and elliptical or oval in shape. Lemons are seedless to seedy. The nipple is prominent and large in some varieties, while others have very small, inconspicuous nipples. Rind color is green to bright yellow at maturity. Fruits are with or without a collar at the neck. Fruits are slightly ribbed to without ribs. Rind is thick to thin and rough to smooth among varieties. Some varieties, such as Meyer lemons, are quite juicy while others have less juice.

4. Acid Lime

Acid limes are very small (3 cm diameter) to medium (5 cm in diameter) in size and round, obovate, or oblong in shape. They have very small necks, a flat base, and a small nipple at the apex. They have a thick to very thin and papery rind and are green to yellow in color. They are seedy to seedless. The rind surface is smooth and the flesh is tender, juicy, and yellowish-green. Fruits of *Citrus aurantifolia* Swingle are small (30–45 g) while fruits of *Citrus latifolia* are large (80–100 g).

5. Grapefruit

Fruits are medium to large in size (10–12 cm in diameter), oblate to spherical in shape, and slightly depressed from stylar end and flat on the stem end. The peel is medium-thick, yellow to pink-blushed, and smooth. Flesh is white and tender/melting with a central core that is usually open. Fruits are borne in clusters and are seeded to seedless. Pink- and red-fleshed cultivars are available. Flavor is strong.

6. Pummelo

Fruits are large to very large (15–20 cm in diameter or even larger), round to obovate in shape (some varieties are pyriform), and borne singly. The rind is thick to very thick (3–4 cm) with a smooth, green- to yellow-colored surface when mature. Flesh is firm and crisp, juice vesicles are separable, and core is open and hollow. Pink-fleshed cultivars are available. Fruit is seeded to seedless and flavor is mild to strong.

II. FRUIT ANATOMY

Citrus fruit arises through the growth and development of an ovary and consists of 8–16 carpels clustered around and joined to the floral axis, which forms the core of the fruit. The carpels form locules, or segments, in which seeds and juice sacs (vesicles) grow. The pericarp (rind or peel) is divided into exocarp, or *flavedo*, and mesocarp, or *albedo*. The flavedo consists of the outermost tissue layers, which have cuticle-covered epidermis and parenchyma cells. The flavedo is the outer, colored part and the albedo is the inner, colorless (white) or sometimes tinted part (as in red grapefruit or blood oranges) (Fig 5.1).

The flavedo consists of the epicarp proper, the hypodermis, the outer mesocarp, and oil glands. Above the epicarp is a multilayered protective skin or

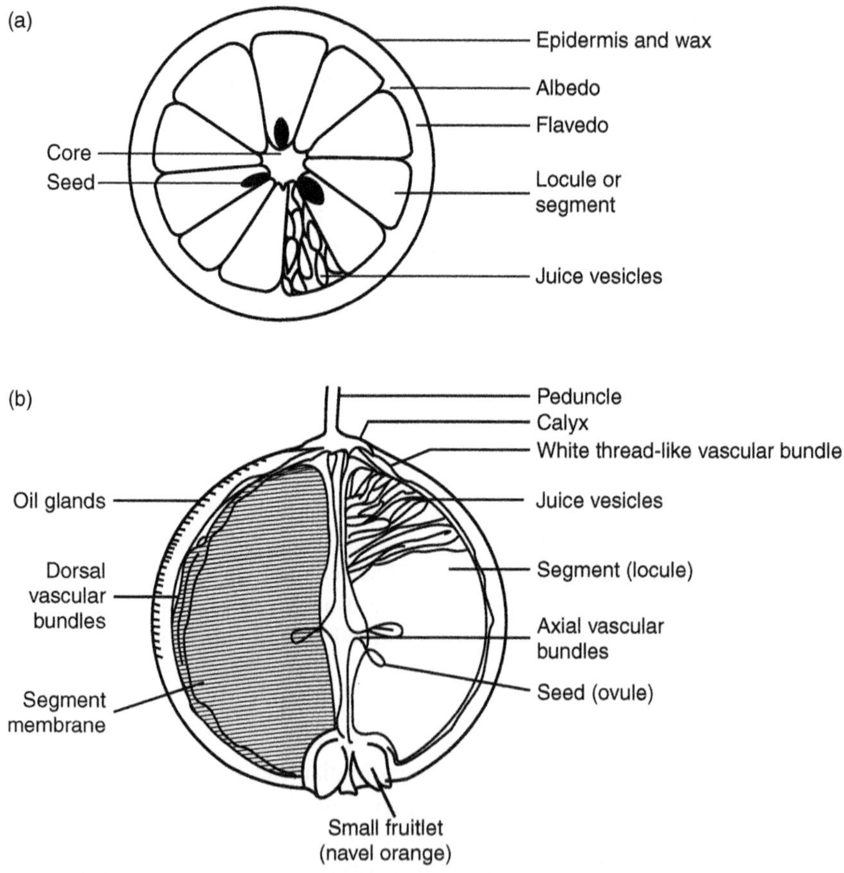

FIGURE 5.1 (a) Transverse Section of Citrus Fruit Showing Various Anatomical Parts of Citrus Fruits and (b) Longitudinal Section Showing Vascular Bundles and Fruitlets of Navel Fruit.

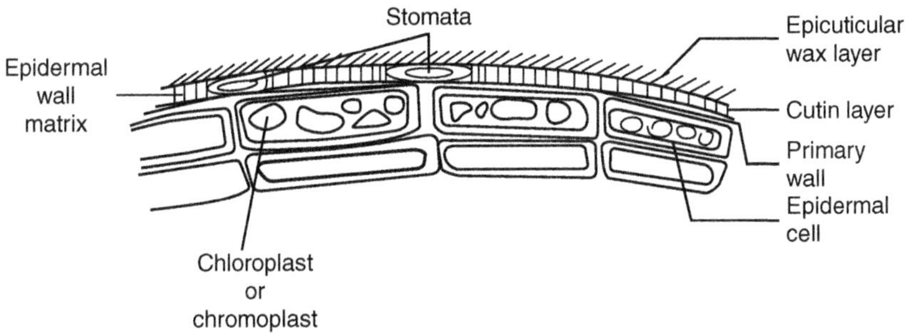

FIGURE 5.2 Schematic Diagram of Epidermal Layer, Cutin, and Wax on Rind of Citrus Peel.

cuticle that is quite complex in origin, structure, and development. The cuticle consists of an inner layer of cutin, which is a heterogeneous polymer of fatty acids and cellulose, and an outer layer consisting of cutin (Baker et al., 1975). In all, the cutin matrix is formed with cutin, wax, and a cell-wall material. Wax deposition continues as fruit grows; the wax hardens and develops breaks naturally. Epidermal cells are known to synthesize lipids and waxes for depositing on the cutin layer (Fig 5.2).

Wax in the form of platelets, rods, and other shapes is embedded within and over the cuticular surface (Albrigo, 1972a; Freeman, 1978). Epicuticular waxes, which are mainly responsible for restricting water loss from the peel, are complex in nature and formed of alcohols, paraffin, aldehydes, ketones, etc. (Freeman, 1978). The wax layers of Pineapple and Navel oranges, Dancy tangerines, and Eureka lemons have been observed to be initially amorphous. Small protrusions and isolated regions of upright platelets develop thereafter (Freeman et al., 1979). All surfaces eventually crack and are uplifted to form large, flat, irregular plates.

This outer surface can be rubbed gently to give shining polish. By restricting water loss through evaporation, the cuticle plays an essential role in maintaining high water content within tissue that is necessary for normal metabolism. One estimate claims that cuticle reduces the rate of evaporation from living plant cells from about 3.6 to 0.14 $mg/cm^2/Pa/h$. Water loss varies with the type of fruit or commodity, its anatomy, and surrounding conditions. This coefficient of transpiration (the rate of water loss) is 42 for apple and 7400 for lettuce (Wills et al., 1998). Numerous *stomata* are scattered over the surface of the epidermal cell layer over the parenchymatous tissues between the oil glands (Turrell and Klotz, 1940). The number of stomata is greater in the stylar half of the fruit than at stem end, and there are few or no stomata around the stem and the calyx (Albrigo, 1972b). Stomata are plugged with wax as the fruit matures. These stomata remain functional even after harvest unless they are plugged by applied wax. Uptake of exogenously applied PGRs through stomata is less and mostly occurs through cracks in wax layers and cutin. Immature fruits do not have a well-developed wax layer; the uptake of applied PGRs is relatively high in these fruits.

The epicarp also has cells with plastids containing chlorophyll (in other words, chloroplasts) which gradually change into chromoplasts as fruit color changes (Thompson, 1969). Fruits change color during the seasons from green to yellow and then to green again (in regreening). Photosynthetic activity is detected only in young green and regreened fruits. In yellow fruits a small amount of chlorophyll is detected. Electron microscopy showed that changing fruit color was the result of a typical transformation of chloroplasts into globular chromoplasts and vice versa (Ljubesic, 1984).

Colorless cells, called hypodermis and outer mesocarp, lie immediately below the epicarp and contain oil glands. The size of oil glands range from 10 to 100 μm or more. The terpenes (mainly *d*-limonene) and sesquiterpenes of oils in these glands give characteristic aroma and flavor to fruits of different citrus species. If the glands are ruptured by impact, oleocellosis develops – a characteristic lesion on the rind that can give fruit an unsightly appearance, as the oil is injurious to other cells of the rind (Wardowski et al., 1976). The rind of most citrus fruits is generally inedible, largely because of the oil. However, the rind of kumquats is sweet and can be eaten along with the pulp.

Albedo consists of inner mesocarp, which consists of parenchymatous cells with large air spaces (Scott and Baker, 1947). This is an extremely effective cushioning material against pressure and impact to fruits. The albedo is 1–2 mm thick in limes and tangerines, 2–5 mm thick in sweet oranges, and up to 20 mm thick in pummelos. Albedo is attached to flavedo on the outer side and connected with segment membrane from the inner side. The exocarp, or flavedo, and white spongy mesocarp, or albedo, are blended together. The endocarp is the inner side of the pericarp and a portion of the locular membrane. When the peel/rind is stripped, the entire exocarp and all but the inner portion of the mesocarp are removed.

The segments surrounding the central axis form the edible pulp of a mature citrus fruit. Each segment is surrounded by a continuous endocarp membrane. The juice in the fruit is contained in closely compacted, club-shaped multicellular sacs, also called juice vesicles, which completely fill the segments and are attached to a thin wall called the carpellary septum surrounding the segments. Each juice sac also has a very minute oil gland at the center. The seeds (ovules) are also attached to segment walls (toward the central column) by means of axial placentation.

The *central core* is composed of the same type of colorless or tinted, loose, spongy network of cells as the albedo. The core is connected with the albedo by membranes between each segment. The central axis of all citrus fruits is solid in the immature stages of development and also in mature fruit, particularly sweet oranges, grapefruits, lemons, and limes. In overmature stages, the central axis may open up in these fruits. The central axis of mandarins and their hybrids, and of pummelo, is normally open, with an air space in the center.

As the rind or peel is removed, the white, thread-like, vascular bundles forming the network in albedo and running parallel to the fruit axis along the outside of the segments are also removed. These vascular bundles carry water and food to the juice vesicles during fruit growth and maturity.

The vascular system extends down the central axis of the fruit, reaching the blossom end (stylar or distal) first, then ramifies surrounding segments and back up the carpels to the stem end (calyx, proximal) of the fruit. Consequently, the distribution of photosynthates and sugars is higher in the blossom end than the stem end. The vascular bundles in citrus fruit provide nutrition to developing fruit and these form a highly ramified network of main and subsidiary traces whereby every cell in the various tissues is connected directly or is adjacent to a cell in contact with a particular section of the vascular system.

The vascular system of the peduncle (fruit stem) is similar to that of other young stems on the tree. It consists of concentric cylinders of phloem, lateral meristem or cambium, and xylem surrounding a central core of pith.

Studies (using labeled CO_2) on movement of photosynthates indicate that photosynthates accumulate in the peel. Most of them enter the fruit via dorsal vascular bundles and are partially hydrolyzed during slow transfer through non-vascular segment epidermis and juice stalks (Koch, 1984). Anatomical changes in the structures of the cells of the Persian lime have been studied during fruit maturation by Garcia and Rodriguez (1992). Chloroplast structure did not change during fruit development and ripening. Fruits are dark green when young and turn light green when mature and ripe. Gradual dissolution of the middle lamella on ripening occurred in the mesocarp. This is linked to polygalacturonase enzyme action and the ripening process. Juice vesicles had an epidermal cell layer and were covered by a cuticle. Vesicles had cell organelles initially; as the fruit developed, all organelles decreased. Juice vesicles showed senescent symptoms earlier than other tissues of the fruit.

III. FRUIT PHYSIOLOGY

As citrus fruit develops, several physiological changes occur. There is a relationship between various functions of different organelles, tissues, organs, and the system as a whole. With growing age changes occur in functioning of the tissues and systems. Biochemical changes occurring with growth are discussed in Chapter 6. Some important physiological aspects with respect to postharvest life of the fruit are discussed here.

A. Respiratory Activity

Climacteric is defined as a period in the ontogeny of certain fruits during which a series of biochemical changes is initiated by the autocatalytic production of ethylene marking the change from growth to senescence and involving an increase in respiration and leading to ripening (Biale, 1950; Rhodes, 1980). Citrus fruit are non-climacteric, hence their respiration rate and ethylene production do not exhibit remarkable increase along with changes related to maturity and ripening as in mango or banana. These fruits do not ripen after harvest. Internal quality of citrus fruits is at its best when fruits are at optimum maturity on the tree.

Subramanyam et al. (1965) observed that respiratory rate of 450 mg CO_2/kg/h 30 days after fruit-set in acid lime fruit decreased to 200 mg after 60 days, 100 mg after 90 days and then increased up to 140 mg after 120 days. At 150 and 180 days, respiration rate was 50 and 40 mg CO_2/kg/h, respectively. Respiration was very high during rapid cell division. Spurt in respiration was observed at 120 days with a decline at later stages, indicating the possibility of climacteric peak between 90 and 120 days with maximum biochemical activity. Relationship among fruit age, epicuticular wax, weight loss, internal atmosphere composition, and respiration were investigated in maturing Washington Navel fruits by El-Otmani et al. (1986). Fruit epicuticular wax, internal CO_2 and internal ethylene increased, while with advancement of season, weight loss, and respiration decreased during storage. Concomitantly fruit conductance to CO_2 was reduced. Respiration rate of grapefruit peel was shown to be higher than pulp but it decreased after harvest while the rest of the fruit respiration remained constant (Vakis et al., 1970). Aharoni (1968) reported that respiratory climacteric can be detected if fruit are picked well prior to normal harvest time.

Response to exogenous ethylene of citrus fruit is reversible, hence they are not considered a climacteric fruit (Eaks, 1970). In Mosambi sweet oranges, respiration increased from initial rate of nearly 35 mg CO_2/kg/h to 80 mg CO_2/kg/h in ethylene-exposed fruit by the end of 48-h ethylene treatment. This reaction slowly declined (Fig. 5.3) after removal of the fruit from the ethylene atmosphere (Ladaniya, 2001).

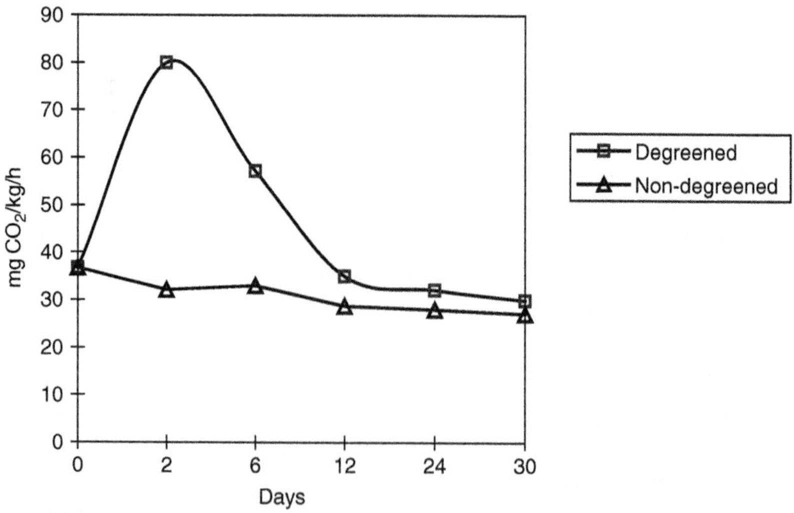

FIGURE 5.3 Respiration Rate of Mosambi Orange Fruit as Affected by Ethylene Application During Degreening.

Respiration of citrus fruits is affected by temperature, humidity, air movement, atmospheric gases, and handling practices. Increasing the temperature increases respiration rate; lowering the temperature restores the original respiration rate with no evidence whatsoever of a climacteric (Vines et al., 1968). Eureka lemons produced 80.5 mgCO_2/kg/h at 37.7°C and 22.7 mgCO_2/kg/h at 21.1°C (Murata, 1997). The respiratory rate increases at higher storage temperatures. This significantly affects storage life because the heat of respiration, or vital heat, generated is also higher (Table 5.1). Heat evolution can be computed from the respiration rate.

TABLE 5.1 Respiratory Activity (CO_2 Evolution in mg/kg/h) of Citrus Fruits as Influenced by Temperatures

Fruit	0°C	5°C	10°C	15°C	20°C	25°C	30–32°C
Sweet oranges	2–4	4–8	6–9	12–24	22–34	25–40	33–46
'Mosambi' orange	–	7–8	11–12	18.5–19	27–28	38–40	–
Kinnow mandarin	–	5–6	10–11	12–14	17–18	28–30	48–50
'Nagpur' mandarin	–	7–9	10–13	15–18	25–30	32–38	40–46
Grapefruit	–	–	7–10	10–18	13–26	19–34	–
Lemons	–	–	11	10–23	19–25	20–28	–
Limes	–	–	4–8	6–10	10–19	15–40	40–55

The heat evolution in Kcal/ton/24 h can be computed by multiplying above given respiration rates with a factor of 61.2.

Respiration of citrus fruit respond differently at temperatures above and below critical temperatures that correspond closely with the temperature at which chilling injury occurs. Citrus fruits exhibit cumulative time-temperature influence of chilling temperature on carbon dioxide evolution (Eaks, 1960). The CO_2 evolution is greater after exposure to 0°C than at 10°C. The chilling temperature also increases the production of ethylene and volatile components (ethanol and acetaldehyde) in fruit after a return to normal temperature (Eaks, 1980). Storage at 1°C for 14 days resulted in elevated respiration in Lisbon and Eureka lemons, with a peak occurring during the first 24 h. After 28 days at 1°C, peak respiration increased to 51 mg/kg/h for Lisbon lemons and 34 mg/kg/h for Eureka lemons. Respiration increased significantly, consistent with extensive chilling injury (Underhill et al., 1999).

The lowest safe temperature minimizes the respiration rate, which prolongs normal metabolism of the fruit and thereby its postharvest life. The 'Nagpur' mandarin has relatively low respiration rate: nearly 40–45 mgCO_2/kg/h at 25–30°C and 50–60 percent RH. Waxing reduced this rate by 34–35 percent. At low temperature (6–7°C), the respiration rate is 12–15 mgCO_2/kg/h. Waxing further reduces this rate by 10–11 percent. During storage, the respiratory activity of Shamouti and Valencia oranges declined and the internal CO_2 rose from a range of 2–4 percent to 5–10 percent, while the O_2 declined from 17–19 percent to 10–12 percent (Ben-Yehoshua, 1969).

If fruits are stored at the lowest safe temperature (above freezing) that significantly suppresses respiration and fungal growth, the storage life of fruits can be increased significantly. While chilling temperature adversely affects fruit quality, the higher temperatures also reduce the storage life and quality. The severity of adverse effects depends on the time–temperature relationship under such conditions.

At lower humidity, the respiration rate of citrus fruits is lower than that at higher humidity (Murata and Yamawaki, 1989). Higher concentrations of oxygen (34.1–99.1 percent) increase the respiration rate of citrus fruits, while lower concentrations reduce the rate. Very low concentration of O_2 (0.5–5 percent) tend to increase the respiration rate. Citrus fruits produce higher quantities of ethanol and acetaldehyde at low O_2 levels and higher N_2 levels, indicating unaerobic respiration.

Particular care is needed during harvesting, storage, transportation, and marketing of citrus fruits because rough handling causes wounding, stimulation of respiration and ethylene production, and can induce physiological disorders and fungal rot. In 'Ponkan' (*C. reticulata*) fruits, respiration rate and superoxide dismutase (SOD) activity in both peel and flesh has been shown to increase during storage when fruits have compression bruising. (Increases were larger following compression at 6.0 kg, compared with compression at 4.5 kg.) Respiration rate and SOD activity decreased rapidly at 40 and 65 days after bruising respectively, to reach levels similar to those of the control fruits. Peel conductivity increased with the degree of bruising, indicating increased membrane permeability (Xi et al., 1995).

Higher respiration has also been recorded in 'Nagpur' mandarins during excessive shriveling, rough handling, and rotting. If the fruit is rotting from inside without any external symptom (as in *Alternaria* core rot), respiration rises, which can be interpreted as deteriorating quality and decay. Internal disorders such as granulation also affect respiratory activity. Climacteric peaks of 'Ougan' and 'Hongju' mandarins reached a maximum after storage for 120 and 45 days, respectively. The respiration rate of the peel of the 'Ougan' was higher than that of its pulp. On the other hand, the pulp of 'Hongju' had a higher respiration rate than its peel until full granulation only; thereafter the peel was higher than the pulp (Ye et al., 2000).

Gas-exchange properties and respiration seem to remain unaffected by insect damage to citrus peels. Regions damaged by rust mite, wind scar, and pitting from physical damage had similar gas-exchange properties as undamaged regions on the same fruit (Petracek, 1996).

B. Biochemistry of Respiration

As the respiration rate is low and steadily declines after the harvest of citrus fruits, the available amount of sugar and organic acids is slowly converted into

CO_2, water, and heat. Since there is no starch, the sweetness of citrus fruits does not increase after harvest except for a slight increase in total soluble solids because of the activity of hydrolytic enzymes, or concentration effect, caused by rapid loss of water under dry storage conditions. At higher temperatures citric acid content also drops rapidly.

Detached fruits require energy for carrying out metabolic reactions to transport metabolites, to maintain cellular organization and membrane permeability, and to synthesize new molecules. This energy comes from aerobic respiration that is the oxidative breakdown of organic compounds such as sugars, organic acids (citric and malic acids present in vacuoles), lipids, and in extreme cases even proteins. The most common substrates in respiration of citrus fruit are glucose and fructose. One molecule of glucose produces energy equivalent to 686 kcal on complete oxidation. This chemical energy is stored in the form of adenosine 5'triphosphate (ATP), nicotinamide adenine dinucleotide (NADH), and $FADH_2$.

$$C_6H_{12}O_6 + 6O_2 \rightarrow 6CO_2 + 6H_2O + \text{energy}$$

In respiration, when sugars are consumed the ratio of oxygen utilized is equal to CO_2 produced: the respiratory quotient $RQ = 1$. ($RQ = CO_2$ produced ml/O_2 consumed ml.) In this case, respiration rate can be measured as either O_2 consumed or CO_2 evolved.

In cytoplasm, glycolysis takes place in which glucose is converted to pyruvate by the enzymes of the EMP (Embeden-Meyerhof-Parnas) pathway. The pyruvate is finally converted to CO_2 and energy in TCA (tricarboxylic acid cycle) by enzymes in mitochondria, the 'powerhouses' of the cells. During the initial fruit growth period, starch is broken down in to glucose by amylase. Phosphorylase enzyme converts glucose to glucose-1-phosphate. Sucrose is broken down to glucose and fructose by invertase. Sucrose synthase is also involved in formation of UDP-glucose (Uridine 5'-diphosphate) and then it is converted to glucose-1-phosphate and fructose-1-phosphate, which are then fed into the EMP pathway. Organic acids are directly utilized in the TCA cycle in mitochondria. Acids consume more O_2 for each CO_2 generated and hence RQ is 1.3. In case of fatty acids, the RQ is 0.7. If we measure both O_2 consumed and CO_2 evolved, we can know the substrate utilized in the respiration by the fruit. However, it is quite possible that several substrates are being utilized at a time and a correct picture may not be available simply on the basis of O_2 consumed and CO_2 produced.

C. Transpiration

When water loss occurs from plant parts in the form of evaporation it is called transpiration. The stomata and cuticle on the epidermal layer of cells (outermost layer of cells) offer the least resistance to moisture loss. Stomata are mostly closed and covered with wax. The wax platelets on the fruit surface overlap; they are strongly hydrophobic. The spaces between these wax platelets are often filled with air. The soft wax component made of alcohols, aldehydes, esters, and fatty acids determine the rate of transpiration. Citrus fruits contain 80–85 percent water. The loss of water has a greater consequence since it affects appearance and also weight. In citrus peels, water exchange was found to be approximately four to six times greater at the stem-end than in other regions (Petracek, 1996).

Usually, fruit peel loses water more rapidly than the flesh during storage under low-humidity conditions, and also becomes thinner. The fruit juice content (percentage) shows an increase (although erroneously) as it is recorded on a fresh-weight basis.

Softening of peel is due to flaccidity of the cell or hydrolysis of intercellular pectic compounds during long refrigerated storage. Under dry ambient conditions with low relative humidity, peel dries rapidly, thus becoming tough and leathery, and hinders normal gas exchange. This causes anaerobic conditions and an increase in alcohol levels inside the fruit. Citrus fruits have relatively long postharvest life if protected from water loss and decay causing microorganisms, mainly fungi. If the peel remains turgid and healthy, normal gas exchange can occur without accumulation of CO_2 or ethylene in and around fruit.

The loss of commercial value of Shamouti and Valencia orange fruit under various storage conditions is caused by transpiration, which lead to shriveling of the peel. During storage, respiratory activity declines and the internal CO_2 rises from a range of 2–4 percent to 5–10 percent while the O_2 declines from 17–19 percent to 10–12 percent. Drying of the peel causes a rise in resistance to gas diffusion, which in turn changes the internal atmosphere. The flavedo portion of the peel is the main site of resistance to gas diffusion (Ben-Yehoshua, 1969).

Stomata of harvested citrus fruits are essentially closed. However, ethylene, O_2, and CO_2 still diffuse through the residual stomatal opening (<1 percent), while water evaporates from epidermal cells. Ethylene, O_2, and CO_2 are constrained from using the water-transport pathway because their diffusivity in water is 104 times less than in air. Waxing fruits partially or completely plugs the stomata. This may increase the off-flavors if coating is not done properly by partially restricting O_2 and CO_2 diffusivity. Wax coating inadequately reduces transpiration because the new surface layer it forms has many pits and breaks. Sealing fruits individually in 10-micron-thick, high-density polyethylene film is more effective than waxing for increasing storage life. The film reduces water loss by 10 times without substantially inhibiting gas exchange (Ben-Yehoshua et al., 1983). The region of neck or stem-end loses water rapidly. Among citrus fruits, acid limes do not have cuticle and hence water loss is quite rapid in these fruits.

Loss of water not only affects appearance or esthetic value but also reduces saleable weight, thus causing direct economic loss. Citrus fruits have low surface-area-to-volume ratio and thus lose water more slowly than many leafy vegetables, but even 5–6 percent water loss can result in some change in appearance and firmness of the fruit that can be detrimental to its marketability.

D. Role of Ethylene

Ethylene is also known as stress hormone. (Its level increases in plants and fruit with the application of stress, and it can create a stress-like condition in plants if applied exogenously.) Ethylene has a special role in fruit maturity, ripening, and senescence, and therefore has its own importance in postharvest management of citrus. Ethylene is known to soften the fruit by disintegrating cell membranes and making them leakier, eventually resulting in fruit softening. Chemical composition, flavor, and texture remain more or less unchanged with ethylene action in citrus. Acidity content decreases slightly with the exogenous application of ethylene. Treatment of fruits with ethylene or ethephon increases nootkatone levels in the rind of both harvested and unharvested Star Ruby grapefruit. The nootkatone level in the rind is therefore proposed as an indicator of ripening/senescence in grapefruits (Garcia et al., 1993).

Citrus fruits have a very low rate of ethylene evolution, in the amount of $<0.1\,\mu l/kg/h$. Even this rate can slowly build up ethylene concentration in closed chambers. The higher CO_2 concentration inside the fruit can counteract the ethylene action. If fruits are rotting in the box, the ethylene evolution is very high and can affect physiology of other fruits. Several other stresses such as freezing, excessive drying/shriveling, and even dropping of fruit can increase the ethylene buildup and respiration. Ethylene production per fruit basis is very low (2 nl/h/fruit), and even this low concentration can be effective endogenously to enhance maturation and senescence. Healthy Satsuma is reported to produce $0.16\,\mu l/kg/h$, while those infected with *Colletotricum gloesporioides* produced $11.80\,\mu l/kg/h$ (Hyodo, 1981).

Citrus fruits produce a very low quantity of ethylene after harvest and there is no associated rise in respiration. However, citrus fruits respond to exogenous ethylene by an increase in respiration, chlorophyll loss, calyx drying, and abscission, although they cannot synthesize large amounts of ethylene autocatalytically. Thus, when the supply of ethylene is terminated, the enhanced respiration decreases to the low level that existed before ethylene treatment.

Young, immature citrus fruits produce large amounts of ethylene, and their respiration increases parallel with a rise in ethylene production. This high ethylene production may be responsible for June drop. Wounding of harvested citrus fruit tissues causes a rise in ethylene production and accelerates coloring and related metabolic changes. This wounding could be due to fungal attacks (green and blue mold and other pathogens), insect damage, freezing injury, hailstorms, or postharvest stresses such as chemical injury, mechanical injury, gamma radiation, and chilling temperature. Preharvest injury and consequent microbial attacks also lead to fruit drop.

In citrus tissues, ethylene is produced from the amino acid methionine, as in most other plants, in an enzymatically controlled reaction as follows:

$$\text{Methionine} \xrightarrow{\text{SAM Synthetase (EC 2.5.1.6)}} \text{SAM(S-adenosylmethionine)}$$
$$\text{(S-Ado-Met)}$$

$$\text{SAM} \xrightarrow{\text{ACC Synthase enzyme}} \text{1-aminocyclopropane-1-carboxylic acid (ACC)}$$

The ACC (1-aminocyclopropane-1-carboxylic acid) is then converted to ethylene through the action of the ACC oxidase enzyme. CO_2 and cyanide are generated and cyanide is detoxified by another enzyme. The SAM is also a precursor for production of polyamines.

ACC is a fundamental intermediate in ethylene production. The ACC content of fresh fruit tissues is much less, but ACC content increases after wounding. Ethylene production in aged albedo tissue is markedly reduced by inhibitors of protein synthesis. This suggests that protein is required to maintain continuous evolution of ethylene (Hyodo, 1981). Conversion of SAM to ACC by ACC synthase enzyme is a rate-limiting step and its gene expression is tightly controlled (Wang et al., 2002).

Potential storage life of citrus fruits with fairly good appearance and eating quality can be obtained if fruits are stored under the most optimum conditions after harvest. Postharvest action that reduces the accumulation of ethylene around citrus and other non-climacteric produce during marketing can result in an increase in postharvest life (Wills et al., 1999). It is suggested that the threshold level of ethylene action on non-climacteric produce is well below 0.005 ppm than the commonly considered threshold level of 0.1 ppm. There is a 60 percent extension in postharvest life of the produce when stored at 0.005 ppm than at 0.1 ppm ethylene. Storage life can be linearly extended in oranges with a logarithmic reduction in C_2H_4 level.

Ethylene destroys chlorophyll and hastens color development by increasing carotenoid synthesis. Temperature and storage duration also affect color development. Carotenoid pigment synthesis takes place at 15–20°C without ethylene treatment. Treatment with ethylene increases chlorophyllase activity in the rind and reduces the number and the size of chloroplasts. Ethylene increases the appearance of chilling injury (CI) symptoms, stem-end rot, and the content of volatile off-flavors in the juice and fruit internal atmosphere. The protective effect of a small amount of ethylene during postharvest storage of Shamouti oranges reduced the amount of decay caused by molds (Porat et al., 1999). The small amount of endogenous ethylene produced by the fruit was considered to be required to maintain their natural resistance against various environmental and pathological stresses.

Plant hormones, such as auxins, GA, ABA, and cytokinin, play a role singly or in combination in the induction of ethylene production. At maturity, the fruit of early-maturing Hamlin and Pineapple oranges contained more ethylene and abscissic acid than late-maturing Valencia and Lamb Summer oranges. Both compounds increased most rapidly in Pineapple, resulting in increased cellulase activity and loosening of fruit separation zones at maturity (Rasmussen, 1974). Application of ABA to Shamouti orange peel plugs and to the excised segments or albedo discs of Satsuma mandarin fruit stimulated ethylene production (Hyodo, 1978). Ethylene and ABA play an interactive role in orange fruit maturation.

Ripening 'Pinalate' oranges (ABA deficient mutant) exhibited delayed degreening, developed a yellow color, contained lower concentrations of ABA, and contained lower concentrations of xanthophylls compared with wild-type fruits. Application of ABA to 'Pinalate' fruits accelerated degreening, and exogenous ethylene also influenced ripening. Both ethylene and ABA are involved in citrus fruit ripening, with ethylene possibly regulating the initiation of ripening and ABA the rate of ripening (Alfetrez et al., 1999).

Ethylene antagonists such as silver nitrate and 2.5-norbonadiene have been shown to inhibit chlorophyll breakdown induced by ethylene, thus inhibiting the degreening process (Goldschmidt et al., 1993; Goldschmidt and Galily, 1995: Goldschmidt, 1998). The compound 1-methylcyclopropene (1-MCP) inhibits the binding of ethylene to the ethylene receptor site, the ethylene binding protein (EBP) and thus blocks ethylene action and degreening (Porat et al., 1999). 1-MCP prevented infection-induced degreening, such that treated grapefruits retained their green, immature color compared to yellow, non-treated controls. However, 1-MCP treatment significantly increased whole fruit ethylene production. This suggested that in the presence of a pathogenic stress, blocking the EBPs prevented regulatory control of the ethylene biosynthetic pathway that resulted in an uninhibited expression of the ACS (ACC synthase) stress-associated genes, increased ACS activity, and elevated ACC accumulation and ethylene production. Blocking the EBPs with 1-MCP did not affect progression of the pathogen through the fruit (Mullins et al., 2000). Even in absence of pathogenic infection, 1-MCP increased ethylene synthesis.

E. Color Development and Regreening

During maturation and development, citrus fruits change color from green to yellow or orange or orange-red as per the genetic character of the variety under favorable climatic and growing conditions. This is called natural degreening, or natural color development. In some citrus fruits, if held on the tree beyond maturity, the yellow-orange color again changes to green, which is called regreening. The regreening process has economic significance since regreened fruit, although internally mature, is not marketable. The regreening process occur on the tree and also after harvest. Huff (1984) observed that the regreening process results from a decrease in soluble sugars as observed in Valencia orange fruit. Ultrastructural studies have indicated that regreening takes place as a result of the reversion of chromoplast to chloroplast and not from the formation of new chloroplast (Wrischer et al., 1986). The chromo-chloroplasts are photosynthetically active and required photosynthetic proteins have been found in chloroplasts of regreened fruit. In most citrus fruits, regreening occurs when fruit is on the tree, but pummelo regreening has been observed in harvested fruit stored in natural light or fluorescent light (Saks et al., 1988). The process depends on light intensity and temperature. Electron-microscopy study indicates that globular chromoplasts in peel tissues of pummelo revert to chloroplasts during regreening although only partially (20 percent chlorophyll level obtained after regreening). The proteins of photosysnthetic system have been detected in regreened peel (reconstructed chloroplasts). These reconstructed chloroplasts were not observed in yellow fruit.

F. Fruit Abscission

In citrus fruit, leaves have abscission zones at the leaf lamina-petiole interface and another at the interface of petiole and stem. Similarly, fruit has two abscission zones: one at the fruit – calyx interface and another between fruit button and stem. When citrus fruit naturally abscises from the tree (as the calyx and button-like receptacle at the stem end has an abscission zone at its base) no button or calyx is attached to fruit. The calyx remains attached to the branch or peduncle. Natural abscission is dependent on several exogenous and endogenous factors. In fruit harvested with clippers with a 2–3 mm peduncle attached, the abscission of the calyx and button takes place naturally after some aging and drying of this part takes place during handling and storage.

In early-maturing oranges such as Hamlin and Pineapple, ethylene and abscissic acid levels have been shown to be higher than the late-maturing Valencia and Lamb Summer. Ethylene (up to 95 nl/l in the internal atmosphere) and abscissic acid (50 μg/kg dry weight of flavedo) increased most rapidly in Pineapple oranges, leading to increased cellulase activity and loosening of the fruit. Fruit of the late-maturing cvs contained less than 25 nl/l ethylene and 40 μg abscissic acid/kg dry weight of flavedo at peak maturity. Cellulase activity and loosening of the fruit of these late-maturing cvs was slight (Rasmussen, 1975). In Marsh grapefruit, high gibberellic acid and cellulase activities were noted in zone 'C' of the peduncle (nearest to the fruit) during June, when fruit drop was high and fruit growth was rapid (Pozo et al., 1989).

G. Fruit Hormonal Balance

The physiological role of various endogenous plant growth substances is to regulate the growth, development, maturity, ripening, and senescence of fruit naturally (when there is no intervention by externally applied plant-growth substances) as a predetermined sequential process in normal growing conditions. Understanding the endogenous hormonal balance and physiology of fruit development is essential for exogenous application of PGRs in commercial citriculture.

Gibberellic acid can promote auxin action, while high doses of auxin can increase ethylene production (Burg and Burg, 1968). Auxins are mainly responsible for cell enlargement by increased water uptake and they increase the extensibility of cell walls. Greater activity of pectin methylesterase was observed in auxin-treated tissues (Osborne, 1958) with higher synthesis of pectic substances

(Albersheim and Bonner, 1959). Thus, there is an interaction between these growth regulators; time and concentration of exogenous application determines the result. Whatever may be the exact role and nature of interaction of these growth regulators, it is certain that all of them are important in fruit development, overall physiology, and shelf life. Growth regulators play a deciding role in growth and development of parthenocarpic fruit and their use is imperative in citrus fruit production.

Fruits are entities of plants: they are a portion that surrounds ovules (seeds). Development of fruit is linked to the development of the ovule. Mostly seedless fruits are small in size and this may be the reason that some sort of stimuli, perhaps hormonal, from the seeds (ovules) or pollens are regulating fruit growth. Larger fruit with more seeds is produced if more pollen is used in Valencia orange (Erickson, 1968). In fact, these stimuli start from the stage of flower formation and tissues of fruit are formed as primordial of flower ovary. Even before anthesis, the development of the ovary that is the fruit takes place. After pollination, ovules (seeds) play a leading role in fruit development and this regulation is mostly by hormonal means. In young ovaries of Washington Naval orange, indoleyl-3-acetic acid (IAA) has been reported (Nitsch, 1965). A higher cytokinin content was observed in ovaries of Washington Navel than in those of Navelate, suggesting that a higher flow of nutrients toward the vegetative organs occurs in Navelate, causing low fruit set and poor productivity. When Navelate trees were subjected to water stress, an increase was observed in the abscissic acid content of leaves and fruits (Furio et al., 1982). Cytokinins – Riboxylzeatin, zeatin, and isopentenyladenosine – have been detected in the developing fruits of Salustiana (seedless) and Blanca Comuna (seeded) (Hernandez Minana et al., 1988).

In general, the level of ethylene increases in fruit before maturation and ripening; this is more profound in climacteric fruit and less in non-climacteric citrus. Ripening is controlled by endogenous as well as exogenous level of ethylene in fruits. Ethylene promotes maturity/ripening, abscission, and senescence. Gibberellins and cytokinins have an antagonistic effect and they delay the ripening and senescence process as observed in oranges (Eilati et al., 1969). It can be said that growth substances direct the movement and utilization of nutrients and also stimulate cell metabolism as well as promoting cell division, enlargement, and maturation. Abscissic acid (ABA), which is known to promote abscission of flowers, leaves, small fruitlets, and fruits, was also reported from *Citrus medica* rind and pulp (Milborrow, 1967) in concentration of 0.097 mg/kg fruit weight. Abscissic acid concentration increases as the limes ripen, suggesting cessation of growth and changes leading to senescence and abscission of fruit (Murthi, 1988). The different citrus species varied in free ABA (3–8 μg/g dry weight) and conjugated ABA (10–39 μg/g dry weight). In general, the amount of conjugated ABA was four times that of free ABA. In the lemon flower, the combined style and stigma tissues contained the predominant (>65 percent) amount of both free and conjugated ABA. During fruit development, different tissues showed dynamic changes in ABA content. In the vesicles, free ABA showed progressive increases with development and reached a high level at maturity, whereas the conjugated ABA showed a corresponding decrease. Free ABA increased in the seed with fruit development. The content of ABA in citrus fruit was correlated with the fruit weight (Aung et al., 1991). In Kinnow mandarins, auxin content in the seeds and pulp increased 34 weeks after fruit set, until just before harvest, while other substances, namely cytokinins, gibberellins, and abscissic acid, decreased during later stages of fruit development – although the decrease in the abscissic acid was less abrupt in the pulp (Dhillon, 1986).

Extracts of the seedless Clementine cv. Fino had higher contents of auxin-like compounds than those of the seeded cv. Monreal. These compounds were suggested to be involved in the parthenocarpic fruit set and development of Fino ovaries. Diffusable gibberellin-like compounds produced in the developing seeds were obtained from Monreal fruits after anthesis. Fruit ABA content was low in Monreal, whereas in Fino and in Monreal it increased markedly after anthesis with the seeds removed. Garcia and Garcia-Martinez (1984) suggested that the control of fruit set and development in Clementines is carried out through an equilibrium between auxin-like substances (in seedless Fino) or gibberellin-like substances (in seeded Monreal) and ABA. Applications of GA3 or paclobutrazol (PBZ) to fruitlets influences fruit retention (Turnbull, 1989a). In Valencia orange, 29 percent of GA-treated fruits were retained to maturity compared with 2 percent in untreated controls. Paclobutrazole, which inhibits GA biosynthesis, caused 100 percent fruit drop within 35 days of application. GA1, GA3, GA8, GA19, GA20, GA29, 3-epi-GA1, 2-epi-GA29, and iso-GA3 were identified in tissues of Valencia orange (Turnbull, 1989b). No major differences were found between tissues of immature seeded and seedless fruit, and developing seeds did not contain high levels of any GA. Seed-produced GAs are not considered essential for fruit retention and development in Valencia oranges.

During growth of citrus fruits the mass (mg/g dry matter) of ABA and GAs per fruit increases. Fruit size and retention were greater in leafy fruitlets (fruits near leaves or leafy inflorescence) than in leafless ones (Hofman, 1988), indicating the role of leaves in maintaining hormonal balance. Although the Blanca Comuna orange and its parthenocarpic mutant cv. Salustiana did not differ qualitatively in the GAs present during flower and fruit development, the Salustiana contained greater amounts as observed in their reproductive organ (Talon et al., 1990).

Satsuma is a male sterile cultivar that shows high degree of natural parthenocarpy, whereas Clementine varieties are self-incompatible and hence have a very low ability to set a commercial crop. GA_3 improves fruit set in Clementines and have little effect on Satsumas. Thus, it appears that self-incompatible Clementines do not have sufficient GA for fruit set. This suggests that endogenous GA content in developing ovaries is the limiting factor controlling parthenocarpic development of the fruit in seedless Clementine mandarins. Levels of 13-hydroxy GA_3 have been found to be lower in Clementines. At petal fall fruits of Satsuma and Clementine had 65 and 43 pg of GA_3 respectively (Talon et al.,

1992). In Spain, usually GA is applied at full bloom of Clementine to increase the set.

Changes in citrus fruit rind color on the tree are due to the weather; however, endogenous growth regulators play the role of internal control mechanism (Eilati et al., 1969; Rasmussen, 1973). Low temperature can provide sufficient stress to produce ethylene, which causes the destruction of chlorophyll and development of carotenoides (Grierson et al., 1982). Plant-water relation, stress conditions, and accompanying endogenous hormonal changes are also likely to change fruit physiology and result in the development of disorders. Concentrations of IAA, gibberellin-like substances, and zeatin have been found to be higher in the pulp than in the peel of cracked fruits, but lower in the pulp than in the peel of normal fruits of lemon (Citrus limon (L) Burm) cv. Baramasi. Abscissic acid concentration was higher in the peel than pulp of cracked fruits, and was higher in cracked fruits than in normal fruits (Josan et al., 1995).

REFERENCES

Aharoni, Y. (1968). Respiration of oranges and grapefruit harvested at different stages of development *Plant Physiol.* 43, 99–102.

Albersheim, P., and Bonner, J. (1959). Auxin, pectic substances and pectinmethylesterase. *J. Biol. Chem.* 234, pp. 3105.

Albrigo, L.G. (1972a). Distribution of stomata and epicuticular wax on oranges as related to stem end rind breakdown and water loss. *J. Am. Soc. Hort. Sci.* 97, 220–223.

Albrigo, L.G. (1972b). Ultrastructure of cuticular surface and stomata of developing leaves and fruit of Valencia orange. *J. Am. Soc. Hort. Sci.* 97, 761–765.

Alferez, F., Zacarias, L., and Grierson, D. (1999). Interaction between ethylene and abscisic acid in the regulation of Citrus fruit maturation. In *Biology and biotechnology of the plant hormone ethylene* (A.K. Kanellis, C. Chang, H. Klee, A.B. Bleecker, and J.C. Pech, Eds). Proceedings of the EU-TMR-Euroconference Symposium, Thira (Santorini), Greece, 5–8 September.1998, pp. 183–184. Kluwer, Netherlands.

Aung, L.H., Houck, L.G., and Norman, S.M. (1991). The abscisic acid content of citrus with special reference to lemon. *J. Expt. Bot.* 42, 241, 1083–1088.

Bain, J.M. (1958). Morphological, anatomical, and physiological changes in the developing fruit of Valencia orange (*Citrus sinensis (L)* Osbeck). *Aust. J. Bot.* 6, 1–25.

Baker, E.A., Procopious, J., and Hunt, G.M. (1975). The cuticles of *Citrus* species: Composition of leaf and fruit waxes. *J. Sci. Food Agric.* 26, 1093–1101.

Bartholomew, E.T., and Sinclair, W.B. (1951). *The lemon fruit: its composition and physiology.* Berkeley: University of California Press.

Ben-Yehoshua, S. (1969). Gas exchange, transpiration and the commercial deterioration in storage of orange fruits. *J. Am. Soc. Hort. Sci.* 94, 524–528.

Ben-Yehoshua, S., Burg, S.P., and Young, R. (1983). Resistance of citrus fruit to C_2H_4, O_2, CO_2 and H_2O mass transport. *Proc. 10th. Ann. Meeting of Pl. Growth Regulator Soc. America*, 145–150.

Biale, J.B. (1950). Postharvest physiology and biochemistry of fruit. Annu Rev. *plant Physiol* I, 183–206.

Burg, S.P. and Burg, E.A. (1968). *Biochemistry and physiology of plant growth substances.* (F. Wightman and G. Setterfield, eds.), pp. 1275–1294. Runge Press, Ottawa.

Dhillon, B.S. (1986). Bio-regulation of developmental processes and subsequent handling of Kinnow mandarin. *Acta Hort.* 179, 251–256.

Eaks, I.L. (1960). Physiological studies of chilling injury in citrus fruits. *plant. Physiol.* 35, 632–636.

Eaks, I.L. (1970). Respiratory response, ethylene production, and response to ethylene of citrus fruit during ontogeny. *plant. Physiol.* 45, 334–338.

Eaks, I.L. (1980). Effect of chilling on respiration, and volatiles of California lemon fruit. *J. Am. Soc. Hort. Sci.* 105, 865–869.

Eilati, S.K. Goldschmidt, E.E. and Monselise, S.P. (1969). Hormonal control of colour changes in orange peel. *Experientia* 25, 209–210.

El-Otmani, M., Coggins, C.W., and Eaks, I.L. (1986). Fruit age and Gibberellic acid effect on epicuticular wax accumulation, respiration and internal atmosphere of navel orange fruit. *J. Am. Soc. Hort. Sci.* 111, 228–232.

Erickson, I.C. (1968). The general physiology of citrus. In *"The citrus industry"*, (W. Reuther, L. Batchelor, and H.J. Webber, eds.) Vol. II. Division of Agricultural Science, University of California, Berkeley.

Freeman, B. (1978). Cuticular waxes of developing leaves and fruits of citrus and blueberry: Ultra structure and chemistry). Ph.D. Dissertation, University of Florida, Gainesville.

Freeman, B., Albrigo L.G. and Biggs, R.H. (1979). Ultrastructure and chemistry of cuticular waxes of developing Citrus leaves and fruits. *J. Am. Soc. Hort. Sci.* 104, 6, 801–808.

Furio, J. Calvo, F. Tadeo, J.L., Primo-Millo, E. and Millo, E.P. (1982). Relationship between endogenous hormonal content and fruit set in citrus varieties of the navel group. *Proc. Int. Soc. Citric, Japan* I, 253–256.

Garcia, M.A., and Garcia-Martinez, J.L.(1984). Endogenous plant growth substances content in young fruits of seeded and seedless Clementine mandarin as related to fruit set and development. *Scientia Hort.* 22, 265–274.

Garcia, M.E., and Rodriguez, J. (1992). Main ultrastructural changes during maturing stage of Persian lime fruit. *Proc. Int. Soc. Citric. Italy*, Vol. 1, pp. 475–477.

Garcia, P.D., Ortuno, A., Sabater, F., Perez, M.L, Porras, I, Garcia, L.A., and DelRio, J.A. (1993). Effect of ethylene on sesquiterpene nootkatone production during the maturation-senescence stage in grapefruit (*Citrus paradisi* Macf.). *Proc. Int. Symp. Cellular Molecular Aspects of Biosynthesis* and Action of the Plant Hormone Ethylene, Agen, France, August 31–September 4, 1992, pp. 146–147.

Goldschmidt, E.E. (1998). Ripening of citrus and other non-climacteric fruits: A role for ethylene. *Acta Hort.* 463, 335–340.

Goldschmidt, E.E., and Galily, D. (1995). Role of ethylene in spontaneous degreening of Shamouti and Valencia orange fruit after harvest. *Alon Hanotea* 49, 310–313.

Goldschmidt, E.E, Huberman, M. and Goren, R. (1993). Probing the role of endogenous ethylene in degreening of citrus fruit with ethylene antagonists. *plant. Growth Regul.* 12, 325–329.

Grierson, W., Soule, I., and Kawada, K. (1982). Beneficial aspects of physiological stress. *Hort. Rev.* 4, 247–271.

Guardiola, J.L. (1997). Future use of plant bio-regulators. *Proc.nt. Soc. Citric, Vol. 1., South Africa.*

Hardenburg, R.E., Watada, A.C. and Wang, C.Y. (1986). The commercial storage of fruits, vegetables, and florist and nursery stocks. USDA, ARS. *Agric. Handbook No.* 66, pp. 128.

Hernandez-Minana, F.M., Primo-Milo, E., and Primo-Milo, J. (1988). Isolation and identification of cytokinins from developing citrus fruits. *Proc. 6th Int. Citrus Symp.* Israel, Vol.1, pp. 367–372.

Hofman, P.J. (1988). Abscissic acid and gibberellins in the fruitlets and leaves of the 'Valencia' orange in relation to fruit growth and retention. Sixth international citrus congress, Middle-East, Tel Aviv, Israel, 6–11 March 1988, Vol. 1, 355–362.

Huff, A. (1984). Sugar regulation of plastid inter conversion in epicarp of citrus fruit. *plant Physiology*, 76, 307–312.

Hyodo, H. (1978). Ethylene production by wounded tissue of citrus fruit. *plant. Cell Physiol.* 19, 545–551.

Hyodo, H. (1981). Ethylene production by citrus fruit tissues. *Proc. Int. Soc. Citric.*, Tokyo, Japan, pp. 880–882.

Josan, J.S., Sandhu, A.S. and Singh, Z. (1995). Endogenous phytohormones in the normal and cracked fruits of lemon (*Citrus limon* (L) Burm). *Indian J. plant. Physiol.* 38, 238–240.

Koch, K.E. (1984). The path of photosynthate translocation into citrus fruit. *Plant Cell Environ.* 7, 647–653.

Ladaniya, M.S. (2001). Response of Mosambi sweet orange (*Citrus sinensis*) to degreening, mechanical waxing, packaging and ambient storage conditions. *Indian J. Agric. Sci.* 71, 234–239.

Ladaniya, M.S. and Singh S. (2006). Response of mandarin fruits of different maturity stages to chilling temperature with intermittent warming and post-storage holding. *Tropical Agriculture.* (In press)

Ljubesic, N. (1984). Structural and functional changes of plastids during yellowing and regreening of lemon fruits. *Acta Botanica Croatica* 43, 25–30.

Milborrow, B.V. (1967). Abscissic acid in *Citrus medica* fruit. *Planta*, 76, 93.

Mullins, E.D., McCollum, T.G., and McDonald, R.E. (2000). Consequences on ethylene metabolism of inactivating the ethylene receptor sites in diseased non-climacteric fruit. *Postharvest Biol Technol.* 19, 155–164.

Murata, T. (1997). Citrus. In *Post-harvest physiology and storage of tropical and sub-tropical fruits*, (S.K. Mitra, ed.). CAB International, pp. 21–47.

Murata, T., and Yamawaki, K. (1989). Respiratory changes of several varieties of citrus fruits during and after conditioning with two different humidities. *J. Jpn Soc. Hort. Sci.* 58, 723–729.

Murthi, G.S.R. (1988). Changes in abscissic acid content during fruit development of acid lime. plant *Physiol. Biochem.* 15, 138–143.

Nitsch, J.P. (1965). *Encyclopaedia of plant physiology* (W. Ruhland, ed.) 15(1), 1537 Springer Verlag, Berlin.

Osbourn, D.J. (1958). Pectinmethylesterase activity in auxin treated tissues. *J. Expt. Bot.* 9, 446–448.

Petracek, P.D. (1996). A technique for measuring gas exchange through the peel of intact citrus fruit. *Proc. Fla Sta. Hort. Soc.* 108, 288–290.

Porat, R., Weiss, B., Cohen, L., Daus, A., Goren, R., and Droby, S. (1999). Effects of ethylene and 1-methylcyclopropene on the postharvest qualities of 'Shamouti' oranges. *Postharvest Biol. Technol.* 15, 155–163.

Pozo, L., Oliva, H., and Perez, M.C. (1989). The relationship between gibberellin and cellulase activities and fruit abscission and development in the grapefruit cultivar Marsh grafted on *Citrus macrophylla* under Cuban conditions. *Agrotecnia de Cuba.* 21, 1–8.

Rasmussen, G.K. (1973). The effect of growth regulators on degreening and regreening of citrus fruit. *Acta Hort.* 34, 473–478.

Rasmussen, G.K. (1974). The relation of cellulase activity to endogenous abscissic acid and ethylene in four citrus fruit cultivars during maturation. *plant Physiol.* 53, 18.

Rasmussen, G.K. (1975). Cellulase activity, endogenous abscisic acid, and ethylene in four citrus cultivars during maturation. *plant Physiol.* 56, 765–767.

Rhodes, M.J.C. (1980). The maturation and ripening of fruits. In *Senescence in plants* (K.V. Thimann, ed.). CRC Press, Boca Raton, Fl, pp. 157–204.

Saks, Y., Weiss, B., Chalutz, E., Livne, A., and Gepstein, S. (1988). Regreening of stored pummelo fruit. *Proc. 6th Int. Citrus Congress*, Tel Aviv, Israel, Vol. 3, 1401–1403.

Scott, F.M., and Baker, K.C. (1947). Anatomy of Washington navel orange rind in relation to water spot. *Bot. Gaz.* 108, 459–475.

Sinclair, W.B. (1961). *The orange: its biochemistry and physiology*. University of California Press, Berkeley.

Sinclair, W.B. (1972). *The grapefruit: its composition, physiology and products.* Division of Agricultural Science University of California Press, Berkeley.

Subramanyam, H., Narasimhan, P., and Srivastava, H.C. (1965). Studies on the physical and biochemical changes in limes (*C. aurantifolia* Swingle) during growth and development. *J. Indian Bot. Soc.* 44, 105–109.

Talon, M., Zacarias, L., and Primo-Millo, E. (1992). Role of gibberellins in parthenocarpic development of seedless mandarins. *Proc. Int. Soc Citric.*, Italy, Vol. 1, pp. 485–488.

Talon, M., Hedden, P., and Primo-Millo, E. (1990). Gibberellins in Citrus sinensis: a comparison between seeded and seedless varieties. *J. Pl. Growth Regulation.* 9, 201–206.

Thompson, W.W. (1969). Ultrastructural studies on the epicarp of ripening oranges, *Proc. lnt. Soc. Citricul.* Vol. 3, 1163–1116.

Turnbull, C.G.N. (1989a). Gibberellins and control of fruit retention and seedlessness in Valencia orange. *J. plant Growth Regulation.*, 8, 270–272.

Turnbull, C.G.N. (1989b). Identification and quantitative analysis of gibberellins in Citrus. *J. plant. Growth Regulation* 8, 273–282.

Turrell, F. M., and Klotz, L.J. (1940). Density of stomata and oil glands and incidence of water spot in the rind of Washington navel orange. *Bot. Gaz.* 101, 862–870.

Vakis, N., Soule, J., Biggs, R.H. and Grierson, W. (1970). Biochemical changes in grapefruit and 'Murcott' citrus fruit, as related to storage temperature. *Proc. Fla. State Hort. Soc.* 83, 304–310.

Vines, H.M., Grierson, W., and Edwards, G.J. (1968). Respiration, internal atmosphere and ethylene evolution of citrus fruit. *Proc. Am. Sac. Hort. Sci.* 92, 227–234.

Wang, Kelvin L.C., Li, H., and Ecker, J. (2002). Ethylene biosynthesis and signaling network. The plant. Cell. Supplement, 131–151 (www.plantcell.org).

Wardowski, W.F., McCornack, A.A., and Grierson, W. (1976). Oil spotting (oleocellosis) of citrus fruit. Fla Ext. Circular. 410.

Wills, R., McGlasson, B., Graham, D., and Joyce, D. (1998). Postharvest: an introduction to the physiology and handling. CAB International, pp. 262.

Wills, R.B.H., Ku,V.V.V., Shohot, D. and Kim, G.H. (1999). Importance of low ethylene levels to delay senescence of non-climacteric fruit and vegetables. *Australian J. Exptl. Agric.* 39, 221–224.

Wrischer, M., Ljubesic, E., Marcenko, E., Funst, L., and Hlousek-Rodjeie, A. (1986). Fine structural studies of plastids during their differentiation and dedifferentiation (a review). *Acta Botanica Croatica* 45, 43–54.

Xi, Y. F., Dong, Q. H., Lu, Q. W., Wang, L. P., Ruan, J. H., and Mao, L. C. (1995). Influence of compression on postharvest physiology and quality of ponkan fruit. *Acta Agric Zhejiangensis.* 7, 24–26.

Ye, M.Z., Chen, Q.X., Xu, J.Z., Xu, X.Z., Zhao, M., and Xia, K.S. (2000). Some physiological changes and storability of citrus fruits. *Plant Physiol.* commun. 36, 125–127.

6

FRUIT BIOCHEMISTRY

For a long time, citrus fruits have been valued by man for their attractive appearance, refreshing flavor, and nutritional qualities. Biochemical compounds and secondary metabolites such as proteins, amines, polyamines, complex carbohydrates, organic acids, lipids, phenols, flavonoids, terpenoids, aromatic compounds, mineral elements, hormones, and vitamins play a very important role in the physiology and metabolism of citrus plants and fruits. Many of these compounds are part of vital nutrition to human beings. Photosynthesis and respiration are primary metabolic processes leading to the formation of basic compounds that lead in turn to the formation of secondary metabolites. Respiration provides energy for synthetic and breakdown processes. Before harvest, fruit stores different metabolites, which are later utilized, and some more metabolites are formed after harvest as the development takes place towards maturity and senescence. Stored carbohydrates, acids, and amino acids are utilized for synthesis of proteins, sugar derivatives, flavors, and volatiles in fruits.

Carbohydrates, either in a free state or as derivatives, play an important role in formation of fruit properties, such as color, texture, and flavor, which appeal to consumers. Flavor is from the balance of sugars and acids; in addition, specific flavor constituents are often glycosides, terpenoids, and lipid compounds. The colors of many citrus fruits are due to sugar derivatives, and structural polysaccharides contribute to textural properties. Sugar contents vary widely in different citrus fruits; acid lime juice contains 0.7–0.8 percent sugars, whereas mandarin and sweet orange juice contains 8–9 percent total sugars. Composition varies with the stage of maturity also. In citrus fruits, amino acids such as alaine, asparagine, aspartic acid, glutamic acid, proline, serine, aminobutyric acid, and arginine occur in substantial amounts. The characteristic aroma of citrus fruits as a whole is attributed to the relatively high-oxygenated terpenes that are representative of the peel oil. The oils vary in composition with the source and the method of preparation. Citrus oils contain more than 90 percent d-limonene. Methyl *N*-methylanthranilate and thymol, present to the extent of 0.85 and 0.08 percent, respectively in the oil are suggestive of mandarin flavor, whereas valencene contributes to fresh orange flavor.

The pronounced bitterness of some citrus flavonoids and limonoids should be mentioned despite their somewhat restricted distribution. One of the chief flavonoids of the grapefruit and the pummelo and also to some extent of oranges is naringin, which is intensely bitter. It is detectable at 10–100 ppm concentrations depending on the taste of an individual. It is composed of aglycone naringenin (a phenolic) and disaccharide neohesperidose. It is interesting to note that when separate, these compounds are not bitter. Another interesting fact is that a derivative of neohesperidose, dihydro-chalcone, which is readily formed by ring splitting in alkali followed by hydrogenation, has 20 times the sweetness of an equal amount of saccharin.

The available literature on biochemical composition and metabolism of citrus is fairly vast and has been reviewed time to time by various research workers.

In this chapter an attempt is made to briefly outline biochemical composition that contributes to the quality (external appearance, eating quality, and nutritional/medicinal attributes) of fresh fruit.

I. CARBOHYDRATES

Carbohydrates are organic compounds composed of carbon, hydrogen, and oxygen. This group consists mainly of monosaccharides, disaccharides, and polysaccharides. Carbohydrates play a major role in citrus fruit physiology when the fruit is attached to the tree and also after its harvest. Citrus fruits have attractive color, texture, and flavor. In all these properties, carbohydrates, particularly sugars, either in a free state or derivatives, play a very important role. A fine balance of sugars and acids of fresh citrus fruit causes the appealing flavor. Specific flavors of different citrus are often due to glycosides. Attractive colors of many citrus fruits are due to sugar derivatives of anthocyanidins. Texture is governed by complex structural polysaccharides. Ascorbic acid, which is commonly considered to be a sugar derivative, is found widely and abundantly in citrus fruits. Total sugar contents vary widely in different kinds of citrus fruits, and they are present in free form mostly as monosaccharides and disaccharides in the juices.

A. Monosaccharides

D-glucose and D-fructose (Fig. 6.1) are monosaccharides (sugars having only one molecule). These are also major reducing sugars in citrus (Table 6.1) and are *hexoses* (sugar molecules with six carbon atoms). Hexoses other than glucose and fructose are rarely found in these fruits. Ting and Deszyck (1961) reported that free sugars in citrus juices are predominantly glucose, fructose, and sucrose, although xylose is present in trace amounts. Sugars are distributed in ratio of sucrose, glucose, and fructose 2:1:1 in mandarins. These are the major carbohydrates present (Kefford, 1959). Mannose has also been found in sweet oranges. Among *pentoses* (sugar molecules with five carbon atoms), arabinose has been reported in limes and grapefruits (Wali and Hassan, 1965). Xylose, another pentose sugar, has been found in trace amounts. Heptuloses (with seven carbon atoms) have also been reported from oranges and peels of grapefruit (Williams et al., 1952).

B. Oligosaccharides

Oligosaccharides are sugars formed by two or more molecules of monosaccharides. Sucrose is a disaccharide (formed with two molecules of monosaccharides, D-fructose and D-glucose) and a major non-reducing sugar in citrus fruits (Table 6.2). Sucrose is a main sugar of translocation in plants. In oranges, sucrose (Fig. 6.2) concentration is less than that of monosaccharide glucose as compared to

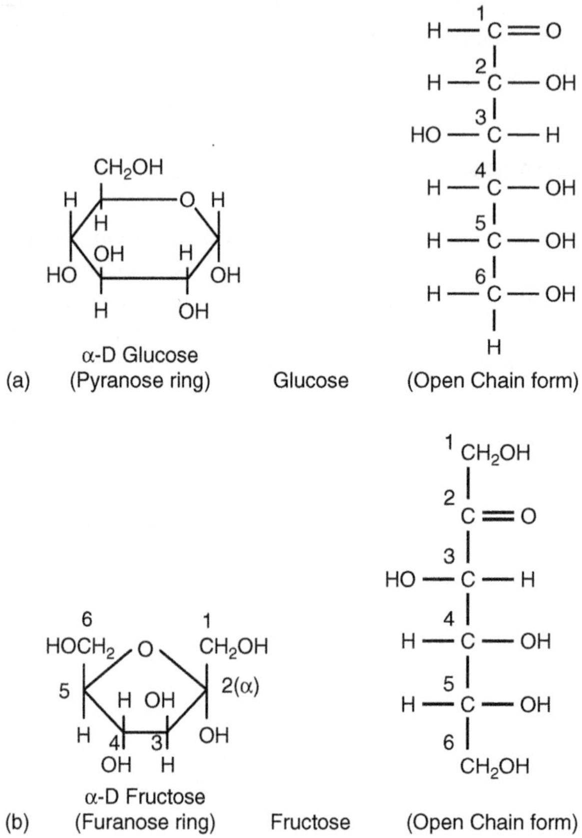

FIGURE 6.1 Chain and Ring Structure of (a) Glucose Sugar and (b) Fructose Sugar.

TABLE 6.1 Sugar Content of Fresh Citrus Fruits

Fruit	Glucose (%)	Fructose (%)	Maltose (%)	Sucrose (%)	Total
Sweet orange, Mosambi	4.05	4.55	Traces	<0.5	8.59
Mandarin, Nagpur	4.00	1.98	–	6.80	–
'Kagzi' lime (light green)	0.39	0.19	–	–	–
Dark green	0.37	0.13	–	–	–
Acid lime (full yellow)	0.61	0.23	–	–	–
Lemon	0.52	0.92	Traces	–	–
Grapefruit	2.97	3.08	<0.5	1.26	7.31

Swisher and Higby (1961); Veldhuis (1971); Hurst et al. (1979); Selvaraj and Edward Raja (2000); Ladaniya and Mahalle (2006).

tangerines and mandarins. Limes and lemons have very small amount of sucrose. Maltose and lactose are not major sugars in citrus.

C. Sugar Derivatives

Among sugar derivatives, sugar acids, sugar phosphates, sugar nucleotides, glycosides, and polyoles are important. D-galacturonic acid, D-Glucuronic acid, D-Gluconic acid, L-Ascorbic acid, and dehydro-ascorbic acid are sugar acids found in citrus fruits. Derivatives are not usually found in a free state. L-Ascorbic acid is found in a free state in citrus juices. Sugar alcohols such as myo-inositol have been reported in oranges and grapefruits in the range of 88–170 mg/100 g of juice (Krehl and Cowgill, 1950). Lemons had 56–76 mg myo-inositol/100 g of juice (Swisher and Higby, 1961). Sugar phosphates are the esters of sugars and active intermediates in sugar metabolism. D-glucose-6-phosphate is an ester found in large amounts. Glycosides are the compounds in which a sugar molecule is linked to a non-sugar molecule (aglycone) such as alcohol or phenol. The term glycoside extends to compounds in which *N* instead of *O* links the sugar and the aglycone. Sugar nucleotides are formed by an ester linkage between a sugar or sugar derivative and the terminal phosphate residue of a nucleotide-5-diphosphate.

TABLE 6.2 Reducing, Non-reducing, and Total Sugar Contents of Citrus Fruits

	Sugars (% of fresh weight basis)		
Fruit	**Reducing**	**Non-reducing (sucrose)**	**Total**
Mandarin	2.50	5.81	9.31
Kinnow	3.20	3.84	8.84
Sathgudi sweet orange	3.57	3.47	7.93
Valencia orange	4.99	4.73	9.72
Navel orange	5.17	5.21	10.38
Bitter orange	N.A.	N.A.	5.49
Grapefruit	4.13	2.19	6.74
Lemon	1.67	0.18	2.19
California lemon	1.09	0.09	N.A.
Lime	0.72	0.14	0.72
Acid lime 'Kagzi'	0.84	0.82	1.76

Note: Sugars in the fresh edible portion of fruit.
Burdick (1961), Birdsall et al. (1961), Siddappa et al. (1962), Money and Christian (1950), Swisher and Higby (1961), Veldhuis (1971), Sandhu et al. (1990), Selvaraj and Edward Raja (2000).

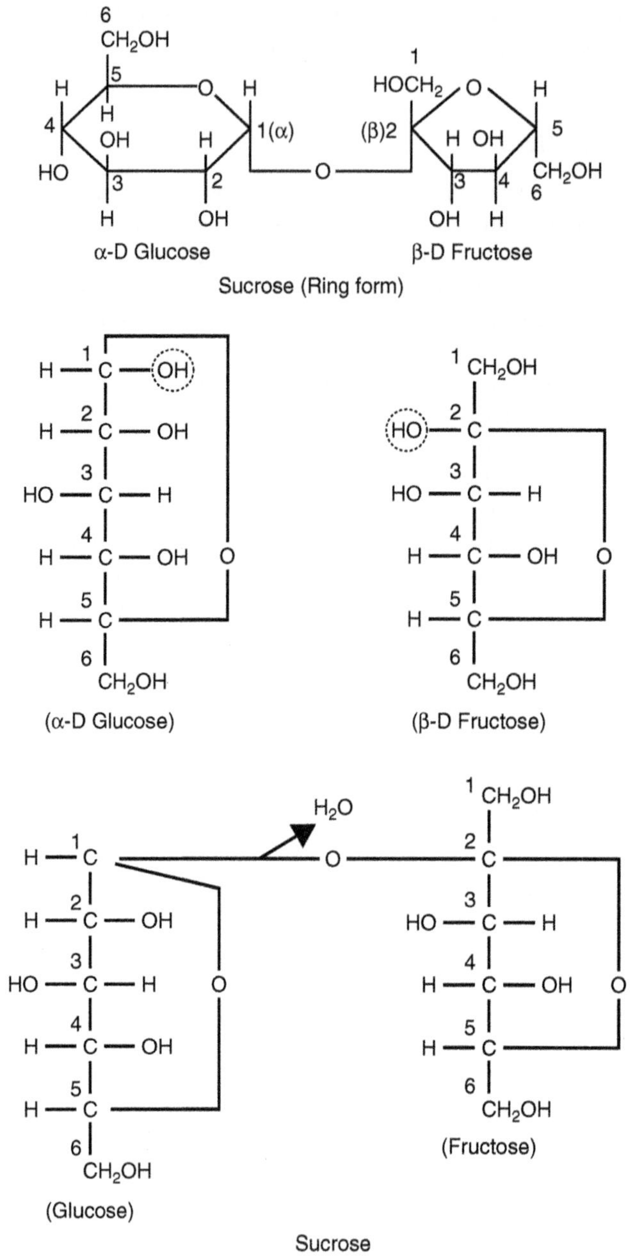

FIGURE 6.2 Sucrose, the Important Disaccharide and Non-reducing Sugar.

These nucleotides play an important role in the interconversion of sugars and polysaccharide synthesis.

D. Changes in Sugars during Fruit Growth and Storage

Nearly 75 to 85 percent of the total soluble solids of orange juice are sugars. The reducing, non-reducing, and total sugars increase as fruit continues to ripen on the tree. This trend is observed in almost all citrus fruits except in acid fruits. Small immature fruits also photosynthesize. The main source of nutrition is leaves. It is generally thought that the main sugar transported from leaf to fruit is sucrose. Sucrose is further utilized for synthesis of various polysaccharides including pectin. Starch is found in the outermost cells of the fruit. Immature oranges have a fairly high sucrose content, but during maturation this decreases slightly (Sawyer, 1963). α- and β-glucose, fructose, sucrose, and a small amount of galactose have been reported in Valencia orange juice (Alberala et al., 1967). As the early and mid-season oranges and tangerines ripen on the tree, the total sugars in the juice increase rapidly due to an accumulation of sucrose. Using 14-C labeled compounds, Sawamura and Osajima (1973) found that translocation from leaf to fruit occurs in the form of glucose and fructose which are converted to sucrose in the fruit.

In a similar way to the sugars in juices, the sugars of the peel also accumulate with maturity; in general, the increases are due mainly to reducing sugars. In grapefruit peels, sucrose increases in the early part of the ripening period but falls during the later part. There is generally an increase in reducing sugars with maturity. As the alcohol-soluble fractions of the peel increase, there is a corresponding decrease of the alcohol-insoluble solids, which are mainly the cell wall and cytoplasmic constituents. Sucrose showed little change in the flavedo of Fortune mandarins (*Citrus reticulata*) during the growing season, but fructose and glucose increased, in nearly equal amounts, throughout the autumn and winter. Starch was less abundant than soluble carbohydrates (Holland et al., 1999).

Changes in sugar content of the flavedo tissue of Satsuma mandarins grown in Japan have been studied by Kuraoka et al. (1976). Concentrations of sucrose in addition to total and reducing sugars were low during August to September but rapidly increased in October. Reducing sugars were 22.7 percent, and non-reducing sugars were 11.4 percent (on a dry weight basis) at harvest. In albedo tissues, reducing sugars were 19.3 percent, and non-reducing sugars were 7.5 percent. Sucrose content declined at harvest in the cvs Ikeda and Iyo but not in Wase. Puffy fruit had higher sugar content in flavedo and albedo than non-puffy fruit. The predominant sugars in Okitsu Wase (early maturing) and Silverhill are sucrose, fructose, and glucose in the juice and peel (Daito and Sato, 1985). Sucrose increased in juice from 1.08 to 7.53 mg and from 1.23 to 6.90 mg/100 ml in Okitsu Wase and Silverhill respectively from September to March. Similarly, fructose increased from 0.81 to 2.06 mg and from 0.75 to 1.84 mg/100 ml, whereas glucose remained relatively constant (0.8 to 1.8 mg/100 ml) in both

TABLE 6.3 Soluble Sugars of 'Mosambi' Orange During Maturation

Days (after fruit set)	Fructose (%)	Glucose (%)	Sucrose (%)	Xylose (%)	Maltose (%)	Total
180	1.30	1.10	1.70	0.52	0.14	4.76
190	1.90	1.30	1.80	0.05	–	5.05
200	2.30	2.40	1.50	–	–	6.20
210	3.50	2.90	2.35	0.025	–	8.77
220	3.82	2.60	2.40	–	–	8.82
230	4.18	2.61	3.00	–	0.16	9.95
240	1.98	4.81	3.61	–	–	10.40
250	1.80	4.70	4.00	–	–	10.50

Ladaniya and Mahalle (2006); percent sugars by weight (edible portion).

cultivars. Sucrose and glucose are the predominant sugars in early stages of peel maturation. All three main sugars increased gradually later on.

Soluble sugars in Mosambi sweet oranges increased as the fruit matured (Table 6.3). Fructose, glucose, and sucrose were prominent sugars, whereas xylose and maltose were found in trace amounts (Ladaniya and Mahalle, 2006). Ribose was not detected. Glucose increased continuously from 1.1 percent at 180 days to 4.70 percent at 250 days, whereas fructose increased up to 230 days. Sucrose, a primary non-reducing sugar, increased considerably between 230 and 250 days resulting in increased sweetness of juice. A fall in fructose at 240 and 250 days could be due to its utilization in synthesis of sucrose, which increased with fruit maturity. Glucose also recorded a sharp increase between the intervals of 230 and 240 days. Xylose content was 0.52 percent after 180 days but declined to 0.025 percent after 210 days. Maltose was detected at 0.14 percent at 180 days and 0.16 percent at 230 days. Ting and Attaway (1971) reported the ratio of fructose:glucose:sucrose as 1:1:2 in Valencia oranges. The difference in the sugar content can occur due to differences in ripening stage of fruit at the time of analysis or other factors including agro-climatic conditions and variety.

E. Polysaccharides

Polysaccharides are compounds produced by the combination of many molecules of simple sugars and sugar derivatives. The examples are starch, cellulose, hemicellulose, gum, and pectic substances. Starch is a basic photosynthate and consists of a chain of glucose units in the form of amylose and amylopectin. Cellulose is a fundamental constituent of cell walls, and it is a long chain polysaccharide made up of β-glucose units (Fig. 6.3). Hemicelluloses are

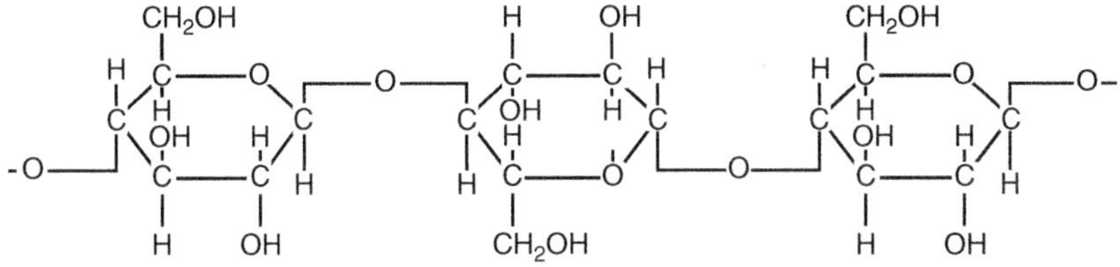

Part of cellulose polysaccharide molecule

FIGURE 6.3 Cellulose Polysaccharide.

a heterogeneous group of long chain polysaccharides. The basic units in hemicelluloses are arabinose, xylose, mannose, or galactose, which are called arabinans, xylans, mannans, or galactans, respectively. These hemicelluloses are found especially in lignified tissues. Pectic substances are polygalacturonides with non-uronide carbohydrate covalently bound to an unbranched chain of 1–4 linked α-galacturonic acid units. The carboxyl groups of polygalacturonic acids may be partly esterified by the methyl group and partly or completely neutralized by one or more bases. Important pectic substances include protopectin, pectinic acid, pectin, and pectic acid. These compounds have varying degrees of methyl ester content and neutralization. These are capable of forming gels (jellies) with sugar and acid or if suitably low in methoxyl content, with certain metallic ions. Pectic substances are deposited in the cell wall and middle lamella. The meristematic and parenchymatous tissues are rich in pectic substances. Pectic substances as a combination of hemicellulose and pectin gel are a part of cell structure. Pectin in the middle lamella can be called *cement* or *adhesive* between cells. Polysaccharides also form the important part of cell sap. The polysaccharides and pectic substances together with the proteins make alcohol-insoluble solids. Nearly 19–25 percent of the alcohol-insoluble solids of the juice of Valencia oranges are proteinaceous, but only 5 percent of that of the peel are. The main polysaccharides present in the peel tissue are polygalacturonic acid, glucosan, araban, and galactan with smaller amounts of xylan (Ting and Deszyck, 1961). Grapefruit contained a higher proportion of glucosan than the oranges. Hemicellulose fraction of both orange and grapefruit peel showed it to contain xylan, glucosan, galactan, and araban in the ratio of 4:3:3:1, with trace amounts of uronic acids. The α-cellulose fraction after hydrolysis yields about 80–90 percent glucose in the total monosaccharides. Arabinose, xylose, and galacturonic acid contribute the remainder of the sugars. Traces of mannose and galactose are also found in this fraction.

Pectic substances in citrus juice are important to the processing industry because of their function as a 'cloud' formation/stabilization in the juice. Because the tissues of citrus fruits have high contents of pectic substances, they are used in commercial pectin production. The increases in °Brix and solubilized sugars in the juice as observed during postharvest storage is also due to cell wall hydrolysis by various enzymes such as pectinase, cellulase, hemicellulase, or pectinesterase (Echeverria et al., 1989).

F. Changes in Polysaccharides during Fruit Growth and Maturation

With changes in polysaccharides, the physiology of fruit also changes. The changes are in pectin content, degree of solubility, and structure. The changes are enzymic and chemical in nature and affect fruit texture. During the course of the development of orange fruit, starch is found in all components of the fruit including the juice vesicles (Webber and Batchelor, 1943). Starch is especially abundant in the albedo, but is also found in the flavedo when green. As the fruit grows older, the starch begin to disappear. When horticulturally mature, starch is usually absent from the fruit. Ting and Deszyck (1961) observed that there is no marked trend in the changes of concentration of polysaccharides (polygalacturonic acid, glucosan, araban, and galactan with smaller amounts of xylan) present in the peel tissue throughout the maturation of oranges and grapefruits. As Valencia oranges ripened, the water-soluble pectin in both the albedo and the pulp increased to a peak and then decreased while the acid-extracted pectic material continued to decrease. The total pectin showed only a slight decrease when the fruit was fully ripened. The total pectin in the juice shows a similar trend to that of the albedo and the pulp. A rapid initial increase of both total and water-soluble pectic substance in the pulp and the peel during rapid growth of the Valencia fruit has been reported (Sinclair and Joliffe, 1961). The percentage of methylation of the carboxyl groups of the pectic compounds in the peel rose rapidly to approximately 80 percent and remained relatively constant during the rest of the period. In maturing oranges, a decrease in total pectin and water-soluble pectic substances in the peel and pulp is observed (Sinclair and Jolliffe, 1958).

Polysaccharides constitute the insoluble fiber in citrus fruit tissues. Because these fibrous materials have good water-holding capacity, they play an important role in the human diet in preventing digestive disorders. The content of dietary fiber in fruit changes with maturity (Larrauri et al., 1997). In Marsh grapefruit, dietary fiber content was higher in September (686 g/kg) than in January (586 g/kg). The constituents of soluble fiber were: uronic acids (172–233 g/kg), arabinose (13–41 g/kg), galactose (4–11 g/kg), glucose (5–10 g/kg), and xylose (2–3 g/kg). Main constituents of dietary fiber were Klason lignin (29–37 g/kg), uronic acids (33–70 g/kg), and neutral sugars. These findings indicate that it is healthier to eat grapefruit in the early harvest season because they have higher dietary fiber content.

Cell wall polysaccharides and their composition also contribute to fruit firmness. The sugar concentration of cellulose in a soft rind fruit (Satsuma cv Aoshima) was lower than a firm rind fruit (*Citrus hassaku*). In flavedo tissue, sugar concentration was highest in cellulose (Muramatsu et al., 1999).

II. ORGANIC ACIDS

Organic acids have acidic properties because they have a carboxyl group (COOH) in a free state. These acids are an important source of acidic taste in fruit and also form a source of energy in plant cells. Organic acids are dissolved in cell sap either free or in combined form with salts, esters, or glycosides. These acids have been found in sap moving from roots to stem and fruits and are produced in leaves also. Juice vesicles of the fruit are also active sites of acid synthesis. Most of the acid is probably present in the vacuole of the cell. Organic acids are water soluble particularly when the carbon chain is short and are weak acids with a dissociation constant of about 10^{-5} at 25°C. Citric is an important and abundant acid in citrus fruits.

The free acidity or titratable acidity of the juice of most citrus fruit is due largely to citric acid, and it is measured by neutralizing fruit extracts with the base (0.1 NaOH). Citrus fruits also contain considerable amounts of cations, mainly potassium, calcium, and magnesium. The acids also combine with cations and form salts. The total acidity represents the sum of all acids (free and those combined with cations). A cation exchange column can be used to remove cations, and total acidity can be determined by neutralizing with a base. The free acid together with the salts forms a very effective buffer system (Sinclair, 1961b). Yamaki (1989) studied free acids, total acids, and combined acidity of the peel of several citrus cvs. Combined acidity was between 3 and 11 me/100 g but exceeded 11 me/100 g in Funadoko (*Citrus funadoko*), Yuzu (*C. junos*), Dancy tangerine, Jimikan, Rusk citrange, Mediterranean mandarin, and Miyauchi Iyo (*C. iyo*). Free acidity of the rind was generally below 17 me/100 g fresh weight, and total acidity of the rind was in the range 4–23 me/100 g. A high positive correlation was found between total acidity and free acidity except for the acid-less species (Yamaki, 1988). One milliequivalent of citric acid is 0.064 g (the equivalent weight of citric acid is 64) whereas 1 milliequivalent of malic acid is 0.067 g (equivalent weight of malic acid is 67).

The pH of citrus juices also provides an idea about the acidity of fruit, and it could be one way of expressing acidity. The pH is an important parameter from a processing point of view. Generally pH varies from about two for lemons and limes to about 4–4.5 in over-mature tangerines. The pH of the juice of Valencia and Washington navel oranges vary between 2.9 and 3.9. In Palestine sweet limes, citric acid content of 0.08 percent was recorded with a pH 5.7 (Clement, 1964a).

Citric acid is the principal acid of the endocarp of all citrus fruit except the sweet lemon and the acidless orange. The peel has less acid than the pulp. The main acids of the citrus peel are oxalic, malic, malonic, and some citric. Together they account for 30–50 percent of the anions present (Clements, 1964a). L-quinic acid was found in the peel and pulp of various citrus fruits (Ting and Deszyck, 1959). Tartaric, benzoic, and succinic acids have also been reported to be present (Braverman, 1933). In the juice of lemons, citric acid may account for 60–70 percent of the total soluble solids. Major acids in some commercially important

TABLE 6.4 Organic Acid Content in Juice of Commercial Citrus Fruits

Fruit	Citric (%)	Malic (%)	Succinic (%)
Clementine	1–1.1	0.10	–
Kinnow	0.8–0.9	N.A.	N.A.
'Coorg' mandarin	0.7–0.8	N.A.	–
'Nagpur' mandarin	0.4–0.7	N.A.	–
Tangerine	0.86–1.2	0.18–0.21	–
Satsuma mandarin	0.97	0.08	–
Washington navel orange	0.72–0.93	0.12–0.15	0.05
Valencia orange	0.30–0.80	0.05–0.15	0.13–0.17
Sathgudi orange	0.6–0.7	N.A.	–
Mosambi orange	0.35–0.45	N.A.	–
Marrs	0.10–0.15	0.10	0.05–0.06
Hamlin	0.20–0.30	0.15–0.23	0.12–0.14
Pineapple	0.30–0.32	0.17–0.26	0.80
Marsh grapefruit	0.42–1.70	0.03	0.40
Villafranca lemon	5.0	0.50	–
Italian lemon	4.76	0.40	–
Asom lemon	6.10–6.43	–	–
Eureka lemon	4.38	0.26	–
Natsudaidai	1.02	0.17	0.09
Acid lime 'Kagzi'	5.56–6.60	0.46	0.01

Chaliha et al. (1963), Mukherjee et al. (1964), Siddappa et al. (1962), Clements (1964a), Kubota et al. (1972), Ting and Vines (1966), Kakiuchi and Ito (1971), Bhalerao and Mulmuley (1992), Selvaraj and Edward Raja (2000).

fruits are given in Table 6.4. Verma and Ramakrishnan (1956) found succinic acid to be the predominant acid in limes (*Citrus acida*) with size less than 1.5 cm in diameter. Acid content of the pulp and peel varied.

A. Changes in Organic Acids During Fruit Growth and Maturation

In oranges and grapefruits, free acid per fruit increased in early growth and then became more or less constant (Sinclair and Ramsey, 1944). The decrease in titratable acidity was considered to be due to dilution as the fruit increased in size and in juice content. A decrease in the concentration of acid with the gradual increase in the ratio of total soluble solids to acidity determines the legal maturity of the fruit as well as their palatability.

As fruit developed, the major acid in Valencia oranges was observed to be citric acid with a concentration of 55 meq/100 g dry weight in the peel of young oranges during May (Rasmussen, 1964). The total acidity in the pulp rose to a peak of 172 meq/100 g in September while remaining fairly constant in the peel. Malic was the major acid in the peel with little citric acid. Clements (1964a, b) reported that oxalate was predominant in the peel of citrus fruits with malonic acid and malate at considerably lower concentration. Only citric and malic acids were determined in the juice, and no malonic acid was detected. In both the flavedo and the albedo of Navel oranges, oxalate decreased and malonate increased with advancing maturity. In over-mature oranges, the malonate concentration of the flavedo is nearly 4 times that of the albedo, whereas oxalate concentration decreases in both component parts.

The concentration of quinic acid in the young fruit is high, especially in the juice vesicles, approaching about one-fifth of that of citric acid and about one-half of that of malic acid in the vesicles (Ting and Vines, 1966). As citric acid increases in concentration, that of malic and quinic acid declines rapidly. Malic acid concentration rises again at the end of the maturing period of the fruit. Quinic acid has been considered a precursor in the synthesis of certain aromatic compounds in plants. It is apparently high in very actively growing tissues of the young fruit and seems to follow a similar trend as the flavonoids in the fruit (Kesterson and Hendrickson, 1957; Hendrickson and Kesterson, 1964). In Shamouti sweet oranges, citric is the main acid in juice and malonic and malic in peel tissues. Succinic, adipic, isocitric, oxalic, lactic, aconitic, α-ketoglutaric, and benzoic acids have also been recorded (Sasson and Monselise, 1977). With tissue aging, malonic, succinic, and adipic acid concentrations increased in peel and juice, whereas citric and malic acids decreased in juice. Malonic acid (the competitive inhibitor of acid respiration) has been suggested as a possible indicator of fruit tissue senescence because it accumulates in the peel at fruit maturity and increases several fold after harvest in storage. This acid component is observed to be most closely linked with aging of fruit tissues as determined by softness and deformation.

In juices of Hamlin, Parson Brown, Pineapple, and Valencia oranges, and Duncan, Marsh, and Triumph grapefruits, malic acid remained fairly constant as the fruit matured whereas citric acid gradually decreased (Shaw and Wilson, 1983). The succinic acid levels were less than 0.01 percent in some of these fruits. In Okitsu Wase (early maturing) and Silverhill satsumas, Daito and Sato (1985) recorded glucuronic, lactic, acetic, pyruvic, malic, citric, succinic, and isocitric acids. The citric acid decreased from 2022 to 802 mg and from 2148 to 896 mg/100 ml juice in Okitsu Wase and Silverhill respectively during maturation. Malic and isocitric acids also decreased markedly. α-ketoglutaric acid was not detected at maturity in these cultivars.

B. Physiological Role of Organic Acids

Organic acids are respiratory substrates in the fruit. Higher (more than 1.00) respiratory quotient (CO_2 produced/O_2 consumed) indicates utilization of acids, mainly citric and malic acids through the TCA (tricarboxylic acid) cycle, in which acids are oxidized and ATPs are formed for synthesis of new compounds. Several metabolites are formed during the process. Organic acids are utilized during formation of many flavor and aromatic compounds.

III. NITROGENOUS COMPOUNDS

Nitrogen-containing compounds have special significance due to their extensive presence in structural and productive organs of the plant including fruit and also in the metabolism. Amino acids, amines, peptides, proteins, and nucleic acids are major nitrogenous compounds. In citrus fruits, nitrogenous compounds are mainly present in the form of free amino acids (Rockland, 1961), and protein content is very low excluding seeds (Townsley et al., 1953). These compounds play a very important role in growth and yield of trees and fruits' nutritive quality. Total nitrogen in citrus juices varies from 50 to 200 mg/100 ml (Ting, 1967). In whole Valencia orange fruit, nitrogen varies from 0.83 to 1.14 g/100 g dry weight (Kefford and Chandler, 1970). Nitrogen content of the fruit varies with maturity. In early stages of fruit growth and development, protein nitrogen is predominant and as the fruit matures, soluble nitrogen (mainly free soluble amino acids) increases and is almost equal to protein nitrogen (Bain, 1958). Nitrogen content of the fruit and growth of the tree depend on the nitrogen status of the tree. Nearly 70 percent of the total nitrogen in citrus juices is present in the form of amino acids (Clements and Leland, 1962a). In Navelinas and Washington navels, total N content in juice increased during the season and was 60–100 mg/100 ml whereas in Navelate, it was about 80 mg/100 ml. Amino N increased from about 30 to 60 mg/100 ml in these varieties (Tadeo et al., 1988). Total N content in the rind decreased with time from 1 to about 0.6 g/100 g dry weight (50–60 percent corresponding to the protein N).

A. Amino Acids

Amino acids are fundamental constituents of living matter and are also coded in to deoxyribonucleic acid (DNA). Amino acids contain both basic amino (NH_2) and acidic carboxyl (COOH) groups. Free amino acids are usually in metabolic equilibrium with the process of protein synthesis and degradation. Amino acids are building blocks of proteins and play a significant role in growth and development of fruit. Free amino acids are water soluble, and they exist in citrus juices. Amino acids are important from a processing point of view also as juices are fermented and made into wine. In several processed products, amino acids are involved in browning reactions. Total amino acids in citrus juices can be used as a means of detecting adulteration in commercial products (Van der cook et al., 1963). Phenols and flavonoids are also synthesized from aromatic amino acids such as tryptophan, tyrosine, and phenylalanine. Amino acids are important in human nutrition. Plants can synthesize all amino acids, but animals need some amino acids (called *essential amino acids*) from plants or other animals fed on plant/plant parts.

The range of amino acids and other nitrogenous compounds in oranges, grapefruits, and lemons is given in Table 6.5. The concentration ranges are very wide. Underwood and Rockland (1953) found that different amino acids are characteristics of the variety of orange. However, environmental and physiological factors such as fruit maturity and tree growth pattern can modify any characteristic varietal

TABLE 6.5 Range of Amino Acids and Soluble Nitrogenous Compounds in Citrus Fruits

	mg/100 ml			juice
Compound	Orange	Mandarin	Grapefruit	Lemon
γ-aminobutyric acid	4–73	–	–	4–20
Arginine	23–150	–	76	25–106
Asparagine	20–188	18–36	–	–
Alanine	3–26	8–15	–	1–31
Aspartic acid	7–115	24–50	470	19–60
Glutamic acid	6–71	17–34	280	6–35
Glutamine	3–63		–	–
Glycine	5		–	–
Proline	6–295		40–59	27–53
Serine	4–37	12–26	310	12–28
Valine	10	3–6	24	–
Histidine	–	–	14	–
Leucines	–	5–11	24	–
Lysine	–	4–12	16	–
Phenylalanine	–	6–15	12	–
Threonine	–	–	10	–
Tryptophan	–	–	4	–
Tyrosine	–	–	6	–
Cysteine	0.3–0.8	–		–
Cystine	–	–	0.18	–
Glutathione	2.8–7.8	–		–
Methionine	–	–	0.35	–
Betaine	39–63	–	–	–
Choline	7–16	–	–	–

Wedding and Sinclair (1954); Wedding and Horspool (1955); Rockland and Underwood (1956); Burdick (1954); Rockland (1958); Townsley et al. (1953), Burroughs (1971).

influence on the amino acid content. Most of the essential amino acids (valine, leucine, phenylalanine, tryptophan, lysine, isoleucine, methionine, threonine, and histidine), which a human being must get from food are present in oranges, lemons, and grapefruits. Mandarins/tangerines provide eight of the nine essential amino acids required by the human body with the exception of tryptophan (Vandercook, 1977).

In navel and Valencia oranges, Eureka and Lisbon lemons, Marsh grapefruit, and Dancy tangerines, the amino acids that are present in substantial amounts are: alanine, asparagine, aspartic acid, glutamic acid, proline, serine, γ-amino butyric acid, and arginine. Proline is the most prominent amino acid in citrus fruit except grapefruits and was especially prominent in Valencia oranges. Aspartic acid was predominant in grapefruit.

Concentrations of amino acids changed with fruit maturity. Asparagine, aspartic acid, and serine remained fairly constant throughout growth, but proline, arginine, and γ-amino butyric acid increased progressively; the last three were predominant in mature Valencia oranges and increased further with overmaturity (Clements and Leland, 1962a). Ting and Deszyck (1960) reported that the concentration of proline was 0.49 m mole/100 ml with a mean total amino acid content of 0.9 m mole/100 ml (range 0.61–1.20) in several sweet orange samples. On average, proline, aspartic acid, asparagine, and arginine in decreasing order comprised 75 percent of the total amino acids (254 mg/100 ml).

High aspartic acid and low arginine content are reported in lemon juice (Vandercook et al., 1963). These amino acids averaged 32 and 3 percent, respectively. Of the total free amino acids, serine (24 percent) and proline (18 percent) were second and third in prominence. Bogin and Wallace (1966) reported that alanine, proline, and glutamine were the main amino acids, whereas valine, serine, and cysteine were present in lesser amounts, and aspartic acid was present in traces in lemons. No aspartic acid was found in Indian lemons (*Citrus medica*), shaddocks (*Citrus decumana*), and oranges (*Citrus sinesis*), whereas a remarkably high content of γ-amino butyric acid was reported (Datta, 1963). Aspartic acid and serine were found to be particularly prominent in Israeli Shamouti oranges (Coussin and Samish, 1968), Alanine, γ-amino butyric acid, aspargine, glutamic acid, and glycine were reported in Assam lemons (Chaliha et al., 1964). In Mosambi sweet oranges, arginine, aspartic acid, proline, serine, threonine, Alanine, γ-amino butyric acid, aspargine, glutamic acid, and glycine were reported (Srivastava and Tandon, 1966). In lime juice, proline was found to be absent, while most other amino acids reported in lemons were present (Alvarez, 1967).

Findings of Clement and Leland (1962b) indicated that in citrus juices, 90 percent of the total amino acids are accounted for by aspartic and glutamic acids, alanine, γ-amino butyric acid, arginine, asparagine, proline, and serine. The relative proportions of the free amino acids, however, vary widely in different cultivars of citrus fruits (Table 6.6). The total free amino acid content in juices and peel of Fina, Oroval, and late cv Hernandina increased during ripening, and the total protein amino acid content of the peel decreased (Martin et al., 1984).

TABLE 6.6 Amino Acids in Some Commercial Cvs of Citrus Fruits (mg/100 ml)

Amino acid	Valencia orange	Navel orange	Shamouti orange	Eureka lemon	Lisbon lemon	Marsh grape fruit	Dancy tangerine
Aspartic acid	31	27	115	36	32	81	36
Asparagine	36	67	–	16	17	42	85
Glutamic acid	19	12	28	19	18	22	16
Serine	19	18	70	17	19	15	19
Glycine	2	2	–	1	1	2	2
Threonine	–	–	–	–	–	–	–
α-alanine	12	12	36	9	10	9	7
γ-amino butyric acid	25	24	–	7	7	19	18
Valine	–	2	12	1	1	2	2
Leucine(s)	–	–	6	–	–	–	–
Proline	239	107	–	41	47	59	100
Arginine	73	54	45	3	3	47	84
Lysine	6	3	10	1	1	3	4
Phenyl-alanine	–	–	–	2	3	3	5

Clements and Leland (1962a, b), Coussin and Samish (1968), Vandercook et al. (1963), Rockland (1958), Ting and Attaway (1971).

The composition of the total free amino acid content varied during ripening with proline and arginine accumulating most. Proline represented at maturity around 50 percent of the total free amino acids in the rinds of Navelina, Navelate, and Washington Navel oranges. Values were about 1.5 m mole/100 ml in juice and over 5.0 m mole/100 g dry weight in rind (Tadeo et al., 1988). Total protein amino acids in rinds decreased during maturation, reaching final values of about 20 m mole/100 g dry weight.

B. Amines

Amines structurally resemble ammonia where one or more hydrogen atoms are replaced by organic substituents (alkyl or aryl group). Amines combine with other organic compounds. Several amines have been reported in citrus fruits. The range of major phenolic amines reported by Stewart and Wheaton (1964) and Wheaton and Stewart (1965a, b) in the juice of some citrus fruits are: Octapamine (4 mg/liter in Meyer lemons), Synepherine (15–43 mg/l in Hamlin, Navel, Valencia, Temple, and Pineapple oranges, and 50–280 mg/l in Murcott, Dancy, and Cleopatra tangerines), Tyramine (25 mg/l in Meyer lemons), Ferulolyte putreiscine (15–41 mg/l in Duncan, Marsh, and Ruby grapefruit). These compounds can be used as index compounds for estimation of citrus juice content in beverages and for detecting adulteration.

Polyamines have two or more amino groups and function as growth factors. Polyamines are present in most mature plant tissues in very low concentrations (about 10 μmoles to 10 mmoles). The major polyamines known for their beneficial effects are putricine, spermine, and spermidine. The production of polyamines is induced by a variety of stresses in plants and fruits. Free radicals and superoxides promote senescence of plants and their parts by targeting cell walls and enzymes, nucleic acids, and cell membranes (Leshem et al., 1986). Polyamines are involved in free radical scavenging and stabilizing DNA because the polyamines are cationic in nature (Slocom et al., 1984). Polyamines can delay senescence in plant tissues. These compounds can stabilize DNA and cell membranes, scavenge free radicals, suppress ethylene production, and inhibit RNAs and proteases (Drolet et al., 1986; Galston and Kaur-Sawhney, 1990). The antisenescent activity of polyamines is generally attributed to their ability to stabilize and protect membranes by associating with negatively charged phospholipids. The suppression of ethylene production by polyamines is shown to be the result of interference with levels of ACC synthase and ACC (1-aminocyclopropane 1-carboxylic acid) (Apelbaum et al., 1981).

Polyamines are supposed to inhibit conversion of ACC to ethylene by reducing synthesis of ACC synthase and scavenging oxygen free radicals involved in catalytic conversion of ACC to ethylene. Usually when fruit matures or leaves turn yellow, polyamine, particularly putriscine, levels drops down thus more S-adenosylmethionine (SAM) is available and ethylene is produced. SAM is a substrate for SAM-decarboxylase in ethylene synthesis, and SAM is also a substrate in polyamine biosysnthesis. As the trees or fruit age, more SAM is available for ethylene synthesis. ACC is formed from SAM in a synthesis of ethylene (Adams and Yang, 1977). Polyamines have also been shown to inhibit softening by reducing activity of the cell-wall degrading enzyme polygalacturonase (Kramer et al., 1989).

Production of polyamines can be increased in plant/fruit tissues by a variety of treatments to derive beneficial effects. These treatments include hot water dips, vapor heat treatments, UV-C radiation, etc.

Mechanical damage to the mandarin cultivars Fortune and Clementine with forces of 0, 10, 20, and 30 N increased polyamine levels (Valero et al., 1998). Spermidine levels increased slightly in Fortune and showed a prominent peak in Clementine 6 h after storage. In general, polyamine levels increased in the peel following bruising. Putrescine levels were positively correlated with the applied force in Fortunes, whereas in Clementines only the 20 N force was effective. Spermidine levels increased in Fortune following injury, whereas 10 N force increased its levels in Clementines.

C. Proteins

Protein constituents are important not only as components of nuclear and cytoplasmic structures but also in metabolism during growth and development till senescence.

If proteins are to be classified on the basis of their functions, they can be grouped into (1) structural proteins; (2) storage proteins; and (3) enzymes. All enzymes are protinous in nature, but all proteins are not enzymes. Many enzymes are also components of cell membranes. hence there is overlapping of functions. Membrane proteins such as lipoproteins and cell-wall proteins are structural proteins. The proteins in grains such as cereals and pulses are storage proteins and are as such a source of food to humans. Proteins are synthesized through genetic signals, and the basic feature of synthesis is DNA→RNA→Proteins. Citrus fruits are not considered a major source of protein; however, proteins contribute to the cloud of citrus juices. The insoluble solids that make the fine cloud contain 45 percent protein (Baker and Bruemmer, 1970) and nucleotide and phospho-proteins (Vandercook and Guerrero, 1969). Protein content of fruit is lower as compared to leaves, twigs, and seeds in citrus. Protein content is generally expressed by multiplying total nitrogen by a factor of 6.25, but many times it does not represent true protein content. In Valencia orange whole fruit, protein N content has been reported as 125 mg/100 g, whereas in pulp, it was 60 mg/100 g. Protein as a percentage of total N was found to be 52 percent in whole fruit and 42 percent in pulp (Bain, 1958). Protein content in fruit (expressed as total nitrogen of fruit × 6.25) of grapefruits, lemons, limes, oranges, and tangerines have been reported as 0.5, 1.2, 0.7, 1.0, and 0.8 percent respectively (Watt and Merrill, 1963).

Protein content was higher in developing fruits, and as the fruits matured, almost equal amounts of proteins were present in pulp and peel (Bain, 1958). Amino acid composition of proteins of Valencia and navel oranges indicated that Valencia oranges had γ-amino butyric acid but navels had none. Amino acids found common in both cultivars were aspartic acid, proline, serine, threonine, alanine, aspargine, glutamic acid, leucine, lysine, phenylalanine, tyrosine, valine, and glycine (Wedding and Sinclair, 1954).

In citrus fruits, seeds are very rich in protein. Citrus seeds contain 18.2 percent protein on a dry weight basis (in whole seeds). Kernels contained 19.5 percent protein, and hulls had 6.1 percent (Ammerman et al., 1963). In lime seeds, 9.8 percent proteins were reported (Kunjukutty et al., 1966).

D. Nucleotides and Nucleic Acids

Nucleic acids are long-chain molecules formed from a large number of nucleotides. These nucleotides contain five-carbon sugar, phosphoric acid, and nitrogen containing bases (purine or pyrimidine). The nucleotides are found in free form in cell cytoplasm such as ATP and as part of various co-enzymes. Deoxyribonucleic acid (DNA), a material of inheritance, and ribonucleic acid (RNA), a molecule concerned with translating DNA information/structure into protein, also contain large number of nucleotides. In citrus leaves, RNA/DNA increased with protein-N during growth. Protein-N content is correlated with RNA content (Monselise et al., 1962). In Calamondin fruits, ribosomal-RNA (rRNA) were higher initially during growth, but as fruit attained its maximum size, rRNA declined, and soluble-RNA (sRNA) were in equal proportion (Ismail, 1966). Gibberellic acid treatment maintained high rRNA, whereas ethylene caused loss of rRNA and increase of sRNA. Acid-soluble nucleotides of citrus fruits declined after 20 days of storage (Biggs, 1972).

IV. ENZYMES

Enzymes are proteins with a certain amino acid sequence and occur in very small quantities. Enzymatic proteins are involved in metabolic processes/biochemical reactions. Fruits regulate their metabolism through synthesis of enzymes, and that is how it adjusts to environmental changes around it. Genetically controlled metabolic activities such as maturation and ripening also lead to enzyme synthesis followed by anabolic and catabolic reactions.

Enzymes catalyze the reactions, and based on their functions can be grouped into: (EC1) Oxidoreductases; (EC2) Transferases; (EC3) Hydrolases; (EC4) Lyases; (EC5) Isomerases; and (EC6) Ligases. Malate dehydrogenase can be grouped into oxidoreductase enzymes, and amylase is a type of hydrolase that catalyzes hydrolyzis by adding water. Transferases catalyze the transfer of a specific group from one molecule to another. There could be classes and further subgroups within the broad classification of enzymes as above. The Nomenclature Committee of the International Union of Biochemistry and Molecular Biology has developed classifications on the basis of reactions catalyzed by the enzymes, and it includes numbers. *EC* stands for Enzyme Commission. Group EC1 of oxidoreductases includes dehydrogenases, oxidoreductases, reductases, oxidases, catalases, peroxidases, hydroxilases, etc. Group EC2 includes hexokinases, transferases, phosphofructokinases, phosphoglucomutases etc. Group EC3 includes esterases, hydrolases, phosphatases, peptidases, etc. Group EC4 includes carboxilases, decarboxilases, aldolases, oxaloacetatelyses, phenylalanine ammonialyases, etc. Group EC5 includes phospho-hexoisomerases, chalcone-flavonone-isomerases, etc. Group EC6 includes synthetases (Vandercook, 1977).

Enzymes are located in cell walls and plasmalemma and in cytoplasmic fluid of the cell (within cell organelles such as peroxisomes, glyoxisomes, lysomes, chloroplasts, mitochondria, ribosomes, vacuoles, membranes, etc.). These enzyme systems operate in a highly coordinated manner in the metabolism of plants and fruits and play an important role in physiology of the plant as a whole. Enzymes are organic catalysts produced within cells and are highly specific in respect to the nature of the reaction catalyzed and substrate utilized. Distribution and action of various enzymes are very complex. Some of them have been studied in detail, but much is yet to be studied. In conventional methods, the total activity of enzymes is measured; however, variants, also known as isoenzymes, may also be present but with different amino acid sequence.

Enzymes carry out metabolic reactions mostly as directed by the genetic makeup of the plant. RNA synthesis is shown to be required for development of proteins of a ripening enzyme complex (Frankel et al., 1968). This has shown that protein synthesis is DNA dependent at least in the beginning. It may be a sort of a signal from the plant at the appropriate time for the onset of the ripening process in fruits. The enzymes related to maturity are characteristically hydrolytic in nature and are involved in the breakdown of large molecules. Increases in ribonucleases, invertases, acid phosphatases, α-1,3-glucanases, α-1,4-glucanases and cellulases have been observed. Enzymes take part in both catabolic and anabolic processes. Increased protein synthesis indicates a rise in enzyme activity when fruits mature and ripen. Investigations have indicated that if protein synthesis is inhibited, the ripening process is adversely affected (Richmond and Biale, 1966). During fruit maturation and ripening, total protein concentration has been shown to be essentially unchanged in navel orange rinds (Lewis et al., 1967). However, changes in protein synthesis that occurs during maturation and ripening are quite dynamic in nature, and while determining status of protein synthesis, the changes taking place in nitrogen content, protein, and other fractions and uptake in tissues need to be taken into account. The types of proteins synthesized also change as fruit ripens. Certain proteins that are synthesized at a high rate early in the climacteric are synthesized at a lower rate later, whereas others that are synthesized at a lower rate in the beginning of the ripening process are increased thereafter. Enzyme synthesis, which is regulated by hormones and ethylene treatment of grapefruits, resulted in increased activity of phenylalanine ammonia lyase activity in flavedo (Riov et al., 1969). Growth regulators, either endogenous or exogenous, can delay or hasten fruit ripening and senescence through enzyme-mediated action. It is confirmed that proteins synthesized during the early stages of ripening are enzymes required to catalyze the ripening process. Enzymes of Pentose Phosphate Pathways (PPP), mainly glucose-6-phosphate, phosphogluconate-dehydrogenase, phophoribose-isomerase, phosphohexose-isomerase, phosphoribose-epimerase, and transketolase are active during the early phase of fruit ripening when protein synthesis is required for the synthesis of enzymes (Wang et al., 1962). Ribose synthesis is considered necessary to sustain RNA synthesis that occurs during ripening. Synthesis of phenols and anthocyanins also take place deriving precursors from the PPP cycle.

Some important specific enzyme complexes of citrus fruits are discussed next.

A. Polysaccharides and Pectic Enzyme Complex

Cellulase is known as one of the cell wall–softening enzymes in fruit, and cellulose hydrolysis structurally weakens the cell wall. Cellulase is known to complement the activity of pectic enzyme complex in fruit softening (Rouse 1953, Rouse et al., 1965). Cellulase activity also leads to fruit loosening from the stem and can be helpful in mechanical harvesting. Cellulase activity in the separation zones of Pineapple and Valencia oranges increased when abscissic acid (ABA) was introduced through the stems. Cellulase activity was greatest in a normal atmosphere in which ethylene accumulated in the ABA-treated fruit. Sprays of ABA were not effective and did not increase ethylene production or cellulase activity or lower the fruit removal force (Rasmussen 1974, 1975).

Citrus fruits have negligible starch content, which is observed in initial stages of fruit growth. Amylases and phosphorylases degrade starch. α-amylase hydrolyses the α-1-4 glucosidic linkage whereas β-amylase hydrolyses alternate α-1,-4 glucosidic bonds thus resulting in maltose formation.

Pectic enzyme complex is involved in textural changes and softening of citrus fruit. Severalfold increases in activity of pectin esterase or pectin methylesterase (PE or PME) and polygalacturonase or PG (pectinase) during maturation and ripening in relation to fruit textural changes have been shown. The PE is also called pectin-pectyle-hydrolase (3.1.1.11). This enzyme is specific in activity and splits the methyl ester group of polygalacturonic acid. Polygalacturonase (PG) or polymethylgalacturonase (PMG) or pectinase is a depolymerizing enzyme that splits the α-(1-4) glycosidic bond between galacturonic monomers in pectic substances. These pectic enzymes also degrade middle lamella pectins of the cell wall. MacDonnell et al. (1950) studied properties of purified pectinesterase of citrus fruit and found that orange pectinesterase is quite specific for pectin and methyl and ethyl esters of polygalacturonic acid. Pectin esterase has an important role in cloud loss in citrus juices, and it is adsorbed on pulp particles. Two polygalacturonases are generally present. Endopolygalacturonase attacks the pectin molecule within the molecule chain at different sites, while exopolygalacturonase sequentially removes galacturonic acid at the end. The endopolygalacturonase is more important because it weakens the molecule by cleaving it in the middle thus increasing its solubility and causing softening. PE de-esterifies the protopectin (Rouse, 1953). During juice processing, orange PE is inactivated with heat treatment at 92°C after 23 s. Thermolabile (TL) and thermostable (TS) pectinesterases (PE) are present in multiple forms in pulp of fresh Valencia oranges (Hou et al., 1997).

Although polygalacturonase activity has been reported in citrus fruits, the correlation between fruit firmness and the activity of polygalacturonase and cellulase was very low (Riov, 1975; Ben-Yehoshua et al., 1981). An exo-polygalacturonase has been isolated in Satsuma mandarin fruits stored at 5°C for 5 months (Naohara and Manabe, 1993). Enzyme activity was 137.3 units/g fresh weight fruit peel. The optimal pH and temperature for the activity of the enzyme were 4.5°C and 40°C, respectively. During ripening, softening enzymes such as PE, PG, and cellulase in general increased but β-galactosidase decreased in acid lime fruits (Selvaraj and Edward Raja, 2000).

B. Sugar Metabolizing Enzymes

Build-up and breakdown of sugars (sucrose) in citrus fruit tissue affect the quality most. Enzymes such as sucrose phosphate synthase (SPS) and sucrose synthase

(Ssy) are known to exist in the fruit and are part of the metabolic complex. Activity of these enzymes increases during development in all the tissues of the fruit. Activity of Ssy was 4–5 times higher than SPS (Tzur et al., 1992):

$$\text{Fructose-6-phosphate} + \text{UDP-glucose} \rightleftharpoons \text{Sucrose-6-phosphate} + \text{UDP}$$

Subsequently sucrose-6-phosphate is hydrolyzed by phosphatase, and free sucrose is released.

Presence of sucrose in juice sacs is derived not only from sucrose import, but juice sacs also have the capacity to synthesize and degrade sucrose independently.

The pyrophosphatase and ATP-dependent phosphofructokinase (Ppi-PFK and ATP-PFK, respectively) and fructose-1,6-biphosphatase have been found to be active in endocarp, and that suggests that juice sacs act independently in both the glycolysis and gluconeogenesis directions (Tzur et al., 1992).

Activities of SPS and Ssy (both synthesis and hydrolysis direction) and acid invertase have been recorded in all stages of acid lime (*C. aurantifolia*) development from green to full yellow stage (Selvaraj and Edward Raja, 2000). Acid invertase increased from dark green to full ripe stage while SPS and Ssy increased until color break and then declined until the full ripe stage indicating that these enzymes were not related to sucrose levels. Acid limes are reported to possess a system of sucrose breakdown that involves enzymes and also non-enzymatic reactions (Echeverria, 1990, 1992).

The invertase enzyme plays an important role in cleaving sucrose into fructose and glucose. It is also known as β-fructofuranosidase. This enzyme has a role in changing composition of important sugars in citrus fruits during growth, ripening, and storage and is also capable of transglycosidation (transfer of hexose to the primary alcohol group of mono- or disaccharides thereby forming trisaccharides).

Fructose-1,6-bisphosphatase (the regulatory enzyme of the gluconeogenic pathway) activity has been observed in stored Valencia orange fruits (Echeverria and Valich, 1989). Activities of the enzymes of acid metabolism (malic enzyme, isocitrate dehydrogenase/aconitase, and alcohol dehydrogenase) either increased during the first 3 weeks (malic enzyme) or remained constant during storage. Activity of enzymes involved in sugar catabolism (hexokinase, sucrose synthase, UDPG pyrophosphorylase, and Ppi-dependent phosphofructokinase) increased during storage. These enzymes are necessary for organic acid use and for the subsequent oxidization of sugars in harvested fruits.

The vacuole was found to contain 70 percent of the malic acid, 75 percent of the fructose and glucose, 89 percent of the citric acid, and 100 percent of the sucrose present in the protoplasts of Valencia orange juice sacs (Echeverria and Valich, 1988). α-mannosidase, phosphohexoisomerase, and phosphoglucomutase showed activity in the vacuole fraction. The activities of acid and neutral invertases, UDPG pyrophosphorylase, and both ATP and Ppi phosphofructokinases were present only in the cytoplasmic fraction. The sucrose synthase, hexokinase, fructokinase, and aconitase showed no activity in protoplast, vacuole, or cytoplasm. The vacuolar sugars are the major form of carbohydrate supplying energy to the mature juice sac cells. In the juice vesicles of Valencia oranges α- and β-galactosidase have been observed to be very active, but α- and β-glucosidase were also present during storage (Burns, 1990).

Total sugar content of fruits doubled during development due to the accumulation of sucrose in Valencia fruits. A significantly positive correlation has been observed between soluble solids, total sugar and sucrose contents, and key enzymes in sucrose synthesis, that is, sucrose synthase (Ssy) activity on the fresh weight of juice (Song and Ko, 1997). Liu and Li (2003) have also reported that sucrose synthase (Ssy) and invertase play a role in sucrose metabolism in citrus fruit juice sacs.

C. Other Important Enzymes

1. Chlorophyllase, Peroxidases, and Catalases

Chlorophyllase is the enzyme that catalyzes chlorophyll by removing the phytol group, which result in chlorophyllide formation. This enzyme is present in chloroplast, and this organelle undergoes degradation before and during maturation and color change in the fruit's rind.

Peroxidases catalyze oxidative reactions. For example, IAA oxidase (a type of peroxidase) catalyzes interconversion and catabolism of IAA. The peroxidases and chlorophyll oxidases are involved in chlorophyll changes and color changes at ripening. Catalase decomposes hydrogen peroxide (H_2O_2) to water and oxygen. Peroxidase in the presence of H_2O_2 catalyzes the oxidation of substrates such as phenols, amines, aromatic compounds, ascorbic acid, etc. Catalases and peroxidases are closely related in structure and function. Catalases utilize H_2O_2 in oxidation of alcohols, phenols, and other H-donors in a manner similar to peroxidase. Catalases also dispose of excess H_2O_2 produced during oxidative metabolism. These enzymes are constituents of peroxisome-like particles within the cells. Peroxidase is considered to be a membrane-bound, or soluble, enzyme; several peroxidase forms are known (Shannon et al., 1966). Catalases and peroxidases with prosthetic group of iron oxidize phenolics.

Increased activity of peroxidases and catalases is generally thought to be indicative of degradative changes and senescence. However, the correlation between peroxidase activity and fruit storage life has been found to be negative (Lin, 1988). The total activity of peroxidase enzymes in peel and pulp tissues of Kinnow mandarins decreased with maturation, whereas catalase activity increased up to color break and declined slightly thereafter (Lallan Ram and Godara, 2005)

Uronic acid oxidase is present at relatively high levels in *Citrus sinensis* peels. The enzyme is a flavoprotein with a molecular weight of 62 000 and an optimum pH of 8.5 (Pressey, 1993). It oxidizes galacturonic acid to galactaric acid in the presence of O_2 and produces H_2O_2. It also oxidizes oligogalacturonides and polygalacturonic acid. The rate of oxidation increases with the degree of polymerization of the substrate.

Catalase, ascorbate peroxidase, and glutathione reductase are O_2-scavenging enzymes, and activity of these enzymes has been found to be higher in chilling-tolerant Clemenules and Clementines. Nova and Fortune are susceptible to chilling with lower activity. Changes in activity of oxygen-scavenging enzymes, superoxide dismutase (EC 1.15.1.1), catalase (EC 1.11.1.6), ascorbate peroxidase (EC 1.11.1.11), and glutathione reductase (EC 1.6.4.2) during low-temperature storage have been studied by Sala (1998). Superoxide dismutase activity increased during cold storage in both chilling-susceptible and chilling-tolerant cultivars, which indicated that oxidative stress may be involved in cold-induced peel damage to harvested citrus fruits, and chilling-tolerant cultivars may have a more efficient antioxidant enzyme system. Superoxide dismutase (SOD, EC 1.15.1.1) is an enzymatic system shown to be functional in citrus leaves, and it catalyzes the potentially harmful superoxide free radicals (O_2^-) generated in cells by univalent reduction of molecular oxygen to H_2O and O_2. The SODs contain the metal prosthetic group Cu, Zn, Mn, and Fe (Sevilla et al., 1988).

2. Phenol/Flavonoid and Limonoid Enzyme System

Phenolases have a role in browning of fruit tissues and browning of processed products resulting in phenol oxidation. The quinines are polymerized to give tannins or brown-colored compounds. These are copper-containing enzymes. Polyphenol oxidases catalyze oxidation of *O*-dihydric phenols such as catechole and chlorogenic acid. The polyphenoloxidase activity in leaves peaked in December and January, and it was highest in Satsuma and lowest in the lemon. Species with the highest polyphenoloxidase activity had the highest frost resistance. Species with a high leaf-polyphenoloxidase activity also showed a high activity of this enzyme in the fruits and generally a low vitamin C content (Kutateladze, 1973).

1-2-Rhamnosyltransferase catalyzes the production of disaccharide-flavonoids in citrus fruits. Rhamnosyltransferase activity in early development of pummelo fruit coincides with an accumulation of naringin. The concentration of naringin in leaves, petals, receptacles, filaments, albedo, and flavedo drops drastically during development and correlates directly with a decrease in the activity and amounts of this enzyme. High 1-2-rhamnosyltransferase activity and naringin concentrations were observed in both young and mature ovaries and in young fruits. The possibility of transport of naringin from leaves to fruit is also indicated. The glycosyltransferases are involved in anthocyanin biosynthesis and genes express late in their production as the color develops during maturity. On the contrary, rhamnosyltransferases are active only in juvenile stages of fruit development (Bar-Peled et al., 1993).

Leaves appear to be the major source of limonoids present in fruit and seeds. Initial steps of the biosynthetic pathway of the B-ring of naringin have been established in grapefruit (Hasegawa and Maier, 1981). Enzymes leading from phenylpyruvate to *p*-coumarate and also naringin chalcone cyclase were identified in the fruit. Phenylalanine ammonia-lyase (PAL) is a probable regulatory enzyme in the biosynthesis of flavonoids. Naringinase is reported to have great potential for reducing naringin bitterness of citrus juices.

Changes in PAL activity in young fruits of *Citrus unshiu* indicated that very high initial activity decreases gradually as the fruit grows and after the June drop declines markedly (Hyodo and Asahara, 1973). PAL behavior appeared to be closely related to fruit tissue differentiation. The protein content of the fruit showed a pattern of change similar to that of PAL activity. The amount of hesperidin was high in immature fruit, and the accumulation of hesperidin increased as fruit growth progressed. When the PAL activity had almost disappeared, this accumulation ceased.

The PAL (EC 4.3.1.5) is also considered as a marker of environmental stress in plant tissues. In chilling-sensitive Fortune mandarins, PAL transcript and PAL activity were restricted to the flavedo tissue in and around the necrotic regions. Exposure to a low non-chilling temperature produced an early, moderate, and transient increase in PAL mRNA levels and PAL activity that declined after 1 day (Sanchez-Ballesta, 2000).

3. Ethylene Metabolizing Enzymes

Enzymes involved in synthesis of ethylene have great bearing on maturity, shelf-life, and senescence of citrus fruits. The ripening process can be controlled if inhibitors of these enzymes are regulated by some means. Shimokawa (1983) reported ethylene-forming enzymes in Satsuma mandarins. The purified enzyme catalyzed ethylene formation from 1-aminocyclopropane-1-carboxylic acid (ACC) in the presence of pyridoxal phosphate, IAA, Mn^{2+}, and 2,4-dichlorophenol.

4. Nutrient Metabolizing Enzymes

Pyruvate kinase, PK (EC 2.7.1.40) (ATP pyruvate phosphotransferase) is closely related to mineral nutrient status in fruit and leaves. High PK activity is recorded in Ca^{2+} deficient leaf and fruit (flavedo tissues of Valencia and Shamouti oranges). Activity of this enzyme is suggested to be useful as an indicator of calcium deficiency in citrus fruit (Lavon et al., 1988).

V. LIPIDS, WAXES, AND OTHER RELATED COMPOUNDS

Lipids are generally defined as biochemical compounds which contain one or more long chain fatty acids to glycerol and are soluble in organic solvents such as chloroform, ether, etc. The term *fat* is usually used for a large store of triglycerides, which is a reservoir of carbon and energy such as in oil seeds and some fruits (avocados e.g.). Although terms such as *fat* and *oil* are used interchangeably, fats generally have saturated fatty acids, and at room temperature, these are in solid form. *Oils* normally have unsaturated fatty acids and are in liquid form at room temperature. These oil/fat reserves can be extracted from dried plant or fruit samples by using non-polar solvents such as petroleum ether, hexane, and others, while lipids in phopholipids, glycolipids, and lipoproteins that are combined with other molecules are extracted by using polar solvent mixtures such as chloroform and ethanol or ether and isopropanol. These complex lipids are amphipathic, that is, they contain a polar 'head' and a non-polar 'tail.' Citrus fruits are not a rich source of fats. Citrus juices contain generally less than 0.1 percent lipids (Swift and Veldhuis, 1951). The lipids/fats are found either in seeds or rinds.

The oil that is found in oil glands of flavedo is essential oil that is volatile in nature and responsible for the characteristic aroma. These oils are covered separately under volatiles in this chapter. The classes of compounds under these volatiles are esters (which give fruity aromas), alcohols, acids, ketones, aldehydes, and hydrocarbons.

The membrane lipids are phospholipids and glycolipids. Lipoprotein membranes of fruit cells and sub-cellular organelles such as mitochondria, lamellae of chloroplast, etc. are rich in phospholipids and glycolipids (Goldschmidts, 1977). Principal accumulation of fats in citrus fruits occur in seeds that contain glycerides of unsaturated fatty acids. These unsaturated fats can be utilized in human consumption. However, this seed oil is bitter due to the presence of limonoid bitter principals and must be refined to remove the bitterness. As the orange fruit matures, seed moisture content decreases to a level of 50–55 percent of the fresh weight, and oil content increases to 35–45 percent (Hendrickson and Kesterson, 1963a).

The dried seeds of oranges contain 30–45 percent lipid, and those of grapefruit contain 29–37 percent (Kesterson and Braddock, 1976). Citrus seed oil is a mixture of glycerides of various fatty acids. The major fatty acids present in citrus seed oils are palmitic, stearic, oleic, linoleic, and linolenic. Different citrus varieties are often similar in fatty acid composition. Lemon and lime oils are high in linolenic acid and show the highest refractive indices and iodine number. Mandarin seed oils have high linoleic acid. Orange seed oils have the least refractive indices and iodine numbers. There was a significant correlation between refractive indices and the degrees of unsaturation as measured by the iodine values (Hendrickson and Kesterson, 1963b, 1964a). The unsaturated fatty acids (palmitoleic, oleic, linoleic, and linolenic) occupy a large percentage of the total fatty acids of the citrus seed oils. This fact makes the citrus oil a desirable dietetic substitute for other unsaturated fats in food. The oil (28 percent recovery) from fresh Eureka lemon seeds contained palmitic acid (41.2 percent) and oleic acid (33.5 percent) as the main constituent acids (Sattar et al., 1987). The other acids present were lauric acid or C12:0 (1.8 percent), myristic acid or C14:0 (0.5

percent), palmitoleic acid or C16:1 (5.1 percent), unknown (3.1 percent), stearic acid or C18:0 (7.2 percent), linoleic acid or C18:2 (5.1 percent), and linolenic acid or C18:3 (1.0 percent).

Fresh citrus fruit flesh/juice also contain some amount of lipids. The lipid content of juice of *Citrus hassaku* and *C. tamurana* fruits decreased as maturity progressed, whereas in *C. unshiu* fruits, the total lipid content decreased initially and increased thereafter (Kadota et al., 1982). The lipid content did not vary in maturing *C. natsudaidai* fruits. The phospholipids were about 50 percent of the total lipids. In all species, the main components of neutral lipids were sterols, free fatty acids, triglycerides, and pigments. The polar lipids were phosphatidylethanolamine, phosphatidylcholine, and glycolipids. The palmitic, oleic, linoleic, and linolenic acids were major fatty acids. The unsaturated fatty acids increased during fruit maturation.

In Satsuma mandarins, total lipids and phospholipids in the pulp increased during the early stages of maturation and then decreased. Glycolipids rose during maturation and storage. Palmitic, oleic, linoleic, and linolenic acids were the major acids of pulp lipids (Kadota and Miura, 1983). In Naveline oranges grown in Sicily, palmitic and palmitoleic acids decreased during maturation, whereas palmitic acid increased in Valencia Late at maturity. Linoleic and linolenic acid values were higher in the overripe stage particularly in the blood-red cultivars such as Tarocco, Moro, and Sanguinello where the fruit juice develop unpleasant odors and taste (Nicolosi-Asmundo et al., 1987).

A. Cutin

Cutin is a major part of the fruit cuticle. It is poorly soluble in most organic solvents. Long chain constituents of cutin are cross-polymerized with membranes. The hydroxyacids of cutin are polymerized (Kollattukudy, 1986). A cuticle is formed of the lipids from outer epidermal cells of the fruit peel and plays an important role in controlling water loss/transpiration. It also protects fruit from insect-pests. The cuticle lipid can be categorized as waxes and cutin. Waxes, triterpenoids, and steroids are similar in chemical properties because they solubilize in lipid solvents and can be found in residue after distillation of citrus peel oils.

B. Waxes

These are a particular kind of lipids and esters of high molecular weight (long chain) formed from a combination of fatty acids and mono alcohol. In broad terms, natural waxes contain not only wax esters but also hydrocarbons such as paraffins and olefins, fatty acids, ketones, alcohols, and aldehydes (Kollattukudy, 1968). These are water repellent in nature, melting at 40–100°C and are crystalline. Wax acids of grapefruit have a molecular weight corresponding to $C_{32}H_{64}O$. Valencia orange peel waxes contain C_{26} acids, stearic acids, and many other fatty acids. Distillation residues of cold-pressed Valencia orange oil contained paraffin waxes. With advancing maturity of fruit, epicuticular wax of oranges increases. The walls of juice vesicles of citrus contain suberin, a high-molecular weight saturated fatty acid (Ting and Attaway, 1971). Wax acts as the adhesive agent between the juice sacs and contributes to the ability of the segments to withstand disintegration (Shomer et al., 1980). The wax in the juice sacs of Marsh Seedless grapefruit, Shamouti and Valencia oranges, and Lisbon lemons grown in Israel contained 33.8–42.3 percent saponifiable matter, 65.8–59.3 percent unsaponifiable matter, hydrocarbons, and primary and secondary alcohols.

Squalene (a terpenoid) present in the wax is considered as a possible natural protectant against chilling injury (CI) of citrus fruits. An inverse relationship was found between CI and levels of squalene in the epicuticular wax. The optimum temperature for biosynthesis of squalene in grapefruit is 15°C, which is the temperature previously reported as optimum for conditioning grapefruit against CI (Nordby and McDonald, 1990). Cold storage affects the amounts of alkanes, squalene, and long-chain aldehydes in the epicuticular wax of grapefruit (Nordby and McDonald, 1991). Fruits dewaxed prior to temperature conditioning at 15°C failed to produce squalene and the aldehydes. The role of C24–C26 aldehydes, C25–C27 alkanes, and squalene, separately or in combination, was suggested in plugging the holes that develop in the wax coating of the grapefruit placed in cold storage. CI is found to be more in exterior-canopy. Marsh grapefruit stored at 5°C. Higher levels of wax in exterior-canopy fruits indicated that CI is not directly related to the level of total epicuticular wax (Nordby and McDonald, 1995). Triterpenes were suggested to interact with sunlight particularly in exterior-canopy fruit and possibly alter the susceptibility of grapefruits to CI.

C. Terpenoids and Steroids

Terpenoids are water-insoluble cyclic compounds chemically similar to lipids. Important compounds are carotenoids, hormones such as gibberellic acid and abscissic acid, and limonoids. Sterols are derivatives of lipids and are grouped under steroids. The acetate-mevalonate pathway leads to formation of terpenoids. Peel oil of grapefruit contained 22-dihydrostigmasterol (β-sitosterol) and a triterpenaoid ketone, friedelin (Weizmann et al., 1955). β-sitosterol D-glucoside has been isolated from Valencia orange juice (Swift, 1952). Citrostadienol has been identified from the peel oils of both oranges and grapefruit (Mazur et al., 1958). The sterol fraction of grapefruit peel was separated into three groups, namely 4-4′-dimethyl sterol, 4-α-methyl sterols, and desmethyl sterols (Williams et al., 1967). Each group of sterols could be further fractionated into several components. Among the desmethyl sterols, three components similar to those of β-sitosterol, stigmasterol, and campesterol were found. The sterols may play role in resistance of some citrus cultivars to diseases and frost. The lemon cultivar Dioskuriya, which is malsecco-resistant and moderately frost-resistant, contained sitosterol, campesterol, stigmasterol, cholesterol, and 24-ethylidenelophenol in the peel and flesh. Free, glycosylated and esterified sterols have been extracted by Zambakhidze et al. (1989).

VI. PIGMENTS

Major pigments that give color to citrus fruits are chlorophylls (green), carotenoids (yellow, orange, and deep orange), anthocyanins (blood red), and lycopenes (pink or red). During growth and maturation, especially in the immature stage, chlorophylls predominate in the peels of all citrus fruits. Due to the presence of chlorophyll, immature fruits are capable of photosynthesis but cannot make a significant contribution to the own nutrition (Todd et al., 1961). Chlorophyll in citrus consists mainly of two pigments: chlorophyll-a (C_{55} H_{72} O_5 N_4 Mg) and chlorophyll-b (C_{55} H_{70} O_6 N_4 Mg). Chlorophyll-c, d, and e are not reported in citrus and are mainly present in algae and certain sea weed. Chlorophyll-a and b are present in a ratio of 2:1. There is a rapid synthesis of carotenoids in the chromoplast during ripening, which is accompanied by a simultaneous loss of chlorophyll. Chloroplasts change into chromoplasts. Lewis et al. (1964) reported a decrease in the concentration of chlorophyll-a from 4.1 to 1.0 $\mu g/cm^2$ and chlorophyll-b from 1.2 to 0.3 $\mu g/cm^2$ in navel orange flavedo 2 months after color-break. Carotenoids are long chain compounds (tetra terpenes) and include carotenes, (α, β, etc.) and xanthophylls (luteins, flavoxanthins, Leuteoxanthin, zeaxanthin, violaxanthin, etc.). β-carotenes of the carotenoids are nutritionally important because of the vitamin A activity. Xanthophylls are esterified with fatty acids and accumulate in lipid globules of chromoplast in peels (Eilati, 1970). Xanthophylls are oxygenated compounds and are methanol soluble, whereas hydrocarbons, monols, and esters are petroleum ether soluble. The petroleum ether fraction of carotenoids increases with maturity. Total carotenoid, chlorophyll, and lycopene content varies greatly in the peel and pulp of various citrus fruits (Table 6.7).

With ripening, total carotenoids increase in the peel as well as in the pulp. Rind of fruit is the region of higher carotenoid concentration, and 50–75 percent of the total carotenoids of oranges exists in the peel (Curl and Bailey, 1956). In the fruit of early season Valencia oranges, there is a qualitative difference between carotenoids in pulp and peel (Table 6.8), the later containing relatively much more violaxanthin (Curl and Bailey, 1956) The disappearance of chlorophyll is not, however, essential for the development of chromoplasts with the synthesis of additional carotenoids. In grapefruit, active carotenoid synthesis occurs before chlorophyll begins to disappear (Yokoyama and White, 1967). In the peel of Marsh seedless grapefruit, there is no further synthesis of colored carotenoids during ripening after all the chlorophyll has disappeared, but there is a marked synthesis of phytoene. In the red grapefruit varieties, flesh (pulp) is of red color due to the pigment lycopene. The lower intensity of lycopene gives a pink color to flesh of the Thompson variety of grapefruit. The carotenoid pigments of the pulp of colored grapefruit are mainly lycopene and β-carotene.

TABLE 6.7 Chlorophyll, Carotenoids, β-Carotene and Lycopene Content in Various Citrus Fruits

	Chlorophyll (mg /100 g)			Total carotenoids (mg /100 g)		β-carotene (mg/100 g)		Lycopene (mg /100 g)	
Citrus fruit	'a'	'b'	Total (peel)	Peel	Pulp	Peel	Pulp	Peel	Pulp
Valencia orange			0.3–0.4	9.8	2.4–3.4	00.02	00.02–0.15	–	–
Navel orange			0.4	6.3–8.0	2.3	00.01–0.05	0.05	–	–
Mosambi orange	0.5	0.3	0.8	2.61					
Parson Brown orange			4.35	5.2					
Pineapple orange			5.00	8.5					
Kagzi lime (green, color turning)			3.5	0.48					
Kagzi lime (yellow)			0.69	0.88					
Lemon				0.3	Traces	0.006	0.005		
Grapefruit Ruby Red				1.4	0.82	0.07	0.2	0.4–0.9	
Dancy tangerine				18.6	2.7	0.07	0.11		
Nagpur mandarin				3.8–5	1.44	0.04	0.001		
Kinnow mandarin				7.6	1.51–3.8				
Clementine			0.3	16.0					

Subramanyam and Cama (1965); Khan and MacKinny (1953); Curl and Baily (1957); Curl (1964); Curl and Bailey (1965); Benk and Bergamann (1973); Mehta and Bajaj (1983); Bower et al. (1997); Selvaraj and Edward Raja (2000); Ladaniya and Mahalle (2006).

There are several factors that govern the formation of carotenoids in maturing citrus fruits. Synthesis of xanthophylls (bright orange to orange) is encouraged by a daytime low temperature (20°C) followed by cool nights (7°C) with soil temperature at 12°C, whereas the synthesis of other carotenoid pigments was not influenced in this way (Young and Erickson, 1961). Rootstock have also been found to affect the pigments in citrus fruit peels. Valencia and Joppa (Jaffa) oranges grown on trifoliate and sweet orange stocks had significantly higher concentration of total carotenoids in the juice than on rough lemon rootstock (Bowden, 1968). Ripening in the absence of oxygen but in the presence of nitrogen, carbon dioxide, or ethylene inhibits pigment formation in various citrus fruits. Carotenogenesis is an endergonic process, and in the later stages an

TABLE 6.8 Distribution of Various Carotenoids in Orange Pulp, Peel, and Juice

	% of total carotenoids		
	Valencia orange		Shamouti orange
Carotenoid	Pulp	Peel	juice
Phytoene	4.0	3.1	5.70
Phytofluene	1.3	6.1	2.40
α-carotene	0.5	0.1	0.30
β-carotene	1.1	0.3	1.26
δ-carotene	5.4	3.5	1.40
OH-α-Carotene-like	1.5	0.3	0.50
Cryptoxanthin epoxide-like	–	0.4	–
Cryptoxanthin	5.3	1.2	12.88
Cryptoflavin-like	0.5	1.2	0.70
Cryptochrome-like	–	0.8	–
Lutein	2.9	1.2	5.20
Zeaxanthin	4.5	0.8	6.20
Capsanthin-like	–	0.3	–
Antheraxanthin	5.8	6.3	6.83
Mutatoxanthins	6.2	1.7	15.13
Violaxanthins	7.4	44.0	3.00
Luteoxanthins	17.0	16.0	6.92
Auroxanthin	12.0	2.3	0.58
Valenciaxanthin	2.8	2.2	–
Sinensiaxanthin	2.0	3.5	–
Trollixanthin-like	2.9	0.5	9.64
Valenciachrome	1.0	0.7	–
Sinensiachrome-like	–	0.2	–
Trollichrome-like	3.0	0.8	1.46

Ting and Attaway (1971); Gross et al. (1971).

oxidative process. However, ethylene at very low concentrations stimulates carotene synthesis as does increasing the oxygen content of the atmosphere. Light is not necessary for carotenoid synthesis because more pigment appears to be synthesized in the dark. Temperatures above 30°C are reported to inhibit lycopene synthesis and its congeners (Tomes et al., 1956). The optimal temperature for lycopene synthesis is 16–21°C. Exposure of fruit to elevated temperatures for a short time has no permanent deleterious effect on lycopene synthesis because fruits held at a high temperature and subsequently transferred to a lower temperature rapidly begin to synthesize the pigment. In general, in pink grapefruit, temperatures up to 30–35°C have no deleterious effect on lycopene synthesis, but carotene synthesis is adversely affected at these temperatures. Carotenes are normally synthesized best at 10–15°C.

Citrus hybrids have a number of naturally occurring apo-carotenoids in considerable quantity. Sintaxanthin, citranaxanthin, and reticulataxanthin occur in large amounts in the Sinton citrangequat, a trigeneric hybrid of the oval kumquat (*Fortunella margarita*) with the Rusk citrange (*Poncirus trifoliata* × *Citrus sinesis*), and are mainly responsible for the deep red color of the flavedo of the fruit. These pigments appear in significant amounts only in hybrids. For example, reticulataxanthin constitutes over 49 percent of the carotenoids in the Sinton citrangequat, whereas it occurs only in traces in the Rusk citrange (Yokoyama and White, 1966). It is only a minor component of the peel of *Citrus sinensis* (Curl and Bailey, 1961) and *C. reticulata* (Curl and Bailey, 1957) but occurs in greater amounts in the peel of the hybrid Minneola tangor (*Citrus reticulata* and *Citrus sinesis*) (Yokoyama and White, 1965). β-Citraurin and β-apo-carotenal have been reported in peels of Clementines and Dancy tangerines (Curl, 1965). Reticulataxanthin and tangeraxanthin (carbonyl carotenoids), which give the typical orange-red color, have been found in tangerine peels (Curl, 1962).

Anthocyanins are only slightly affected by rootstock. Cyanindin-3-glucoside and delphinidin-3-glucoside are the major anthocyanins (Chandler, 1958; Harbourne, 1965). Maccarone et al. (1983, 1985) used the HPLC technique and separated 10 anthocyanins in Moro, Tarocco, and Sanguinello blood oranges grown in Italy. Cyanindin-3-glucoside and some new anthoyanins acylated by hydroxycinnamic acid have been reported.

Maximum anthocyanin content has been reported in Moro sweet oranges followed by Sanguinello-moscato and Tarocco at the same degree of maturity. The positive correlation has been found in anthocyanin pigments and TSS/acidity ratio. The Nucellar line of Moro had 274.2 mg/l of total anthocyanins, Sanguinello Moscato had 174.5 mg/l, and the nucellar Tarocco had less than 50 mg/l of total anthocyanins (Rapisarda and Giuffrida, 1992). Development of pigments and color is mainly dependent on weather conditions and citrus variety. In warm and humid areas, abundant red anthocyanins are produced, whereas in dry areas, color intensity is lower. The concentration of anthocyanins increases rapidly in fruit as they near maturity.

VII. PHENOLS, FLAVONOIDS, AND LIMONOIDS

A. Phenols

Phenolics are widely distributed in plants. It is a class of organic aromatic compounds containing one or more hydroxyl (OH) groups attached to a benzene ring. These are called secondary metabolic products, and they are seemingly not produced for growth and development of plant cells initially like other essential metabolic products i.e. carbohydrates, proteins, lipids (fats), and nucleic acids. Nevertheless, phenolics play important role in plant defenses, fruit and flower coloring, flavor, and hormonal balance (GAs and ABA are terpenoids that are secondary products). The phenols are formed via production of some aromatic amino acids such as tryptophan, tyrosine, and phenylalanine and by-products of their metabolism (Neish, 1964). Solubility of phenolics depends on the presence of sufficient proportions of polar or hydrophilic groups. With higher hydroxyl groups, solubility increases, and when hydroxyl groups are methylated, solubility decreases.

The simple phenols and aromatic amino acids are synthesized via a shikimic acid pathway that further leads to formation of aromatic compounds, tannins, coumarines, and lignins. In this pathway, cinnamic acids are formed (including derivatives of cinnamic acid: caffeic acid, sinapic acid, and ferulic acid). Cinnamic acid is considered to be formed from quinic acid through various steps in which shikimic acid, prephenic acid, and L-phenylalanine are formed. From tyrosine (through oxidative deamination) *p*-coumeric acid is formed, and the process is mediated by PAL (phenyl ammonia lyase enzyme). Coumerins (phenolic compounds) are physiologically active and known to regulate endogenous IAA by regulating, enhancing, or inhibiting IAA oxidase activity. On the other hand, cinnamic acids are bound to sugar either as esters or glycosides. When the cinnamic acid or its derivatives form glycocides, the flavonoids are formed.

There are three classes of phenols: (1) monocyclic or simple phenols (single phenol ring, for example, catechol, hydroquinone, and *p*-hydroxycinnamic acid); (2) dicyclic phenol (two benzene rings, for example, flavanones); and (3) polycyclic (for example, pigments such as cyanins and petunidin). Phenolics determine color and flavor of fruits and also actively play a role in the mechanism of resistance of fruits and plants to diseases and insects. Phenols are generally tasteless compounds, but phenolic acids are sour whereas condensed flavans are astringent. Phenols are generally not found in a free state in the cell, and they are mostly conjugated with other molecules as in the case of flavonoids.

Phenols attached to monosaccharides or disaccharides are glycosides. The flavonoids are glycosylated and are bitter in taste. Similarly, phenols are also esterified to organic acids, amino acids, and other organic molecules.

The common phenolic compounds reported in higher plants are: (1) Cinnamic acid and its derivatives (e.g. chlorogenic acid); (2) Flavans: for example catechines and epicatechins; (3) Anthocyanidins/anthocyanins: these flavonoid compounds are usually red, blue, and purple in color, soluble in water, and easily detectable. The common anthocyanidins are pelargonidin, cyanidin, peonidin, delphinidin, petunidin, and malvinidin. Normally, these compounds are β-glycosides with sugars (glucose, galactose, rhamnose, and rutinose) at the 3 and 5 positions, and thus anthocyanins are formed. The red or blue color of anthocyanins depends on pH. In contrast to the cinnamic acid and flavans, the concentration of anthocyanins increases rapidly at fruit maturity. Changes in the color of the juice of blood group oranges take place on heating and processing due to changes in anthocyanins; (4) flavanones, flavonols, and flavonol glycosides: flavanones are usually in glycoside form because at 3 and 7 positions, D-glucose and L-Rhamnose or rutinose (diasccharide) are the common sugars forming glycosides. Naringin is a glycoside that is composed of aglycon naringenin and the disaccharide neohespespiridose (2-0- ά-lrhamnopyranosyl-D-glucosepyranose). The aglycones and neohespiridose are not bitter when present separately in free form. Thc sugarpresent at position 7 is crucial. Neohespiridose imparts bitterness, glucose gives much less, and rutinose imparts no bitterness. Solubility of flavonoids depends on the presence of a sufficient proportion of polar or hydrophilic groups (hydroxyl groups). With glycosylation, solubility of flavonoids increases. Hydroxyl groups at positions 3 and 5 and carbonyl groups at position 4 form strong complexes with metals such as copper, iron, lead, and aluminum. This property of fixing metals, together with the ability of polyphenols to terminate oxidations mediated by free radicals, gives many flavonoids antioxidant properties (Harbourne, 1967); (5) Flavones: glycosides with sugars attached at position 7 or 5 (epigenin, luteolin, and tricin).

The naringenin is a common flavanone in citrus fruits. Some flavanones, including naringenin, give rise to bitter taste. In grapefruits, neohespiridose (leading to formation of naringin) accumulates rapidly during early growth (Maier and Metzler, 1967a), whereas in lemons and sweet oranges, the rutinose accumulates (Horowitz, 1961). Other phenols also accumulate along with naringin in grapefruit; however, the quantity is much less. The aglycone form of the naturally occurring glycosides and esters of the phenols have been isolated and identified (Maier and Metzler, 1967b, c). The naringenin (the aglycone of naringin) or its glycosides is an early and central intermediate in flavonoid metabolism in grapefruit. The possible interrelationship of citrus phenolics is shown in Fig. 6.4.

Plant phenolic levels have been found to increase after infection by pathogens. Polyphenol oxidases, which oxidize phenols, lead to production of oxidized phenols that are more potent antifungal agents than non-oxidized phenols. Oxidized polyphenols are potent inhibitors of the pectolytic enzymes necessary for the invasive ability of fungal pathogens. Susceptibility of plants to disease also depend on oxidation–reduction balance of the tissues. Presence of high activity of polyphenol oxidase changes this balance in favor of plants as opposed to the pathogen. Citrus fruits protect themselves by lignification of wounds. Hydroxylated cinnamic acids (phenols) are considered to be a precursor of lignin. Lignin-like polymers were shown to be formed from ferulic acid and *p*-coumeric acid (Stafford, 1960). Hermann (1989) reviewed the phenolics in fruits and reported that feruloyl and sinapoyl glucose are major phenolics in citrus fruit. Under high humidity conditions, wounds of citrus fruits on the surface have been found to be healed. Free phenolics increased, conjugated phenolics decreased and a lignified layer was found on injured cells of Valencia oranges at 30°C and 96–98 percent RH. Lignin formation provided a mechanical barrier

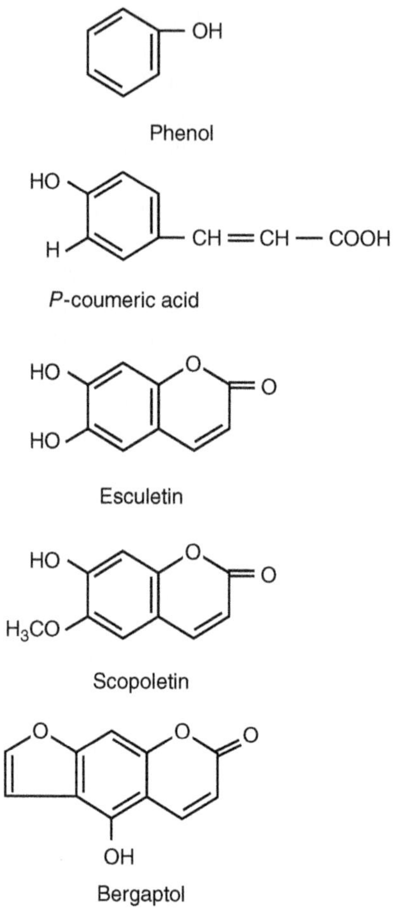

FIGURE 6.4 Phenolic Compounds Present in Citrus Fruit.

that retarded or inhibited penetration of injured tissue by *P. digitatum* (Ismail and Brown, 1975).

The role of phenols and activity of polyphenoloxidase in tolerance to chilling is also a matter of study and further research. Valencia orange fruits with an initially high phenol content in the rind, higher chlorophyll content, and the least polyphenoloxidase activity had the least low temperature injury at 6°C after 2.5 months (Santana et al., 1981).

1. Coumarins: These are derivatives of O-coumeric acid (a phenolic compound) that have a specific odor and are found in essence oils of citrus peels. In lemon oil, coumerin substances such as limettin and bergamotin have been reported, whereas in grapefruit oil, 7-geranoxy-coumarin was identified. The coumarins identified in orange and lime oil are auraptene, meranzine, bergaptol, and isopentenyl psoralenes (Stanley 1964; Stanley and Vannier, 1967).

B. Flavonoids

Anthocyanins, flavones, flavonols, and flavanones are grouped under flavonoids. Flavonoids contain a C_6—C_3—C_6 carbon skeleton with sugar moity (in glucosides). The major glycoside flavonoid in citrus are hespiridin, naringin, and neohespiridin. In general, concentration of flavanones decreases as fruit matures. Hesperidin is a main flavonoid of oranges. It is not bitter in taste. Hesperidin is the 7-β-rutinoside of hesperetin (Hendrickson and Kesterson, 1964b). In hesperidin, the rhamnose and glucose are in the form of rutinose as a disaccharide moity, and because of the rutinose, they are not bitter. Hesperidin is also found in mandarins, lemons, limes, and hybrids. The cloudiness of juice and marmalade made from oranges is due to precipitation of hesperidin, which is less soluble in water. Hesperidin can be found in the segment membrane in the form of white spots/crystals in freeze-damaged oranges and in the form of white specks in frozen concentrated orange juice.

The main flavonoid compound of the grapefruit and shaddock is naringin; it has a bitter taste and is soluble in water, so some fresh grapefruits and pummelos taste slightly bitter. It is a rhamnoglucoside of the aglycone, naringenin. The bitter taste of naringin was found to be due to the structure of the disaccharide moiety (Horowitz and Gentili, 1961, 1963). In grapefruit, rhamnose and glucose sugars link to form neohesperidose (2-0-α-L-rhamnopyranosyl-D-glucose), which is the sugar moiety of naringin and neohesperidin. Citrus flavanones containing neohesperidose as disaccharide moity are bitter, whereas, the flavanones containing the isomeric disaccharide rutinose (6-0-α-L-rhamnopyranosyl-D-glucose) are tasteless (Horowitz and Gentili, 1961, 1963) (Fig. 6.5).

The dihydrochalcone glucosides of naringin, neohespiridin, and hespiridin are 300, 1100, and 300 times as sweet as sucrose, respectively, on a weight basis and are potential sweeteners without calories (Horowitz and Gentili, 1963, 1969). These sweeteners have potential for industrial production because the synthetic sweetener saccharin has some side effects.

Flavonoid content of citrus fruits increases to a maximum during the early stage of fruit development and then remains constant. With an increase in fruit size, the flavonoid concentration decreases. Maximum hesperidin can be obtained in early season fruit that is about 2.5–5 cm in diameter (Hendrickson and Kesterson, 1964b). The concentration of flavonoids decreases with increase in fruit size and with maturity. Only 10 percent of the naringin of the whole fruit is found in the juice of grapefruit while 20 and 30 percent of the hesperidin of the whole fruit was found in the juice of the orange and mandarin (Hendrickson and Kesterson, 1964). Hesperidin content of citrus fruits vary with the species (Omidbaigi et al., 2002). The highest yield of crude hesperidin was obtained (1.67–1.73 percent) from Clementines, and the lowest (0.74 and 0.77 percent) from navel oranges. Grapefruits are extremely bitter when immature due to the high concentration of naringin in the juice vesicles. As the fruit ripens, the bitterness decreases with a concomitant decrease of naringin. The absolute amount

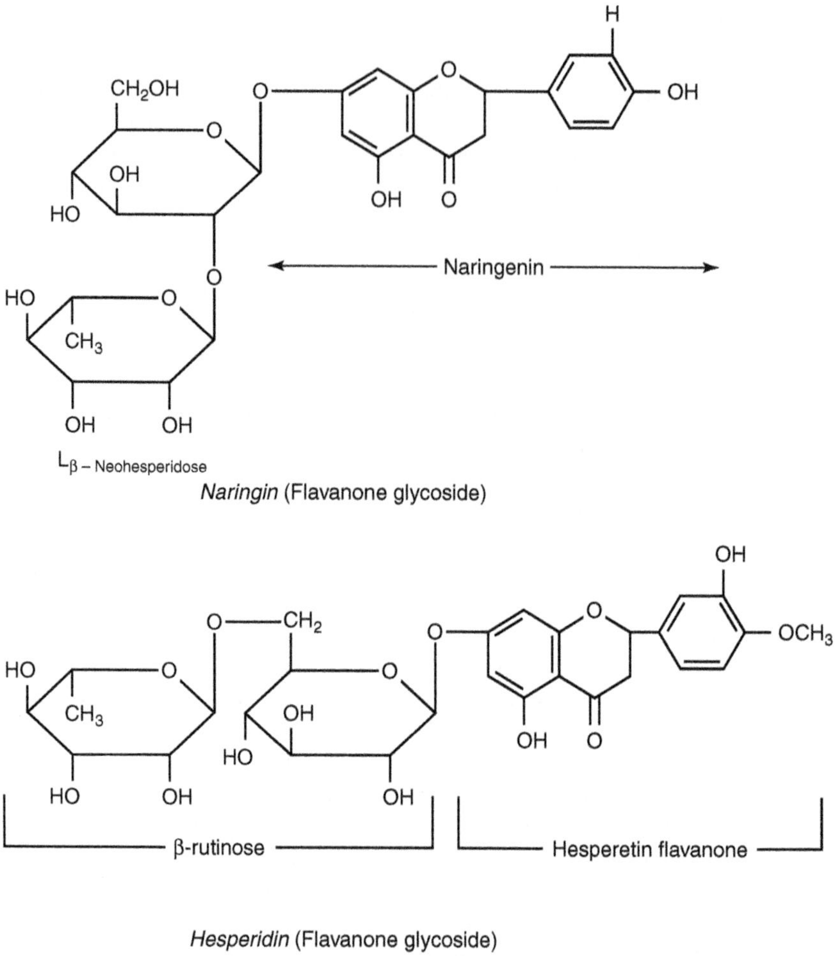

FIGURE 6.5 Important Flavonoids (Flavanone Glycosides) of Citrus-Naringin and Hesperidin.

of flavanone glycoside did not change. The decrease in concentration was due to an increase in fruit size (Mizelle et al., 1967). In Shantian pummelos during October and November, naringin content was 2.25 and 3.18 mg, respectively, per 100 g of peel. In Jincheng oranges, hesperidin content was 12.1 and 11.0 mg/100 g of peel during October and November, respectively. Total flavonoids in this orange decreased from 25.7 to 23.9 mg during the same period as the fruit matured (Houjin et al., 1990).

Diosmin (the flavone analogue of hesperidin) and eriodicitrin are the other flavonoids in lemon peels (Horowitz and Gentili, 1960a) next to hesperidin. Limocitrol, isolimocitrol, and limocitrin were reported in lemon flavonoid extracts

(Gentili and Horowitz, 1964). In Ponderosa lemon citronin, 2-methoxy-5 and 7-dihydroxyflavanone-7-rhamnoglucoside were reported (Horowitz and Gentili, 1960b).

Methylated flavones and flavonols have also been found in citrus fruit. Tangeritin ($C_{20}H_{20}O_7$) has been reported as a pentamethoxyflavone (Goldsworthy and Robinson, 1957) from tangerine oil. From the orange peel extracts nobiletin, tangeretin, a heptamethoxyflavone, sinensetin, and tetra-O-methyl scutellarcin have been reported (Swift, 1965). Wollenweber and Dietz (1981) reviewed the occurrence and distribution of free flavonoid aglycones in plants. They reported 462 flavonoids present in a free state and listed their plant sources, particularly *Citrus* spp.

1. Possible Uses of Flavonoids

Phenolics have certain medicinal properties that have not yet been thoroughly studied. These compounds have antioxidant properties, the ability to form chelates with metals, and complexes with proteins, all of which have potential for use in treating disorders. Compounds such as hesperidin, quercitin, rutin, and eriodictyel have been found to decrease the fragility of capillaries in animals and could have therapeutic value. The antioxidant action of flavonoids can be useful in protecting adrenaline, and ascorbic acid helps to relax muscles. Flavonones can be used as a raw material for medicinal production and to treat eye, kidney, and rheumation diseases. Fluorescent wood dye, benzoic acid, and phlorogluci-nol are manufactured from flavonones (Kefford and Chandler, 1970).

C. Limonoids

Limonoids are the C_{26} oxidized triterpenes found in citrus fruits and are responsible for bitter taste of juice and seeds. The highest concentration of limonoids is in seeds. Seeds of citrus fruits taste intensely bitter if inadvertently crushed under the teeth, and hence seedless mandarins, oranges, and grapefruits are preferred in fresh fruit markets. Limonin, which is a widely found limonoid in citrus fruits, is formed from a basic tri-terpene structure. After shortening of the side chain and formation of a furan ring with cleavage of A and D rings accompanied by oxidization, limonin which is bitter in taste, is formed (Fig. 6.6). The structure of limonin is known to contain one furan ring, one ketone group, one epoxide group, and two lactone rings (Arigoni et al., 1960). Limonin content decreases with the ripening of fruit. Bitterness develops after juice stands for several hours and/or after heating. Bitterness obviously lowers the quality and value of juices. In navel oranges, the monolactone form of limonin, which is non-bitter and stable at neutral pH, is present in albedo and endocarp tissues of fruits, but there is no or very little limonin present (Maier and Beverly, 1968). After extraction of juice, in an acidic pH, the non-bitter monolactone gets transformed to bitter dilactone, which is bitter limonin. Maier and Dreyer (1965) reported that

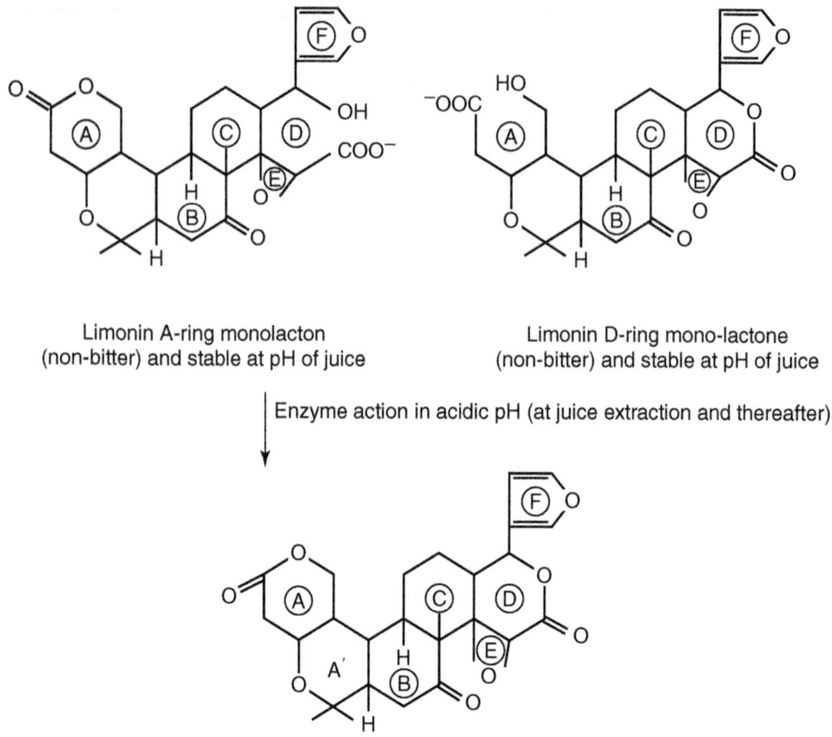

FIGURE 6.6 Non-bitter Mono-lactone and Bitter Dilactone Structures of Limonin.

late season navel oranges and grapefruits did not contain monolactone and did not develop bitterness as the juice was acidified or heated. This indicated that a monolactone is a precursor and slowly disappears (by being metabolized by the enzyme system) from the tissues as fruit matures, so juice of mature fruits develop less or no bitterness. The monolactone is characterized as limonoic acid A-ring lactone (Maier and Beverly, 1968; Maier and Margileth, 1969). In fruit stored at a high temperature (27–32°C), rapid loss of limonin was recorded (Rockland et al., 1957). Navel oranges that degreened in storage with ethylene had no bitterness (Ball, 1949); however, treatment of the entire tree with ethylene did not show any such decrease (Emerson, 1949). In Australian navel and Valencia oranges, limonin and limonoic acid were reported in the peel, juice, and seeds (Chandler and Kefford, 1951, 1953). Samish and Ganz (1950) detected limonin in the juice of Shamouti oranges. In the central part of the grapefruit, limonin content up to 140 ppm was recorded. The other bitter principles in Valencia orange are nomilin and obacunone. Structurally, obacunone, nomilin, deacetylenomilin, and limonin are closely related. Nomilin has also been separated from the grapefruit juice vesicle. The concentration was higher in early season rather than in riper fruit, and in Marsh seedless than in Duncan (Rouseff, 1982). The conversion of obacunone to obacunoate and further to limonin has been demonstrated by Herman and Hasegawa (1985) with a radioisotope tracer technique. Nomilin is considered to be the precursor of all other limonoids accumulated in *Citrus* and related species. It is biosynthesized from acetate via the terpenoid biosynthetic pathway in the phloem region of stems and then translocated to the leaves, fruit tissues, peels, and seeds where it is further metabolized to other limonoids. The citrus limonoid aglycones are then glucosidated by limonoid UDP-D-glucose transferase in maturing fruit tissues and seeds (Hasegawa and Miyake, 1996). These limonoid glucosides are one of the major secondary metabolites in citrus fruit tissues, and they play an important role in fruit quality and possibly in human health.

Regulation of nomilin biosynthesis using phytohormones have been found to be possible. This needs to be done when fruit is growing. IAA, IBA, NAA, and 2,4,5-T (10–15 ppm) inhibited nomilin biosynthesis in lemon seedlings by 82–97 percent compared with controls. GA_3 is not effective. When applied separately, ABA and BA (3 ppm) caused 71 and 37 percent inhibition, respectively (Hasegawa et al., 1986). Orme and Hasegawa (1987) showed inhibition of nomilin accumulation in *Citrus limon* fruit by a naphthaleneacetic acid application. NAA (20 ppm), when fed to stems on mature lemon trees at a distance of 2 cm from the base of the fruit (5–9 mm diameter), it inhibited nomilin accumulation in the stems and fruits by 50–100 and 80–100 percent, respectively, compared with untreated controls.

Limonoid bitterness is also related to fruit maturity and fruit cultivar. When Miyagawa Wase and Okitsu Wase Satsuma fruits were harvested at an early ripening stage, the juice was more bitter and had a higher limonoid content than that of more mature fruit. In fruits held for 3 weeks before juice extraction, bitterness and limonoid contents decreased slightly. Juice from fruits that matured fully on the tree was almost limonoid-free and without bitterness (Izumi et al., 1981). The limonin content of Thai mandarin fruit decreased from 21.56 mg/l 6 months after fruit set to 4.2 mg/l 9 months after fruit set. A small decrease in limonin was recorded between 9 and 10 months. Naringin content decreased from 181.0 mg/l 6 months after fruit set to 140.0 mg/l 10 months after fruit set (Noomhorm and Kasemsuksakul, 1992). Development of delayed bitterness is associated with acid content in fruit. Limonin and acid contents were strongly correlated in Washington Navel oranges (Rodrigo et al., 1985). Navel orange juice has relatively more bitterness than Valencia orange juice. The maturity stage of fruit and bitterness development indicate the role of acids in reaction in which bitterness develops. The mature and degreened Nagpur mandarin fruit had minimum bitterness (Sonkar and Ladaniya, 1995).

I. Possible Uses of Limonoids

Limonoids have been shown to inhibit the formation of certain chemically induced carcinogenesis in laboratory animals by inducing glutathione S-transferase activity. Some of the limonoids act as repellants and possess antifeedant activity against insects and can be used in plants for pest control. The citrus limonoids are unique for many species and varieties and make excellent taxonomic markers (Hasegawa and Miyake, 1996).

VIII. VITAMINS

The main contribution of citrus fruits in human nutrition is undoubtedly their supply of vitamins, especially ascorbic acid (vitamin C). Daily vitamin C intake of 5 mg is sufficient to prevent symptoms of scurvy in an adult (Mapson, 1967). The intake of 30–60 mg is estimated to be required for full grown adults, and an orange a day can fulfill this requirement and ensure good health. Citrus fruits contain many other vitamins too (Table 6.9). Vitamin C content of juice of different citrus fruits varies considerably. Oranges generally contain 40–70 mg vitamin C/100 ml juice, whereas grapefruits, tangerines, and lemons provide 20–50 mg/100 g. Since fruits are eaten raw, the required amount of 30–60 mg per person per day of vitamin C is accomplished by one orange per day. Ascorbic acid is usually high in immature oranges and grapefruits. As fruit ripens and increases in size, the concentration decreases. When calculated on a per-fruit basis, the total ascorbic acid usually increases. Vitamin C is higher in the stem-half of the pulp of oranges and grapefruit than in the stylar-halves and higher at the central axis than on the outside (Ting and Attaway, 1971). The peel is rich in ascorbic acid. Concentration of ascorbic acid in the juice of oranges is one-fifth of that of the flavedo and one-third of that of the albedo on a fresh weight basis (Atkins et al., 1945). In grapefruit, the juice had one-seventh, and one-fifth of the vitamin C concentration of that in flavedo and albedo, respectively. On a whole fruit basis, the juice contained about 25 percent of the total vitamin C of the orange and only 17 percent of that of grapefruit.

A positive correlation exists between vitamin C and the soluble solids of the juice of Valencia oranges from the same tree (Sites and Reitz, 1951). Fruits at the top and on the outside of the tree contain more vitamin C than those in the inside and at the lower level. Green-colored oranges have less vitamin C than the orange-colored ones when harvested at the same time. Fruits from the north side of the tree have a considerably lower concentration than those from the south side. There is no difference in vitamin C content of fruits picked from inside the tree. A significant correlation between ascorbic acid content and the contents of reducing sugars (hexose sugars) suggest that these constituents are associated in ascorbic acid synthesis. Satsuma (*Citrus unshiu*), Hayashi and Miyagawa, and hassaku (*C. hassaku*) fruits from the exterior (sunlit) part contain more vitamin C and sugars with significant correlations between sucrose and ascorbic acid contents in the flavedo and juice (Izumi et al., 1988).

Compared with other foods, citrus juice may supply larger amounts of several vitamins on a per-calorie basis. Citrus juices, on a concentration basis, are higher in vitamin A, thiamine, and Nicotinic acid (niacin) than milk but lower in

TABLE 6.9 Vitamin Content in Some Citrus Fruits

Vitamin per 100 g	Tangerine	Grape fruit	Valencia orange		Navel orange		Eureka lemon	
	juice	juice	peel	juice	peel	juice	peel	juice
Ascorbic acid, (mg)	30	40	137	44	222	59	129	44
Total carotenoids, (mg)	–	–	9.9	2.8	12	1.4	0.3	0.04
β-Carotene, (mg)	0.23	0.01	0.3	0.2	0.2	0.03	0.0	0.0
Choline, (mg)	–	–	23	8	23	6	11	6
Folic acid (B complex), (μg)	1.2	0.8–1.8	12	3	9	2	5	1
Inositol, (mg)	135	88–150	257	159	185	156	216	66
Nicotinic acid (B complex), (μg)	–	–	888	376	665	429	356	71
Pantothenic acid (B complex), (μg)	–	290	40	207	303	187	319	104
Pyridoxine (B_6), (μg)	–	–	176	57	102	48	172	51
Riboflavin (B_2), (μg)	30	20–100	91	27	95	34	79	12
Thiamine (B1), (μg)	95	40–100	120	100	90	100	58	31
Biotin (B complex), (μg)	0.5	0.4–3.0	5	0.8	3	0.6	2	0.3

Joseph et al. (1961); Birdsall et al. (1961); Ting and Attaway (1971); Kefford and Chandler (1970).

riboflavin (Ting and Attaway, 1971). Rakieten et al. (1951) reported that inositol and tocopherol were present in fairly large amounts in the juice of citrus fruit, but the other vitamins also occurred in amounts appreciable enough to be of dietetic importance.

Losses of vitamin C are lower in acidic media. In intact fruit, an enzymic system of oxidation of vitamin C is controlled, but when juice is extracted and processed, some losses do occur. Enzymes such as ascorbic acid oxidase, phenolase, cytochrome oxidase, and peroxidase can oxidize ascorbic acid. Phenolic compounds, flavonoids, and acids can inhibit enzymic and non-enzymic oxidation. In a non-enzymic oxidation process, copper and iron salts catalyze the oxidation. The mono dehydro ascorbic acid is formed first and then dehydroascorbic acid in the second stage. Dehydro ascorbic acid (DHA) in oxidized form is also biologically active as L-ascorbic acid (reduced form). In intact fruit, the changes from reduced to oxidized form and vice-versa continue taking place depending on hydrogen ion balance (Mapson, 1967). The DHA level increases during storage of citrus fruits. The DHA present is less (1.0–4.6 mg/100 g) initially and contributed less than 10 percent of total vitamin C. During storage, DHA increased about 3.0–6.0 mg/100 g. The proportion of vitamin C present as DHA is 10–20 percent in lemons and oranges (Wills et al., 1984).

Citrus fruits are also a relatively good source of vitamin B complex and vitamin A. Vitamin A as such is not present in citrus fruits. α- and β-carotene in the juice vesicles are the precursors of vitamin A. Vitamin A is usually reported in international units (IUs), which are equal to 0.3 μg of vitamin A (Salunkhe et al., 1991). Considering that each molecule of β-carotene yields one molecule of vitamin A (about half of its molecular weight), the 0.6 μg β-carotene or 1.2 μg of α-carotene are equivalent to 1 IU. Tangerines and mandarins are the best source of carotenes among citrus fruits.

Citrus fruits are rich in polyphenolic compounds with some medicinal properties earlier called vitamin P. Orange and tangerine juices contain 0.02 and 0.03 percent hesperidin, respectively. The precipitate in the juice of both contained 11.8 percent hesperidin on a dry weight basis (Romanenko et al., 1973).

IX. INORGANIC CONSTITUENTS

The inorganic substances of citrus fruit are all found in the ash. These are mainly minerals (Table 6.10). Besides their importance in human nutrition, the minerals play vital roles in the biochemical reactions within the tree and the fruit. Mineral analysis of the leaves is used as a tool to determine the status of particular elements in tree nutrition. Many of these elements are associated with enzyme systems in the fruit and hence are extremely important in the metabolism of fruit. Phosphorus is a part of nucleic acids and phospho-proteins. Phosphorus is determined to be inorganic phosphorus, lipid phosphorus, and ethanol-insoluble phosphorus, and the ratio of these fractions can be utilized as an index for characterization of citrus juices and may have its application in trade (Vandercook and Guerrero, 1969). Potassium, calcium, and magnesium occur in fruit in combination with organic acids such as citric, malic, and oxalic. Calcium is also usually associated with the pectic substances and found as calcium pectate. High potassium is related to high total acidity in the fruit. Copper is known to have a destructive effect on ascorbic acid as it catalyzes oxidation.

Inorganic constituents of the fruit have been found to be influenced by fertilizer applications (Labanouskas et al., 1963). Considerable variations can be found in the mineral content of fruit from different positions of the same tree (Koo and Sites, 1956). Juice of citrus fruit contains about 0.4 percent ash. Ash content of orange juice was generally the highest in immature fruit and gradually decreased as the fruit maturity progressed (Harding et al., 1940). Citrus juices are a good source of potassium in human nutrition as potassium is the most abundant element in orange, grapefruit, and tangerine juices. Due to a favorable ratio of sodium to potassium, citrus juices are generally prescribed in diets to balance the body

TABLE 6.10 Mineral Content in Juice/pulp of Various Types of Citrus Fruits

Fruit	N	K	Ca	Mg	Fe	Cu	P	S	Cl	Na
Orange juice, (mg/100 ml)	151 – 181	179 – 320	5.4 – 15.6	7.3 – 15.3	0.2 – 0.4	0.05 – 0.06	17.8 – 21.7	4.0 – 4.6	2.1 – 5.6	0.2–2.0
Tangerine, (mg/100 g)	69– 80	155 – 178	18 – 41.5	7 – 11.2	0.2 – 0.27	0.7 – 0.09	14 – 16.7	8.2 – 10.3	2.4	1 – 2.2
Grapefruit juice, (mg/100 ml)	32 – 94	78 – 234	10 – 12.1	7.9 – 15	0.06 – 0.10	0.06 – 0.07	7 – 19	5.1 – 6	0.6 – 0.7	0.8 – 2.6
Lemon, (mg /100g)	35– 84	94 – 193	3.1 – 8.4	1 – 6.6	0.16 – 1.0	0.13	3.2 – 14.4	1 – 2.0	1 – 2.6	1 – 2.9
Lime, (mg/100 g)	48 – 112	104	4.5 – 10.4	–	0.19 – 0.92	–	9.3 – 11.2	–	–	1 – 1.1
Acid lime, (Kagzi) (mg/100 g)	133	140	44	6.6	1.83	0.16	16.8	–	–	–

Benk (1968); Watt and Merill (1963); Birdsall et al. (1961); Dawes (1970); Kefford and Chandler (1970); Nagy (1977); Selvaraj and Edward Raja (2000).

electrolytes in summer months. In Valencia and navel oranges from California, sodium, silicon, iron, manganese, and boron are found. Strontium and aluminum were found to be less than 1 percent in ash, whereas copper, lithium, titanium, lead, tin, nickel, cobalt, and silver were less than 0.01 percent (Birdsall et al., 1961). In orange juice, anions such as phosphates, sulfates, and chlorides are present. Bromine and iodine have also been reported (Rakienten et al., 1952; Stevens, 1954).

Apart from the nutritional value of mineral elements, there have been efforts to find out whether minerals play any significant role in overall fruit quality and their susceptibility to disorders. It is well known that excesses and deficiencies of elements also cause disorders. Mineral concentrations of Marsh grapefruit peels were observed to be K (2.79–3.52), Ca (0.55–0.96), Mg (0.18–0.28), Na (0.003–0.034), and Zn (0.001–0.002) mg/g fresh peel weight. Freezing temperatures resulted in higher peel K but lower Ca and Mg concentrations (Nagy et al., 1984). Mineral nutrition of the tree consequently gets reflected in fruit quality, and fruits grown at different locations on the tree also have different mineral composition and quality. Fruit weight, total juice per fruit, peel fresh weight and dry weight, and rind thickness of Kinnow mandarins, Red Blush grapefruits, Valencia oranges, and Lisbon lemons from internal canopies were significantly higher than those of external fruits. Fruits from internal canopies of all cultivars had generally higher peel concentrations of N, P, and K on a percent dry weight basis (Fallahi et al., 1989). Peel Mg and S concentrations from external fruits were higher in all cultivars when expressed as percentages of either dry weight or fresh weight. Nitrogen content of mandarin and orange juices and Ca content of grapefruit and lemon juices from external fruits were significantly higher than those of juices from internal fruits. In maturing fruit of Kinnow, Ca, Mg, K, and N decreased in the peel, while N and P increased in juice tissues. The Zn, Cu, Fe, and Mn content fluctuated in the peel and rag while decreasing in juice tissues. The juice of mature fruit had higher K content (Lallan Ram and Godara, 2006). The changes in these minerals are related to physical and biochemical changes as a decrease in Ca is related to softening, and N content increased in juice as amino acids and protein content increased (Tadeo et al., 1988). Takagi et al. (1989) studied the effects of sugar and nitrogen content in peels on color development in Satsuma mandarin fruits. N content in the peel declined and total sugars increased as the peel color began to develop. In an experiment conducted by these workers, when peel segments were cultured on an agar medium containing various concentrations of sucrose and N, a high concentration of sucrose observed to promote degreening, whereas N inhibited degreening. This suggested that color development in Satsuma mandarins is caused by low N concentrations in the peel.

X. CITRUS OILS AND VOLATILE FLAVORING COMPOUNDS

Citrus fruits emanate distinctive aromas as they release small quantities of volatiles into the atmosphere. The characteristic aroma of citrus fruits can be ascribed to high-boiling sparingly–water soluble oils. The amounts of volatiles emanated from fruit increase with increasing maturity and storage temperature. The release of volatiles also increases greatly if the peel is injured and oil sacs are ruptured. Volatiles of citrus fruit are associated with their characteristic flavor and aroma. Chemically, these include terpene-hydrocarbons (aliphatic monoterpenes, sesquiterpenes), alcohols, esters, carbonyl components (aldehydes and ketones), and volatile organic acids. These compounds are generally associated with the peel oil in the flavedo but are also found in oil sacs embedded in the juice vesicles. The volatiles of citrus comprise more than 150 compounds. In most cases these volatiles have to be isolated and concentrated by distillation and/or solvent extraction followed by drying. The essence is obtained and then analyzed/resolved into its components by gas chromatography, and components are identified by infrared, mass, and/or nuclear magnetic resonance spectrometry. Essence oils are a by-product obtained during the concentration of frozen orange juice, and they are important as flavoring agents (Moshonas and Shaw, 1979). The peel oils from limes, navel oranges, grapefruits, and lemons have been observed to be highly toxic to the insect species – *Tribolium confusum* and *Sitophilus granaries* affecting stored products (Abbassy et al., 1979). Lime oil is the most effective. Oils from mandarins, navel oranges, lemons, and limes were found to be synergistic in insecticidal activity with pyrethroids such as SH 1479. Volatile sulfur compounds – hydrogen sulfide, dimethyl sulfide, methanethiol, and dimethyl disulfide have also been identified (at ppm concentrations) in headspace gases above fresh citrus fruits. The quantity of hydrogen sulfide present was decreased more by frost damage than by normal maturation during a season. Dimethyl sulfide may be an important contributor to off-flavor in citrus juices (Shaw et al., 1981).

Monoterpene hydrocarbon (+)-limonene (d-limonene) accounts for 80–95 percent by weight of all citrus oils. Oxygenated terpenes, representing about 5 percent of the oil, provide species-specific aromas (Stanley, 1962). There is a significant variation in both the qualitative and quantitative composition of other volatiles from citrus species. Other hydrocarbons include the monoterpenes α-pinene, α-thujene, camphene, β-pinene, sabinene, myrcene, Δ-3-carene, α-phellandrene, α-terpinene, β-terpinene, p-cymene, terpinolene, p-isopropenyltoluene, and 2,4-p-menthadine. The sesquiterpenes are cubebene, copaene, elemene, caryophyllene, farnescene, α-humulene, valencene, and Δ-cardinene (Hunter and Brogden, 1965a). Valencene is one of the important sesquiterpenes (Hunter and Brogden, 1965b).

Quantitatively, esters account for only a small fraction of citrus oils but they impart a characteristic aroma. Important esters identified are: ethyl formate, ethyl acetate, ethyl butyrate, ethyl isovalerate, ethyl caproate, ethyl caprylate, linalyl acetate, octyl acetate, nonyl acetate, decyl acetate, terpinyl acetate, geranyl acetate, ethyl 3-hydroxyhexanote, citronellyl butyrate, geranyl butyrate, methyl anthranilate, and methyl N-methylanthranilate. Ethyl butyrate is found in substantial quantity in orange essence (Wolford et al., 1963; Ikeda and Spitler, 1964; Shaw, 1979).

The carbonyl compounds (aldehydes and ketones) make important contributions to citrus flavors (Stanley et al., 1961). Terpene aldehydes, neral and geranial, provide the characteristic flavor of lemons; and the sequiterpene ketone, nootkatone, is a major factor in the flavor of grapefruit juice (MacLeod and Buigues, 1964). The conversion of valencene to nootkatone is by oxidation of the sesquiterpene with tetrabutyl chromate to form the sesquiterpene ketone (Hunter and Brogden, 1965b). Important aldehydes and ketones identified in orange flavor are 2-hexenal, *n*-octanal, *n*-decanal, and geranial. Traces of acetaldehyde, acetone, *n*-butyraldehyde, *n*-hexanal, methyl ethyl ketone, *n*-heptanal, *n*-nonanal, furfural, methyl heptenone, citronellal, *n*-undecanal, *n*-dodecanal, neral, carvone, perillaldehyde, piperitenon, e and β-sinensal have also been reported.

Alcoholic compounds are also present among the volatile flavor materials of citrus. The predominant alcohol of orange oil and essence are linalool and octanol. Significant amounts of 4-terpinenol and α-terpineol have also been found. Methanol, ethanol, *n*-propanol, isobutanol, *n*-butanol, isopentanol, *n*-pentanol, *n*-hexanol, 3-hexenol, *n*-heptanol, methyl heptenol, 2-nonanol, *n*-nonanol, *n*-decanol, citronellol, nerol, geraniol, carveol, undecanol, and dodecanol are present in traces (Attaway et al., 1962; Hunter and Moshonas, 1965). Essentially the same alcohols, but in different proportions, are found in lemon, grapefruit, and tangerine oils (Hunter and Moshonas, 1966).

Volatile organic acids of orange juice essence are acetic, *n*-propionic, *n*-butyric, caproic, and capric (Attaway et al., 1964). Volatile acids present in trace amounts include isovaleric, valeric, isocaprioc, and caprylic.

A. Tangerine and Mandarin

Mandarin oil contains mainly terpene hydrocarbons. The distinctive notes in the aroma of Mediterranean mandarin oil are due to the presence of methyl *N*-methyl-anthranilate (0.85 percent) and thymol (0.08 percent). In the peel oil of Satsuma mandarins, sesquiterpenes, notably β-elemene and β-sesquiphellandrene, and the acetate esters of geraniol, nerol, citronellol, *p*-mentha-1,8-dien-7-ol, were found to give characteristic aroma of the fruit (Yamanishi et al., 1968; Kita et al., 1969). Methyl *N*-methylanthranilate and thymol were not found in Satsuma mandarins. In Kovano-Wase, Miagava Wase, and Owari, limonene predominated followed by γ-terpinene, linalool, and β-elemene (Kekelidze and Dzhanikashvili, 1985). Thymol, dimethylanthranilate, and two monoterpenes (γ-terpinene and β-pinene) were important to aroma in mandarin peel oil and flavor in juice (Shaw et al., 1981).

With a maturity of fruit, the d-limonene increased from 20 to 87 percent, whereas the linalool dropped from 62 to 4 percent in tangerine oil. Oxygenated terpenes also decreased in concentration during the season. Thymol, a constituent found only in tangerine oil, decreased from 3.4 to 0.2 percent (Attaway et al., 1967). The terpene hydrocarbon, which increases with maturity, is myrcene, which increased from 0.46 to 1.4 percent in tangerine oil.

Differences in peel oil of tangerines and mandarins have also been found. The tangerine oils contained more d-limonene than the mandarin oils and less γ-terpinene, α-pinene, and camphene (Ashoor and Bernhard, 1967).

Florida tangerine oils were distinguished by the presence of the sesquiterpenes α- and β-elemene (Hunter and Brogden, 1965b). Thymol was also identified; however, *N*-methylanthranilate was not present. The peel oil of the Murcott tangor resembled tangerine oil more than orange oil (Kesterson and Hendrickson, 1960).

B. Orange

When boxes of ripe oranges are opened after storage, a fruity aroma spreads all around. The components identified from volatiles that emanated from whole Valencia oranges were d-limonene, β-myrcene, α-pinene, acetaldehyde, octanal, ethanol, and ethyl acetate (Norman et al., 1967). In more than 100 constituents of orange oil, qualitative differences on a varietal basis are observed. Several esters have been recovered from Hamlin oranges, including ethyl acetate, butyrate, hexanoate, and octanoate (Attaway and Oberbacher, 1968). In the cuticular wax extract of undamaged Hamlin orange fruit with methylene chloride, a sesquiterpene, valencene, and other components including sabinene, β-elemene, β-carbophyllene, farnecene, humulene, and δ-cadinene have been found to contribute to the aroma of fresh oranges. Over-ripe Valencia oranges produce a considerable amount of valencene, a typical flavor-related compound (Moshonas and Shaw, 1979).

Pino et al. (1981) identified as many as 60 volatiles contributing to flavor of Valencia orange juice. Terpene hydrocarbons, alcohols, aldehydes, ketones, and esters contribute most to the overall flavor. New compounds such as 2-hexanal, beta-terpineol, ethyle anthranilate, and tetradecanol in Valencia juice have been reported. Acids and bases are also reported. Tamura et al. (1996) found that a mixture of 11 compounds, namely limonene, linalool, octanal, decanal, dodecanal, geranial, neral, myrcene, α- and β-sinensal, and citronellal duplicated the aroma of naval oranges very closely.

1. Orange Peel Oils

Orange peel oil is high in monoterpenes, and d-limonene is the major component it makes up about 90 percent of this oil (Table 6.11). Orange oils are distinguished by the presence of valencene as the principal sesquiterpene. It is observed that California Valencia peel oil is higher in hydrocarbons than Florida Valencia oil and is especially high in valencene. Hydrocarbons such as tetradecane and pentadecane were also found in California Valencia peel oil. Linalool is the principal compound of oxygenated fraction. Its concentration declines with maturity. The major aldehydes are the even-carbon saturated aliphatic members, which include octanal, decanal, and dodecanal. The concentrations of total aldehydes and

TABLE 6.11 Percentage of Important Constituents in Peel Oils of Some Citrus Fruits

Constituent	Orange	Mandarin Satsuma	Dancy	Grapefruit	Lemon	Lime (Cold-pressed)	Lime Distilled
Monoterpene (Total)	89–91 (% of oil)	98 (% of oil)	–(% of oil)	88 (% of oil)	81–85 (% of oil)	69 (% of oil)	77 (% of oil)
d-limonene	83–90	65–68	87–93	88–90	72–80(% of terpene)	64	60
Hydrocarbons							
α-pinene	0.5	0.8	1.00	1.6	2.00	1.2	0.8
β-pinene	1.00	–	0.4	–	7–13	1.2	0.8
Myrcene	2.00	2.00	1.2	1.9	2.00	–	0.8
γ-Terpinene	0.1	–	3.4	0.5	10.00	22.00	0.6
p-cymene	–	2.8	0.4	0.4	–	1.9	12.00
Aldehydes	1.8% of oil	–	–	1.2–1.8(% of oil)	–	–	–
Heptanal	3.00 (% of aldehyde)	–	–	4.00 (% of aldehyde)	1.00 (% of aldehyde)	–	–(% of aldehyde)
Octanal	39.00	–	4.00	16–35	4.00	–	0.3
Nonanal	5.00	–	–	7.00	6.00	–	–
Decanal	42.00	5.00	–	43–54	3.00	–	0.09
Citral	0.05–0.2 (% of oil)	–	–	0.06 (% of oil)	1.9–2.6 (% of oil)	3.1–5.3 (% of oil)	0.3 (% of oil)
Alcohols	0.9 (% of oil)	–	–	0.3–1.3 (% of oil)	–	–	–
Octanol	2.8		–		1		
Decanol	–	–	–				
Linalool	5.3 (% of oil)	2		0–3 (% of terpene less fraction)			
Esters	2.9 (% of oil)	–	–	3–4 (% of oil)	–	–	–

Yokoyama et al. (1961); Attaway et al. (1967, 1968); Kesterson and Hendrickson (1962, 1963, 1964); Ikeda and Spitler (1964); Hunter and Brogden (1965); Stanley (1962); Norman et al. (1967); Shaw 1979; Kefford and Chandler (1970).

octanal increases with fruit maturity but declines with overmaturity (Kesterson and Hendrickson, 1962). The saturated aliphatic aldehydes have a sweet pungent fatty aroma. Citral is a minor component of orange oil (Stanley, 1962). α- and β-unsturated aldehydes, α-sinenal, contributes significantly to orange aroma because it has a sweet pungent penetrating aroma. It is present in cold-pressed orange oil (Stanley, 1965). Ketones, piperitenone, and 6-methyl-5-hepten-2-one have also been identified in orange oil (Moshonas, 1967). Out of 41 oxygenated compounds identified, the major constituents are linalool, octanal, and decanal in Fukuhara oranges grown in Japan (Uchida et al., 1984). The aliphatic aldehydes (comprising 27.8 percent of the oxygenated fraction) are of major importance in the characteristic aroma of this cultivar.

Climatic factors influence the components and their concentration in the peel oils (Kesterson and Hendrickson, 1966). The aldehyde content rose with increased rainfall and fell in the drier years. However, Scora and Newman (1967) reported that the composition of the oil was little influenced by climate provided that samples were examined at equivalent maturities. The content of peel oil increases with fruit maturity. Kesterson and Hendrickson (1962) reported that yield of peel oil from Valencia oranges grown in Florida increased with maturity from 1.68 kg per metric ton in immature fruits (at 3 months of age) to 2.63 kg per metric ton at 8 months. The oil content achieved true orange character at 12 months and at the same time declined to about 2.27 kg per metric ton.

Changes have been found to occur between the terpene hydrocarbons, particularly (+)-limonene, and the terpene alcohols, particularly linalool during maturity of oranges (Attaway et al., 1967). In Hamlin oranges, the flavedo oil contained 52 percent d-limonene and 27 percent linalool in May, but by October the d-limonene level increased to 95 percent and the linalool decreased to 0.6 percent. Geranial decreased from 3.5 to 0.2 percent, geraniol from 0.5 to 0 percent, citronellal from 0.55 to 0 percent, terpinen-4-ol from 5.5 to 0.6 percent, and α-terpineol from 4.3 to 0.2 percent. The terpene hydrocarbon, myrcene, increased from 0.69 to 1.76 percent.

C. Grapefruits and Pummelo

The peel oil of grapefruit has high content of the monoterpene fraction, which is mainly (+)-limonene (Hunter and Brogden, 1964, 1965). The quantity of oil recovered also changes with maturity. The yield of oil from Marsh grapefruit in Florida declined with advancing maturity (Kesterson and Hendrickson, 1963). Only the mature fruit yielded oil of satisfactory organoleptic quality. Grapefruit oil develops optimal quality after holding the oil at 10–15°C for at least 6 months (Kesterson and Hendrickson, 1967).

Total carbonyls in oil, which were principally octanal, nonanal, and decanal, increased as grapefruit ripened. Oils from white grapefruit varieties in Florida had higher aldehyde contents than those from red varieties, and the relative properties of the constituent aldehydes were different. In the white varieties, the decanal content was 1.1–1.4 times the octanal content, whereas in the red varieties, this relation was reversed (Kesterson and Hendrickson, 1964). In Arizona grapefruit, however, decanal was the dominant aldehyde in both white and pink varieties (Stanley et al., 1961).

The alcohols in grapefruit oil decrease in concentration with fruit maturity, and the linalool content, in particular, falls to undetectable levels particularly in mature grapefruit of the white varieties. In red varieties, linalool was up to 3 percent of the terpeneless fraction (Kesterson and Hendrickson, 1964). The limonene concentration was 83 percent in May and increased to 93 percent as linalool decreased from 3 to 0.4 percent (Attaway et al., 1967).

Some of the linalool is converted into *cis*-and *trans*-linalool oxides (Hunter and Moshonas, 1966). During the 'curing' of grapefruit oils, linalool decreases considerably (Kesterson and Hendrickson, 1967). Difference in linalool content is the major difference in the composition of grapefruit oil and orange oil.

The sesquiterpene ketone, nootkatone, which gives the characteristic aroma to grapefruit, has the same skeleton as valencene, the major sesquiterpene of oranges (MacLeod, 1965). Valencene is readily converted to nootkatone by oxidation with t-butyl chromate (Hunter and Brogden, 1965b). Nootkatone is stable in air, in acid juice, and at high temperatures in the absence of air. The nootkatone content of grapefruit oil increases with advancing maturity. In California grapefruit oil, nootkatone levels as high as 1.8 percent have been recorded (MacLeod, 1966). In general, oils most preferred for grapefruit character have the highest aldehyde contents, around 1.8 percent decanal and 0.5–0.7 percent nootkatone (Kesterson et al., 1965). Small amounts of nootkatone are present in grapefruit juice that is free from peel oil. Only traces of nootkatone have been found in oranges, tangerines, lemons, limes, and bergamots (MacLeod and Buigues, 1964). Nootkatone is one of the characteristic components of the pummelo's aroma, similar to that of grapefruit (Sawamura and Kuriyama, 1988).

Nootkatone has an astringent taste and an odor variously described as pungent, aromatic, and musty. Its flavor threshold in water is about 1 ppm, and in grapefruit juice, it is 4–6 ppm (Berry et al., 1967). At 6–7 ppm, nootkatone improved the flavor of grapefruit juice, but at levels about 8 ppm, it imparted an unpleasant bitter taste.

Del Rio et al. (1992) suggested that nootkatone as an indicator of ripening of grapefruit. An increase in nootkatone concentration in grapefruit increased with the maturity index (TSS/acid ratio). It increased rapidly between 33 and 48 weeks from anthesis. As the nootkatone increased, the valencene level dropped. Marsh grapefruit had nootkatone levels up to 1356 μg/100 g fresh weight of peel.

D. Lemon

The principal components detected in lemon peel oil are γ-terpinene, terpinolene, d-limonene, and citral. In yellow lemon fruit, production of volatiles was greater in quantity and variety than from green fruit (Norman and Craft, 1968).

In comparison with orange, mandarin, and grapefruit oils, lemon oil has lower proportions of monoterpenes. Relatively higher proportions of β-pinene and γ-terpinene are observed in lemon oil (Table 6.11). The γ-terpinene in lemon oil contributes to flavor deterioration because it is degraded to *p*-cymene (Ikeda et al., 1961).

Tetradecane and pentadecane are prominently present as a hydrocarbon fraction of lemon oil. The major sesquiterpenes are α-bergamotene, β-bisabolene, and caryophyllene (Hunter and Brogden, 1965).

The composition of the oxygenated fraction in lemon oil differs from the oils of the orange, mandarin, and grapefruit. Citral is the major aldehyde, and the two isomers geranial and neral are present in the proportions of 75 and 25 percent, respectively. Normal saturated aliphatic aldehydes from C-6 to C-17 are present but only in small amounts (Ikeda et al., 1962a). Contrary to the position in orange and grapefruit oils, nonanal is present in a higher proportion than octanal or decanal.

In lemon oil, composition of alcohol and ester fractions is variable. Aliphatic alcohols are present only in small proportions, but large proportions of α-terpineol, 4-terpinenol, and nerol have also been reported (Hunter and Moshonas, 1966). It is likely that the tertiary terpene alcohols are formed to a greater or lesser extent by hydration of terpenes during contact with acidic juice and water during extraction (Ikeda and Spitler, 1964). The primary alcohols are present mainly as acetate esters.

Lemon oils of high quality tend to have low citral contents (around 2 percent) but contain a wide variety of components, especially the high-boiling members geraniol, geranyl acetate, neryl acetate, and bergamotene (Guenther, 1968). Oils of moderate quality have higher citral contents, and low-quality oils lack high-boiling components.

Climatic conditions affect composition of lemon volatiles. Ikeda et al. (1962b) found that oils from lemons grown on the California coast had a total monoterpene content of about 80 percent, whereas oils from desert-grown lemons had about 85 percent monoterpenes and only about half the β-pinene content of the coastal lemon oils. Coastal lemons had high citral content (Yokoyama et al., 1961).

In Moroccan lemons, linalool and linalyl acetate contents were low, whereas Corsican lemons had higher amounts of these compounds (Huet and Dupuis, 1968). A high content of oxygenated compounds in bergamot oil is favored by atmospheric humidities above 70 percent and low temperatures during the later part of fruit development.

Essential oil content was reported to be 6.5 μl/g fruit in Monachello and 2.8 μl/g in Meyer lemons. The essential oil of Meyer is characterized by a high content of thymol, which is almost absent in Monachello and Gruzinskii but lacked components that imparted a lemon aroma such as neral and geranial (isomers of citral). Gruzinskii had a higher content of neral and geranial than those of Monachello (Sardzhveladze et al., 1987).

Lemon oils undergo changes immediately after extraction, and in freshly extracted oils, there was some loss of aldehydes and esters and some oxidation and isomerization of terpenes. During storage, these changes get accentuated.

E. Lime

In composition, peel oils of limes resemble those of lemon oil with respect to the distribution and constituents of terpene, sesquiterpene, and oxygenated fractions (Hunter and Moshonas, 1966). Among the oxygenated constituents, α-terpineol, 1,4-cineole, and 1,8-cineole are common. Citral, present in the natural oil, is degraded during processing and largely disappears by transformation to *p*-cymene (Attaway et al., 1966). Srivas et al. (1963) reported that acid limes yielded 0.5–0.6 percent oil when fresh, but the oil was almost completely lost during storage at room temperature or at 8–10°C. D-limonene is the main component of *C. limon* and *C. latifolia* oils. Other compounds are α- and β-pinene, linalool, L-menthone, nerol, carvone, geraniol, citral, and decanone (Lancas and Cavicchioli, 1990).

Long-chain hydrocarbons have been detected in the juice sacs of acid limes (Nagy and Nordby, 1972). The dominant hydrocarbon in the saturated fractions for all limes is linear C25. C29 predominated in monoene fractions in both Key and Persian limes, but C25 predominated in Columbia sweet limes.

F. Other Citrus and Related Fruits

In sesquiterpene fractions of Natsudaidai oil, α-selinene, aromadendrene, calamenene, and α-calacorene are reported. The characteristic aroma of the Natsudaidai is due to perillaldehyde, carvone, and decyl acetate (Ohta and Hirose, 1966). Linalool, nonanol, and citronellal are the principal oxygenated constituents in trifoliate oranges (Scora et al., 1966). The distilled peel oil of the trifoliate orange is lower in d-limonene content than orange, lemon, and grapefruit oils, and the terpene fraction is made up of larger proportions of β-myrcene and γ-terpinene. In the essential oil of kumquat fruit, 120 compounds were found and 71 volatile compounds have been identified (Koyasako and Bernhard, 1983). Thirteen sesquiterpenes, eight terpenes, eleven alcohols, one ketone, eight aldehydes, and thirteen esters were identified. Limonene was the most abundant compound comprising 93 percent of the whole oil. Many aldehydes and esters have been found to contribute to the aroma of *Citrus iyo* fruit. Aldehydes and neryl, geranyl, and citronellyl-acetate have been found to increase during storage (Korayashi et al., 1981).

REFERENCES

Abbassy, M.A.A., Hosny, A.H., Lamaei, O., and Choukri, O. (1979). Insecticidal and synergistic citrus oils isolated from citrus peels. Mededelingen van de Faculteit-Landbouwwetenschappen, *Rijksuniversiteit-Gent.* 44, 21–29.

Adams, D.O., and Yang, S.F. (1977). Methionine metabolism in apple tissue. Implication of S-adenosylmethionine as an intermediate in the conversion of methionine to ethylene. *Pl. Physiol.* 60, 892–896.

Alberala, A.J., Casas, A., and Primo, E. (1967). Detection of adulteration in citrus juices. Identification of sugars in orange juices and commercial sucroses by gas liquid chromatography. *Rev. Agroquim. Technol. Aliment.* 7, 476–482.

Alvarez, B.M. (1967). Detection of adulteration of fruit juices by thin layer chromatography. *Analyst*, 92, 176–179.

Ammerman, C.B., Van Wallengham, P.A., Easley, J.F., Arrington, L.F., and Shirley, R.L. (1963). Dried citrus seeds – nutrient composition and nutritive value of protein. *Proc. Fla. Sta. Hort. Soc.* 76, 245–249.

Apelbaum, A., Burgon, A.C., Anderson, J.D., Lieberman, M., Ben-Arie, R., and Mattoo, A.K. (1981). Polyamines inhibit biosynthesis of ethylene in higher plant tissue and fruit protoplasts. *P. Physiol.* 68, 453–456.

Arigoni, D., Bartan, H.R., Corley, E.J., and Jeager, O. (1960). The constitution of limonin. *Experientia* 16, 41–49.

Ashoor, S.H.M., and Bernhard, R.A. (1967). Isolation and characterization of terpenes from *Citrus reticulata* Blanco and their comparative distribution among other citrus species. *J. Agric. Fd. Chem.* 15, 1044–1047.

Atkins, C.D., Wiederhold, E., and Moore, E.L. (1945). Vitamin C content of processing residue from Florida citrus waste. *Fruit Products J. Am. Fd Manuf.* 24, 260–262, 281.

Attaway, J.A., Wolford, R.W., and Edwards, G.J. (1962). Isolation and identification of volatile carbonyl components from orange essance. *J. Agric. Fd. Chem.* 10, 102–104.

Attaway, J.A., Wolford, R.W., Alberding, G.E., and Edwards, G.J. (1964). Identification of alcohols and volatile organic acids from natural orange essance. *J. Agric. Fd. Chem.* 12, 118–121.

Attaway, J.A., Pieringer, A.P., and Barabas, I.J., (1966). Origin of citrus flavour components. II. Identification of volatile components from citrus blossoms. *Phytochem.* 5, 1273–1279.

Attaway, J.A., Pieringer, A.P., and Barabas, I.J., (1967). Origin of citrus flavour components. III. A study of percentage variation in peel and leaf oil terpenes during one season. *Phytochem.* 6, 25–32.

Attaway, J.A., Pieringer, A.P., and Buslig, B.S. (1968). Origin of citrus flavour components. IV. The terpenes of Valencia leaf, peel and blossom oils. *Phytochem.* 7, 1695–1698.

Attaway, J.A., and Oberbacher, M.F. (1968). Studies on the aroma of intact Hamlin oranges. *J. Fd Sci.* 33, 287–289.

Bain, J.A. (1958). Morphological, anatomical and physiological changes in the developing fruit of Valencia orange. *C. sinensis*, L, Osbeck. *Aust. J. Bot.* 6, 1–24.

Baker, R.A., and Bruemmer, J.H. (1970). Cloud stability in the absence of various orange juice soluble components. *Citrus Industry.* 51, 6–11.

Balls, A.K. (1949). Enzyme problems in citrus industry. *Fd. Technol.* 3, 96–100.

Bar-Peled, M., Fluhr, R., and Gressel, J. (1993). Juvenile-specific localization and accumulation of a rhamnosyltransferase and its bitter flavonoid in foliage, flowers, and young citrus fruits. *Pl. Physiol.* 93, 1377–1384.

Benk, E. (1968). Natural sodium, potassium, calcium and chloride content of orange and lemon juices. *Riechst. Aromen Koerperpflegem*, 18, 126, 128, 132, 134.

Benk, E., and Bergamann, A. (1973). The red fleshed grapefruit and its juice. *Ind. ObstGeumuserverwert.* 58, 437.

Bennett, R.D., Herman, Z., and Hasegawa, S. (1988). I. Changensin: a new citrus limonoid. *Phytochem.* 27, 1543–1545.

Ben–Yehoshua, S., Shapiro, B., and Even-Chen, Z. (1981). Mode of action of individual fruit wrapping. *Proc. Int. Soc. Citric.*, Japan, pp. 718–721.

Berry, R.E., Wagner, C.J., and Moshonas, M.G. (1967). Flavour studies of nootkatone in grapefruit juice. *J. Fd. Sci.* 32, 75–78.

Bhalerao, S.D. and Mulmuley, G.V. (1992). Studies on bitterness in orange. *Indian Fd. Packer*, 46: 23–29.

Biggs, R.H. (1972). Effect of storage and processing on citrus ribonucleotide. *Proc. Florida Sta. Hort. Soc.* 85, 204–206.

Birdsall, J.J., Derse, P.H., and Teply, L.J. (1961). Nutrients in California lemons and oranges. II. Vitamins, minerals and proximate composition. *J. Am. Dietet. Assos.* 38, 555–559.

Bogin, E., and Wallace, A. (1966). CO_2 fixation in preparations from Tunisian sweet lemon and Eureka lemon fruits. *Proc. Am. Soc. Hort. Sci.* 88, 298–307.

Bowden, R.P. (1968). Processing quality of oranges grown in the near North Coast area of Queensland. *Queensland J. Agric. Animal Sci.* 25, 93–119.

Bower, J.P., Lesar, K., Farrant, J., and Sherwin, H. (1997). Parameters relating to citrus chilling sensitivity. *Citrus J.* 7, 22–24.

Braverman, J.S. (1933). The chemical composition of orange fruit. *Hadar* 6, 62–65.

Burdick, E.M. (1954). Grapefruit juice In *The chemistry and technology of fruit and vegetable juice production*' (D.K. Tressler, and M.A. Joslyn, eds.) pp. 381–410. AVI Publishing Co., New York.

Burdick, E.M. (1961). Citrus juice In *Fruit and vegetable juice processing Technology* (D.K. Tresseler, and M.A. Joslyn, eds.) pp. 874–902. AVI Publishing Co. Inc., Westport, Connecticut.

Burns, J.K. (1990). α- and β-Galactosidase activities in juice vesicles of stored Valencia oranges. *Phytochem.* 29, 2425–2429.

Burroughs, L.F. (1970). Amino acids. In *The bio-chemistry of fruits and their products.* (A.C. Hulme Ed.). Academic Press, N.Y. pp. 119–146.

Chaliha, B.P., Sastry, G.P., and Rao, V.R. (1964). Chemical studies on Assam citrus fruits. III. Examination of Assam round lemons (*Citrus jambhiri*). *J. Proc. Inst. Chem.*(India). 36, 208–210.

Chaliha, B.P., Berma, A.D., and Siddappa, G.S. (1963). Studies on utilization of Assam lemon. *Ind. Fd. Packer* 17(3), 5–7.

Chandler, B.V. (1958). Anthocyanins of blood oranges. *Nature*, 182, 933–940.

Chandler, B.V., and Kefford, J.F. (1951). The chemistry of bitterness in orange juices. I. An oxidation product of limonin. *Aust. J. Sci.* 13, 112–113.

Chandler, B.V., and Kefford, J.F. (1953). The chemistry of bitterness in orange juices. IV. Limonoic acid. *Aust. J. Sci.* 15, 28–29.

Clements, R.C., and Leland, H.V. (1962a). An ion exchange study of free amino acids in the juices of six var. of citrus. *J. Fd. Sci.* 27, 20–25.

Clements, R.C., and Leland, H.V. (1962b). Seasonal changes in free amino acids in Valencia juice. *Proc. Am. Soc. Hort. Sci.* 80, 300–307.

Clements, R.L. (1964a). Organic acids in citrus fruits. I. Varietal differences. *J. Fd. Sci.* 29, 278–280.

Clements, R.L. (1964b). Organic acids in citrus fruits. II. Seasonal changes in the orange. *J. Fd. Sci.* 29, 281–286.

Coussin, B.R., and Samish, Z. (1968). The free amino acids of Israel orange juice. *J. Fd. Sci.* 33, 196–199.

Curl, A.L. (1962). Reticulataxanthin and tangeraxanthin, two carotenoids from tangerine peel. *J. Fd. Sci.* 27, 537–543.

Curl, A.L., and Baily, G.F. (1956). Orange carotenoids. I. Comparison of Valencia orange peel and pulp. *J. Agric. Fd. Chem.* 4, 156–159.

Curl, A.L. (1965). The occurance of β-citraurin and β-apo-8-carotenal in the peels of California tangerines and oranges. *J. Fd. Sci.* 30, 13–18.

Curl, A.L., and Baily, G.F. (1957). The carotenoids of tangerines. *J. Agric. Fd. Chem.* 5, 605–608.

Curl, A.L., and Baily, G.F. (1961). The carotenoids of navel oranges. *J. Fd. Sci.* 26, 424.

Daito, H., and Sato, Y. (1985). Changes in the sugar and organic acid components of satsuma fruits during maturation. *J. Japanese Soc. Hort. Sci.* 2, 155–162.

Das, S., and Thakur, S. (1989). Constituent acids of *Limonia acidissima* leaf cutin. *Phytochem.* 28, 509–511.

Datta, S.C. (1963). Citrus. Bulletin of the Botanical Society of Bengal No. 17, 89.

Dawes, S.N. (1970). Composition of New Zealand fruit juices. II. Grapefruit, orange and tangelo juice. *New Zealand J. Sci.* 13, 452–459.

Del-Rio, J.A., Ortuno, A., Garcia-Puig, D., Porras, I., Garcia-Lidon, A., and Sabater, F. (1992). Sesquiterpene nootkatone as an indicator of ripening in *Citrus paradisi* Macf. *Proc. Int. Soc. Citric.* Vol. 1. pp. 420–422.

Dreyer, D.L. (1965). Citrus bitter principles III. Isolation of deacetyl nomilin and deoxylimonin. *J. Org. Chem.* 30, 749–751.

Drolet, G., Dumbroff, E.B., Legge, R.L., and Thompson, J.E. (1986). Free radical scavenging properties of polyamines. *Phytochem* 25, 367–371.

Echeverria, E. (1990). Developmental transition from enzymatic to acid hydrolysis of sucrose in acid lime. *Pl. Physiol.* 92, 168–171.

Echeverria, E. (1992). Activities of sucrose metabolizing enzymes during sucrose accumulation in developing acid lime fruit. *Pl. Sci.* 85, 125–129.

Echeverria, E., and Valich, J. (1988). Carbohydrate and enzyme distribution in protoplasts from Valencia orange juice sacs. *Phytochem.* 27, 73–76.

Echeverria, E., and Valich, J. (1989). Enzymes of sugar and acid metabolism in stored Valencia oranges. *J. Amer. Soc. Hort. Sci.* 114, 445–449.

Echeverria, E., Burns, J.K., and Wicker, L. (1989). Effect of cell wall hydrolysis on Brix in citrus fruit. *Proc. Florida State Hort. Soc.* 101, 150–154.

Eilati, S.K. (1970). The changes in the ripening orange fruit as affected by exogenous and endogenous factors. Ph.D. Thesis. The Hebrew Univ. of Jerusalem.

Emerson, O.H. (1949). The bitter principle of navel oranges. *Fd. Technol.* 3, 248–250.

Fallahi, E., Moon, Jr. J.W., and Mousavi, Z. (1989). Quality and elemental content of citrus fruit from exposed and internal canopy positions. *J. Pl. Nutrition.* 12, 939–951.

Frankel, C., Klein, I., and Dilley, D.R. (1968). Riboneucleic acid synthesis and proteins. *Pl. Physiol.* 43, 1146–1153.

Galston, A.W., Kaur-Sawhney, R. (1990). Polyamines in plant physiology. *Pl. Physiol.* 94, 406–410.

Gentili, B., and Horowitz, R.M. (1964). Flavonoids of citrus-VII. Limocitrol and isolimocitrol. *Tetrahedron* 20, 2313–2315.

Goldschmidt, E.E. (1977). Galacto-lipids and phospholipids of orange peel and juice chromoplasts. *Phytochem.* 16, 1046–1047.

Goldsworthy, L.J., and Robinson, R. (1957). A correction respecting the structure of tangeritin. *Chem. Ind.* 47.

Gross, J., Gabai, M., and Lifshitz, A. (1971). Carotenoids in juice of Shamouti orange. *J. Fd. Sci.* 36, 466–473.

Guenther, H. (1968). Gas chromatographic and infra-red spectroscopic studies of lemon oils. *Deut. Lebensm Rundscheu* 64, 104–111.

Harbourne, J.B. (1967). *Comparative biochemistry of the flavonoids*. Academic Press, London and New York.

Harding, P.L., Winston, J.R., and Fisher, D.F. (1940). Seasonal changes in Florida oranges Technical Bulletin 753USDA.

Hasegawa, S., and Maier, V.P. (1981). Some aspects of citrus biochemistry and juice quality. *Proc. Int. Soc. Citric.* Vol. 2. 1983, pp. 914–918.

Hasegawa, S., and Miyake, M. (1996). Biochemistry and biological functions of citrus limonoids. *Fd. Rev. Int.* (USA). 12: 413–435.

Hasegawa, S., Maier, V.P., Herman, Z., and Ou, P. (1986). Phytohormone bioregulation of nomilin biosynthesis in *Citrus limon* seedlings. *Phytochem.* 25, 1323–1325.

Hendrickson, R., and Kesterson, J.W. (1963a). Seed oils from *Citrus sinensis* oranges. *J. Am. Oil Chemists Soc.* 40, 746–747.

Hendrickson, R., and Kesterson, J.W. (1963b). Florida lemon seed oil. *Proc. Fla. State Hort. Soc.* 76, 249–253.

Hendrickson, R., and Kesterson, J.W. (1964a). Seed oils from Florida mandarins and related varieties. *Proc. Fla. State Hort. Soc.* 77, 347–351.

Hendrickson, R., and Kesterson, J.W. (1964b). Hespiridin in Florida oranges. Technical Bulletin 684. Univ. of Florida, Agric. Expt. Stn., Gainesville, Florida.

Hermann, K. (1989). Occurrence and contents of phenolic acids in fruit. Vorkommen und Gehalte der Phenolcarbonsauren. *Obst. Erwerbsobstbau.* 31, 185–189.

Herman, Z., and Hasegawa, S. (1985). Limonin biosynthesis from obacunone via obacunoate in *Citrus limon. Phytochem.* 24, 2911–2913.

Holland, N., Sala, J.M., Menezes, H.C., and Lafuente, M.T. (1999). Carbohydrate content and metabolism as related to maturity and chilling sensitivity of cv. Fortune mandarins. *J. Agric. Fd. Chem.* 47, 2513–2518.

Horowitz, R.M. (1961). The citrus flavonoids. In *The orange, its biochemistry and physiology.* (W.B. Sinclair Ed.). Univ. Calif., Berkeley, California.

Horowitz, R.M., and Gentili, B. (1960a). Flavonoid compounds of citrus. III. Isolation and structure of eriodictyole glycoside. *J. Am. Chem. Soc.* 82, 2803–2806.

Horowitz, R.M., and Gentili, B. (1960b). Flavonoids of citrus. IV. Isolation of some aglycones from the lemon. (*C. limon*). *J. Org. Chem.* 25, 2183–2187.

Horowitz, R.M., and Gentili, B. (1961). Phenolic glycosides of grapefruit. A relation between bitterness and structure. *Arch. Biochem. Biophysic.* 92, 191–192.

Horowitz, R.M., and Gentili, B. (1963, 1969). US Patents. Dihydrochalcone derivatives and their uses as sweeteners, Patent No. 3087826, 3429873.

Horowitz, R.M., and Gentili, B. (1963). Flavonoids of citrus. VI. The structure of neohespiredose. *Tetrahedron* 19, 773–782.

Hou, W.N., Jeong, Y., Walker, B.L., Wei, C.I., and Marshall, M.R. (1997). Isolation and characterization of pectinesterase from Valencia orange. *J. Fd. Biochem.* 21, 309–333.

Houjin, W.U., Calvarano, M., and Giacomo, A.D. (1990). Some flavonones in the peel of ten *Citrus* spp. and varieties in China. *Proc. Int. Citrus Symp.*, China, pp. 844–849.

Huet, R., and Dupois, C. (1968). Essential oil of bergamot in Africa and in Corsica. *Fruits* (Paris) 23, 301–311.

Hunter, G.L.K., and Brogden, W.B. (1964). A rapid method of isolation and identification of sesqueterpene hydrocarbons in cold-pressed grapefruit oil. *Anal. Chem.* 36, 1122–1123.

Hunter, G.L.K., and Brogden, W.B. (1965a). Analysis of terpene and sesqueterpene hydrocarbon in some citrus oils. *J. Fd. Sci.* 30, 383–387.

Hunter, G.L.K., and Brogden, W.B. (1965b). Conversion of valencene to nootkatone. *J. Fd. Sci.* 30, 876–878.

Hunter, G.L.K., and Moshonas, M.G. (1965). Isolation and identification of alcohols in cold-pressed Valencia orange oil by liquid–liquid extraction and gas-chromatography. *Anal. Chem.* 37, 378–380.

Hunter, G.L.K., and Moshonas, M.G. (1966). Analysis of alcohols in essential oils of grapefruit, lemon, lime and tangerine. *J. Fd. Sci.* 31, 167–171.

Hurst, W.J., Martin, R.A., and Zoumas, B.L. (1979). Application of HPLC for characterization of individual carbohydrates in foods. *J. Fd. Sci.* 44: 892–895.

Ikeda, R.M., and Spitler, E.M. (1964). Isolation, identification, and gas-chromatographic estimation of some esters and alcohols of lemon oil. *J. Agric. Fd. Chem.* 12, 114–117.

Ikeda, R.M., Rolle, I.A., Vannier, S.H., and Stanley, W.L. (1962a). Isolation and identification of aldehydes in cold pressed lemon oil. *J. Agric. Fd. Chem.* 10, 99–102.

Ikeda, R.M., Stanley, W.L., Rolle, I.A., and Vannier, S.H. (1962b). Monoterpene hydrocarbon composition of citrus oils. *J. F. Sci.* 27, 593–596.

Ikeda, R.M., Stanley, W.L.Vannier, S.H., and Rolle, I.A. (1961). Deterioration of lemon oil, formation of p-cymene from gamma-terpinene. *Fd. Technol.* 15, 379–380.

Ismail, M.A. (1966). Polymeric nucleic acids from citrus fruits and leaves in various stages of growth and development, including chemically modified senescence. Ph.D. Dissertation, Univ. of Florida.

Ismail, M.A., and Brown, G.E. (1975). Phenolic content during healing of Valencia orange peel under high humidity. *J. Am. Soc. Hort. Sci.* 100, 249–251.

Izumi, Y., Araki, C., and Goto, A. (1981). Relationships between bitterness of juice and fruit maturities of Wase satsuma mandarin (*Citrus unshiu* Marc. var. Praecox Tanaka). *Proc. Int. Society of Citric.*, Vol. 2, 1983, 877–879.

Izumi, H., Ito, T., and Yoshida, Y. (1988). Relationship between ascorbic acid and sugar content in citrus fruit peel during growth and development. *J. Japanese Soc. Hort. Sci.* 57, 304–311.

Joseph, G.H., Stevens, J.W., and MacRill, J.R. (1961). Nutrients and California lemons and oranges. I. Source and treatment of samples. *J. Am. Dietet. Assoc.* 38, 552–554.

Kadota, R., and Miura, M. (1983). Studies on lipids of citrus fruits. III. Changes in lipids of satsuma fruits (*Citrus unshiu* Marc.) and their fatty acid composition during maturation and storage. Bull. Faculty of Agriculture, Miyazaki Univ. 30(2), 95–104.

Kadota, R., Tokunaga, Y., and Miura, M. (1982). Studies on lipids of citrus fruits. I. Changes in the lipid content and fatty acid composition of juice from citrus fruits during maturation. Bull. Faculty Agric. Miyazaki University. 29(1), 213–222.

Kakiuchi, N., and Ito, S. (1971). Fundamental studies on citrus juices. II. Seasonal changes of organic acids and sugars of Natsudaidai and Fukuhara orange. Enge Shikenjo Hokok, Series B, 11, 101–117.

Kefford, J.F. (1959). The chemical constituents of citrus. In *Advances in food research*, Vol. 9, E.M. Mrak, and G.F. Stewart, (eds). Academic Press, New York.

Kefford, J.F., and Chandler, B.V. (1970). The chemical constituents of citrus fruits. In *Advances in food research*, Supplement 2, Chichester, C.O. (E.M. Mrak, and G.F. Stewart, eds.), Academic Press, New York.

Kekelidze, N.A., and Dzhanikashvili, M.I. (1985). Essential oils of the early-ripening cultivars of *Citrus unshiu*. Khimiya Prirodnykh Soedinenii. No.4, 572–573.

Kesterson, J.W., and Hendrickson, R. (1957). Naringin: a bitter principle of grapefruit. Univ. of Florida Expt Sta. Bull. 511A.

Kesterson, J.W., and Braddock, R.J. (1976). By-products and specialty products of Florida citrus. Univ. Fla, Agr. Expt. Stn. Bulletin 784. Gainesville, Florida.

Kesterson. J.W., and Hendrickson, R. (1960). Florida cold-pressed Murcott oil. *Am. Perfumer. Aromat.* 75(11), 35–37.

Kesterson, J.W., and Hendrickson, R. (1962). Composition of Valencia orange oil as related to fruit maturity. *Am. Perfumer Cosmet.* 77(12), 21–24.

Kesterson, J.W., and Hendrickson, R. (1963). Evaluation of cold-pressed Marsh grapefruit oil. *Am. Perfumer Cosmet.* 78, 32–35.

Kesterson, J.W., and Hendrickson, R. (1964). Comparison of red and white grapefruit oils. *Am. Perfumer cosmet.* 79, 34–36.

Kesterson, J.W., and Hendrickson, R. (1966). Aldehyde content of Valencia orange oil as related to total rainfall. *Am. Perfumer Cosmet.* 81, 39–40.

Kesterson. J.W., and Hendrickson, R. (1967). Curing Florida grapefruit oils. *Am. Perfumer Cosmet.* 82, 37– 40.

Kesterson, J.W., Hendrickson, R., Seller, R.R., Huffman, C.E., Brent, J.A., and Griffith, J.T. (1965). Flavor of expressed Duncan grapefruit oil as related to fruit maturity. *Proc. Fla. State Hort. Soc.* 78, 207–210.

Khan, M.U.D., and McKinney, G. (1953). Carotenoids of grapefruit. *Pl. Physiol.* 28, 250–252.

Kita, Y., Nakatani, Y., Kobayashi, A., and Yamanishi, T. (1969). Composition of peel oil from *Citrus Unshiu. Agric. Biol. Chem.* (Tokyo) 33, 1569–1565.

Kollattukudy, P.E. (1968). Biosynthesis of surface lipids. *Science* 159, 498–505.

Kolattukudy, P.E. (1984). Biochemistry and function of cutin and suberin. *Canadian J. Bot.* 62, 2918–2933.

Koo, R.C.J., and Sites, J.W. (1956). Mineral composition of citrus leaves and fruit as associated with position on tree. *Proc. Am. Soc. Hort. Sci.* 68, 245–252.

Korayashi, A., Matsumoto, M., Uchida, K., and Yamanishi, T. (1981). Aroma development in *Citrus iyo* by pre-treatment and storage. *Proc. Int. Soc. Citric.* Vol. 2, Tokyo, Japan, pp. 909–910.

Koyasako, A., and Bernhard, R.A. (1983). Volatile constituents of the essential oil of kumquat. *J. Fd. Sci.* 48, 1807–1812.

Kramer, G.F., Wang, C.Y., and Conwey, W.S. (1989). Correlation of reduced softening and increased polyamines levels during low oxygen storage of McIntosh apples. *J. Am. Soc. Hort. Sci.* 114, 942–946.

Krehl, W.A., and Cowgill, G.R. (1950). Vitamin contents of citrus products. *J. Fd. Sci.* 15, 179–191.

Kubota, S., Fukui, H., and Anao, S. (1972). Free amino acids, organic acids, and sugars in citrus juices. Shikoku Agric. Expt Sta. Bulletin 24, 97–107.

Kunjukutty, N., Sankunny, T.R., and Menarchery, M. (1966). Chemical composition and feeding values of lime (*C. aurantifolia*) and lemon (*C.limon*). *Ind. Vet. J.* 43, 453–455.

Kuraoka, T., Iwasaki, K., Hino, A., Kaneko, Y., and Tsuji, H. (1976). Studies on the peel puffing of satsumas. IV. Changes in sugar content during the development of the fruit rind. *J. Japanese Soc. Hort. Sci.* 44, 375–380.

Kutateladze, D.S. (1973). The activities of polyphenol oxidase and ascorbic acid oxidase enzymes in citrus. *Subtropiecheskie Kaltury*. No. 2. pp. 67–72.

Labanauskus, C.K., Jones, W.W., and Embleton, T.W. (1963). Effects of foliar applications of manganese, zinc, and urea on yield and fruit quality of Valencia oranges, and nutrient concentration in the leaves, peel and juice. *Proc. Am. Soc. Hort. Sci.* 82, 142–153.

Ladaniya, M.S., and Mahalle, B. (2006). Changes in 'Mosambi' orange (*Citrus sinensis* Osbeck) fruit during maturation under sub-humid tropical climate. *Tropical Agriculture*.

Lallan Ram and Godara, R.K. (2003). Activities of hydrolytic enzymes in peel and juice tissues of Kinnow mandarin during fruit ripening. *Ind. J. Hort.* 62, 227–230.

Lallan Ram and Godara, R.K. (2006). Mineral distribution pattern in peel, rag and juice of Kinnow mandarin at different stages of fruit ripening. *Ind. J. Hort.* 65, 8–11.

Lancas, F.M., and Cavicchioli, M.J. (1990). Analysis of the essential oils of Brazilian citrus fruits by capillary gas chromatography. *High Resolution Chromat.* 13, 207–209.

Larrauri, J.A., Ruperez, P., Borroto, B., and Saura-Calixto, F. (1997). Seasonal changes in the composition and properties of a high dietary fibre powder from grapefruit peel. *J. Sci. Fd. Agric.* 74, 308–312.

Lavon, R., Soloman, R., and Goldschmidt, E. (1988). Pyruvate kinase a potential indicator of Ca level in citrus leaves and fruit. *Proc. 6th Int. Citrus Congr. Int Soc Citric.* Vol. 1, Israel, pp. 541–545.

Leshem, Y.Y., Halevy, A.H., and Frenkel, C. (1986). Processes and control of plant senescence. Elsevier, Amsterdam.

Lewis, I.N., Coggins, C.W., and Garber, M.J. (1964). Chlorophyll concentration in the navel orange rind as related to potassium giberellate, light intensity and time. *Proc. Am. Soc. Hort Sci.* 84, 177–180.

Lewis, I.N., Cogins, Jr. C.W., Labanauskas, C.K., and Dugger, Jr. W.M. (1967). Biochemical changes associated with natural and gibberellin A_3 delayed senescence in the navel orange rind. *Pl. Cell Physiol.* 8, 151–160.

Lin, S.D., and Chen, A.O. (1995). Major carotenoids in juices of Ponkan mandarin and Liucheng orange. *J. Fd. Biochem.* 18, 273–293.

Lin, W.Z. (1988). An experiment on the peroxidase activity and isoenzyme patterns of major citrus fruits growing in Guangdong. *J. South China Agric. Univ.* 9, 11–16.

Liu, Y.Z., and Li, D.G. (2003). Sugar accumulation and changes of sucrose-metabolizing enzyme activities in citrus fruit. *Acta Hort. Sinica*. 30(4), 457–459.

Maccarone, E., Maccarone, A., and Rapisarda, P. (1985). Acylated anthocyanins from oranges. *Anali di Chimica*, Rome, 75, 79–86.

Maccarone E., Maccarone A., Perrini, G., and Rapisarda, P. (1983). Anthocyanins of Moro orange juice. *Anali di Chimica*, Rome, 73, 533–539.

MacLeod, W.D. (1965). Constitution of Nootkatone, Nootkatene and valencene. *Tetrahedron Letters*. pp. 4779–4783.

MacLeod, W.D. (1966). Nootkatone, grapefruit flavor and the citrus industry. *Calif. Citrog.* 51, 120–123.

MacLeod, W.D., and Buigues, N.M. (1964). Sesquiterpenes. I. Nootkatone, a new grapefruit flavor constituent. *J. Fd. Sci.* 29, 565–568.

Maier, V.P., and Margileth, D.A. (1969). Limonoic acid A-ring lactone: a new limonin derivative in citrus, *Phytochem.* 8, 243.

Maier, V.P., and Metzler, D.M. (1967a). The phenolic constituents of *C. paradisi* and their biosynthetic significance. *Seventh Ann. Meet. Phytochem Soc. North Amer. Madison*, Wisconsin, August 23–26.

Maier, V.P., and Metzler D.M. (1967b). Grapefruit phenolics. I. Identification of dihydrokaemphenol and its co-occurrence with naringenin and karmpherol. *Phytochem.* 6, 763–765.

Maier, V.P., and Metzler D.M. (1967c). Grapefruit phenolics. II. Principal aglycones of endocarp and peel and their phenolic biosynthetic relationship. *Phytochem.* 6, 1127–1135.

Maier, W.P., and Beverly, W. (1968). Limonin monolactone, a non bitter precursor responsible for delayed bitterness in certain citrus juices. *J. Fd. Sci.* 33, 488.

Maier, W.P., and Dreyer, D.L. (1965). Citrus bitter principle. IV. Occurrence of limonin in grapefruit juice. *J. Fd. Sci.* 30, 874–875.

Mapson, L.W. (1967). *The vitamins.* Vol. I, pp. 385. Academic Press, London and New York.

Martin, B., Tadeo, J.L., and Ortiz, J.M. (1984). Amino acid changes in the juice and peel of clementines (*Citrus clementina* Hort. ex. Tan.) during ripening. Anales Inst. Nacional Invest. *Agr. Agric.* 25, 107–114.

Mazur, Y., Weizmann, A., and Sondheimer, F. (1958). Steroids and triterpenoids of citrus fruits. III. The structure of citrostadienol, a natural 4α-methylsterol. *J. Am. Chem. Soc.* 80, 6293–6296.

McDonnell, L.R., Jang, R., Jansen, E.F., and Lineweaver, H. (1950). The properties of orange pectinesterase. *Arch. Biochem.* 28, 260–275.

Mehta, U., and Bajaj, S. (1983). Physico-chemical characteristics of Kinnow, Blood Red, and Villa Franca. *Punjab Hort J.* 23, 40–44.

Mitcham, E.J., and McDonald, R.E. (1993). Changes in grapefruit flavedo cell wall noncellulosic neutral sugar composition. *Phytochem.* 34, 1235–1239.

Mizelle, J.W., Dunlap, W.J., and Wender, S.H. (1967). Isolation and identification of two isomeric naringenin rhamnoglucosides from grapefruit. *Phytochem.* 6,1305–1307.

Money, R.W., and Chriestian, W.A. (1950). Analytical data of some common fruits. *J. Sci. Fd. Agric.* 1, 8–12.

Monselis, S.P., Cohen, A., and Kessler, B. (1962). Changes in RNA and DNA in developing orange leaves. *Pl. Physiol.* 37, 572–578.

Mookerjee, K.K., Tandon, G.L., and Siddappa, G.S. (1964). Varietal suitability of lemon for making lemon squash. *Indian Fd. Packer* 18(1), 1–3.

Moreno-Alvarez, M.J., Gomez, C., Mendoza, J., and Belen, D. (2000). Total carotenoids of orange peel (*Citrus sinensis* L. var. Valencia). *Revista Unellez de Ciencia-y-Tecnologia, Produccion-Agricola.* 171, 92–99.

Moshonas, M.G. (1967). Isolation of piperetenone and 6-methyl-5-hepten-2-one from orange oil. *J. Fd. Sci.* 32, 206–207.

Moshonas, M.G., and Shaw P.E. (1979). Composition of essence oil from overripe oranges. *Agric. Fd. Chem.* 27, 1337–1339.

Muramatsu, N., Takahara, T., Ogata, T., and Kojima, K. (1999). Changes in rind firmness and cell wall polysaccharides during citrus fruit development and maturation. *Hort Science.* 34, 79–81.

Nagy, S. (1977). Inorganic elements. In *Citrus science and technology* (S. Nagy, P.L. Shaw, M.K. Veldhuis, eds.), Vol. 1, pp. 479–495. AVI Publishing Co. Westport, Connecticut, U.S.A.

Nagy, S., and Nordby, H.E. (1972). Saturated and monounsaturated long-chain hydrocarbons of lime juice sacs. *Phytochem.* 11, 2865–2869.

Nagy, S., Dezman, D., and Rouseff, R. (1984). Mineral composition of Marsh grapefruit peel during maturation. *Hort Science.* 19, 654–655.

Naohara, J., and Manabe, M. (1993). The properties of pectate depolymerizing enzyme isolated from Satsuma mandarin fruits (*Citrus unshiu* Marc.). *J. Japanese Soc. Hort. Sci.* 40, 485–489.

Neish, W. (1964). In *Biochemistry of phenolic compounds* (J.B. Harbourne ed.), pp. 295–359, Academic Press, London.

Nicolosi-Asmundo, C., Cataldi-Lupo, C.M., Campisi, S., and Russo, C. (1987). Lipids in citrus fruit juice. 1. Lipid content during orange ripening in eastern Sicily. *J. Agric. Fd. Chem.* 35, 1023–1027.

Noomhorm, A., and Kasemsuksakul, N. (1992). Effect of maturity and processing on bitter compounds in Thai tangerine juice. *Int. J. Fd. Sci. Technol.* 27, 1, 65–72.

Nordby, H.E., and McDonald, R.E. (1990). Squalene in grapefruit wax as a possible natural protectant against chilling injury. *Lipids*. 25, 807–810.

Nordby, H.E., and McDonald, R.E. (1991). Relationship of epicuticular wax composition of grapefruit to chilling injury. *J. Agric. Fd. Chem.* 39, 957–962.

Nordby, H.E., and McDonald, R.E. (1995). Variations in chilling injury and epicuticular wax composition of white grapefruit with canopy position and fruit development during the season. *J. Agric. Fd. Chem.* 43, 1828–1833.

Norman, S., and Craft, C.C. (1968). Effect of ethylene on production of volatiles by lemons. *Hort Science*. 3, 66–68.

Norman, S., Craft, C.C., and Davis, P.I. (1967). Volatiles from injured and uninjured Valencia oranges at different temperatures. *J. Fd. Sci.* 32, 656–659.

Ohta, Y., and Hirose, Y. (1966). Constituents of cold pressed peel oil of *Citrus natsudaidai. Hayata Agric. Biol. Chem.* (Tokyo) 30, 1196–1201.

Omidbaigi, R., Nasiri, M.F., Sadr, Z.B., and Kock, O. (2002). Hesperidin in *Citrus* species, quantitative distribution during fruit maturation and optimal harvesting time. *Acta-Horticulturae*. 576, 91–97.

Orme, E.O., and Hasegawa, S. (1987). Inhibition of nomilin accumulation in Citrus limon fruit by naphthaleneacetic acid. *J. Agric. Fd. Chem.* 35, 512–513.

Pino, J., Tapanis, R., Rosado, A., and Baluja, R. (1981). Quality control of sweet orange juice by gas liquid chromatography of volatile constituents. *Proc. Int. Soc. Citric.* 12, 936–939.

Pressey, R. (1993). Uronic acid oxidase in orange fruit and other plant tissues. *Phytochem.* 32, 1375–1379.

Purvis, A.C. (1989). Soluble sugars and respiration of flavedo tissue of grapefruit stored at low temperatures. *Hort Science* 24, 320–322.

Rakienten, M.L., Newman, B., Falk, K.B., and Miller, I. (1952). Comparison of some constituents in fresh frozen and freshly squeezed orange juice. *J. Am. Diet. Assoc.* 28, 1050–1053.

Rapisarda, P., and Giuffrida, A. (1992). Anthocyanin level in Italian Blood oranges. *Proc. Int. Soc. Citric.* Vol. 3, 1120–1133.

Rasmussen, G.K. (1964). Seasonal changes in organic acid of Valencia orange fruit in Florida. *Proc. Am. Soc. Hort. Sci.* 84, 181–186.

Rasmussen, G.K. (1974). Cellulase activity in separation zones of citrus fruit treated with abscisic acid under normal and hypobaric atmospheres. *J. Am. Soc. Hort. Sci.* 99, 229–231.

Rasmussen, G.K. (1975). Cellulase activity, endogenous abscisic acid and ethylene in four citrus cultivars during maturation. *Pl. Physiol.* 56, 765–770.

Rekieten, M.L., Newman, B., Falk, K.B., Miller, J. (1951). Comparison of some constituents in fresh, frozen and freshly squeezed orange juice. *J. Am. Diet. Assoc.* 27, 864–868.

Richmond, A., and Biale, J.B. (1966). Protein synthesis during the climacteric rise in respiration. *Pl. Physiol.* 41, 1247–1253.

Riov, J., Monselise, S.P., and Kahan, R.S. (1969). Ethylene induced phenyl alanine ammonia lyase activity in grapefruit. *Pl. Physiol.* 44, 631.

Riov, K. (1975). Polygalacturonase activity in citrus fruit. *J. Fd. Sci.* 40, 201–202.

Rockland, L.B. (1958). Free amino acids in citrus and other fruit and vegetable juices. *Food Res.* 24, 160–164.

Rockland, L.B. (1961). Nitrogenous constituents. In *The orange: its biochemistry and physiology.* (W.B. Sinclair ed.) Univ. Calif. Press.

Rockland, L.B., and Underwood, J.C. (1956). Rapid procedure for estimation of amino acid by direct photometry on filter paper chromatograms. Estimation of seven free amino acids in orange juice. *Anal. Chem.* 28, 1679–1684.

Rockland, L.B., Beavens, E.A., and Underwood, J.E. (1957) Debittering of citrus fruits. U.S. patent 2816835. December 17.

Rodrigo, M.I., Mallent, D., and Casas, A. (1985). Relationship between the acid and limonin content of Washington navel orange juices. *J. Sci. Fd. Agric.* 36, 1125–1129.

Romanenko, E.V., Tylibtseva, N.N., and Katsitadze, M.G. (1973). Compounds possessing vitamin 'P' properties in citrus fruit. *Konserv.Ovoshchesushil'naya Promyshlennosti'*. 12, 21–22.

Ronneberg, T.A., Hasegawa, S., Suhayda, C., and Ozaki, Y. (1995). Limonoid glucoside β-glucosidase activity in lemon seeds. *Phytochem.* 39, 1305–1307.

Rouse, A.H. (1953). Distribution of pectinesterase and total pectin in component parts of citrus fruits. *Fd. Technol.* 7, 360–362.

Rouse, A.H., Atkins, C.D., and Moore, E.L. (1964). Evaluation of pectin in component parts of Pineapple orange during maturation. *Proc. Fla. Sta. Hort. Soc.* 77, 274–278.

Rouse, A.H., Atkins, C.D., and Moore, E.L. (1965). Seasonal changes occurring in pectin esterase activity and pectic constituents of component parts of citrus fruits. *Fd. Technol.* 19, 673–676.

Rouseff, R.L. (1982). Nomilin, a new bitter component in grapefruit juice. *J. Agric. Fd. Chem.* 30, 504–507.

Sala, J.M. (1998). Involvement of oxidative stress in chilling injury in cold-stored mandarin fruits. *Postharvest Biol. Technol.* 13, 255–261.

Sala, J.M., and Lafuente, M.T. (2000). Catalase enzyme activity is related to tolerance of mandarin fruits to chilling. *Postharvest Biol. Technol.* 20, 81–89.

Sala, J.M., Lafuente, T., and Cunat, P. (1992). Content and chemical composition of epicuticular wax of Navelina oranges and Satsuma mandarins as related to rind staining of fruit. *J. Sci. Fd Agric.* 59, 489–495.

Salunkhe, D.K., Bolin, H.R., and Reddy, N.R. (1991). *Storage, processing and nutritional quality of fruits and vegetables.* 2nd edition, Vol. 1, CRC Press, Boca Raton, Florida, pp. 115–145.

Samish, Z., and Ganz, D. (1950). Bitterness in Shamouti oranges. *Canner.* 111-0. No. 23:7–9. No. 24:36–37, No. 25:22–24.

Sanchez-Ballesta, M.T., Lafuente, M.T., Zacarias, L., and Granell, A. (2000). Involvement of phenylalanine ammonia-lyase in the response of Fortune mandarin fruits to cold temperature. *Physiol. Plantarum.* 108, 382–389.

Sandhu, K.S., Bhatia, B.S., and Shukla, F.C. (1990). Effects of treatments on quality of Kinnow. *Indian J. Hort.* 47: 55–59.

Santana, M.C., Slutzky, B., and Alauso, A. (1981). Total phenol content and polyphenoloxidase activity in Valencia orange flavedo during storage. *Proc. Int. Soc. Citric. Japan.* Vol. 2, 740–741.

Sardzhveladze, G.P., Tsilosani, M.V., and Kharebava, L.G. (1987). Study of the chemical composition of the fruit of three lemon varieties. *Subtropicheskie-Kul' tury.* (5), 115–118.

Sasson, A., and Monselise, S.P. (1977) Organic acid composition of 'Shamouti' oranges at harvest and during prolonged postharvest storage. *J. Am. Soc. Hort. Sci.* 102, 331–336.

Sattar, A., Mahmud, S., and Khan, S.A.(1987). The fatty acids of indigenous resources for possible industrial applications. Part XIV. Fatty acid composition of the seed oil of *Citrus limon* var. Eureka. *Pakistan J. Sci. Ind. Res.* 30, 710–711.

Sawamura, M., and Kuriyama, T. (1988). Quantitative determination of volatile constituents in the pummelo (*Citrus grandis* Osbeck forma Tosa-buntan). *J. Agric. Fd. Chem.* 36, 567–569.

Sawamura, M., and Osajima, Y. (1973). Studies on the quality of citrus fruits. II. Studies on the translocation of 14C-labelled compounds from leaves to fruit in satsumas. *Nippon Nogei Kagaku Kaishi.* 47, 733–735.

Sawyer, R. (1963). Chemical composition of some natural and processed citrus juices. *J. Sci. Fd. Agric.* 14, 302–310.

Schultz, T.H., Black, D.R., Bombden, J.L., Mon, T.R., and Teranishi, R. (1967). Volatiles from oranges. VI. Constituents of the essence identified by mass spectra. *J. Fd. Sci.* 32, 698– 701.

Scora, R.W., and Newman, J. E. (1967). A phenological study of the essential oil of the peel of Valencia orange. *Agr. Meteorol.* 4, 11–26.

Scora, R.W., England, A.B., and Bitters, W.P. (1966). Essential oils of *Poncirus trifoliata* and its selection in relation to classification. *Phytochem.* 5, 1139–1146.

Selvaraj, Y., and Edward Raja (2000). Biochemistry of ripening of kagzi lime (*Citrus aurantifolia* Swingle) fruit. *Indian J. Hort.* 57, 1–8.

Seymour, T.A., Preston, J.F., Wicker, L., Lindsay, J.A., and Marshall, M.R. (1991). Purification and properties of pectinesterases of Marsh White grapefruit pulp. *J. Agric. Fd. Chem.* 39, 1080–1085.

Shannon, L.M., Kay, E., and Lew, J.Y. (1966). Peroxidases. *J. Biol. Chem.* 241, 2166–2172.

Shaw, P.E. (1979). Review of quantitative analysis of citrus essential oils. *J. Agric. Fd. Chem.* 27, 246–256.

Shaw, P.E., and Wilson, C.W. (1983). III. Organic acids in orange, grapefruit and cherry juices quantified by HPLC using neutral resin or propylamine columns. *J. Sci. Fd. Agric.* 34(11), 1285–1288.

Shaw, P.E., Wilson, C.W., and Berry, R.E. (1981). Some important flavor compounds in mandarin, grapefruit and orange juices and peel essential oils. *Proc. Int. Soc. Citric.*, Vol. 2, 1983, 911–914.

Shimokawa, K. (1983). An ethylene-forming enzyme in *Citrus unshiu* fruits. *Phytochem.* 22, 1903–1908.

Shomer, I., Ben-Gira I., and Ben-Shalom N. (1980). Epicuticular wax and its hydrocarbons from inter-juice-sac spaces in citrus fruit segments. *J. Agric. Fd. Chem.* 28, 1158–1163.

Siddappa, G.S., Bhatia, B.S., and Indira, K. (1962). Day today and seasonal variation in the quality of orange juice. *Indian J. Hort.* 19:155–161.

Sinclair, W.B. (1961). *The orange: its biochemistry and physiology.* Univ. of Calif. Press, Berkeley.

Sinclair, W.B. (1972). *The Grapefruit: Its composition, physiology and products.* Div. Agric. Sci. Univ. of California Press. Berkeley.

Sinclair, W.B., and Jollife, V.A. (1958). Changes in pectic constituents of 'Valencia' oranges during growth and development. *Bot. Gaz.* 119, 217–223.

Sinclair, W.B., and Jollife, V.A. (1961). Pectic substances of Valencia oranges at different stages of maturity. *J. Fd. Sci.* 26, 125–130.

Sinclair, W.B., and Ramsay, R.C.(1944). Changes in the organic acids of 'Valencia' orange during development. *Bot Gaz.* 106, 140–148.

Sites, J.W., and Reitz, H.J. (1951). The variation in individual Valencia oranges from different locations of the tree as a guide to sampling methods and spot-picking for quality. III. Vitamin C and juice content of the fruit. *Proc. Am. Soc. Hort. Sci.* 56, 103–110.

Slocom, Kaur-Sawhney, and Galston, A.W. (1984). The physiology and biochemistry of polyamines in plants. *Arch. Biochem. Biophysics.* 225, 283–303.

Song, K.J., and Ko, K.C. (1997). Relationship between sugar content and sucrose synthase activity in orange fruits. *J.Korean Soc. Hort. Sci.* 38, 242–245; 20.

Sonkar, R.K., and Ladaniya, M.S. (1995). Effect of fruit maturity and degreening on bitterness of 'Nagpur' mandarin juice. *Indian Fd. Packer* November/December, pp. 11–16.

Srivas, S.R., Pruthi, J.S., and Siddappa, G.S. (1963). Effect of stage of maturity of fruit and storage temperature on the volatile oil and pectin content of fresh limes (*C. aurantifolia*) *Fd. Sci.* 10, 340–343.

Srivastava, M.P., and Tandon, R.N. (1966). Free amino acid spectrum of healthy and infected Mosambi fruit. *Naturewissenscaften,* 53, 508–509.

Stafford, H.A. (1960). Lignin biosynthesis. *Pl. Physiol.* 35, 612–618.

Stanley, W.L. (1962). Citrus oils: analytical methods and compositional characteristics. *Intern. Fruchtasaft-Union,* Ber. Wiss. Tech. Komm. 4, 91–103.

Stanley, W.L. (1964). Recent developments in coumarin chemistry. Aspects of plant phenolic chemistry. *Proc. 3rd Int. Symp. Univ. Toronto,* 1963, pp. 79–102.

Stanley, W.L. (1965). Recent progress in analysis (characterization) of citrus juices. *Intern. Fruchtasaft-Union,* Ber. Wiss. Tech. Komm. 6, 207–220.

Stanley, W.L., and Vannier, S.H. (1967). Psoralene and substituted coumarins from expressed oil of lime. *Phytochem.* 6, 585–596.

Stanley, W.L., Ikeda, R.M., Vannier, S.H., and Rolle, I.A. (1961). Determination of the relative concentration of the major aldehydes in lemon, orange and grapefruit oils by gas chromatography. *J. Fd. Sci.* 26, 43–48.

Stevens, J.W. (1954). Preparation of dehydrated agar media containing orange juice serum. *Food Technol* 8, 88–92.

Stewart, I., and Wheaton, T.A. 1964. l-Ocatapamine in citrus. Isolation and identification. Science. 145:60–61.

Swift, L.J., and Veldhuis, M.K. (1951). Constitution of the juice of Florida Valencia oranges. *Fd. Res.* 16:142–146.

Swift, L.J. (1952). Isolation of β-sitosteryl-D-glucoside from the juice of Florida Valencia orange (*C. sinensis* L). *J. Am. Chem. Soc.* 74, 1099–1100.

Swift, L.J. (1965). Tetra-O-methylscutellarein in orange peel. *J Org. Chem.* 30, 2079–2080.

Subramanyam, C., and Cama, H.R (1965). Carotenoids in Nagpur orange pulp and peel. *Indian J. Chem.* 3, 463–465.

Swisher, H.E., and Higby, W.K. (1961). Sugars in citrus juices. In *Fruit and vegetable juice processing technology.* D.K. Tresseler and M.A. Joslyn eds., pp. 903–932, AVI Publishing Co. Inc., Westport, Connecticut.

Tadeo, J.L., Ortiz, J.M., Martin, B., and Estelles, A. (1988). Changes in nitrogen content and amino acid composition of navel oranges during ripening. *J. Sci. Fd. Agric.* 43, 201–209.

Takagi, T., Masuda, Y., and Ohnishi, T. (1989). Effects of sugar and nitrogen content in peel on color development in satsuma mandarin fruits. *J. Japanese Soc. Hort. Sci.* 58, 575–580.

Tamura, H., Fukuda, Y., Padrayuttawat, A., and Kobayashi, A. (1996). Characterization of citrus aroma quality by odor threshold values. In *Biotechnology for improved foods and flavors* G.R. Takeoka, R. Teranishi, P.J. Williams, ed., ACS Symp. Series No. 637 Am. Chem. Soc., Washington, DC, USA, pp. 282–294.

Ting, S.V. (1967). Nitrogen content of Florida orange juice and Florida orange concentrate. *Proc. Fla. State Hort. Soc.* 80, 257–261.

Ting, S.V. (1969). Distribution of soluble components and quality factors in the edible portion of citrus fruits. *J Am. Soc Hort. Sci.* 94, 515– 519.

Ting, S.V., and Attaway, J.A. (1971). Citrus fruits. In 'The biochemistry of fruits and their products.' Vol. 2. A.C. Hulme ed., pp. 107–179, Academic Press, New York.

Ting, S.V., and Deszyck, E.J. (1959). Isolation of L-quinic acid in citrus. *Nature.* 183, 1404–1405.

Ting, S.V., and Deszyck, E.J. (1960). Total amino acid content of chilled orange juice and frozen orange concentrate. *Proc. Fla. Sta. Hort. Soc.* 73, 252–257.

Ting, S.V., and Deszyck, E.J. (1961). The carbohydrates in the peel of oranges and grapefruit. *J. Fd. Sci.* 26, 146–152.

Ting, S.V., and Vines, H.M. (1966). Organic acids in the juice vesicles of Florida 'Hamlin' orange and Marsh seedless grapefruit. *Proc. Am. Soc. Hort. Sci.* 88, 291–297.

Todd, G.W., Bean, R.C., and Propst, B. (1961). Photosynthesis and respiration in developing fruits. II. Comparative rates at various stages of development. *Pl. Physiol.* 36, 69–73.

Tomes, M.L., Quackenbush, F.W., and Kargill, T.E. (1956). Carotenoid synthesis in citrus. *Bot. Gaz.* 117, 248.

Townsley, P.M., Joslyn, M.A., and Smith, C.J.B. (1953). The amino acids in various tissues of citrus fruits and orange protein. *J. Fd. Sci.* 18, 522–531.

Tzur, A., Goren, R., and Zehavi, U. (1992). Carbohydrate metabolism in developing citrus fruit. *Proc. Int. Soc. Citric.* Vol. 1, pp. 405–411.

Uchida, K., Kobayashi A., and Yamanishi T. (1984). The composition of oxygenated compounds in the peel oil of Fukuhara oranges. *J. Agric. Chem. Soc.* Japan. 58, 691–694.

Underwood, J.C., and Rockland, L.B. (1953). Nitrogenous constituents in citrus fruits. I. Some free amino acids in citrus juices by paper chromatography. *J. Fd. Sci.* 18, 17–29.

Valero, D., Martinez, D., Riquelme, F., and Serrano, M. (1998). Polyamine response to external mechanical bruising in two mandarin cultivars. *Hort Science.* 33, 1220–1223.

Vanderkook, C.E. (1977). Nitrogenous compounds. In '*Citrus science and technology*,' Vol. 1. AVI Publishing Co., Inc., Westport, Connecticut, pp. 229–265.

Vandercook, C.E., and Guerrero, H.C. (1969). Citrus juice characterization. Analysis of the phosphorus fraction. *J. Agric. Fd. Chem.* 17, 626–628.

Vandercook, C.E., Rolle, L.A., and Ikeda. R.M. (1963). Lemon juice composition. I. Characterization of California–Arizona lemon juice by its total amino acid and L-malic acid content. *J. Asso. Off. Agric. Chem.* 46, 353–358.

Veldhuise, M.K. (1971). Orange and tangerine juices. In '*Fruit and vegetable juice processing Technology*' (D.K. Tresseler and M.A. Joslyn eds.), pp. 838–873, AVI Publishing Co. Inc., Westport, Connecticut.

Verma, N.S., and Ramakrishnan, C.V. (1956). Succinic acid in *Citrus acida*. *Nature,* 178, 1358–1359.

Wali, Y.A., and Hassan, Y.M. (1965). Simple sugars in citrus fruits. *Proc. Am. Soc. Hort. Sci.* 87, 264–269.

Wang, C.H., Doyle, W.P., and Ramsey, J.C. (1962). Proteins in food ripening. *Pl. Physiol.* 37, 1–7.

Watt, B.K., and Merill, A.L. (1963). Composition of foods: fresh and processed. USDA Handbook 8. Washington DC., 190.

Webber, H.J., and Batchelor, L.D. (1943). *The Citrus Industry*. Vol. 2, Univ. Calif. Press. Berkeley and Los Angeles.

Wedding, R.T., and Sinclair, W.B. (1954). A quantitative paper chromatographic study of the amino acid composition of proteins and juice of orange. *Bot. Gaz.* 116, 183–188.

Weizmann, A., Meisels, A., and Mazur, Y. (1955) Steroids and triterpenoids of grapefruit (*C. paradisi*, Macf). *J. Org. Chem.* 20, 1173–1177.

Wedding, R.T., and Horspool, R.P. (1955). Compositional changes in juice of Valencia and Washington navel oranges during fruit development. *Citrus leaves*. 35, 12–13.

Wheaton, T.A., and Stewart, I. (1965a). Quantitative analysis of phenolic amines using ion exchange chromatography. *Anal. Biochem.* 12, 585–592.

Wheaton, T.A., and Stewart, I. (1965b). Feruloyleputriscine: Isolation and identification from citrus leaves and fruit. *Nature* 206, 620–621.

Williams, K.T., Potter, E. F., and Bevenue, A. (1952). A study by paper chromatography of the occurance of non-fermentable sugars in plant materials. *J. Asso. Agric. Chem.* 35, 483–486.

Williams, B.L., Goad, L.J., and Goodwin, T.W. (1967). The sterols of grapefruit peel. *Phytochem.* 6, 1137–1145.

Wills, R.B.H., Wimalasiri, P., and Greenfield, H. (1984). Dehydroascorbic acid levels in fresh fruit and vegetables in relation to total vitamin C activity. *J. Agric. Fd. Chem.* 32, 836–838

Wollenweber, E., and Dietz, V.H. (1981). Occurrence and distribution of free flavonoid aglycones in plants. *Phytochem* 20, 869–932.

Wu, T.S., (1988). Alkaloids and coumarins of *Citrus grandis*. *Phytochem.* 27, 3717–3718.

Yamaki, Y.T. (1988). Differences in juice acidity values of Citrus species and relationships between types of acidity and between acidity values and potassium concentration. *J. Japanese Soc. Hort. Sci.* 56, 457–469.

Yamaki, Y.T. (1989). Variation in acidity and acid content in rind among citrus fruits, and their relationship to fruit juice acidity. *J. Japanese Soc. Hort. Sci.* 57, 568–577.

Yokoyama, H., and White, M.J. (1965). Citrus carotenoids. II. Structure of citranaxanthin, a new carotenoide ketone. *J. Org. Chem.* 30, 2481–2482.

Yokoyama, H., and White, M.J. (1966).Citrus carotenoids. VI. Carotenoid pigments in the flavedo of cinton cintrangequat. *Phytochem.* 5, 1159–1173.

Yokoyama, H., and White, M.J. (1967). Carotenoids in the flavedo of 'Marsh' grapefruit. *J. Agric. Fd. Chem.* 15, 693–696.

Yokoyama, F., Levi, L., Laughton, P.M., and Stanley, W.L. (1961). Determination of citral in citrus extracts and citrus oils by conventional and modern chemical methods of analysis. *J. Assoc. Offic. Agric. Chemists*. 44, 535–541.

Young, I.B., and Erickson, I.C. (1961). Influence of temperature on color change in Valencia oranges. *Proc. Am. Soc. Hort. Sci.* 78, 192–200.

Zambakhidze, N.E., Museridze, T.T., and Amashukeli, I.D. (1989). Sterols in the peel and flesh of lemon cultivar Dioskuriya. *Fiziologiya-i-Biokhimiya Kul'turnykh-Rastenii*. 21, 590–593.

7

GROWTH, MATURITY, GRADE STANDARDS, AND PHYSICO-MECHANICAL CHARACTERISTICS OF FRUIT

I. CITRUS FRUIT AND CLIMATE

Citrus fruit growth and quality are dependent on climatic conditions in addition to soil type, water availability, cultural practices, and nutrient supply. The most important condition is climate, which includes temperature, relative humidity, rainfall, wind velocity, and sunshine. Frost, fog, and hailstorms are occasional or regular phenomena in some areas and are detrimental to fruit growth and quality. These factors can be controlled to a very limited extent under open field conditions. Optimum temperature, relative humidity, sunshine, and well-distributed rainfall can produce heavy yields and excellent quality fruit, provided that the nutrient supply is optimum. Citrus species are considered to be native to tropical climates, but excellent quality can be produced in a subtropical climate, which is mostly available 23.5–40° north and south to the equator. The tropics lie between 23.5° latitude from equator toward north and south. Tropical climate is characterized by steady temperatures with little diurnal variation, particularly in the lowlands (not much elevation from mean sea level). Average annual temperatures are around 20–30°C. Variations sometimes occur from elevation (altitude) from sea level or nearness to mountain ranges. Subtropical climate generally has 15–20°C average temperature with mild to severe winters and occasional frosts. Minimum temperatures are as low as 5–10°C or sometimes even below 0°C, while higher temperatures go up to 35–40°C in interior areas away from the seashore. Citrus tree vegetative growth is adversely affected below 12.5°C; thus the heat-unit summation for fruit growth is calculated considering this temperature as a base. Heat units (H.U.) can be calculated from the average temperature of the region minus 12.5°C multiplied by time (days). For example, if a growing region has an average temperature of 15°C, the heat unit would be ($15-12.5 = 2.5$; $2.5 \times 30 = 75$ for a month). If a growing region has an average temperature of 20°C, heat summation units would be 225 in a 30-day month, indicating faster growth of the fruit. Very high heat units of lowland tropics (e.q., 5000–6000 units during fruit growth period in some parts of Kenya and Sri Lanka) lead to faster growth but produce poor-quality fruits because the respiration is high and therefore net carbohydrate accumulation is less. On the contrary, in subtropical climates, with frost, fog, and cloudy weather, growth takes longer with a slower accumulation of carbohydrates (sugars), since such climate reduces temperature, light intensity, and CO_2 assimilation. Lower heat units delay growth and result in higher acids and lower sugar content. 'Valencia' fruit mature within 6–7 months under lowland tropics (Cartagena, Colombia) while it take 14 months under the arid subtropical and coastal climate at Santa Paula, California, USA (Reuther and Rios-Castano, 1969). The climatic conditions of 'Nagpur' and the surrounding region in Central India lead to faster growth because of higher temperatures and the subhumid climate. Fruit maturation is faster with accumulated temperatures of more than 3000 H.U. summation. The region normally receives around 800–1000 mm of precipitation with mean monthly temperatures of 21–28°C from June to February. The monsoon-blossom crop fruit grows within 8 months, while spring-blossom crop fruit takes 9 months for acceptable internal maturity. As per heat unit concept, the minimum range (1000–1400 H.U.) results in poor growth rate, and the rate increases until above 6000 H.U. (Mendel, 1969). In temperate regions, citrus is not grown commercially, although Satsuma mandarin on trifoliate (*Poncirus*) rootstock can tolerate temperatures below zero for quite some time.

Climatic conditions have a very significant effect on fruit quality and this effect is more predominant than any other factor, including soil, cultural practices, and even genetic factors. Navel oranges (same variety) grown in Florida and California produce different quality fruit. In cool winters with a subtropical climate, fruit color is intense in mandarins and oranges. Higher respiration rate of fruit in lowland tropics results in faster maturity because acids decline very rapidly. The TSS:acid ratio increases very quickly; however, fruit becomes insipid as the acid content decreases. Tropically grown oranges tend to be less orange in color and peel less easily.

In the cool climate of interior subtropics or coastal areas of subtropics, TSS accumulates slowly, hence maturity takes a longer time. In these areas with the semi-arid to arid climate of interior desert and coastal regions (with cool climates) acidity is greater. In tropical climates, continuous vegetative growth also competes with fruit growth and accumulation of solids and therefore TSS is less. In subtropical climates with lower temperatures at the time of fruit maturity, vegetative growth slackens and hence TSS (sugars) increases slowly in fruit. In cool, dry climates with ample sunlight and irrigation water, excellent fruit can be grown.

The best-quality fruit is grown in the Mediterranean-type climate, which is characterized by relative dryness (low rainfall), hot summers, and cool, wet winters (winter rains) at fruit maturity. The eating quality and appearance is good with beautiful color because of fewer insect-pest problems when fruit is young.

II. GROWTH AND DEVELOPMENT

Flowering in citrus lasts for a month or so under subtropical climate. There are several flushes under tropical climates and citrus trees flower year-round. Out of several thousand flowers that open, very few (2.5–5.4 percent) set fruit, as observed in Valencia oranges (Nogueira and Franco, 1992). 'Nagpur' mandarins and many other commercially grown citrus fruits are similar. Setting also depends on climatic factors, as high to very high temperatures (about 38–40°C) and dry spells during and after set may lead to drop of fruitlet.

The duration of growth and maturation varies with variety. The growth and development of a citrus flower's ovary into a fruit ready to harvest takes 6–18 months or more depending upon the type of fruit, particular cultivar, and climate. For early varieties such as Hamlin and Navels, harvesting commonly starts 6–7 months after bloom, whereas for the late Valencia variety, harvesting starts about 12 months after bloom. Harvesting also continues during a 'tree storage' period that lasts several months after the maturity of fruit. The late oranges and grapefruits have two crops on the tree at the same time: both the fruitlets and the bloom of the new crop and the mature fruits of old crop can be seen.

The growth of citrus can be shown as a single sigmoid curve: two stages of slow growth with a period of rapid growth in between (Bain, 1958; Subramanyam et al., 1965; Dhillon, 1986; Garcia-Luis et al., 2002). Usually early growth is marked by cell division followed by cell expansion. Washington Navel fruit completes growth within 245 days from anthesis to maturity, while Valencia oranges take 413 days. The growth rate (g weight/day) was recorded 1.22 and 0.36 in Washington Navel and Valencia respectively (Bain, 1958; Bouma, 1959).

From fruit set, stage I lasts for 30–40 days, during which cell division is extremely rapid but fruit enlargement is negligible (Fig. 7.1). At this stage the cuticle is not developed so small fruitlets are extremely vulnerable to superficial damage by winds and insects. This problem is relatively greater in areas such as Brazil, USA (Florida), India, and elsewhere where rains in the post-bloom period facilitate superficial infection and insect attack. Windscars also provide entry to waterborne spores of the melanose fungus (*Diaporthe citri*). Although most cell division takes

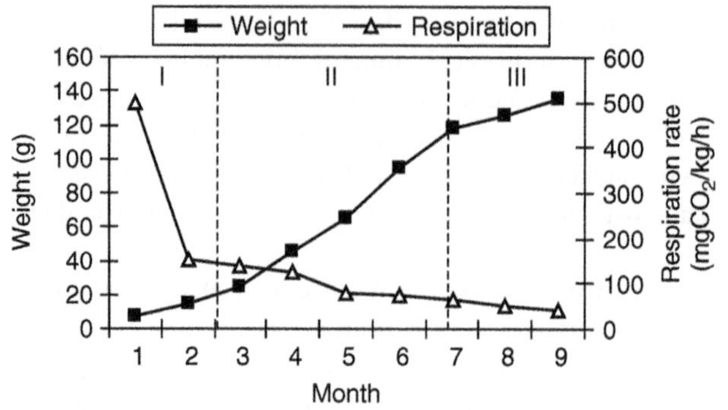

FIGURE 7.1 Growth and Development Stages of Mandarin Fruit.

place in this period, some cell division can continue in the peel until maturation, particularly with Navel oranges, making such fruit very vulnerable to damage.

Developing citrus fruits contain chlorophyll and they can assimilate CO_2 in light (Bean and Todd, 1960; Todd et al., 1961). The rate of photosynthesis to respiration was reported to be unity (no net gain of energy or photosynthates) and with maturity the fruit photosynthesis dropped (Bean et al., 1963). Simple sugars, sucrose, and organic acids are known to be transported from leaves to fruits and fruits are also known to synthesize compounds in their own tissues. Tissues from vesicles of lemon fruits growing in sterile culture with sucrose as the sole source of carbon accumulated citric acid as much as the tissues of intact fruit (Bollard, 1970). In most cases, leaves supply nutrients and nearby leaves are the most potent suppliers of required nutrients to fruit growth. Growing fruits and leaves compete for nutrients and water and act as sinks where inorganic ions or organic metabolites move preferentially. Nutrients move preferentially to growing fruits. As far as water is concerned, fruits act as reservoir and whenever moisture is in short supply leaves draw water from fruits. Growth substances either naturally produced in the fruit or exogenously applied can affect the movement of nutrients in to fruit tissues. In Satsuma mandarin trees, it was noticed that 3–7 cm-long developing shoots were the most powerful sinks. However, as the young fruits developed they became the most powerful sinks. When root growth was restricted, translocation of C^{14} (labeled carbon) to the other parts of the tree decreased. If the older leaves were removed when the trees were in flower, the growth of the young fruitlets was poor, indicating that the new leaves had not developed sufficiently to become a source of photosynthates for other organs. Kinetin or GA_3 applications, each at 200 ppm to the fruit surface, increased the translocation of photosynthates to the fruits and stimulated fruit respiration, whereas the application of chlormequat at 200 ppm had the opposite effect. Under conditions of water stress the amount of C^{14} incorporated in the fruit juice decreased, although the total sugars in the juice were higher than normal (Kadoya, 1974). Externally applied plant growth regulators play an important role not only in controlling the abscission mechanism but also on subsequent development and quality of fruit. 2,4,5-T (Trichlorophenoxy acetic acid) at 10–50 ppm applied as aqueous spray on 1-month old fruitlets of 'Mosambi' oranges (*C. sinensis*) resulted in retention of higher number of fruits till harvest. NAA (10 ppm) and 2,4-D (10 ppm) were also effective (Das, 1983).

In stage II, cell enlargement (and hence fruit enlargement) is rapid. As the fruit expands, CO_2 output per fruit also increases. However, CO_2 evolution per unit weight (the usual way of expressing respiration) declines gradually as the fruit develops (Fig. 7.1). During this period, the juice sacs are enlarging and developing their distinctive solutes. Such solutes are initially high in organic acids and low in sugar. As the orange matures, sugars increase steadily while acids decline. Fruits usually contain water up to 85 percent; this may vary from 80 to 90 percent in fruits of various citrus species. Reducing sugars, sucrose, and citric acid accumulation is faster during growth stage II, which is marked by cell expansion and maturity. Sugar and acid contents, which contribute to fruit flavor, vary greatly depending on fruit species (within citrus) and variety. The sour lemon (*Citrus limon*) contains 4–7 percent acids, while sweet lime (*Citrus limettioides*) contains less than 1 percent (Bogin and Wallace, 1966).

Stage III mostly constitutes changes related to maturation, and these changes also influence maturity. In mature Washington Navel fruits, epicuticular wax, internal CO_2, and internal ethylene increased, whereas as the season advanced, weight loss (during storage) and respiration decreased. Simultaneously, fruit conductance to CO_2 was also reduced (El-Otmani et al., 1986).

The hybrid 'Kinnow' mandarin fruit also matures following a single sigmoid curve. After fruit set in spring (April), fruit weight and diameter increases rapidly from July to November. A rapid increase in carotenoids in the peel becomes evident by late November when the fruit attains full size and volume. The level of chlorophyll -'*a*' and -'*b*' decreases by that time. Total soluble solids increase and acidity decreases until the end of January (Dhillon, 1986). The three distinct phases of Kinnow fruit growth are: May 15th to July 15th, August 15th to November 15th, and November 29th to February 8th. The TSS content increases as sugars increase while juice acid decreases during the entire period of fruit growth. Ascorbic acid content increases initially and decreases with the advancement of fruit maturity. Three periods of fruit drop were also observed. The first extensive fruit drop occurred during mid-May (29.16 percent); the second from May to mid-June (50.84 percent); and the third between mid-September and mid-October, which is a pre-harvest drop (17.33 percent). The chlorophyll content of the fruit peel reduces to a minimum, whereas total carotenoids show a continuous increase as the fruit approaches ripening.

Respiratory pattern of the fruit also varies with growth as it is higher initially when energy requirement is higher for fast-growing fruit. The respiration rate of 'Kagzi' acid lime (*C. aurantifolia*) fruit was quite high (159.65 mgCO_2/kg/h) initially around 120 days when the fruits were small, indicating very high metabolic activity in rapidly growing fruit (Fig. 7.2). After 160 days, as the fruit

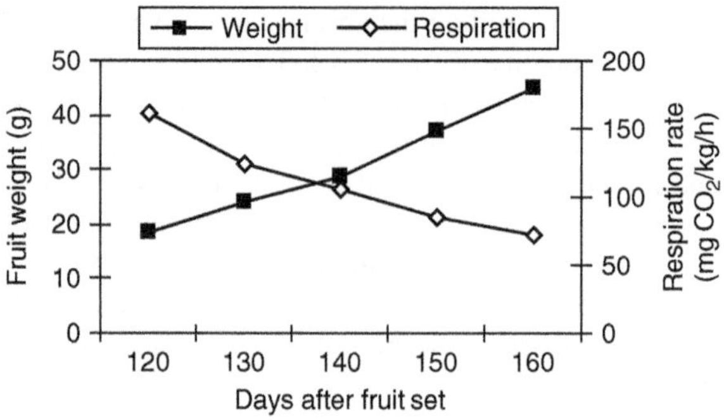

FIGURE 7.2 Acid Lime Fruit Growth and Respiration.

matured, respiration rate per unit weight reduced to about 80 mgCO_2/kg/h at 32°C (room temperature in summer), indicating completion of growth. In harvested mature limes, respiration rate decreased gradually to 40 mgCO_2/kg/h at 29–31°C during a span of 8–10 days (Ladaniya and Shyam Singh, 2000). Subramanyam et al. (1965) reported a similar trend of respiration (450, 140, and 40 mgCO_2/kg/h after 30, 120, and 180 days, respectively) in developing acid limes at Mysore, India.

A. Mineral nutrition and fruit growth

Application and availability of various nutrients applied to trees also affect fruit growth and development and overall final quality. The role of potassium (K) and nitrogen (N) is worth mentioning in this regard. Koo and Reese (1977) reviewed the role of nitrogen and potassium on citrus fruit quality. Higher doses of nitrogen and potassium beyond certain limits that are required to achieve normal yield are detrimental to fruit quality. Higher N results in thicker and greener peel. The effects of K on citrus peel, fruit growth, and overall size of fruit are similar to those produced by gibberellic acid. These plant growth substances are known to increase K uptake by meristemetic tissue. By increasing N and K nutrition, peel growth increases and thus results in thicker peel. K application increases fruit size when applied, up to a certain point. Increasing K also increases titratable acids and reduces the TSS:acidity ratio (Embleton et al., 1975). For increased fruit size, the N:K ratio is required to be 2.4 and 3.0, with higher N and K levels (Miller and Hofman, 1988). Fertilizers and growth regulators such as monopotassium phosphate (2 percent), diammonium phosphate (2 percent), Potassium nitrate (2 percent), 2,4-D, and GA_3 have been used to improve fruit growth in 'Nagpur' mandarins (*C. reticulata* Blanco). The sprays of MPP (2 percent) + GA_3 (10 ppm), followed by DAP (2 percent) + 2.4-D (10 ppm) at 15-day intervals 8–12 weeks after fruit set improved fruit size (Huchche, 2005). Chapter 4 explains the effect of nutrients and growth regulators on fruit quality in great detail.

B. Fruit Maturation

Citrus fruits are non-climacteric: they ripen on the tree. They do not ripen after harvest and do not show any respiratory rise accompanied by major changes in flavor and biochemical composition after harvest in relation to ripening. Citrus fruits of different species differ in time taken for maturity depending on soil and climate; internal composition also varies accordingly. Under temperate to subtropical conditions in North India, citrus flowers only in spring, whereas in the tropical conditions of Central and South India, trees bloom during spring, summer or monsoon season, and also autumn, or before the onset of winter (October–November). In monsoon season and in autumn, trees are forced to flower by withholding irrigation and then breaking the water stress after some time. The maturity and harvesting periods are different in these crops. A similar practice is followed in other parts of the world, such as the Mediterranean region, Africa, and Central and Latin America to achieve the desired crop.

The use of terminology such as 'ripening' for citrus can be controversial, as the changes related to edible quality development are not as distinct and rapid as they are in mangos and bananas. The term ripening appears repeatedly in the citrus literature, yet it is erroneous because there is no starch, oil, and so on capable of being converted to sugars or other soluble products present in citrus fruit flesh during maturation (Grierson and Ting, 1978). Citrus fruits gradually become edible on the tree and remain so (beyond normal harvest time) for several months: 1–2 months for mandarins/tangerines or tangerine hybrids, 5–6 months for oranges, and up to 8 months for grapefruits. This duration of maturity varies with cropping season as mentioned above.

As in most of the fruits, the maturity of citrus heralds the accumulation of sugars and the loss of acidity and host of other biochemical changes. In Spain, Clementine cultivars Fina, Oroval, and Hernandina mature from September to March; the orange cultivars Navelina, Washington Navel, and Navelate mature from September to May. Gradual sugar accumulation has been observed in all cultivars during ripening; this is mainly due to an increase of sucrose in the juice and reducing sugars in the rind. The late cultivar Hernandina showed a slower rate of sugar increase in the rind than the other cultivars (Tadeo et al., 1987). Nagy et al. (1978) studied changing lipid class patterns during maturation of sweet oranges. Phospholipid P declined at rates of 1.57, 1.02, and 0.66 μg of P/mg of lipid/month for Hamlin, Pineapple, and Valencia oranges respectively. While ripening occurred fastest in Hamlin (early-maturing fruit), the rate of senescence (as measured by phospholipid loss) was also fastest. Temperatures also affect maturation of citrus fruit (Utsunomiya et al., 1982). After 50 days, whole fruit and peel weights in Satsuma mandarins were greatest at 23°C and lowest at 30°C. Fruit-specific gravity was lowest at 15°C and lower at 23°C than at 30°C. Abscisic acid levels were lowest at 30°C, but GA-like activity was unaffected. TSS and sugar contents in the juice were highest at 23°C. The lower the fruit temperature, the earlier that the chlorophyll degradation and carotenoid accumulation occurred. In Satsuma mandarins, flavonoid content decreased rapidly with maturity. Limonoid contents were high at the first harvest and then decreased rapidly. Flesh maturity differed between fruits harvested with the same rind color (Kozaki et al., 1984). The activity of several enzymes also changes during fruit maturation. Polygalacturonase activity in peel tissue increased, whereas in juice tissue it fluctuated with the advancement of fruit maturity in Kinnow mandarins. In peel tissue, β-D-galacturosidase enzyme showed higher activity up to 50 percent color-change in fruit surface and then decreased slightly at mature stage, whereas in juice tissue, its activity fluctuated during fruit maturation. Cellulase activity in peel and juice tissues followed the increasing trend. However, its activity was higher in juice tissue than in peel tissue. Pectinmethyl esterase (PME) activity was faster after color-break stage (Lallan Ram et al., 2002).

III. INDICES OF MATURITY, FRUIT GRADES, AND STANDARDS

A. Indices of Maturity (Internal Standards)

Sweet oranges, mandarins, grapefruits, and pummelos are considered mature when their juice content and total soluble solids:acidity ratio have attained certain minimum limits for palatability. Total soluble solids constitute about 80 percent sugars, 10 percent acids, and 10 percent nitrogenous compounds. An increase in sugars is accompanied by an increase in TSS; there is a very strong correlation between TSS and acidity (Bartholomew and Sinclair, 1943).

In citrus fruits that are used for table purposes (such as fresh fruits) and that are processed into juices, maturity is determined mainly on the basis of the ratio of total soluble solids to titratable acidity. This ratio is called the maturity index. The reliance on this ratio alone can be deceptive: 10:1 TSS:acidity ratio means 10 percent TSS and 1 percent acidity, or 5 percent TSS and 0.5 percent acidity. In the latter example, fruit would be of poor quality because of insufficient sugars. Hence, a minimum sugar or TSS content should also be the part of maturity indices like juice content and color break, since peel color and juice content also correlate with palatability (Sites and Reitz, 1949, 1950). Juice content is usually determined as a percentage by weight or volume of fruit.

The content of TSS is very practical guide for harvest in rural areas where growers do not have any other means to measure maturity index objectively. A handheld refractometer, which is inexpensive, can give an almost exact picture of fruit maturity in the orchard. The change of fruit color from light green to yellow-orange in most of the fruits on a tree is also very simple criterion for judging maturity. The size of fruit cannot be a reliable criterion for maturity since fruit size may remain small depending on the bearing and nutritional status of the plant (Goldweber et al., 1957). The fruit size and weight do not bear any definite relationship to the percentage of juice and acidity content.

1. Mandarin

For mandarins, the TSS:acid ratio of 14 was considered necessary for good eating quality, while a ratio higher than 22 produced flat taste (Siddappa, 1952). There is always a variation in this ratio depending on growing regions and acceptability by the consumers, as certain markets prefer slightly sour fruit. In Florida, the minimum maturity standard for 'Temple' orange fruit is 8.5:1 for the fresh market. In India, sweet fruits are preferred and hence the TSS:acid ratio of 14–18 is quite acceptable. In Central India there are two crops of 'Nagpur' mandarins in a year. The maturity indices for harvesting the spring blossom crop of 'Nagpur' mandarins grown at Nagpur (Central India) have been set at a minimum of 10 percent TSS content and a minimum TSS:acid ratio of 14 for acceptable flavor. Fruit diameter, length/height, and weight increase up to 280 days from fruit set in this cultivar (Ladaniya, 1996). In the monsoon-blossom crop season, fruits mature after 240 days from anthesis, when a minimum TSS: acid ratio is 14 (Ladaniya and Shyam Singh, 1999). Fruits of the monsoon-blossom crop are less prone to decay during storage. Fruits of spring blossom crop grow during rainy season and hence are more prone to stem-end rots. Direction had no effect on fruit maturity except on the northeast side of the canopy, where juice content and fruit color were better (Kunte and Deshpande, 1970). For 'Nagpur' mandarins grown at Allahabad, Uttar Pradesh (North India), optimum time for harvesting has been suggested as first 2 weeks of November, when TSS content is 8.3 percent and titratable acidity 0.8 percent (Mazumdar, 1976).

The growth and maturation of 'Coorg' mandarins, grown in Karnataka state, India, harvested in December–February (main crop) and June-August (monsoon) continues for 8 and 9 months respectively. The maturity indices for main crops are recommended as TSS:acid ratios of 13.2. For the monsoon crop, the ratio is set at 9.1 because the fruits of the monsoon crop have relatively poor keeping quality (Ramana et al., 1980). Darjeeling mandarins (grown in Kalimpong) mature between 214 and 245 days from fruit set when juice content is maximum (Roy et al., 1999). Mandarins (Santra) grown under UP Hill conditions matured after 240 days from anthesis, while Kinnows matured after 270 days (Vinay et al., 1987). Seminole and Sunshine tangelos and Futrell's Early mandarin grown at Allahabad (UP) matured from the first to the third week of December (Ukalkar and Shankar, 1979).

In Himachal Pradesh (Dhaulakuan), India, a TSS:acid ratio of 13–14, a TSS content of 13–14 percent and titratable acidity of nearly 1 percent are reported optimum for mature Kinnow fruit (Bhullar, 1982). Joolka and Awasthi (1980) suggested February 4th as normal harvest time for Kinnow in Himachal Pradesh. Under Hissar conditions (Haryana), the third week of December is recommended for harvesting of Kinnow with 10–11 percent TSS and a TSS:acid ratio of 10–11 (Surinder Kumar and Chauhan, 1989).

Wilking mandarins grown at Abohar (the semi-arid plain of Punjab in Western India), take 9 months to attain maturity, which occurs by the first week of January (Josan et al., 1988). In the sub-mountainous region of Hoshiarpur in Punjab, Kinnow mandarins matured during the first week of February; in the semi-arid irrigated condition of Abohar, they matured during January–February when their TSS:acid ratio, TSS content, and acidity were 13.70, 11 percent and 0.8 percent respectively with maximum juice content and palatability (Jawanda et al., 1973; Devkota et al., 1982). Juice content decreases with delayed harvest and so does the fruit's keeping quality. Under Ludhiana conditions (Punjab), the best harvesting time of Kinnow fruits falls between mid-January and the first week of February (Singh et al., 1998).

In California, maturity indices for mandarin/tangerine fruits are set at TSS: acid ratio of 6.5 or higher with yellow or orange or red color on 75 percent of the fruit surface (Arpaia and Kader, 2000).

In Pakistan, the optimum harvesting time for Feutrells' Early mandarin is considered to be early November and for Kinnow early January (Ullah and Haque, 1984).

Liu et al. (1998) reported acceptable harvesting periods as mid-November to mid-January for Ponkan mandarin at Miaoli, early January to late March for Tankan tangor at Miaoli, and early February to April for Tankan tangor at Yangmingshan.

2. Sweet Orange

In citrus regions with high humidity (tropical climate), fruits cannot compete on the basis of appearance and so have to rely principally on standards based on the high soluble solids. Fruits do not develop an attractive color in tropics. TSS content and TSS:acid ratio are also more reliable indices than rind color in sweet oranges. For example, maturity standards in Florida are based on TSS content and also on the ratio of total soluble solids (TSS, mainly sugars) to acids (titratable, such as citric acid). The scale is sliding throughout the season as TSS increases and acidity decreases. At the beginning of the season, Florida oranges must have 8 percent TSS with a TSS:acid ratio of 10.5:1. By the end of the season, this ratio may exceed 20:1, but with the provision that (for fresh fruit sale) acid cannot be below 0.4 percent so that the oranges do not taste too insipid. Regardless of growing district, consistent gradients occur within a citrus fruit, particularly in terms of sugar content. The TSS:acid ratio of 12.0 (a reliable indicator of legally acceptable juice in the USA) is achieved on November 27th, January 12th, and March 28th for Hamlin, Pineapple, and Valencia oranges respectively, and agreed to be reasonably accurate with field maturation in Florida (Nordby and Nagy, 1977). Maturity indices for oranges as reported from California are minimum 10 percent TSS with a TSS:acid ratio of 8 or higher with yellow-orange color on at least 25 percent of the fruit surface (Arpaia and Kader, 2000).

For 'Mosambi', Pineapple, Jaffa, Blood Red, and Valencia sweet oranges grown at Abohar in Punjab (India), TSS content of 13.1, 11.8, 12.5, 12.1, and 12.6 and TSS:acid ratios of 30:1, 14:1, 14:1, 14:1, and 10:1 have been considered as optimum maturity indices respectively (Jawanda, 1961). Fruit size, weight, TSS, and titratable acidity content declines if Valencia Late fruits are harvested by March 1st and fruit becomes unpalatable by the end of March under Punjab conditions (Bakhshi et al., 1967). In Himachal Pradesh, Mosambi, Jaffa, Pineapple, and Blood Red oranges mature by November 15th, December 1st, December 1st, and December 15th, respectively, when corresponding TSS:acid ratios are 22:1, 12.5:1, 16:1, and 17.5:1 (Bhullar, 1983; Khokhar and Sharma, 1984). Under Delhi conditions, early November is suggested as harvesting time for Hamlin (to avoid granulation) and February for Valencia Late (Sinha et al., 1962). Mosambi orange was earliest to ripen in late October with a TSS:acid ratio of 20.36:1 followed by Pineapple, Jaffa, and Valencia Late (February) at Bhatinda, Punjab (Aulakh and Mehrotra, 1999). Sathgudi oranges grown in the Andhra Pradesh state of India matured after 240 days from fruit set when juice content, TSS, and TSS:acid ratio were highest (Reddy et al., 1984). The standards for good-quality Sathgudi oranges are suggested as juice content 42 percent and above, titratable acidity 0.4–0.7 percent, and Brix:acid ratio of 16–20 (Satyanarayana and Ramasubba Reddy, 1994).

In the Jaguey Grande valley in Cuba, Valencia oranges are harvested from November (minimum TSS:acid ratio of 8) until February (TSS:acid ratio of 12). Fruit is generally harvested when its age is 270–330 days with a weight of 180–205 g, juice content 50–53 percent and diameter 68–71 mm (Aranguren et al., 1992).

3. Lime and Lemon

TSS:acid ratio is not considered a suitable index for determining the maturity of limes and lemons. These fruits with desired juice content and size are preferred for fresh salads and garnishing curries, for example. A juice volume of 30 percent (v/v) is the sole internal quality standard for US lemons. Limes and lemons for processing should have higher juice, acids content, and peel oil.

In North Karnataka (India), acid limes (*C. aurantifolia* Swingle cv. Kagzi) take 195 days from fruit set, and fruit with 51.1 percent juice and 9 percent acidity is considered to be mature (Rao et al., 1983). At Bangalore (South Karnataka) it takes 165 days for maximum growth, acidity, and ascorbic acid content (Mustafa and Kumar, 1997). Tahiti limes (*Citrus latifolia* Tanaka) mature within 255 days with 6.7 percent acidity in North Karnataka conditions (Rao et al., 1983). Acid limes (Kagzi) mature after 6 months from anthesis around Mysore (Subramanyam et al., 1965). Under agroclimatic conditions of Nagpur (Central India) the summer crop (flowering in mid-January) of acid lime (Kagzi) matures after 160 days from anthesis (Ladaniya and Shyam Singh, 2001) when acidity and juice content is 8.06 percent and 47.93 percent respectively. In Andhra Pradesh, the acid lime fruit maturity standard is suggested as a minimum of 45 percent juice and 6–7 percent acidity (Satyanarayana and Ramasubba Reddy, 1994). In sweet limes (*C. limettioides* Tanaka), 50 percent juice content was considered optimum for harvesting (Singh and Samaddar, 1962).

Harvesting of Hill and Columbia lemons is recommended by the end of December and November, respectively, under Delhi conditions (Soni and Randhawa, 1969a).

4. Grapefruit and Pummelo

In the US there are per-state maturity standards for grapefruit. The yellow color should be on two-thirds of the fruit surface with minimum of TSS 6–7 percent. Minimum juice percentage is set at 35 percent for some areas. Maturity index for Marsh Seedless and Pink grapefruits grown near Delhi (India) is reported as minimum TSS:acid ratio of 7:1 with optimum harvest time as the third week of December (Randhawa et al., 1964). For Red Blush grapefruit grown in Punjab, TSS content of 17.70 percent and acidity 1.5 percent were found optimum (Chohan et al., 1984).

In Pakistan, grapefruit cultivars Thompson, Marsh Seedless, and Foster exhibited maximum sugar-to-acid ratio during November and December and considered mature by that time. Ascorbic acid content is maximum in all cultivars by December 30th (Ahmed et al., 1992).

The optimum harvest time for pummelo fruits (*C. grandis* or *C. maxima*) of cultivar Ma-tou grown in Taiwan is 23–25 weeks after anthesis, when the fruits showed higher total soluble solids (TSS) and juice contents (Chang, 1987).

B. Fruit Grades (External Standards)

Citrus fruit grades are mostly related to size, appearance, extent of defects, shape, and color of the fruit. European citrus-growing countries, South Australia, California, and other places with Mediterranean-type climates (cool winter nights, bright days, and low rainfall) can rely almost entirely on external standards to sell their fruit. As Grierson and Ting (1978) put it, the real basis for fresh citrus fruit grades and standards is economics. What is economically justified under one situation may not be so in another. Hence almost all the countries have their own standards for domestic markets and these standards also vary as per early-, mid- and late-season crop fruit. The same variety of citrus also performs differently in different climatic conditions and this also leads to setting of different standards.

The grade standards of citrus fruits are published by international bodies and national governments of different countries. The Organization for Economic Cooperation and Development (OECD) introduced standards for marketing fruit between countries. The Economic Commission for Europe (ECE) also publishes standards for grades of fruits and vegetables including citrus. For fruit to be palatable, all grades of fruit must meet minimum internal maturity standards.

Besides international standards published by the FAO Codex committee, several countries have their own fruit-quality standards. The Philippines have the following standards for Valencia oranges: a TSS:acid ratio of 10:1 with minimum solids of 8.5 percent, a juice content minimum of 50 percent by weight, and a color break of 25 percent. For Ponkan mandarins, standards are 50 percent color break, 8.5 percent solids, and 10:1 TSS:acid ratio (Pantastico, 1975). In the U.S., standards for grades of oranges such as Fancy, No. 1, 2, and 3 are described by the states of California, Florida, Texas, and Arizona. The Codex (FAO) Extra class is equivalent to US Fancy (Superior quality); Class 1 is equivalent to US No. 1 (good quality) and Class II is equivalent to US No. 2 (marketable quality). In the US, fruits are graded in packinghouses and also at the time of official grade inspection. Fruits of each size are inspected for defects in a given grade. In Florida, the maturity and grade standards are rigidly enforced by the Florida Department of Citrus and Florida state statutes. The internal quality of Florida citrus is higher than for most other areas, while appearance is lower because of pale color, windscar, mite damage, and so on. Florida shippers are usually directed to select only crops grown for optimum external and good internal quality.

In India, 'Nagpur' mandarins are presently size-graded as Grade 1 (Medium) 6.40–6.70 cm (140–160 count packed in wooden containers of 47 × 35 × 35 cm size), Grade 2 (extra large) 7.30–8.0 cm (80–120 count), and Grade 3 (small), 5.50 to 6.00 cm (168–200 count). Because the 'Nagpur' mandarin fruit gets puffy, the large fruit is not preferred by the trade and therefore medium-sized fruit is assigned Grade 1. Kinnows are marketed with grading on the basis of size (Sandhu, 1990). The size grades for Sathgudi sweet oranges are suggested as Grade I = 6.5 cm and above, Grade II = 5–6.5 cm, and Grade III = 5 cm and less. In India there is a need for grade standards (based on size as well as internal and external quality) and their statutory enforcement in all citrus for orderly marketing. The Directorate of Marketing and Inspection (Ministry of Agriculture, Government of India) is in the process of revising, drafting, and adapting standards for citrus fruit to be marketed in India.

To meet Indian grade standards of 'Extra Special', acid lime (*C. aurantifolia* Swingle) fruit should have minimum diameter of 44 mm (1.75 in.), which resembles grade 'I' recommended for fruit with diameter of 45 mm in Andhra Pradesh, India (Rajput and Haribabu, 1993). For 'Special' and 'A' grades of Indian standards corresponding to grade II and III of acid limes in Andhra Pradesh recommended sizes are 40 and 35 mm diameters, respectively.

1. Some International and National Standards for Citrus

The Codex Commission of FAO prepares the draft of standards for fresh fruits, including citrus (FAO, 2004, 2005). Recently, standards along similar lines were also being drafted for fresh citrus grown in India (APEDA, Directorate of Marketing and Inspection, Government of India, personal communication). Brief guidelines of these standards are given below. These standards indicate minimum quality requirements only; fruit entering trade generally has better quality than the stipulated one.

(i) Citrus Species and Varieties: The citrus fruits included are: (1) oranges grown from *Citrus sinensis* (L.) Osbeck (2) Mandarins and tangerines grown from species of *Citrus reticulata* Blanco, Satsumas, Clementines, common mandarins (*Citrus deliciosa* Ten.), and their hybrids, including 'Kinnow' (*Citrus nobilis* × *Citrus deliciosa*) (3) grapefruits (4) limes grown from the species of *Citrus aurantifolia* Swingle (small-fruited) known as 'Key', 'Mexican', and

'Kagzi' lime and its hybrids, and limes from *Citrus latifolia* (large-fruited) or Persian lime and its hybrids (5) lemon and its hybrids.

(ii) Minimum Requirements: The fruit should be wholesome, firm, intact, and free of bruises or cuts or disorders. It should be sound with characteristic shape of the variety without any rotting or deterioration. Fruit should be clean and free of any foreign matter or smell. There should not be internal shriveling or damage by frost.

(iii) Maturity Requirements: Orange fruits with a light green color are allowed, provided that it does not exceed one-fifth of the total surface area. The degreening is permitted only if the natural oganoleptic characteristics are not modified. Oranges produced in tropics can be of a green color exceeding one-fifth of the total surface area, provided that they satisfy the criteria mentioned below. Completely green fruit is excluded. Minimum juice content in blood oranges should be 30 percent. Navels with 33 percent juice are allowed. Other varieties shall have 35 percent juice, while Mosambi and Sathgudi with more than one-fifth green color shall have 33 percent juice. Other varieties with more than one-fifth green color shall have 45 percent. Minimum TSS content should be 10 percent or 10° Brix with minimum TSS:acidity ratio 12:1.

In mandarins, color must be typical of the variety on at least two-thirds of the surface of the fruit. Mandarins produced in areas with high air temperatures and high relative humidity conditions during the developing period can be of a light green color exceeding two-thirds of the total surface area, provided that they satisfy the criteria for juice, TSS, and TSS:acidity contents. Minimum juice content for Satsumas and their hybrids should be 33 percent. Clementines must have 35 percent juice. 'Nagpur,' 'Khasi,' and 'Coorg' with two-thirds green surface must have 35 percent juice while 'Kinnow' hybrids must have 40 percent. Minimum TSS content for 'Nagpur,' 'Coorg,' 'Khasi,' and 'Kinnow' should be 9 percent with minimum TSS:acid ratio of 12:1.

Grapefruit must have the characteristic color of the variety. However, fruits of a light greenish color with color break are allowed, provided that they comply with the minimum internal quality requirements. Red-pulp varieties may have reddish patches on the rind. Minimum juice content for Marsh seedless and other varieties is set at 33 percent with minimum TSS content 8 percent or 8° Brix.

In lemons, minimum juice content must be 25 percent with typical varietal color. No lemon fruit shall be entirely dark green. The fruit should be light green and shining.

In limes, minimum juice content should be 42 percent. The fruit should be light green and shining but may show yellow patches on up to 30 percent of its surface.

(iv) Fruit Classification: For fruit entering in international trade three classes are defined: 'Extra' Class, 'Class 1', and 'Class II'. Fruits in the Extra class must meet minimum maturity criteria and should be of superior quality. Very slight superficial defects that do not affect the general appearance of the produce, its quality, its keeping quality, and its presentation in the package are allowed. In coloring, mandarins must be characteristic of the variety. Fruits in Extra class shall not be smaller than minimal size standards as specified below. Fruits of Class I must be of good quality, must meet minimum maturity criteria, and should be characteristic of the variety. Slight defects in shape, coloring, and skin that occur during the formation of the fruit, such as silver scurfs, russets, and rust mite damage, are allowed. Slight skin discoloration from rust mite or melanose on less than one-fifth of the surface of grapefruit is also allowed. Skin defects not exceeding more than 1 sq. cm surface of lemons are also permitted. The defects must not affect the pulp of the fruit. Class II must satisfy the minimum quality and maturity requirements. The defects may be allowed, as in Class I. Rough skin, superficially healed skin alterations, and slight and partial detachment of the pericarp are permitted in oranges. Defects must not, in any case, affect the pulp of the fruit. Slight puffing of fruit is allowed in mandarins. In limes and lemons, skin defects should not exceed more than 2 cm^2. The small-fruited (Key or Mexican) limes in this class shall not be smaller than 33 mm in diameter and lemons shall not be smaller than 45 mm in diameter.

(v) Fruit Size: Fruit size is determined by the maximum diameter at the equatorial section of the fruit. 'Key' or 'Kagzi' limes of a diameter below 33 mm are excluded. The suggested size codes for small-fruited limes are 1 (41–45 mm), 2 (30–44 mm), 3 (37–42 mm), 4 (35–39 mm), and 5 (33–37 mm). For lime fruit arranged in layers in the package the maximum difference between the smallest and the largest fruit must not exceed 5 mm for size codes 1 and 2, and 4 mm for size codes 3–5.

Oranges with a minimum dimension of less than 53 mm are not permitted. Oranges packed by count must be uniform in size. The size code is 0 for 92–110 mm diameter fruit, 1 for 80–100 mm, 2 for 84–96 mm, 3 for 81–92 mm, and 4 for 77–88 mm. There are 13 such size codes. For fruit arranged in regular layers in packages, the maximum difference between the smallest and the largest fruit for size codes of 0–2 must not exceed 11 mm. For size codes of 3–6 the difference should not be more than 9 mm.

Mandarin with sizes of 78 mm and above are given a code of 1-XXX, while those with sizes of 67–78 mm have a size code of 1-XX. The size codes of 1-X and 2 are given to fruit of 63–74 mm and 58–69 mm diameter respectively. Size codes 3, 4, and 5 are given to fruit with 54–64 mm, 50–60 mm, and 46–56 mm diameters respectively. Satsumas and 'Nagpur', Coorg, and Khasi mandarins less than 45 mm in size are not permitted. Clementines with sizes of 35 mm, and less and Kinnow with sizes of 50 mm and less are not permitted.

Size codes for grapefruit are 0 for fruit of 139 mm or more, 1 for 109–139 mm, 2 for 100–119 mm and 3 for 93–110 mm. Nine such size codes are prescribed. Fruits of less than the minimum size of 70 mm are excluded.

The suggested size codes for lemons are 0 for 79–90 mm, 1 for 72–83 mm, 2 for 68–78 mm, 3 for 63–72 mm, 4 for 58–67 mm, and 5 for 53–62 mm. Lemons of a diameter below 45 mm are not permitted. For fruit arranged in layers in the package the maximum difference between the smallest and the largest fruit in the same package of size codes 1 and 2 should not be more than 7 mm; for size codes 3–7 not more than 5 mm.

(vi) Tolerances: Tolerances with respect to quality and size are allowed in the package for fruit not satisfying the requirements of the class indicated on the box. Up to 5 percent fruit (by number or weight) not satisfying the requirements of the Extra class, but meeting those of Class I is allowed. In fruit of Class I, up to 10 percent fruit of Class II or, exceptionally, coming within the tolerances of that class is allowed. In Class II, 10 percent fruit satisfying neither the requirements of the class nor the minimum requirements is permitted. Within this tolerance, a maximum of 5 percent fruit showing slight superficially unhealed damage, dry cuts, or softening and shriveling is allowed. Size tolerance up to 10 percent by number or weight of fruit corresponding to the size immediately above and/or below that indicated on the package is allowed. The size minimums as given for each type of citrus fruit must be abided by.

(vii) Presentation: Fruit of each package must be uniform and contain only fruit of the same origin, variety, quality, size, degree of maturity, color, and development. The visible part of the contents of the package must be representative of the entire contents. Fruit of Extra class must be uniform in color. Arrangement of fruit in regular layers in the package is mandatory for Extra class and optional for Classes I and II. The packaging material must be new, clean, and of a quality that does not cause any external or internal damage to the fruit. The use of materials, particularly of paper or stamps bearing trade specifications, is allowed provided that the printing or labeling has been done with non-toxic ink or glue. Packages must be free of all foreign matter and smell.

(viii) Labeling/Marking: The requirements of the Codex General Standard for the labeling of packaged foods should be followed. The name of the produce, variety, country of origin, treatment, and so forth should be labeled on each package. Packages must also bear the name and address of the exporter, packer, and/or dispatcher. The identification code is optional. Class, size code, and net weight for fruit should also be marked on the box.

(ix) Hygiene and Fruit Contaminants: Fruits shall comply with maximum limits for heavy metals and pesticide residues established by the Codex Alimentarius Commission. Fruit should be prepared and handled in accordance with the appropriate sections of the Recommended International Code of Practice: General Principles of Food Hygiene, Code of Hygienic Practice for Fresh Fruit and Vegetables, and other relevant Codex to avoid microbiological contamination.

IV. PHYSICAL AND MECHANICAL CHARACTERISTICS

Physical and mechanical properties of various citrus fruits are important for designing mechanical systems for optimal harvesting and handling capacities with minimum damage to fruit. This is very important in terms of bruising and abrasion, because decay and/or oleocellosis may develop. In general, sweet oranges can bear the impact of mechanical handling (pressures) better than soft fruit (mandarins and tangerines). Vibration, impact of drop, and compression all cause bruises. Various types of bruise can be minimized by using proper cushioning and smooth surfaces in handling machinery. Some physical properties of citrus fruits, such as firmness, density, volume, weight, specific gravity, puncture resistance, and shape, also indicate quality. Physical properties such as weight, volume, and juice content are important from a marketing point of view. Firmer fruit has a greater ability to sustain and resist impact and will deteriorate more slowly than soft fruit. Fruit density is reduced with maturity and also by freeze damage.

Changes occur in physical properties of fruit during development and maturation on the tree as well as during storage. Harvesting at the right stage of maturity is as important for mechanical handling as it is for palatability. Over-mature (ripe or puffy) fruit is difficult to handle on machines. In 'Kagzi' acid limes and 'Mosambi' sweet oranges, fruit volume and weight increases while peel thickness decreases with development and maturity (Tables 7.1 and 7.2). The juice content of acid limes increased rapidly from 28.92 to 47.93 percent in last leg of development, between 120 and 160 days after fruit set (Ladaniya and Shyam Singh, 2001). In Mosambi oranges, juice content almost doubled between 180 and 240 days from fruit set. Fruit firmness dropped from 115.52 Newtons (N) to 74.58 N between 180 and 240 days (Ladaniya, 2004). Rind color changed from deep green to greenish-yellow during maturity. Increasing L* indicated brightness of color, with a change in color from dark green to light green and then to greenish-yellow in acid lime. The increase in C* (Chroma) indicated increased color intensity. Maximum hue angle 115.63° after 120 days indicated deep green

TABLE 7.1 Changes in Physical Quality Attributes During Growth and Maturity of Acid Lime Fruit Cultivar 'Kagzi'

Days after fruit set	Fruit weight (g)	Fruit height (mm)	Fruit diameter (mm)	H/D ratio	Fruit volume (ml)	Peel color			Peel thickness (mm)	Juice (%)
						L*	C*	h°		
120	20.24	33.25	32.25	1.03	18.41	42.92	27.64	115.6	1.57	28.92
130	28.10	39.13	36.25	1.07	25.86	43.29	32.65	114.3	1.41	35.19
140	30.15	40.41	37.38	1.08	28.29	49.82	35.68	110.4	1.30	38.77
150	37.41	43.55	40.87	1.06	37.68	58.17	42.65	100.4	1.24	43.59
160	45.30	45.55	44.09	1.03	44.35	61.66	51.46	90.7	1.19	47.93
SE for ± Means	0.67	2.09	0.33	0.02	0.79	1.20	1.15	1.3	0.05	1.42
CD ($P = 0.05$)	2.03	2.96	1.02	NS	2.39	3.60	3.46	3.90	0.16	4.26

Ladaniya and Shyam Singh (2001); H/D, Height to diameter ratio; L*, Lightness or darkness; C*, Chroma (0 – lowest intensity, 100 – greatest intensity); h^0, hue angle (0° – red purple, 90° – yellow, 180° – green, 270° – blue).

TABLE 7.2 Changes in Physical Quality Attributes During Growth and Maturity of 'Mosambi' Sweet Orange

Days after fruit set	Fruit weight (g)	Fruit height (cm)	Fruit diameter (cm)	H/D ratio	Fruit volume (ml)	Fruit firmness (N)	Peel color 'a'	Peel color 'b'	Peel thickness (mm)	Juice (%)
180	161.62	7.02	7.09	0.99	178.61	115.52	−16.91	30.04	0.82	21.38
190	190.66	7.49	7.52	0.99	205.08	106.97	−18.00	39.42	0.74	23.22
200	194.52	7.54	7.65	0.98	213.27	91.84	−16.00	48.52	0.73	26.50
210	223.86	8.04	8.28	0.97	258.20	87.41	−15.55	45.81	0.74	28.72
220	247.90	8.02	8.38	0.96	267.31	77.22	−14.55	49.59	0.72	31.84
230	249.56	8.01	8.40	0.95	268.00	74.39	−8.50	54.90	0.73	35.81
240	250.40	7.92	8.42	0.94	267.62	74.58	−7.60	56.54	0.69	39.53
250	253.28	7.94	8.36	0.94	269.83	74.64	5.00	62.21	0.70	41.34
SE for Means ±	5.51	0.10	0.11	0.01	7.02	1.90	–	–	0.04	0.49
CD ($P = 0.05$)	15.91	0.28	0.28	0.04	20.27	5.50	–	–	0.12	1.41

H/D, Height to diameter ratio.
Ladaniya (2004).

color along with minimum L* value showing darkness of color. As the peel color changed from light green to yellow, hue angle decreased to 90.71°. Objective measurement of all these properties provides sufficient information about fruit's ability to withstand mechanical handling and at the same time its acceptability in the market. As citrus fruits are non-climacteric, they have to be harvested at optimum maturity so that they are neither raw/immature nor overripe. In both conditions fruit is unfit for mechanical handling and has low keeping quality.

The mechanical characteristic such as detachment force (force required to separate the fruit from stem/peduncle) varies in different citrus fruits. This property has been studied in different orange cultivars to find their amenability and suitability for mechanical harvesting and to design suitable harvesters. Tangerines and mandarins generally get plugged and rind ruptures at the stem end when pulled (if the natural abscission layer is not yet developed). The pulling force for detachment in 'Salustiana' and 'Washington Navel' oranges grown in Spain was observed to be 70–75 N and 55 N, respectively. The torsion force (rotation detachment) was observed to be less than 3 torques for both varieties. The most effective detachment system suggested for fresh market citrus was torsion, because the calyx is not detached in it and losses are least (Juste et al., 1988). In harvesting systems based on pulling of fruit, losses from rupture of the peel were excessive. Fruit detachment force (FDF) of orange cultivars Hamlin,

Pineapple, and Valencia located in the interior of the canopy is lower than on the exterior. The FDF of all these cultivars is positively correlated with fruit weight and negatively correlated with juice content (Kender and Hartmond, 1999).

The rind resistance to penetration or puncture is also an important characteristic of the fruit to consider when designing mechanical systems. The resistance to penetrate 'Salustiana' orange fruit is observed to be 33.5 N in mid-December and 27.9 N in the last week of February at equatorial plane when measured with 4.7 mm diameter punch attached to a Chatillon dynamometer puncture tester. With similar measuring conditions, the resistance to penetration was 26.3 N in mid-December and 24.2 N in February for Washington Navel oranges grown in Spain (Juste et al., 1988).

As far as pressure impact is concerned, the speed rate used to apply pressure has no influence on the deformation. However, the time involved in contact of pressure influenced the deformation and damage (Juste et al., 1988). The pressure resistance of 'Salustiana' and 'Washington Navel' at speeds of 300 and 500 mm/sec using 'Instron' model 4301 has been measured at equatorial plane with circular surface of 1.1 cm^2. The damage is measured in terms of deformation and cell splitting. In December, Washington Navel oranges resisted pressure of about 4.5 kg/cm^2, which declined to about 2.5 kg/cm^2 in March. In Salustiana the decline was from 4.6 to 4.1 kg/cm^2 from February to April. Immature oranges sustain greater oil cell damage and mature fruit could withstand 3.5–4.5 kg/cm^2, stabilizing at these levels (Juste et al., 1988).

Damage resulting from compression of citrus fruit depends on maturity and the environmental conditions at harvesting time. Fruit picked under good environmental conditions show a compression resistance of 4–4.5 kg/cm^2 and no severe damage occurring during subsequent handling (Juste et al., 1990). Under dew and high RH, injuries from oleocellosis appear as a result of the rupturing of oil glands present on the rind. Such injuries are more important and cause concern where the fruit is not fully colored. Drying reduces the damage rate, but is not adequate to allow the safe harvesting of wet fruit.

Storage conditions (ambient or refrigerated) also affect mechanical properties of orange fruit. Tensile strength and cutting energy decreased with storage. Firmness measured as puncture resistance (for a 5 mm cylindrical probe) increased, while compression force required for permanent deformation decreased. Firmness values measured in compression test decreased from 109.6 to 65.4 N and from 135 to 81.9 N under ambient and refrigerated conditions respectively. Peak cutting force, albedo cutting force, and cutting energy of whole fruit increased with storage period (Singh and Reddy, 2006).

REFERENCES

Ahmad, M.J., Muhammad, M., Muhammad, I., and Kayani, M.Z. (1992). Chemical changes in grapefruit (*Citrus paradisi* Macf.) during maturation and storage. *J. Agric. Res.*, Lahore 30, 489–494.

Aranguren, M., Lopez, B., Rodriguez, J., Suarez, C., and Zayas, J. (1992). Maturity calendar and harvest organization for Valencia in Jaguey Grande, Cuba. *Proc. Int. Soc. Citric*, Italy, pp. 1023–1025.

Arpaia, M., and Kader, A.A. (2000). Produce facts (grapefruit, lemon, mandarin, orange) – Recommendations for maintaining postharvest quality. http://postharvest.ucdavis.edu/producefacts/fruit/grapefruit/lemon/orange/mandarin.html

Aulakh, P.S., and Mehrotra, N.K. (1999). Determination of maturity standards in sweet orange cultivars under arid irrigated region of Punjab. *Proc. Nat. Symp. Citric.*, NRCC, Nagpur, pp. 443–445.

Bain, J.M. (1958). Morphological, anatomical, and physiological changes in the developing fruit of Valencia orange (*Citrus sinensis* (L) Osbeck). *Aust., J. Bot.* 6, 1–25.

Bakhshi, J.C., Singh, G., and Singh, K.K. (1967). Effect of time of picking on fruit quality and subsequent cropping of Valencia Late variety of sweet orange (*C. sinensis*). *Indian J. Hort.* 24, 67–70.

Bal, J.S., and Chohan, G.S. (1987). Studies on fruit quality at maturity/ripening of Kinnow mandarin on different root stocks. *Indian J. Hort.* 44, 45–51.

Bartholomew, E.T., and Sinclair, W.B. (1943). Soluble constituents and buffer properties of orange juice. *Pl. Physiol.* 18, 185–206.

Bean, R.C., and Todd, G.W. (1960). Photosynthesis and respiration in developing fruits I. C^{14} O_2 uptake by young oranges in light and dark. *Pl. Physiol.* 35, 425–429.

Bean, R.C, Porter, C.G., and Barr, B.K. (1963). Photosynthesis and respiration in developing fruits III. Variation in photosynthetic capacities during change in citrus. *Pl. Physiol.* 38, 285–290.

Bhullar, J.S. (1982). Determination of maturity standards for mandarins at Dhaulakuan in Himachal Pradesh. *Punjab Hort. J.* 22, 35–40.

Bhullar, J.S. (1983). Determination of maturity standards of sweet oranges in Himachal Pradesh. *Haryana J. Hort. Sci.* 12**(3–4)**, 183–188.

Bogin, E., and Wallace, A. (1966). Sugar and acid content of citrus fruit. *Proc. Am. Soc. Hort. Sci.* 83, 182.

Bollard, E.G. (1970). The physiology and nutrition of developing fruits. In *The biochemistry of fruits and their products* (A.C. Hulme, ed), Vol. I, pp. 387–426.

Bouma, B. (1959). Changes in the developing fruit of Washington navel orange. *Aust. J. Agric. Res.* 10, 804.

Chang, L.R. (1997). Studies on quality of 'Ma-tou' Wentan (*Citrus grandis* Osbeck cv. 'Ma-tou') fruits. *Proc. Symp. Enhancing Competitiveness* of Fruit Industry, Taichung, Taiwan, 20–21 March 1997. Special Publication Taichung District Agric. Improvement-Station, No. 38, 77–91.

Chohan, G.S., Dhillon, B.S. and Josan, J.S. (1984). A note on the performance of Red Blush grapefruit in the arid-irrigated region of Punjab. *Punjab Hort. J.* 24, 46–48.

Das, R.C. (1983). Rate of fruit growth and retention of Mosambi (*C. sinensis* Osbeck). *Proc. Int. Citrus Symp. Hort. Soc. India*, Bangalore, pp. 299–303.

Devkota, R.P., Grewal, S.S., and Dhatt, A.S. (1982). Maturity determination in Kinnow mandarin. *Punjab Hort. J.* 22 **(3–4)**, 131–135.

Dhillon, B.S. (1986). Bio-regulation of developmental processes and subsequent handling of Kinnow mandarin. *Acta Horticul.* No. 179, 251–256.

Embleton, T.W., Jones, W.W., and Platt, R.G. (1975). Plant nutrition and citrus fruit crop quality and yield. *HortScience*. 48, 48–50.

El-Otmani, M., Coggins, Jr., C.W., and Eaks, I.L. (1986). Fruit age and gibberellic acid effect on epicuticular wax accumulation, respiration, and internal atmosphere of navel orange fruit. *J. Am. Soc Hort. Sci.* 111, 228–232.

FAO (2004 and 2005). *Codex standards for fruit and vegetables*. Food and Agriculture Organization, Rome.

Garcia-Luis, A., Oliveira, M.F.N., Bordon, Y., Siqueira, D.L., Tominaga, S., and Guardiola, J.L. (2002). Dry matter accumulation in citrus is not limited by transport capacity of pedicel. *Annals Bot.* 96, 755–764.

Goldweber, S., Boss, M., and Lynch, S.J. (1957). Some effects of nitrogen, phosphorous and potassium fertilization on growth, yield and fruit quality of Persian limes. *Proc. Fla. Sta. Hort. Soc.* 69, 328–332.

Grierson, W., and Ting, S.V. (1978). Quality standards for citrus fruit, juices and beverages. *Proc. Int. Citrus Congress*, Australia, pp. 21–27.

Huchche, A.D. (2005). Fruit size improvement in 'Nagpur' mandarin. Annual Report 2004–2005. National Research Centre for Citrus, Nagpur.

Jawanda, J.S. (1961). Maturity standards for sweet orange in the Punjab. *Punjab Hort. J.* 1, 207–210.

Jawanda, J.S., Arora, J.S., and Sharma, J.N. (1973). Fruit quality and maturity studies of Kinnow mandarin at Abohar. *Punjab Hort. J.* 13, 3–12.

Joolka, N.K., and Awasthi, R.A. (1980). Studies on maturity standards of Kinnow in Himachal Pradesh. *Punjab Hort. J.* 20, 149–51.

Josan, J.S., Monga, P.K., Chohan, G.S., and Sharma, J.N. (1988). Biochemical changes during development and ripening in the fruit of Wilking mandarin. *Indian J. Hort.* 45, 13–17.

Juste, F., Gracia, C., Molto, E., Eranez, R., and Castillo, S. (1988). Fruit bearing zones and physical properties of citrus fruits for mechanical harvesting. *Proc. 6th Int. Citrus Congress*, Israel, Vol. 4, pp. 1801–1809.

Juste, F., Fornes, I., and Castillo, S. (1990). Compression damage on citrus fruits. *Int. Conf. Agric. Mechanization*, Zaragoza, Spain, 27–30 March. Workshop on impact damage of fruits and vegetables. Vol. 2. pp. 103–112.

Kadoya, K. (1974). Studies on the translocation of photosynthates within Satsuma trees during fruit growth. Memoirs College of Agric, Ehime University. 18(2), 193–254.

Kender, W.J., and Hartmond, U. (1999). Variability in detachment force and other properties of fruit within orange tree canopies. *Fruit Varieties J.* 53(2), 105–109.

Khokar, U.U., and Sharma, R. (1984). Maturity indices for sweet orange cv. Blood Red. *Haryana J. Hort. Sci.* 13, 22–25.

Koo, R.C.J., and Reese, R.L. (1977). Influence of nitrogen, potassium and irrigation on citrus fruit quality. *Proc. Int. Soc. Citric.* Vol. 1. pp. 34–38.

Kozaki, I., Hirai, M., Ueno, I., Izumi, Y., Okudai, N., Ishiuchi, D. Oiyama, I., Takahara, T., Shibata Y., Iwakiri, T., Kodama, M., Matsumoto, Y., and Bessho, Y. (1984). Flavouring substances related to the juice quality of the Satsuma (*Citrus unshiu* Marc.) I. Differences in flavouring substances among fruits grown at different locations. Bulletin, Fruit Tree Research Station, Japan, Okitsu, No. 11, pp. 17–39.

Kunte, Y.N., and Deshpande, V.B. (1970). Quality of Santra fruit as affected by aspects – direction and height on the tree. *Nagpur Agric. College Mag.* 42, 33–39.

Ladaniya, M.S. (1996). Standardization of fruit maturity indices in Spring blossom (Ambia) crop of Nagpur mandarin. *J. Maharashtra Agric. Univ.* 21, 73–75.

Ladaniya, M.S. (2004). Standardization of maturity indices for Mosambi sweet orange. Paper presented in National Horticultural Congress, November 2004, New Delhi.

Ladaniya, M.S., and Singh S. (2001). Maturity indices for acid lime (*Citrus urantifolia*) cultivar Kagzi grown in Central India. *Indian J. Agric. Sci.* 70, 292–295.

Ladaniya, M.S. and Singh S. (1999). Determination of optimum maturity indices for 'Mrig' bahar Crop of Nagpur mandarin. *Annual Report for 1998–1999* National Research Centre for Citrus, Nagpur.

L Ram, Godara, R.K., Sharma, R., and Siddiqui, S. (2002). Cell-wall modifying enzyme activities in peel and juice tissues of Kinnow fruits at various stages of maturity. *Indian J. Hort.* 59, 24–30.

Liu, F.W., Wang, T.T. and Pan, C.H. (1998). Maturity characteristics of 'Ponkan' mandarin (*Citrus reticulata* Blanco), Tankan (*C. tankan* Hayata) and 'Liucheng' orange (*C. sinensis* Osbeck) in harvesting seasons. *J. Chinese Soc. Hort. Sci.* 44(**3**), 265–274.

Mazumdar, B.C. (1976). Physico-chemical changes in fruits of 'Kinnow' and 'Nagpur' Santra during development and maturation. *Punjab Hort. J.* 16(**3–4**), 96–100.

Mendel, K. (1969). The influence of temperature and light on the vegetative development of citrus trees. *Proc. First Int. Citrus Symp. Citrus Congress*, Riverside, California Vo. 1, pp. 259–265.

Miller, J.E., and Hofman, P.J. (1988). Physiology and nutrition of citrus fruit growth, with special reference to Valencias: a mini review. *Proc 6th Int. Citrus Symp.* Vol. 1, Israel, pp. 501–503.

Mustafa, M.M., and Kumar, V. (1997). Studies on the growth and development of Kagzi lime fruits. Abstracts, Nat. Symposium Citric. 17–19 Nov. 1997. NRCC, Nagpur pp. 29–30.

Nagy, S., Nordby, H.E. and Smoot, J. M. (1978). Changing lipid class patterns during maturation of sweet oranges. *J. Agric. Food Chem.* 26, 838–842.

Nogueira, A.P., and Franco (1992). Contribution to the study of phenology and biology of sweet orange (*C. sinensis*) in Portugal. *Proc. Int. Soc. Citric.*, Italy, Vol. 1, pp. 443–445.

Nordby, H.E., and Nagy, S. (1977). Relationship of alkane and alkene long-chain hydrocarbon profiles to maturity of sweet oranges. *J. Agric. Fd. Chem.* 25, 224–228.

Pantastico, B. E. (1975). *Postharvest physiology, handling and utilization of tropical and sub-tropical fruit and veg.* AVI Publishing Co., Westport, CT, USA.

Ramana, K.V.R., Moorthy, N.V.N., Radhakrishnaiah-Shetty, G., Saroj, S., and Nanjundaswamy, A., M. (1980). Physiological and chemical changes in the developing fruits of Coorg mandarins (*C. reticulata* Blanco). *Indian F. Packer* 34(**2**), 3–11.

Randhawa, G.S., Khanna, R.C., and Jain, N.L. (1964). Seasonal changes in fruits and bearing shoots of the grapefruit (*C. paradisi* Macf.). *Indian J. Hort.* 21, 21–32.

Rao, M.M., Hittalmani, S.V., and Bojappa, K.M. (1983). A comparative study on the developmental physiology of Kagzi (*C. aurantifolia* Swingle) and Tahiti (*C. latifolia* Tanaka) lime fruits. *Proc. Int. Citrus Symp.*, Bangalore, Horticultural Society of India, pp. 211–216.

Reddy, P.J., Madhavachari, S., and Reddy, P.V. (1984). Studies on sweet orange fruit maturity. *South Indian Hort.* 32(**6**), 367–368.

Reuther, W., and Rios-Castano, D. (1969). Comparison of growth, maturation and composition of citrus fruits in subtropical California and topical Colombia. *Proc. 1st Int. Citrus Symp.*, Riverside, California, Vol. 1, pp. 277–300.

Roy, A., Haque, R., and Gurung, A. (1999). Studies on the physico-chemical changes of Darjeeling mandarin during growth and development to identify the optimum harvest maturity. *Proc. Nat. Symp. Citric.*, 17–19 November, 1997. NRCC, Nagpur, pp. 271–274.

Sandhu, S.S. (1990). Maturity indices and harvesting of citrus fruits with special references to Kinnow. *Proc. Seminar on prospects and problems of Kinnow cultivation.* PAU, Ludhiana, pp. 211–221.

Satyanarayana, G., and Ramasubbareddy, M. (1994). *Citrus cultivation and protection.* APAU and Department of Horticulture, Andhra Pradesh, Hyderabad, Wiley Eastern Ltd. New Delhi, 66pp.

Sidappa, G.S. (1952). Quality standards for South Indian citrus fruits. *Indian J. Hort.* 9(**4**), 7–24.

Singh, K.K., and Reddy, B.S. (2006). Measurement of mechanical properties of sweet orange. *J. F. Sci. Technol.* 42, 442–445.

Singh, H.K.P.P., Singh, S.N. and Dhatt, A.S. (1998). Studies on fruit growth and development in Kinnow. *Indian J. Hort.* 55. 177–182.

Singh, J.P. and Samaddar, H.N. (1962). Seasonal changes in sweet lime (*C. limettioides* Tanaka). *Indian J. Hort.* 19, 42–49.

Sinha, R.B., Randhawa, G.S., and Jain, N.L. (1962). Seasonal changes in Hamlin and Valencia Late orange. *Indian J. Agric Sci.* 32, 149–161.

Sites, J.W., and Reitz, H.J. (1949). The variation in individual Valencia oranges from different locations of the tree as a guide to sampling methods and spot picking for quality. I. Soluble solids in the juice. *Proc. Am. Soc Hort. Sci.* 54, 1–10.

Sites, J.W., and Reitz, H.J. (1950). The variation in individual Valencia oranges from different locations of the tree as a guide to sampling methods and spot picking for quality. III. Vitamin C and juice content of the fruit. *Proc. Am. Soc. Hort. Sci.* 56, 103–110.

Soni, S.L., and Randhawa, G.S. (1969a). Morphological and chemical changes in the developing fruits of lemon *C. limon* (L) Burm. *Indian J. Agric. Sci.* 39, 813–829.

Soni, S.L., and Randhawa, G.S. (1969b). Changes in pigments and ash contents of lemon peel during growth. *Indian J. Hort.* 26, 21–26.

Subramanyam, H., Narasimham, P., and Srivastava, H.C. (1965). Physical and biochemical changes in limes during growth and development. *J. Indian Bot. Soc.* 44, 105–108.

Surinder Kumar, and Chauhan, K.S. (1989a). Standardization of maturity indices of Kinnow mandarin. *Res. Dev. Reporter* 6, 26–30.

Tadeo, J.L., Ortiz, J.M. and Estelles, A. (1987). Sugar changes in clementine and orange fruit during ripening. *J. Hort. Sci.* 62, 531–537.

Todd, G.W., Bean, R.C., and Propst, B. (1961). Photosynthesis and respiration in developing fruits II. Comparative rates at various stages of development. *Pl. Physiol.* 36, 69–73.

Ukalkar, S.S., and Shankar, G. (1979). Measurement of physico-chemical changes in the developing fruits of mandarin hybrids. *Allahabad Farmer* 50(**4**), 447–448.

Ullah, M., and Haq, M.A. (1984). Effect of maturation on the nutritional factors of Kinnow and Feutrells' Early mandarins. *J. Agric Res.*, Pakistan. 22, 347–350.

Utsunomiya, N., Yamada, H., Kataoka, I., and Tomana, T. (1982). Effect of fruit temperatures on the maturation of satsuma fruits. *J. Japanese Soc. Hort. Sci.* 51, 135–141.

Vinayshankar, and Sinha, M.M. (1987). Studies on physico chemical changes of hill orange and Kinnow mandarin under UP hills conditions. *Progressive Hort.* 19(**3–4**), 183–188.

8

HARVESTING

I. METHODS OF HARVESTING

Harvesting is a very crucial operation in which fresh fruit is removed from the plant after completion of its growth and development. This also marks the last cultivation operation for the crop in the orchard and beginning of its postharvest handling. The method of harvesting, injury to fruit during harvesting, and weather conditions during harvest greatly determine the extent of decay losses during subsequent handling and storage. The snap method (twisting the fruit stem and pulling) is very common in manual harvesting. Clipping is done in specialty fruit. Mechanization of fruit destined for the fresh fruit market is yet to be commercialized, although research efforts have been going on in this direction for quite some time.

A. Manual Harvesting

Citrus fruits are non-climacteric and they mature on the tree. Oranges and mandarins do not fall on the ground at optimum maturity. Overripe fruits fall on the ground long after the development of acceptable taste; however, it is not advisable to wait until that time. It is advantageous in that fruit can be kept longer on the tree for fresh fruit market. It depends on the maturity of the fruit and also its stem-wood, which can break while snap-harvesting or can separate at the abscission zone near the fruit. Oranges harvested in early season (November–December) do not separate at abscission zone but the stem breaks at the junction of button and the fruit stem, leaving the button attached to fruit (Fig. 8.1). This also happens in most other citrus fruits, as the natural abscission zones are at two places in

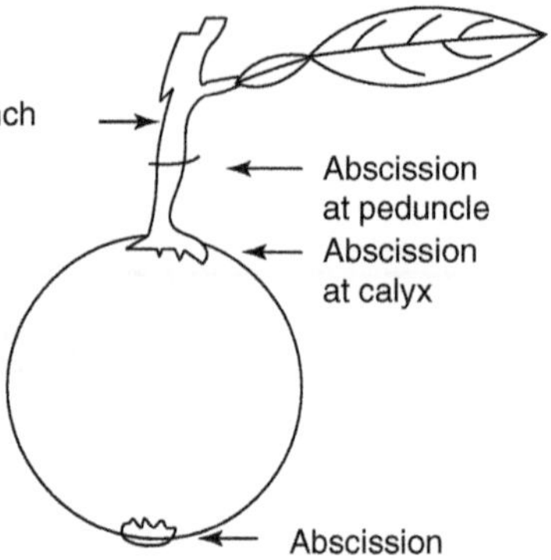

FIGURE 8.1 Abscission Zones (Layers) of Citrus Fruit.

citrus – one between the button and the stem and another between the fruit and the button with the calyx. When harvesting an orange, about 8–9 kg of fruit removal force is required to detach it from stem (Juste et al., 1988). Detachment force decreases as the fruit maturity advances.

Most of the world's citrus crop for the fresh fruit market is harvested manually. If citrus fruits are all harvested at one time, the result is a mix of good- and poor-quality fruits. For the fresh fruit market, fruit has to be spot-picked. Fruits are harvested normally by snap method, either keeping some part of the pedicel attached to the fruit or without any pedicel at all. The peel is torn off while pulling, particularly when mandarin fruits are over-mature and puffy, because the peel becomes brittle. In Florida, the usual method is to snap-pick, which breaks or pulls the stem and sometimes part of the button from the fruit if not properly done (Jackson, 1991). In California and Arizona, citrus fruit are clipped – particularly mandarins – to avoid the danger of infection. While a certain amount of injury to citrus fruits is unavoidable during harvesting, the extent of decay is directly proportional to the amount and severity of injury (Eckert and Eaks, 1989).

The use of ladders and some type of bag or container tied/strapped to the back or waist of the picker is a common practice. In most areas the picker places the fruit into a canvas or plastic sack, which he then empties into a larger wooden, plastic, or metal container kept on the ground. In an improved method of picking, the bag is strapped on over the left shoulder and the picker climbs a ladder and picks fruit between the ladder rungs and on his right side. After reaching the highest point he shifts the bag to the right and picks fruit on left side (Seamount et al., 1972).

The container is then moved from the orchard to the truck for transport to the packinghouse, or the fruit is dumped from the containers into large trailers for transport to a processing plant. The fruit is transported out of the orchard by lift trucks, tractor-drawn trailers, small vehicles on rail systems (Japan), or manually (China). Cable ways or ropeways with trolleys are used to carry fruits on steep slopes in Japan.

South African citrus is harvested traditionally by hand using specially designed clippers. Snap-picking is also common if the crop is mature and results in less rotting. In snap-harvested fruit, *Alternaria* rot was found to increase, but an accompanied benefit was a greater decrease in *Diplodia* rot (Pelser, 1977). Therefore, fruits for export are mostly harvested by snap method. In Morocco, fruit harvest is done by hand, taking maximum care to avoid any damage to fruit that could result in decay (El-Otmani, 2003). In Australia, citrus fruit for the fresh fruit market is hand-harvested, which is a major cost to the grower. Aluminum ladders and mobile power ladders are used to reach the fruit, which is collected in canvas picking bags. Harvested fruit is collected in 0.5–1.5-tons bins and taken to packinghouses in trucks (Gallasch and Ainsworth, 1988). In Spain, harvesting is normally done manually using clippers and baskets. The filled baskets are unloaded into boxes of 18 kg capacity. These boxes are loaded onto trucks for transportation to the packinghouses (Juste et al., 1988).

In Florida, pickers standing on ladders (6–7 m long) generally remove the fruit and collect it in 25 kg-capacity canvas bags strapped over their shoulders. Canvas bags of 20 kg capacity and two shoulder straps are found to be convenient to most pickers regardless of age and sex (Grierson and Wardowski, 1986). Fruit is emptied into pallet boxes with about 400 kg capacity. These pallet boxes are then lifted with forklift trucks and loaded onto trucks for transportation to the packinghouse. In the 1950s, smaller field boxes of 90 lbs capacity were used to collect harvested fruit in the field. In the U.S., harvested citrus fruit is usually estimated in terms of boxes of fruit instead of tons, as in other countries. A box of oranges in Florida has net weight of 40.9 kg (90 lbs), while in California it is 31.8 kg (70 lbs). In California, a lemon box weighs 34.5 kg or 76 lbs. The Florida tangerine harvest unit is a box of 18.1 kg or 40 lbs. Grapefruit in California and Florida are harvested in a box with 36.3 kg (80 lbs) net weight. In Florida, operators other than growers do the harvesting. These operators are cooperatives of growers or private firms. They also do marketing and maintain crews for harvesting, loading, and transporting. Specialized citrus dealers perform the harvesting and deliver the fruit to market.

Pickers are usually paid on a piece-rate basis but also may work at an hourly rate. In Florida, pickers are paid usually for each box rather than on an hourly basis. There is generally a crew foreman who oversees the harvesting and records the number of containers harvested by the picker. The foreman also inspects for improperly harvested fruit, such as off-sized or plugged fruit or fruit harvested with a portion of long stem attached. The long pedicel or stem may cause puncturing of the adjacent fruit. Plugging (rind tearing) occurs in mandarin-type fruit that is not clipped, but poor harvesting leads to plugging of all citrus. In mandarins, economic analysis usually indicates that the higher cost for clipping is justified. Plugging predisposes the fruit to fungal infection and desiccation during transit and packing. Foremen are also responsible for ensuring that the entire crop or the properly sized or colored portion is harvested from a particular orchard or area of an orchard.

Hand-harvesting is arduous work – it is difficult and time-consuming to harvest the fruit of large trees. Out of total harvesting time, a picker spends 75 percent and 60 percent of his time in picking (reaching and detaching) oranges and grapefruits, respectively (Coppock and Jutras, 1960). In California, a picker spends 75 percent and 85 percent of his time picking oranges and lemons, respectively (Ross, 1968).

In some parts of South India, acid limes and mandarins are harvested by pulling off the fruit with the help of a bamboo pole with a scythe attached to one end (Naik, 1948). Harvesting of Kinnows is conventionally done using garden secateurs and even tailors' scissors. 'Kinnow' plucking scissors have also been developed. The sac-bag, basket, bucket with rope, and Israeli auto-empty bags have all been found to be inconvenient. A harvesting device with flexible, wire-reinforced PVC tube with an enlarged feeding mouth has been developed. The picked fruit has to be put into the tube by the picker; the fruit then rolls down to the ground (Jai Singh, 1999). 'Nagpur' mandarins are harvested by the snap method using ladders and bags and sometimes clippers. Pickers collect fruit in a bag or a cloth tied around the waist and collect in round baskets made of bamboo. Fruits are heaped at one place in the orchard on paddy straw and sorted before transportation. Sweet oranges are manually picked by snap method and collected in bags. For 'Sathgudi' oranges, careful clipping with shears, leaving 2 mm of stem, is recommended (Satyanarayana and Ramasubba Reddy, 1994). Sweet oranges and acid limes on the lower part of the canopy, which is easily accessible, are picked by the snap method and collected in bags. To reach acid lime fruit at the top and inside of the canopy, a pole with hook is used and dropped fruits are collected. This is a common practice in some parts of India because acid lime trees are large and thorny. Usually the ground is plowed and soft and most of the time grass or polyethylene or cloth is spread on the ground so that fruit is not damaged (Fig. 8.2, see also Plate 8.1).

The speed of harvesting is slower in clipping because both the hands of picker are engaged in harvesting one fruit. In snapping, picker can harvest two fruits at a time. A picker can harvest 350 'Nagpur' mandarin fruits in an hour by the snap method as opposed to 316 by clipping. The speed of harvesting also depends on the bearing habit of the tree and how much time the picker has to spend to reach the fruit. If mobile, height-adjustable platforms are used instead of ladders, the efficiency of picker is likely to increase because it will be convenient for him to reach the fruit. In Cuba, oranges are harvested using a multi-stand harvesting platform. Harvesting productivity using platforms is 0.48 ton/ha greater and labor requirements and costs are less than conventional manual harvesting (Alonso et al., 1989).

Keeping labor availability and consumption patterns in mind, hand-picking continues to be the only method of harvesting citrus fruits in India and many

FIGURE 8.2 Harvesting of Acid Lime with a Hook as a Practice in Central India. Note the Plain Field Inside the Canopy Covered with Cloth to Avoid Fruit Injury and Contact with Soil.

other underdeveloped and developing countries in the near future. More than 95 percent of produce is consumed fresh and hence mechanical harvesting, which is expensive and most likely to cause bruises and injuries, has little scope in the near future.

1. Injuries to Fruit

Careful harvesting and handling are very important to maintain fruit quality. Although most citrus cultivars have a rigid, tough peel, poor or delayed handling in the field has a significantly deleterious effect on subsequent fruit quality. Reduced fresh marketing is often related to poorer handling, resulting in unreliable delivery condition of the fruit (Grierson, 1981). Fruit for export must be picked with reasonable care, even if a premium has to be paid to ensure this. Oleocellosis is the common hazard and peel turgor should be less. Injury should be minimal to the stem end. The pulling (snapping) method can be preferable to clipping if it is done carefully without plugging. No fruit should ever touch the ground. Care should be taken that pallet boxes are not filled above the top rim. Lemons are to be picked only in rigid framed bags, and once picked, the fruit should be shaded from sun. Oranges and mandarins should be hauled as quickly as possible and never allowed to dry out. Lemons are left undisturbed in the shaded grove for a day or two (for slight water loss/curing) to avoid injury to the peel.

Bruising, plugging, stem-end tears, scratching, and pitting of fruit are common injuries observed in fruits collected from pallet bins (Burns and Echeverria,

1990). Maximum damage is observed at pallet bin collection at the packinghouse. Improper harvesting and handling cause extensive fruit injuries and immediate decreases in pack-out percentages for fresh fruit as well as later increases in decay during storage or transit.

Plugging (rupture of skin at stem-end) or severe button-hole injury and longer peduncle were observed in 'Nagpur' mandarin fruit harvested by conventional methods. Harvesting injuries in the form of button holes at the stem-end have been found to be 8.79–9.14 percent in snap-harvested fruit as compared with negligible (less than 1 percent) injury in clipping. As a result of longer pedicels left on the fruits, particularly during clipping, puncture injuries were recorded in another 1.72 percent of fruits (Sonkar et al., 1999). In the spring-blossom crop, plugging and slight skin rupture at the stem-end (near the collar) accounted for 4.3 percent and 4.49 percent of injured fruit, respectively in snap method. Longer stems/pedicel (10–40 mm and even longer) on 1.58 percent of fruits resulted in injury to 0.25 percent fruits during harvesting and handling operations (Fig. 8.3, see also Plate 8.2). In the monsoon-blossom crop harvested in February–March, the extent of injury or plugging at the stem end has been very low (<0.5 percent). In warm and dry weather – particularly if irrigation is stopped at least 2–3 weeks before harvest – fruit can be readily harvested by snapping. Higher injuries to fruit in the spring-blossom crop harvest season is probably due to more moisture content in peel, resulting in brittleness. This may also result in the slowing of abscission-zone formation. In general, the relative humidity range in spring-blossom (November–December) and monsoon-blossom crop season (February–April) are 60–70 percent and 30–40 percent, respectively. Low humidity in the February–March season results in drying of the stem (a sort of stress) and easy detachment of fruit.

FIGURE 8.3 Injury to 'Nagpur' Mandarin Caused Due to Long Stems of Other Fruit during Handling.

It is observed that at times, 'Nagpur' mandarin fruits with greener stems (where a normal abscission layer is not formed) are harvested by the pickers keeping longer stems/pedicels. In the snap method, the pedicel/button usually does not remain attached to the fruit. However, when the pedicel is still green though the fruit is mature, the picker has to struggle to harvest these fruits. Thus longer stems are left on such fruit, which can injure other fruit. Clippers can be selectively used to harvest such fruit.

In Kinnow mandarins, conventional harvesting methods have been found to cause injuries to fruit. With use of garden secateurs and scissors, a 4–5 mm peduncle is left on the fruit, resulting in puncture damage to 17 percent of other healthy fruit (Jai Singh, 1999).

Christ (1966) showed that the amount of *Penicillium* decay in Navels and Valencias could be directly correlated with the level of injury inflicted during commercial picking and packing. In clipped 'Nagpur' mandarin fruit, whether treated or non-treated with fungicidal wax, decay losses were significantly less as compared with snapping (Sonkar et al., 1999). Although the harvesting rate is slightly less in clipping, the injuries to fruit and subsequent decay can be considerably reduced by this method. Since harvesting injuries are more frequent in the spring-blossom crop, the decay losses are likely to be greater in stored fruit of this season. This corroborates with and proves the popular belief among growers and traders that '*Ambia*' crop (spring-blossom) fruits have low keeping quality (due to decay losses).

The addition of 2.4-D as ethyl ester (500 ppm) in fungicidal wax resulted in a higher percentage of greener pedicels in clipped 'Nagpur' mandarin fruit after storage. In 2.4-D-treated lemons, Stewart et al. (1952) reported delayed button deterioration and a reduced rate of coloration and water loss during storage. The vitality of the fruit button is considered to be the major impediment to development of stem-end rots. To keep buttons green, 2.4-D applications to lemons is a commercial practice in California (Eckert and Eacks, 1989). Wax (highshine wax 2.5 percent) treatment with 2.4-D (500 ppm) and carbendazim (2000 ppm) reduced decay from *Penicillium* rot, particularly in clipped 'Nagpur' mandarin fruit. The carbendazim treatment was ineffective, especially in snap-harvested fruit with slight to severe injury (Sonkar et al., 1999).

2. Cost of Manual Harvesting

Harvesting and handling costs of citrus fruit often equal or exceed total production costs of the crop. The pickers' rate of harvesting varies depending on the kind of fruit (for example, orange or mandarin, difficulty in picking etc.). The total costs of picking, loading, and handling to the plant have been lowest for grapefruit (73.8 cents/56 lit equivalent) and highest for tangerines (1.62 US$/56 lit equivalent) in 1976–77 (Brooke and Spurlock, 1977). Spain is a major producer and exporter of fresh citrus. Fruit harvesting accounted for more than 25 percent of the total production costs, entailing 50–60 percent of hand labor used in cultivation. The high costs and shortage of labor for harvesting poses a major problem (Juste et al., 1992).

In mandarins, economic analysis indicated that the higher cost for clipping (because of a slower rate of harvest) is justified (Davies and Abrigo, 1994), keeping harvesting losses in mind.

B. Mechanical Harvesting

Mechanical harvesting has been tried since the 1970s in Florida for fruits sent for processing. Although most of the world's fresh citrus fruits are harvested manually, research on mechanical harvesting continues to develop the system with fewest bruises and least damage to fruit at a lower cost of picking. Injuries to fruit at the time of harvesting lead to subsequent decay. The percentage of splits, fruits with stems attached, and punctured fruits is higher in mechanically harvested fruits than in manually picked fruits (Whiteny et al., 1973). Fungicide (thiabendazole) treatment has been found to reduce decay in mechanically harvested fruit (Rekham and Grierson, 1971). Keeping in mind labor problems, individual fruit picking using robotics is being considered as a long-term solution in Spain. The mechanical harvesting work for fresh citrus is focussed on this aspect of robot development (Juste et al., 1992).

There are several problems in the mechanical harvesting of citrus fruits. Valencia oranges particularly pose problems because two crops are on the tree (the mature crop of the current season and the green small fruits and even bloom of next year's crop). Mechanical harvesting of fruit destined for processing using mass removal method is usually stopped during peak bloom until immature fruit is 6 mm in diameter (late April to May). Juste et al. (1992) list the major problems encountered in mechanization of harvesting of fruit as follows:

1. The presently used harvesting machinery is very large, while field holdings are small with varying tree sizes, spacing, and age.
2. Irrigation layouts and soil-management practices are different in different fields.
3. In many countries, citrus is grown on slopes, presenting difficulties for mechanization.
4. Breakage of limbs and damage to fruit.
5. Cost/benefit ratio.
6. Some leftover fruit after mechanical harvesting has to be manually removed.
7. Adverse effect of loosening chemicals on tree physiology.

These problems are specifically related to the Spanish citrus industry considering the use of mass-removal machines, but these constraints are true for many other citrus-growing areas.

In mechanical harvesting where a robot arm is used, the detection of fruit and its selective picking under field conditions at the desired rate of picking is also a major challenge keeping in view that fruit is destined for the fresh fruit market.

1. Mechanical Harvesters

Mechanical harvesting machines can be broadly classified as contact machines and mass-removal machines (Coppock, 1978; Coppock et al., 1981). The contact machines consists of (1). the positioning mechanism and (2). the picking hand or arm. The mass-removal machines operate by applying external force, shaking the limb or tree trunk mechanically by holding it or applying force in the form of a jet of water or air to vibrate limbs, foliage, and twigs. In mass-removal machines, fruit drops onto padded catch frames or is allowed to drop on plowed ground. Fruit is collected in pallet bins or open trucks and transported. In general, mass-removal type mechanical harvesting is suitable for fruit destined for processing (Whitney, 1978), while contact machines are useful for harvesting fruit destined for fresh consumption. The contact machines are based on the principle of selective picking and may use mechanical fingers, which are flexible and imitate human fingers (Chen et al., 1982).

For citrus destined for the fresh fruit market, robots or mechanical arms and the torsion method of detachment is considered effective because the calyx is not removed. The distribution of fruit in the canopy has to be considered in designing a robot. The cultivars with maximum fruits in the outer periphery are most suitable since electronic detection and selection is easier. Satsumas and Clementines are the most suitable for this purpose, as 80 percent of the fruit is available 70 and 30 cm from the periphery, respectively in these cultivars. Fruits of Washington Navel and Salustiana oranges are borne 140 and 100 cm inside the periphery, respectively (Juste et al., 1988). According to Juste et al. (1992), some promising results have been obtained in developing prototypes of mechanical arms, vision systems, fruit detectors, and end effectors. Fruit peduncles in 85 percent of fruit have been less than 5 mm and 98 percent of fruits presented intact calices. The injuries to fruit have been negligible to none. The system is very useful for fruit destined for the fresh fruit market.

The use of spectral reflectance and chrominance information to enhance digital color images to control a robotic manipulator for harvesting is possible (Slaughter and Harrell, 1987; Slaughter et al., 1986). The spectral information is useful to differentiate the fruit image from background leaves, tree limbs, soil, and sky. The ability to harvest citrus fruits automatically by means of image processing (vision system) has also been investigated by Kawamura (1985). The field conditions pose difficulties in fruit recognition by automatic mechanical systems or robots for selective harvesting of fresh fruit. The important task for the robot is to recognize fruit against varying backgrounds (green leaves, blue sky, brown branches, and black soil) that have different colors under different light conditions. Fruits are also clustered most of the time. Cerruto et al. (1996) conducted experiments in which images of fruit are taken in RGB (red, green, and blue) system and analyzed under HIS (hue, saturation, and intensity) system because the HIS system offers several advantages, and recognition by this system is very close to the human eye. These researchers have reported real-time fruit recognition by color imaging using video cameras. The automating of system settings and procedures for different light conditions is likely to provide recognition without error. Molto et al. (1998b) reported usefulness of machine-vision technology to properly detect oranges and the exact location of the stem based on image analysis. The electronic arm of the harvester therefore will be able to identify and selectively harvest fruits under the grove conditions.

On the large commercial plantations of Florida and Brazil, where fruit is mostly utilized for processing, mechanization is beneficial as the manual harvesting is becoming uneconomical and laborers are not available when needed. Efficient mechanical harvesting is a key factor in economical fruit production for industrial processing. Mechanical harvesting by a conventional trunk shaker works well in young and uniform orchards, where shaker parameters are pre-adjusted to the uniform trees. In an old or non-uniform orchard, most shakers cannot be operated properly unless the shaker properties are adjusted to the individual tree during the harvesting operation (Galili et al., 1999).

The Florida Department of Citrus has been working with inventors and manufacturers to develop technologies that reduce harvesting costs for processing oranges and increase worker productivity (Brown, 1998). Mechanical harvesting systems are in various stages of development and fruit-abscission compounds are being evaluated. Peterson (1998) has described an experimental, direct-drive, double-spiked-drum canopy shaker to harvest oranges from high-density groves. The drums have horizontal whorls on a vertical shaft and each whorl has nylon rods that penetrate the canopy up to 1 m. Shaking frequency is 4–5 Hz, with maximum horizontal displacement of the rod tip of 250 mm. The shaker can be towed by a tractor along a tree row at travel speeds of 1.4–3.2 km/h. In the canopy space penetrated by the shaking rods, mature fruit removal averaged 71–91 percent. The shaker drums are 3.66 m in diameter with height to harvest trees up to 4 m high. Fruit catching and conveying components are added under the shaker mechanism to collect and transport the detached oranges to the rear center of the shaker unit. A self-propelled bulk transport unit follows the harvest unit at a synchronized speed. The bulk transport unit has a conveying system that receives the oranges from the harvester and transfers them to its rear hopper (with a 6-ton capacity). The system has trash-removal devices also. The grade of fruit received at the processing plant has been as good as hand-harvested fruit.

Trunk shake-catch systems are being commercially used to some extent to mechanically harvest Florida oranges for processing. Whitney and Wheaton (1987) studied the efficiency of air and trunk shakers plus abscission chemicals for orange harvesting. Harvesting efficiencies of the air and trunk shaker averaged 77 percent and 87 percent, respectively. Tree-size management practices did not affect the fruit-removal performance or harvesting efficiencies of the shakers.

Fruits left on the trees by the shakers and those missing the catch frames must be gleaned by hand harvesters. The cost of gleaning reduces or may eliminate the profit for the mechanical harvesting operation. Abscission chemicals to reduce the detachment force of oranges are being tried to increase the removal efficiency of the shakers. Shaker removal efficiencies increase by 10–15 percent when orange detachment force is reduced 50–80 percent (Whitney et al., 1999).

2. Loosening Chemicals

In mass-removal harvesting, some chemical aids (plant growth regulators, particularly abscission agents) are applied so as to loosen the fruit before mechanical harvesting. The effects of these chemicals and the mass-removal method of mechanical harvesting have been studied by several researchers. The important criteria that abscission chemical should meet are (1) selective action on mature fruit within 3–4 days and (2) non-phytotoxic. The chemical should also be inexpensive and eco-friendly.

Ethephon (350 ppm) reduced the fresh fruit quality of Washington Navel oranges and increased wastage, while 'Pick-off' (glyoxal dioxime, 200 ppm) increased wastage in Valencias (El-Zeftawi et al., 1978). Mechanical harvesters reduced fruit quality and increased wastage in all varieties and losses were higher in Washington Navels. Different types of mass mechanical shakers have different impacts and the extent of losses varied. Burns et al. (2005) reported that application of abscission agent 'Release' (5-chloro-3-methyl-4-nitro-1H-pyrazole, CMPN, 17 percent a.i.) ranging from 10–500 ppm at 300 gallons/acre to Hamlin and Valencia trees increased the harvesting capacity of trunk and canopy shakers by reducing the time necessary to harvest each tree while maintaining a high percentage of mature fruit removal. CMPN is selective and non-phyto-toxic.

In various methods of mechanical harvesting (limb shakers, air-blast and trunk shakers, foliage shakers, and robotic arms), when abscission chemicals are not used, a large percentage of fruit comes off with stems that could puncture adjacent fruit during the transporting and handling processes. The de-stemming machine (including roller-cutter assembly) and the oscillating conveyer have been reported to cut stems of oranges at a speed of one fruit per second with destemming of 90–95 percent of the fruit (safe length of 3 mm) fed on the machine (Chen, 1994).

3. Cost of Mechanical Harvesting

The increased necessity of mechanical harvester is mainly due to rising labor costs. According to estimates of Brooke and Spurlock (1977), at least 105 ha of fruit needs to be harvested annually for mechanical harvesting to be cheaper than hand-picking. One trunk shaker and catch system harvests 90–140 trees/h. Adoption of the mechanical harvester eliminates five jobs for each job of equipment operator it creates. Mechanical harvesting is an important tool that enables the Florida citrus industry to be competitive in a global juice market (Futch and Roka, 2004). Growers save 20–50 percent per box compared to conventional harvesting systems. In Sao Paulo (Brazil), the cost of harvesting one box is 0.52 US$, while in Florida it is relatively higher. Therefore, more and more growers are following mechanical harvesting (Neff, 2004).

REFERENCES

Alonso, A., Cardenas, T., Sadourian, S., and Suarez, J.L. (1989). Technical – economic evaluation of multi-stand platform APC – 9 M in citrus fruit harvesting. *Ciencia Y Techica en la Agric.: Mecaniza cion de la Agric.* 12, 15–22.

Brooke, D.L., and Spurlock, A.H. (1977). Production costs – fruit production, harvesting, packing-house, processing. In *Citrus science and technology* (S. Nagy, P.E. Shaw, and M.K. Veldhuis, eds.), Vol. 2. AVI Publishing Co., Westport, CT, USA pp. 141–170.

Brown, G.K. (1998). Florida citrus can be mechanically harvested. *ASAE Annu. Int. Meeting*, Orlando, FL, USA, 12–16 July 1998, 6 pp., ASAE Paper no. 981091.

Burns, J.K., and Echeverria, E. (1990). Quality changes during harvesting and handling of 'Valencia' oranges. *Proc. Fla. State Hort. Soc.* 103, 255–258.

Burns, J.K., Buker, R.S., and Roka, F.M. (2005). Mechanical harvesting capacity in sweet orange is increased with an abscission agent. *Hort. Technol.* 15, 758–765.

Cerruto, E., Manetto, G., and Schillaci, G. (1996). Trials on citrus fruit recognition by colour image. *Proc. Int. Soc. Citric*, Sun City, South Africa, pp. 1122–1125.

Chen, P. (1994). Mechanical de-stemming of oranges. In '*Post-harvesting operations and quality sensing*' (F. Juste, ed.). *IVth Int. symp. fruit veg. prod. Engg.*, Valencia, Spain, Vol. 2, pp. 205–221.

Chen, P., Mehischan, J., and Ortiz-Caflavate, J. (1982). Harvesting Valencia oranges with flexible curved fingers. *Trans. ASAE* 25, 534–537.

Christ, R.A. (1966). The effect of handling on citrus wastage. *S. African Citrus J.* 587, 7, 9, 11, 13, 15.

Coppock, G.E. (1978). Mechanical harvesting and handling citrus fruits. *Proc. Int. Soc. Citric. Int. Citrus Congress*, Australia, pp. 87–91.

Coppock, G.E., and Jutras, P.J. (1960). Mechanizing citrus fruit harvesting. *Trans. ASAE.* 3(2), 130–132.

Coppock, G.E., Sumner, H.R., Churchill, D.B., and Hedden, S.L. (1981). Shaker method for selective removal of oranges. *Trans. ASAE.* 24, 102–104.

Davis, F.S., and Albriago, L.G. (1994). *Citrus*. CAB International, Wallingford, Oxon, UK, 254 pp.

Eackert, J.W., and Eaks, I.L. (1989). Postharvest disorders and diseases of citrus fruits. In '*The citrus industry*'. (W. Reuther, E.C. Calavan, and G.E. Carman, eds.) Vol. V, pp. 179–260. Divinan. Agriculture Science University of California, Berkeley, California.

El-Otmani, M. (2003). Citriculture in Morocco. *Citrus Ind.* 84 (11), 22–23.

El-Zeftawi, S.M., Thornton, I.R., and Gould, I.V. (1978). Effects of mechanical shake-removal on citrus fruit quality. *Proc. Int. Soc. Citric. Int. Citrus Congress*, Australia, pp. 106–109.

Futch, S.H., and Roka, F.M. (2004). Trunk shaker mechanical harvesting system. *Citrus Ind.* 85(7), 20–21.

Galili, N., Rubinstein, D., and Shdema, A. (1999). Adaptive shaker for mechanical harvesting of olives and citrus fruits. *ASAE-CSAE-SCGR Ann. Int. Meeting*, Toronto, Ontario, Canada, July 18–21, 1999, 13 pp., ASAE Paper No. 997061.

Gallasch, P.T., and Ainsworth, N.J. (1988). Developments in the Australian citrus industry. *Proc. 6th Int. Citrus Congress*, Israel, Vol. 4, pp. 1613–1623.

Grierson, W. (1981) Harvesting Florida citrus for overseas export. *Proc. Fla. Sta. Hort. Soc.* 94, 252–254.

Grierson, W., and Wardowski, F. (1986). Transportation to packing house. In *Fresh citrus fruits* (W. Wardowski, W. Grierson, and S. Nagy, eds.), AVI Publishing Co., Westport, CT, USA. pp. 227–242.

Jackson, L. (1991). Citrus growing in Florida. University of Florida Press, Gainesville, FL, USA, 293 pp.

Jai Singh (1999). Equipments to mechanize harvesting and handling Kinnow. *Proc. Nat. Symp. Citriculture*, Nagpur, 17–19, 1997. NRCC, Nagpur.

Juste, F., Gracia, C., Molto, E., Eranez, R., and Castillo, S. (1988). Fruit bearing zones and physical properties of citrus fruits for mechanical harvesting. *Proc. 6th Int. Citrus Congress*, Israel, Vol. 4, pp. 1801–1809.

Juste, F., Ferres, J., Pla, F., and Sevila, F. (1992). An approach to robotic harvesting of citrus in Spain. *Proc. Intn Soc. Citric*, Italy, pp. 1014–1018.

Kawamura, N. (1985). Vision of fruit for the development of fruit harvesting robots. *Proc. Third Int. Conf. on Physical Properties of Agricultural Mater.*, Prague, Czechoslovakia, pp. 445–450. Agricultural Engineering Department, Kyoto Univeristy Japan.

Molto, E., Ruiz, L.A., Aleixos, N., Vazquez, J., and Juste, F. (1998). Machine vision for non-destructive evaluation of fruit quality. *Acta Hort.* 421, 85–90.

Naik, K.C. (1948). South Indian fruits and their culture. P. Varadachari Co. Madras. 335 pp.

Neff, E. (2004). Mechanical harvesting: it is all about the costs. *Citrus Ind.* 85(4), 18.

Pelser, P. (1977). Postharvest handling of South African citrus fruit. *Proc. Int. Soc. Citric., Florida* Vol. 1, pp. 244 249.

Peterson, D.L. (1998). Mechanical harvester for process oranges. *Appl. Engg. Agric.* 14(5), 455–458.

Rackam, R.L., and Grierson, W. (1971). Effect of mechanical harvesting on keeping quality of Florida Citrus for fresh fruit market. *HortScience* 6, 163–165.

Ross, J. (1968). Multiman picking machines and systems in California. *Proc. First Int. Citrus Symp., California*, Vol. 2, pp. 647–651.

Satyanarayana, G., and Ramasubba Reddy, M. (1994). *Citrus cultivation and protection*. APAU and Department of Horticulture, Andhra Pradesh, Walley Eastern Ltd. New Delhi. p. 66.

Seamount, D.T., Nash, P., and Opitz, K.W. (1972). An improved method of ladder and bag picking. *Citrograph.* 58(11), 397–398.

Slaughter, D.C., and Harrell, R.C. (1987). Colour vision in relative fruit harvesting. *Trans. ASAE* 30(4), 1144–1148.

Slaughter, D.C., Harrel, R.C., Adsit, P.D., and Pool, T.A. (1986). Image enhancement in robotic fruit harvesting. *Trans. ASAE* 29, 1137–1152.

Sonkar, R.K., Ladaniya, M.S., and Singhs (1999). Effect of harvesting methods and post-harvest treatments on storage behaviour of 'Nagpur' mandarin (*Citrus reticulata* Blanco) fruit. *Indian J. Agric. Sci.* 69, 434–437.

Stewart, W.S., Palmer, J.E., and Heild, H.Z. (1952). Packing house experiments on the use of 2,4-trichlorophenoxy acetic acid and 2,-4-5 trichlorophenoxy acetic acid to increase storage life of lemons. *Proc Am. Soc. Hort. Sci.* 59, 327–334.

Whitney, J.D. (1978). Air-shakers for removal of oranges in Florida. *Proc. Int. Soc. Citric*, Australia. pp. 91–93.

Whitney, J.D., and Wheaton, T.A. (1987). Shakers affect Florida orange fruit yields and harvesting efficiency. *Appl. Engg. Agric.* 3, 20–24.

Whitney, J.D., Hartmond, U., Kender, W.J., Burns, J.K., and Salyani, M. (1999). Orange removal with trunk shakers and abscission chemicals. *ASAE-CSAE-SCGR Ann. Int. Meeting*, Toronto, Ontario, Canada, 18–21 July 1999, 9 pp. ASAE Paper No. 991078.

9

PREPARATION FOR FRESH FRUIT MARKET

Postharvest treatments are applied to citrus fruits before storage in order to delay senescence, minimize spoilage, and improve appearance and marketability. Degreening is done mainly to improve color. Applications of surface coatings, fungicides, and other chemicals are common before fruits are marketed or stored under ambient or refrigerated conditions. These treatments are also effective in reducing chilling injury in refrigerated storage. Plant growth regulators are used to delay aging/senescence and various fungistats are being used to control rots. Gamma radiations are also being tried to reduce microbial spoilage or disinfest the fruits with fruit flies (see Chapters 15 and 18). These are supplemental treatments and cannot substitute refrigeration for long-term storage. Pre-cooling is done before fruits are stored under refrigerated conditions. Safer and more effective supplementary treatments are being developed throughout the world. Chapter 16 discusses common fungicide treatments and emerging eco-friendly, non-hazardous treatments such as hot water treatment, heat treatment, UV-rays, and bio-agents. Commonly used postharvest treatments such as degreening, curing, wax coating, the application of growth regulators, and packing-line operations are discussed in this chapter, with a focus on packinghouse activities with regard to treatments to fruit.

I. DEGREENING

Consumers prefer brightly colored citrus fruit and are willing to pay a premium for them. Green-colored fruits are considered unripe and fetch lower prices. Hence the color of the rind is important for the aesthetic value and as such it is the most important factor determining marketability. Chlorophyll imparts green color to citrus fruits, while carotenoids give yellow, orange, and reddish-orange colors. The external color of citrus fruit is not a reliable index of internal maturity, but it does indicate maturity to a certain extent when color develops under normal conditions in the field. The color of the juice is also important, as it is used as a criterion for determining grade and hence juice price. In general, juice color can also be related to pro-vitamin A content, especially when the predominant pigment is cryptoxanthin, as with oranges and mandarins/tangerines. In red grapefruit, color is due to lycopene and hence not related to vitamin A. The deep orange color of 'Nagpur' mandarin (*Citrus reticulata* Blanco) pulp is due to cryptoxanthin, while the peel contains violaxanthin, cryptoxanthin, and neoxanthin carotenoids (Subramanyam and Cama, 1965).

In tropical climates, oranges and mandarins may show a color break only when mature. In the subtropics, citrus fruits (depending on their natural color) become orange or red, as with brightly colored mandarin/tangerine hybrids such as Minneola and Osceola. The most attractive color in citrus fruit develops in those grown in dry climates where nights are cool with warmer days during maturation time. Studies in Florida have shown that a night temperature below 13°C initiates the desired color. Carotenoids are temperature-sensitive, and at lower temperatures even very low concentrations produced by fruit are sufficient to induce color.

Less degreening is required for the Florida citrus varieties as the season progresses until it is no longer necessary in December and January except for some early maturing fruit of later varieties such as Valencia and Murcott Honey tangerines. Robinson tangerines can only be degreened for short periods because of peel injury and decay problems (McCornack and Wardowski, 1977).

Development of color is governed by several factors such as fruit maturity, tree nutrition, rootstock, cultivation practices, water availability, and temperature. Fruit color also depends on climatic conditions and groundcover in the orchard. Carotenoid content was higher with lower chlorophyll content in the rind of Valencia fruit from the groundcover plots than in fruit from the bare-ground plots (Eaks and Dawson, 1979).

In the more temperate regions of the world where citrus is grown, the final color of most orange cultivars is produced by decline in chlorophyll pigments and accumulation of carotenoids. In most lemon, lime, and grapefruit cultivars, the final color is produced by decline in chlorophyll but little or no net increase in carotenoids. The fact that certain exogenous growth regulators have a minor impact on coloration supports the speculation that endogenous growth regulators are closely linked to coloration. The development of chloroplast and chromoplast is influenced by endogenous growth regulators. Ethylene causes loss of chlorophyll and produces minor changes in carotenoids, while GA and cytokinin cause a delay in the loss of chlorophyll and produce minor changes in carotenoids. Exogenously applied 2,4-D delays the loss of chlorophyll and has little or no influence on carotenoids. CPTA [2(4–Chlorophenyl thio) triethylamine] promotes the accumulation of carotenoids and lycopene. R-33417 (2,4-Dichloro-1-cyanoethanesul phoncinilide) causes rapid and uniform loss of chlorophyll in the presence of light (Coggins and Jones, 1977).

A. Degreening Conditions

The purpose of degreening treatment is to improve the esthetic value of the fruit – it does not help in extending keeping quality. Early-season citrus varieties in subtropical climates usually become edible when the rind is still green. In tropical climates, oranges and mandarins do not develop an attractive color at the time of maturity and hence require degreening. They therefore have to be treated with ethylene or ethephon to accelerate chlorophyll breakdown and the development of an orange color. Analysis of rind samples revealed changes in the relative concentration of chlorophyll, a, and b and a consequent decrease in the a/b ratio as total chlorophyll levels decrease with fruit maturation and degreening (Jahn and Young, 1976). The use of the dye citrus red #2 is limited to few cultivars. It is allowed in the U.S. and some other countries. For degreening of citrus fruit, treatment with 1–5 ppm ethylene at 20–29°C and 90–96 percent relative humidity is recommended. A ventilation facility in the storage room is essential to reduce ethylene to the threshold level after treatment because excess ethylene stimulates stem-end rot decay. Ethylene also induces carotenoid accumulation in lemons and tangerines

(Young and Jahn, 1972). In general, the color of all harvested citrus fruit improves by holding at 15–25°C temperature with ethylene concentration 1–10 ppm. Ethylene can be substituted with ethephon (2-chloroethyl) phosphoric acid (2-CEPA) at 500–1000 ppm. Application of 2(4-chlorophenylthio) triethylamine (CPTA) can improve rind color of citrus by means of induction of carotenoids (Jahn and Young, 1975).

The role of ethylene in citrus coloring falls into at least two categories. The first is for degreening (degradation of chlorophyll), which takes place at 30°C with 5–10 ppm C_2H_4. The second is for cool coloring, where ethylene is required for biosynthesis of β-citraurin. This carotenoid is temperature-sensitive. At decreasing temperatures decreasing levels of C_2H_4 are required. At 15°C significant amounts of color develop without the addition of ethylene; presumably this is caused by endogenous C_2H_4 (Stewart, 1977). Ethylene application induces the accumulation of carotenoids such as cyptoxanthin, β-citraurin, and some violaxantin in flavedo (Stewart and Wheaton, 1972). At 15°C the highest rate of chlorophyll degradation was observed at 1 μl/l ethylene. Higher ethylene concentration resulted in lower rates of chlorophyll loss. At 1250 μl/l ethylene, the rate of chlorophyll loss was slower than in untreated fruits (Knee et al., 1988).

Gaseous concentration, particularly that of oxygen, also affects the degreening rate. Exposure to C_2H_4 levels at 5–10 ppm caused rapid losses of chlorophyll in Hamlin oranges with two days of degreening. High O_2 (50 percent) alone increased the rate of degreening, but high $O_2 + C_2H_4$ did not produce any further degreening response compared with C_2H_4 alone. Low O_2 (10 percent) reduced the degreening response to C_2H_4 in both Hamlin and Washington Navel oranges. The rate of degreening was slower at 21°C than at 29°C; this is observed particularly at initial stages when chlorophyll content in the rind is higher. Degreening continued after removal from degreening chambers and resulted in modification or elimination of the original treatment differences (Jahn et al., 1969).

The initial rate of chlorophyll loss is more rapid at 30°C than at 21°C, but subsequent changes at 21°C were such that the fruit reached an acceptable color almost as rapidly as at 30°C after 4 days of degreening, Fruit at both temperatures showed the same response to ethylene concentration. Both ethylene concentrations of 5 and 10 ppm resulted in the best color development. Responses at 24°C and 27°C were similar to those at 21°C and 30°C respectively. There was no evidence of seasonal change in the response to ethylene. Stem-end decay generally was greater with increasing ethylene concentration, temperature, and length of the degreening period. Increasing the temperature or the length of the degreening period resulted in greater decay at high rather than at low ethylene levels (Jahn et al., 1973).

Continuously higher temperatures (30–35°C) accelerate the loss of green color, but the fruits become pale yellow. A higher temperature was beneficial only during early degreening when chlorophyll content was high. Different levels of RH (70–90 percent) during degreening do not affect color development (Cohen, 1978). However, it is always desirable to have higher humidity – up to 90 percent – to avoid shriveling and gas burn. Grierson et al. (1986) have reported on an RH control system used in Florida where a lithium-chloride humidity sensor (humistat) is placed in the return air stream and a temperature sensor (thermostat) in the delivery stream. When humidity falls below the set level, the humidifier starts automatically. This system is cheap and satisfactory. Higher temperatures and ethylene concentrations and low rates of ventilation are factors responsible for the buildup of high CO_2 levels in degreening rooms. CO_2 *per se*, even at higher concentration had no inhibitory effects on the color development or physiological response of the degreened fruits. Sufficient O_2 is needed for the degreening process (Cohen, 1977). The ventilation with air change of 0.5–1 air changes per hour is satisfactory, but depends on degreening room design and other parameters. The air circulation within the room should be such that air moves through the fruit load rather than around the fruit in order to achieve uniform ethylene distribution and maintain O_2, CO_2, and RH levels (Grierson et al., 1986). Optimum temperature/ethylene combinations vary with respect to fruits of different citrus species. In some instances, manipulating temperature and increasing ethylene concentration increased the rate of degreening by over 200 percent. The degreening rate is greater at the higher temperature (29.4°C) and at 100–250 ppm ethylene (Ahrens and Barmore, 1987).

The preharvest ethephon spray had unpredictable results, and the concentration for coloring and defoliation greatly overlap. The main drawback to use of ethephon is that is effective for only 3 days (Stewart, 1977). Chapter 4 describes the effect of ethephon and some other chemicals as preharvest applications in greater detail. Preharvest applications of ethephon to improve color is successful to some extent but must be used cautiously.

B. Ethephon, Ethylene, and Other Compounds Enhancing Rind Color

Ethylene is used in gaseous form or in the form of an ethylene-releasing compound that decomposes in fruit tissues, leading to degreening and improving appearance. The method selected for applying ethylene depends on cost, convenience, and safety factors.

Compounds that decompose in or on the fruit to release ethylene have the advantage of easy application. For example, 2-chloroethyl phosphonic acid, which is commonly called ethephon (Ethrel as commercial product), has been used as a source of ethylene for decades. Besides enhancing the degreening of citrus fruits it also initiates ripening in climacteric fruit. Ethephon is hydrolyzed in plant tissue to produce ethylene, phosphate ($PH_2O_4^-$), and chloride (Cl^-).

$$ClCH_2 - CH_2 - PHO_3^- \text{ (ethephon)} \xrightarrow{OH^-} C_2H_4 \text{ (ethylene)} + PH_2O_4^- + Cl^-$$

Ethylene (C_2H_4) can also be released from Ethrel by mixing it with alkali such as sodium hydroxide. Ethephon dip has the same effect as exposing the fruit to ethylene gas for 24–48 h. Since fruits are to be dipped in ethephon solution, its use may be governed by food regulations in different countries. Ethylene, which is a hydrocarbon, is available from petroleum refineries and supplied in large steel cylinders as a compressed gas. Cylinders may vary in capacities of 14–15 kg (nearly 10–12.9 m^3) and 2–3 kg (1.5–2.6 m^3) ethylene. Because ethylene is highly flammable (3 percent and above up to 33 percent) it is usually supplied after diluting with nitrogen. Mixtures are 95 percent nitrogen and 5 percent ethylene or 90 percent nitrogen and 10 percent ethylene. The method of application is to meter the gas into the degreening room containing the fruit. The volume of the room is calculated and the volume of ethylene introduced is controlled with a flow meter and needle valve. Ethylene concentration is calculated from fresh air introduced as the gas is injected along with fresh air using a blower.

Ethylene generators are also available and are placed in degreening/ripening rooms. Ethylene is generated by heating ethanol or some other liquid compound (which is a trade secret) in a controlled manner in the presence of a catalyst. A measured quantity of liquid is poured into the generator to achieve the required concentration over a period of time in a room with a measured/fixed volume. This method offers a better chance of achieving the desired degreening even though room is not perfectly gas-tight.

Other gases, particularly unsaturated hydrocarbons of the ethylene family, have been shown to initiate ripening/degreening. However, these gases are considerably less effective than ethylene (Burg and Burg, 1967). The most commonly used chemical other than ethylene for ripening initiation is acetylene.

Acetylene, or C_2H_2, is generated from calcium carbide (CaC_2) because it is cheaper than ethylene and easier to apply in simple ripening rooms or even fruit containers. Calcium carbide is a byproduct of the iron and steel industry and the material available contains impurities. The gas is released when the calcium carbide is exposed to moisture (H_2O) [$CaC_2 + H_2O = C_2H_2 + CaO$]. It is commonly wrapped in a small amounts (just a few grams) in a pouch of paper and these pouches are placed among the fruit. High humidity reacts with the calcium carbide, creating a slow release of acetylene. It is a common practice of retailers to color fruits in some south Asian countries. If large quantities of acetylene are required, small amounts of calcium carbide can be dropped into a bucket of water in a closed room where fruit is kept.

Care must be taken in handling these chemicals and gases – ethylene and acetylene both are toxic and explosive. Flames, cigarette smoking, and electrical sparks should be strictly prohibited in areas where these chemicals and gases are kept.

1. Use of Ethephon for Degreening

Ethephon dip as postharvest treatment is the most common and simplest method of color improvement in citrus fruits. The results of citrus degreening with Ethrel (ethephon) have been quite encouraging in many countries. The optimum concentration of treating solutions is 500 ppm or more and a dipping time of 5–10 min is satisfactory. Wetting agents have no effect when added to an ethephon solution. The dilute solution of ethephon deteriorates rapidly and hence must be used immediately (Tesson, 1970). Postharvest dips in 500–4000 ppm ethephon resulted in the satisfactory degreening of Cyprus-grown lemons and

Marsh grapefruit (Vakis, 1976). The best concentration was related to the maturity of the fruit. Rates of color change were comparable or better than with ethylene degreening. Subsequent waxing of ethephon-treated fruit drastically arrested color changes in Marsh grapefruit but had little effect on degreening of treated lemons. Less oleocellosis was recorded in ethephon-treated fruit than with ethylene treatments, but button abscission is accelerated by ethephon.

Korean Satsuma fruits treated with ethephon (250 or 500 ppm) dip had a significant change in color – the orange color developed considerably in 5 days and was fully developed in 8 days (Oh et al., 1979). Peel chlorophyll content decreased as the orange color developed. Fruit quality was not adversely affected.

Jahn (1973) tried ethephon up to 8000 ppm concentration as a dip to Bearss lemons, Robinson and Dancy tangerines, Hamlin oranges, and Marsh grapefruits and held the fruit at 10°C, 21°C, or 7°C. Responses differed among cultivars, but were greater and more rapid as temperature increased. Untreated fruit degreened at 16°C rather than at a higher temperature. Degreening was maximum at 21°C and 27°C with concentration of 500–1000 ppm. Robinson tangerines responded to lower concentrations and showed inhibition in color development at 2000–8000 ppm ethephon. Tangerines and grapefruits degreened with ethephon approached the same color as that obtained with ethylene. Ethephon increased stem-end rot, which was greatest at the higher concentration and temperature but lower and later than with ethylene degreening.

Various methods of ethephon application and ethylene gassing in Cuban Valencia oranges (internally ripe with color break) have been compared (Ma de Los et al., 1981). Fruits were packed after the treatments. After 24 days of storage at 24°C, the yellow-orange color developed in washed and ethephon dip-treated (3 min – 2000 ppm) fruits, which was attributed to washing, where natural wax was partially removed and ethephon was absorbed easily. In aqueous emersion and spray application of ethephon, color development was relatively slower. In control, poor color was obtained. In waxed fruit, color changed from greenish-yellow to yellowish-orange, but the final color was only yellow in waxed oranges. Jahn (1976) observed that degreening occurred in waxed Hamlin oranges and Dancy tangerines, but the color was not satisfactory, mainly because of poor carotenoid development. Ethylene degreening was quicker than ethephon degreening.

In Central and South India, mandarins and sweet oranges do not develop attractive peel color, although the desired taste and aroma is attained. In the subtropical climate of North India (cool nights and warmer days when fruit is ripening), lack of desired color is not a problem. The color of 'Coorg' mandarins grown in South India was improved by Ethrel dip (1000 ppm) or smoke treatment for 24 h followed by storage at 24–28°C and 88–90 percent relative humidity (Ramana et al., 1973). Addition of benomyl or thiabendazole (0.1 percent) in ethephon reduced storage rots (Ramana et al., 1979). Similar results have been reported by Singh et al. (1978c) in Coorg mandarins and Hamlin sweet oranges. Pearl tangelos and Dancy mandarins developed color at a lower (750 ppm) concentration of Ethrel (Gupta et al., 1983). Green Valencia and Mosambi sweet oranges developed a yellow color within 5–7 days at room temperature after Ethrel dip (1000–2000 ppm) and lost 6–7 percent weight during degreening (Arora et al., 1973; Chauhan and Parmar, 1978; Purandare et al., 1992). Ethrel concentration of 3000 ppm is the best treatment for bright-yellow color development in Mosambi oranges (Sanghvi and Patil, 1983).

Mature green acid lime fruits (*Citrus aurantifolia* Swingle) can be degreened by ethephon dip (500–2000 ppm) for 1 min at room temperature. In 3 days about 70 percent of the fruits turn yellow following ethephon treatment at 750 ppm and above (Gautam et al., 1977). Rana and Chauhan (1976) tried normal light and dark conditions for holding lime fruit during degreening (after Ethrel dip) but chlorophyll content was not significantly different under these two conditions. TSS and acidity content were higher in fruits held in the dark. In Italian Round lemons, ethephon-dip (1000 ppm) treatment gave the best results (Josan et al., 1981).

The major drawback of ethephon dip application is that effective concentration of ethylene gas is not maintained in the atmosphere around the fruit, and degreening takes 5–8 days after the treatment. Sulikeri et al. (1980) used Ethrel to fumigate mature Tahiti lime fruits in an airtight chamber instead of dipping. After 5 days, Ethrel-treated fruits were degreened. Partial color developed in 33 percent of the fruit, while completely colored fruits were 77 percent. In untreated fruits the corresponding percentages of partially and completely degreened fruits were 27 percent and 5 percent respectively.

Ethephon releases ethylene gas in the presence of a base. One mole of CEPA (2-chloroethyl phosphonic acid) was found to release nearly one mole of ethylene in presence of a base (Warner and Leopold, 1969). Reid (1985) reported that 28.3 ml of ethylene gas can be released from 206 ml of ethephon.

A successful method of degreening fruit with ethylene gas liberated from ethephon has been demonstrated in 'Nagpur' mandarins grown in Central India (Ladaniya, 1998). In the intermittent degreening technique, ethylene was generated from ethephon by adding sodium hydroxide in a closed chamber where temperature and humidity were controlled. A treatment cycle of 12 h of ethylene (75–90 ppm) followed by 1 h of ventilation resulted in orange color development in stage 1 (two-thirds to one-half green surface), stage 2 (one-quarter to one-half green surface) and stage 3 (one-quarter or less green surface) fruits after 4.5, 3.5, and 2.5 days, respectively. This technique was found to be very cost-effective, simple to adopt and effective in terms of color development and ripening because it could also be used for climacteric fruits. As the storage decay increased with degreening period, fruits of stage 3 were considered the best for degreening.

2. Degreening with Ethylene Gas

Although ethephon dip is simple and inexpensive, it has some drawbacks. Rind color was yellow and not orange when green 'Nagpur' mandarins were dip-treated with 2500–5000 ppm ethephon (Kohli and Ladaniya, 1988). Ethephon degreening is slower (Jahn, 1973), and green oranges that responded to ethylene failed to degreen with ethephon. However, the response of mature fruits was better. After ethephon-dip treatment, the fruit required appropriate temperature and ethylene concentration for faster color development (Wheaton and Stewart, 1973). Therefore, degreening with ethylene has become a common commercial practice for large-scale operations worldwide.

Gas is metered into the degreening chamber along with fresh air and simultaneously ventilated in a continuous flow process at a controlled temperature and relative humidity. This method is more than five decades old and several changes/modifications have taken place to improve efficiency and minimize the adverse effect of the treatment on fruit quality and storage ability. In Australia, Jorgenson (1969) compared the standard but old shot method and new trickle method of ethylene application. In the trickle method of ethylene application for degreening, the fruit is kept in a continuously ventilated chamber at a controlled temperature and RH with a continuously maintained (trickling) low concentration of ethylene. This method was more effective than the standard Australian shot degreening system, in which fruit is kept in a closed chamber at a certain concentration of ethylene for 1–12 h with no environmental control. The trickle system with ethylene at 10 ppm was more rapid with respect to color development than the previously used shot method with ethylene at 250–1000 ppm. For optimum results, a temperature of about 29°C and 95 percent RH were required (Jorgensen, 1977).

The operational conditions for degreening with the trickle method of ethylene application have been described by Grierson and Newhall (1960). In three types of degreening rooms – slatted-floor rooms, solid-floor rooms (both for fruit in boxes), and vented degreening bins in which the fruit is handled in bulk – a 29.5°C dry-bulb temperature, 28°C wet-bulb temperature, and ethylene delivery at 1 bubble per minute per 10-box capacity have been recommended. A fan capacity of 0.21–0.28 cum/min per 1-box capacity (1 box = 2.232 bu) was considered necessary for adequate air movement. Ventilation to prevent CO_2 accumulation was recommended either by continual air intake of approximately 2 percent of the total volume of room per minute or else complete airing twice daily.

McCornack and Wardowski (1977) recommended degreening conditions for Florida citrus as 1–5 ppm ethylene concentration, 27.8–29.4°C, 90–96 percent RH, and an air circulation rate of 47.2 lit/s/408 kg fruit or one fresh air change per hour. Fruit that meets legal maturity standards internally and is exposed to these conditions can be normally degreened within 72 h. In the initial stage of degreening, 30°C for rapid chlorophyll loss and thereafter at 20–25°C for carotenoid synthesis are usually recommended for satisfactory results (Cohen, 1991). The trickle method of ethylene application is commonly used in large degreening rooms.

In Japan, Kitagawa et al. (1971) tried different methods of ethylene application. In small-scale tests with Satsuma mandarins, baskets of fruit were wrapped in laminated polyester/polyethylene film (agriculture grade, 0.2 mm, PVC film) and treated with ethylene for 15 h. After exposure to air, the chlorophyll disappeared rapidly and coloring was greatly hastened. Ethylene concentration of 500–1000 ppm was satisfactory. Further scaling up the operation, Kitagawa et al. (1977) developed a method using a 0.2 mm thick plastic (PVC) cover for storing fruit. The ethylene concentration of 1000 ppm and contact time of 15 h were most suitable. Within 3–4 days after removal from the closed plastic cover, chlorophyll degradation occurred. 20–25°C was found to be the most effective temperature.

Cohen (1981) compared three methods of degreening and found that during the 15-h ethylene treatment, ethylene and CO_2 concentration were very high and O_2 decreased considerably inside the Satsuma fruit and in the atmosphere surrounding it. These conditions did not allow color development. When the chlorophyll did not disappear completely after a single treatment, a repeated degreening was required. The intermittent degreening method (fruit was exposed to 5–10 ppm ethylene by trickle method at 25°C for 12 h/day followed by 12-h interruption) gave the best results for variable color. The third method that he tried was fumigating fruits with fumes from a kerosene stove. It was not applicable in large-scale operations and did not produce uniform color; additionally it led to higher decay.

Intermittent degreening with 12-h cycles of ethylene gas has been found to produce similar results as that of continuous flow method, with 75 percent cost savings in ethylene gas and 30 percent cost savings on energy (Cohen, 1991). Ethylene treatment for 12 h followed by heating for twelve hours with higher RH and a low rate of air movement (without ethylene) is followed in Israel. This results in less decay. Once color development starts, it is unaffected by a break in ethylene supply. The duration of degreening in the two methods was the same (48–72 h). During interval ethylene concentration dropped by 0–1 ppm while temperature dropped by only 1–2°C. Under these conditions, no buildup in atmospheric CO_2 was noted.

In India, ethylene treatment was standardized for the 'Nagpur' mandarin for first time during early 1990s, using compressed gas applied through the trickle method in specially designed degreening chambers for 1 ton of fruit. The ethylene concentration of 5 ± 2 ppm by trickle method (continuous flow) was equally effective as 25 ± 2 ppm, 50 ± 2 ppm, and 100 ± 2 ppm with respect to efficacy in color improvement in 'Nagpur' mandarins (Ladaniya, 2000). 'Nagpur' mandarin fruits with initial yellowish-green color (50 percent surface green) needed 48 h for degreening as compared to 24 h for yellowish-orange fruit (25 percent surface green). Longer ethylene exposure resulted in higher incidence of stem-end rots. Complete degreening of green but mature Mosambi orange fruits (TSS 9–10 percent and titratable acidity nearly 0.30 percent) was achieved within 48 hours with 5–10 ppm ethylene at 27–29°C and 90–95 percent relative humidity with four air changes per hour and air circulation of 0.5–0.6 lit/sec/kg fruit. Ethylene-exposed fruit developed a bright-yellow color with a hue angle of less than 90°, while non-exposed fruit remained green with a 103° hue angle after fourtyeight hours (Table 9.1). The 48 h treatment resulted in 1.95 percent and 2.15 percent weight loss in ethylene-exposed and control fruit, respectively. Fruits stored without ethylene treatment also developed yellow-orange color but at the end of 30 days of storage, with insignificant differences in the hue angle of treated and non-treated fruit (Table 9.2) (Ladaniya and Shyam Singh, 2001).

TABLE 9.1 Color Development in 'Mosambi' Sweet Orange During Degreening

	Color								
	'L'			Chroma			Hue angle (θ)		
Treatment	At hours								
	0	24	48	0	24	48	0	24	48
Ethylene	62.69	70.25	76.31	42.29	45.31	59.04	110.7	90.07	89.70
Control (No ethylene)	61.43	63.98	67.04	41.82	42.54	48.94	111.8	108.8	103.8
$P = 0.05$	NS	0.76	1.71	NS	NS	7.93	NS	3.83	1.28

Ladaniya and Shyam Singh (2001) 'L' = lightness (0 = maximum darkness, 100 = maximum lightness), Chroma (0 = lowest intensity, 100 = greatest intensity), Hue angle (0° = red purple, 90° = yellow, 180° = bluish-green, 270° = blue). NS, Non-significant.

TABLE 9.2 Color Values in Degreened and Non-degreened 'Mosambi' Fruits during Ambient Storage

	Color								
	'L'			Chroma			Hue angle (θ)		
	Days								
	0	15	30	0	15	30	0	15	30
Degreened	77.57	77.87	78.28	65.66	66.73	70.03	86.02	85.22	81.64
Control	70.85	74.09	75.48	52.95	62.65	62.58	100.0	89.09	82.24
$P = 0.05$	2.45	1.90	1.32	2.82	1.30	2.69	2.15	1.75	NS

Ladaniya and Shyam Singh (2001).

C. Degreening Rooms

Commercial degreening chambers have a holding capacity of 50–100 tons of fruit or even larger. The ethylene application, air movement, ventilation, temperature, and relative humidity are critically monitored during the commercial operations. Gas concentration, temperature, and air movement must be uniform in every corner of the room and also within pallet bins. Miller and Ismail (1995) reported that a computer-based monitoring and control system in Florida citrus packinghouses would be very effective in optimizing packinghouse operations, thus reducing reliance on human intervention. The system, which included a differential pressure sensor for fan operating conditions, an infrared sensor for CO_2 concentration, thermocouples for temperature, a lithium-chloride humidity probe for RH, and an ethylene solid-state sensor for ethylene concentration, was tried

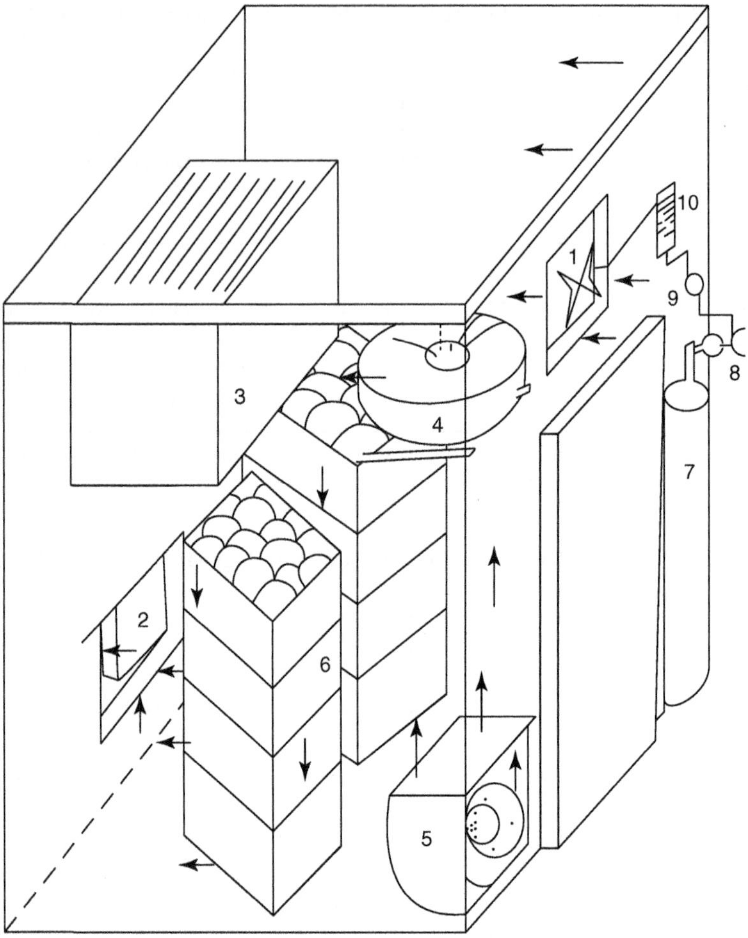

FIGURE 9.1 Degreening Chamber for Continuous Flow Application of Ethylene: (1) Fan to Suck fresh Air with Ethylene, (2) Exhaust, (3) Cooling Unit, (4) Humidifier, (5) Heating Unit, (6) Stacked Vented Plastic Crates with Fruit, (7) Compressed Ethylene Gas Cylinder, (8) Two -Stage Regulator, (9) Needle Valve and (10) Flowmeter.

for automatic monitoring and control under real-time conditions so that packing-house personnel could be alerted before substantial damage occurred.

For degreening of 1 ton of fruit at a time, a degreening chamber has been developed (Fig. 9.1) using polyurethane foam–insulated steel panels (Ladaniya, 2000). This chamber is fitted with humidifying, cooling, and heating units and can also be fitted with suitable sensors for automatic controls of gas concentration. The inlet for fresh air is provided on the front side at the top; an exhaust vent is located on the opposite side at the bottom. The inlet and outlet have shutters to open or close the vent. The chamber, with inner dimensions of 170 cm L ×

170 cm W × 227 cm H and volume of 6.5603 m^3 (6560 l), has a capacity to hold one ton of fruit in vented plastic crates. The ethylene compressed-gas cylinder (10 ± 2 percent ethylene in nitrogen) – with a two-stage pressure regulator, a needle valve, and a flow meter to control and measure the gas flow while injecting ethylene through air in the chamber – is used for trickle application of the gas. The technique can be scaled up for 10 tons of fruit by increasing chamber size and fan capacity for fresh air. Extra fans are required for air movement inside big chambers.

Degreening Chamber

Fan delivery can be calculated by considering carbon dioxide evolution of a maximum of 40 mg/kg/h by the Mosambi sweet orange fruit at 27–29°C. Since 1 ton of fruit would evolve nearly 40 g of carbon dioxide in 1 h, its approximate volume of 22.4 lit at 27°C would result in a concentration of roughly 0.34 percent in the chamber of 6.5603 m^3 volume in 1 h. To keep CO_2 level close to the concentration in fresh air (0.03 percent), a minimum of four changes were considered necessary to reduce CO_2 levels from 0.34 percent to 0.08 percent in 1 h; hence fan delivery of 0.4373 m^3/min = 6.5603 m^3/15 (60/4 min) can be set during degreening.

The ethylene injected at 0.4373 m^3/min air flow is 3.06 ml/min (= 0.4373 × 7, for 7 ppm ethylene considering it is 100 percent in concentration). If a cylinder contains 10 ± 2 percent ethylene (in nitrogen), the gas flow is adjusted accordingly. Sampling of gas inside the chamber is done for monitoring ethylene levels. Concentration varies between 5 and 10 ppm during the treatment. Fan delivery and gas flow are set based on a 1-ton fruit-holding capacity of the chamber.

Fruits placed in vented plastic crates (48 cm L × 31 cm W× 28 cm H internal dimensions with netted four sides and bottom) having open tops can be kept in a degreening chamber maintained at 27–29°C and 90–95 percent RH. Ethylene is injected from the inlet vent and drawn inside with a fresh air current mixed with air movement inside and exhausted from the opposite side. Air from the evaporator coil (an evaporator fan with 12.7 m^3/min air flow), heating unit, and fresh air from inlet vent are mixed with ethylene gas inside the chamber during degreening. The air circulation (velocity) inside the chamber varied from 0.5–0.6 L/s/kg fruit at different locations.

D. Changes in Physico-Chemical Attributes of Degreened Fruit

Ethephon treatment mainly changes rind color. However, some changes in internal composition have also been reported. Total soluble solids, sugar, and juice content increased (Mazumdar and Bhatt, 1976) while chlorophyll, acidity, and ascorbic acid content fell during degreening (Ramana et al., 1973; Singh et al., 1978 c). Juice and TSS content were also higher in fruit treated with Ethrel before harvest (Chauhan and Rana, 1974; Soni and Ameta, 1983).

The difference between fruit firmness, juice percentage, total soluble solids, and ascorbic acid content of degreened and non-degreened 'Nagpur' mandarin fruit was not significant, and ethylene treatment had no effect on internal composition except titratable acidity content. Titratable acidity declined in ethylene-treated fruit (Ladaniya, 1998, 2000; 2001). Ethylene treatment affected the composition of grapefruit to some extent. With early-picked degreened (5 ppm at 30°C for 60 h) or non-degreened Marsh grapefruit kept at 15°C for 12 weeks, ethylene treatment decreased acidity to the same level as that of freshly harvested fruits picked mid-season. Malic acid decreased by as much as 60 percent with a drop in soluble solids content. Fruits from all harvests treated with ethylene contained more ethanol and acetaldehyde than non-treated fruits (Davis et al., 1974).

E. Respiration

The degreening process increases the respiratory rate of citrus fruit. Respiration of Mosambi fruit increased more than two-fold from 35.4 mg CO_2/kg/h to 80.1 mg CO_2/kg/h after 2 days of degreening (Ladaniya, 2001). In fruit not exposed to ethylene, respiration rate did not increase and instead gradually declined. In non-treated stored fruit (1 month under ambient storage conditions), respiration declined from the initial value of 35.4 mg CO_2/kg/h to nearly 30 mg CO_2/kg/h. In treated fruit, as the fruit was removed from the degreening chamber, respiration rate slowly declined and was on par with non-treated fruit at the end of storage. Lemons responded similarly to ethylene as respiration shot up 2–3 fold (Craft, 1970).

F. Fruit Color and Mass Loss in Degreened Fruit

Fruit color continues to improve in ethylene-treated fruit during storage under ambient conditions (24 ± 4°C; 55–65 percent RH). There is considerable residual effect of ethylene treatment even when fruits are removed from an atmosphere containing ethylene. It is observed that once color development is started with ethylene, it remains unaffected by the break in ethylene supply (Cohen, 1991). In non-treated Mosambi fruits wrapped in polyethylene, degreening occurred naturally over a period of 30 days under ambient conditions of storage, as evident from the increase in brightness (L*) and intensity (Chroma) and the decrease in hue angle (Ladaniya, 2001). The storage temperature in the range of 20–28°C and endogenous production of ethylene can lead to degreening in non-exposed fruit. Although it is a slow process, it results in the elimination of the original difference in treated and non-treated fruit (Jahn et al., 1969). Navel sweet oranges held in air for 9 days at 20°C also lost chlorophyll, although slowly compared with ethylene-treated fruits (Eaks, 1977).

The loss of chlorophyll in ethylene-exposed fruits is attributed to increased chlorophyllase activity and reduced size and number of chloroplasts in the peel (Barmore, 1975; Shimokawa et al., 1978). With the change of color from green to yellow, brightness also increased. Lancaster et al. (1997) reported that higher chlorophyll contributed to lower L* value and therefore to a darker shade. A significant linear relationship was detected between chlorophyll content and L*, indicating a logarithmic relationship between increasing chlorophyll with lower L*value. Doubling the chlorophyll content reduced L* by 0.04 units. The hue angle decreased (<90°) as the chlorophyll decreased and yellow-red pigments appeared on the fruit surface.

During degreening, mass loss is negligible as the RH remains high during the process. Degreened fruit lost nearly 3 percent of its mass during storage for 30 days under ambient conditions (Ladaniya, 2001), and a vented polyethylene liner was an effective water-vapor barrier in minimizing water loss. The difference in mass loss of degreened and non-degreened fruit remained insignificant during storage.

G. Decay and Disorders in Degreened Fruit

Ethylene as a senescence promoter weakens the fruit tissues and hence is most likely to increase decay and cause loosening of fruit buttons. Fruits of several orange and tangerine cultivars degreened with ethylene (0–120 ppm) for 48 h have shown higher decay during storage at 21°C for 4 weeks. The incidence of rot from *D. natalensis* increased with ethylene concentration (McCornack, 1971). The U.S. citrus industry uses 1–5 ppm C_2H_4. The temperature used is 22°C in California and 29°C in Florida. The rate of degreening does not increase by ethylene concentration above 10 ppm. Above 20 ppm C_2H_4 causes fruit senescence, loss of button (calyx, disc, and receptacle), and in some instances physiological disorder. Ethylene may increase the susceptibility of some citrus fruits to anthracnose, a superficial infection by *Colletotrichum gloeosporioids* (Eckert and Eaks, 1989). The duration of degreening has a direct effect on decay; the longer the treatment, the higher the number of rotten fruits. Partially colored mandarin cultivar Sunburst fruits subjected to 66 h of degreening (5 ppm ethylene) showed 10 percent decay, whereas fruits degreened for 45 h had <2 percent decay. The main causes of decay were *Penicillium digitatum, Alternaria* sp., and *Collectotrichum gloeosporioides* (*Glomeralla singulata*). Fully colored untreated fruits did not decay after storage at 4°C for 4 weeks (Hatton et al., 1987).

2,4-D (amine form; 200 ppm) with TBZ (500 ppm) in a drench prior to degreening is reported to reduce PME and cellulase activity in abscission zones, minimize button drop and fallout, preserve vitality of buttons, and prevent stem-end decay, which is enhanced by ethylene treatment. The 2,4-D isopropyl ester form is more effective but causes noticeable delay in color development (Cohen, 1991). The incidence of *P. digitatum* was found to be minimally affected by degreening.

Brown and Craig (1989) compared several methods of fungicide applications to reduce decay in degreened citrus fruit in Florida. Aqueous aerosol application (prepared with Tifa Microsol model 202 mechanical aerosol generator, particle size 50–200 μm at generator) of benomyl applied before degreening consistently reduced stem-end rot. TBZ and Imazalil (1000 ppm) are more effective as aqueous preharvest sprays or drench applications (by dipping in aqueous solution for 30 s) before degreening.

To minimize disease incidence in Satsuma mandarins, harvesting the fruits at sufficient color break and exposing them to no more than 36 h of degreening is recommended. The subsequent treatment of fruits with an active postharvest fungicide is recommended. After degreening, fruit has to be dipped for 3 min in a solution of 1000–1500 ppm (a.i.) Imazalil, prochloraz, and fenpropimorph (Tuset et al., 1988). In Florida, TBZ and Imazalil (1000 ppm) as a non-recovery drench of fruit is applied before degreening.

Curing at a higher temperature after degreening treatment can lead to heat damage on citrus fruit surfaces. All ethephon-treated fruits (dipped in ethephon at 2000 or 4000 μl/l) showed heat damage when cured for 72 h. Pummelos showed damage at 33–39°C and lemons at 42°C (Ben-Yehoshua et al., 1990), which indicated that curing of fruit after degreening treatment should generally be avoided.

Pre- and postharvest treatment with carbendazim minimized the stem-end decay (mainly incited by *Botryodiplodia theobromae*) in degreened 'Nagpur' mandarin fruit under ambient conditions after 10 days (Table 9.3), while stem-end rot was higher in fruit without pre- or post harvest fungicide treatment (Ladaniya, 2000). In Mosambi oranges, too, three pre-harvest carbendazim (500 ppm) sprays minimized decay in degreened fruit (Ladaniya and Shyam Singh, 2001).

TABLE 9.3 Weight Loss, Decay and Sensory Qualities of 'Nagpur' Mandarin Fruits as Influenced by Various Treatments and Degreening after Storage for 10 Days at Ambient Conditions

Type of treatment	Weight loss (%)	Decay (%)	Sheen (Scale 1–5)*	Color (Scale 1–4)**	Turgidity (Scale 1–3)#	Flavor (Scale 1–3)#
Degreening only	7.96	6.57	2.11	4.00	1.88	3.00
Waxing only	6.56	3.83	3.06	2.44	2.88	2.55
Degreening and waxing	7.03	15.13	3.44	3.66	2.55	2.00
Degreening and waxing with pre-harvest fungicide treatment	6.42	4.19	4.11	4.00	2.55	2.55
Degreening and waxing with postharvest fungicide treatment	6.52	14.08	3.99	3.22	2.66	2.10
Degreening and waxing with pre- and postharvest fungicide treatment	6.96	2.70	3.88	3.32	2.88	2.22
No degreening, no waxing and no pre- or postharvest fungicide treatment (Control)	8.92	11.46	1.66	2.00	1.33	2.10

Ladaniya (2000),* Sheen scale 1 = unacceptable, 2 = slight, 3 = acceptable, 4 = good, 5 = bright;** Color scale 1 = green, 2 = greenish-yellow, 3 = yellowish-orange, 4 = orange # Turgidity and flavor scale 1 = Poor, 2 = acceptable, 3 = good.

In some cases, lower decay loss is reported after ethylene treatment, which could possibly be attributed to ethylene-induced activities of enzymes, particularly the phenyl-alanine ammonia-lyase (PAL). Kuc (1982) reported that PAL catalyzes the branch-point step reactions of the shikimic acid pathway, resulting in biosynthesis of phenols, phytoalexins, and lignins, which are associated with induced resistance to diseases. The level of resistance to *Geotrichum candidum* has been found to be parallel to the amount of lignin-like deposition (Baudoin and Eckert, 1985).

H. Color Improvement with Other Methods

Besides ethylene and ethephon treatment, several other methods have also been tried to improve the color of citrus fruits. The red-light irradiation increased Hunter a and a/b values and decreased Hunter b values in *Citrus iyo* cultivar Miyauchi fruits. The a/b values of irradiated fruits were higher than those of controls during the whole storage period. Red-light irradiation has been shown not only to accelerate overall color development, but also to enhance red color pigmentation via a specific pathway of carotenoid biosynthesis (Ohishi et al., 1996).

In sweet orange fruits (Hamlin cultivar) harvested shortly before color break, exposure to UV-B radiation for up to 30 days accelerated rind chlorophyll degradation at 25°C in the dark, especially in fruits treated with ethephon (1000 ppm for 1-min dip). Fruit composition was not directly affected by UV-B radiation, as changes in TSS content and acidity are associated with changes in juice content (Basiouny and Biggs, 1975).

II. PACKINGHOUSE OPERATIONS

Packinghouses are an integral part of the citrus industry worldwide. Per one estimate, commercial packages of fresh citrus are being produced in more than 800 packinghouses located in 28 countries. The output of these packinghouses varies from 50 tons to 50 000 tons in a season. In the U.S., lemons are packed on packing lines exclusively designed for this fruit. Packing lines are also designed exclusively for more delicate fruit such as mandarins and tangerines. Care is taken to minimize impact, drops, and vibrations so as to avoid bruises and cuts (Johnson, 1991). A packinghouse is purely a commercial operation, and therefore has to be commercially viable in a highly competitive business. Hence smaller packinghouses are declining in number and larger ones are increasing their capacities with modernization and automation to reduce costs.

A. Mechanized Operations

The commercial packing of citrus begins with careful picking and hauling to the packinghouse. In a modern packinghouse, operations include degreening, mechanized handling for pretreatment, grading, packing, precooling, and cool storage. The standard sequence of operations might be: (1) receiving at the packinghouse, (2) degreening if necessary, (3) dumping and trash removal, (4) sorting or pre-sizing, including rot and small-fruit removal, (5) washing and rinsing, (6) grading, (7) fungicide treatment, (8) waxing, (9) drying, (10) fruit stamping, (11) sizing, (12) carton filling and labeling/marking, (13) palletizing, (14) pre-cooling and storage, (15) railcar or truck loading. The details of these operations follow. Degreening is described in part I of this chapter.

1. Receiving at the Packinghouse

Fruits are brought to the packinghouse in containers such as wooden field bins (1.2 × 1.2 × 0.6 m depth with capacity of 0.86 cum in, or about 500 kg), plastic crates, or even loose in trucks. Field bins and pallet boxes are handled with fork-lift trucks and inverted on the packing line if fruit is to be packed immediately. Some fruit may be treated with postharvest fungicides and stored for some time until ready for packing. On receiving at the plant, trucks are usually unloaded in holding areas (90–95 percent RH). Fungicides are applied as drench to the bulk bins holding the fruits or to the trucks holding loose fruit. As a practice in Florida fruit is drenched with fungicide in the trailer itself before unloading. Pallet boxes are stacked with sufficient space for degreening with ethylene (1–5 ppm) for early-season fruit, which is green externally but has internally achieved minimum maturity. Fruit is usually dumped dry onto the packing line conveyer. The purpose of a wet or dry dump is to remove debris or traces of leaves, dust, and soil coming from the field along with the fruit. Usually this part of the packing line is somewhat away from main packing line operation to avoid fungal spores entering the packinghouse. Prompt movement of fruit from orchard to packinghouse is necessary (Wardowski, 1981) except for lemons, which are moved 24 h after harvest. Trash elimination should be done outside the packinghouse. The water dumps are designed so as to remove water and dirt quickly and again refill it in order to maintain sanitary conditions. The bin is slowly inverted so that fruit is delivered gradually without any injury or crushing/bruises. Fruits with freezing injury symptoms (fruit that has become dry and useless) can be separated on packing line by flotation or X-ray techniques (Hale and Risse, 1974).

Wet dumps or water dumps were earlier used in some packinghouses for Kinnow mandarins in Punjab (India) but have now been discontinued. Wet dumps and soak-tanks have been discontinued in most packinghouses in California and Florida also. Dry dumps have proved better in terms of preventing the spread of decay pathogens through bruises and wounds on fruit peel. Usually water in wet dumps is mixed with chorine (at 50–200 ppm or sometimes an even higher concentration depending on the chlorine-toxicity tolerance of the fruit). The fruits are taken out of the dump by elevator and passed on to sorting conveyer. In case of 'Nagpur' mandarins and Mosambi sweet oranges, disinfection with a spray of water containing 800–1000 ppm chlorine on nylon brushes (contact time 6–8 s) has been found useful for minimizing fungal spore load. Fruit is then rinsed with plain (non-chlorinated) water (8–10 s). Fruit then moves to the sponge roller belt one layer deep to remove excess moisture.

FIGURE 9.2 Manual Sorting of Lemons.

2. Sorting

Commercial-scale manual sorting conveyers made of aluminum rollers are used. Various capacity machinery (from 2 to 5 tons/h or 5 to 6 tons/h capacity with 120–150 cm width of the conveyer for even more fruit-conveying capacity) is available depending on requirements. The rollers rotate on their axes in order to rotate the fruit to reveal the entire surface while the fruit moves forward. The removed fruit or cull needs to be removed quickly from the area, so such sorting conveyers need extra space around them for fruit handling. In manual sorting, the fruit-sorting personnel have to be sufficiently trained to identify culls on the basis of color, size, shape, blemishes, and so on (Fig. 9.2, see also Plate 9.1). The sorting conveyer should have sufficient white light/illumination (180 foot candle) focused on the conveyer and fruit. Light should not fall on the eyes of the sorting personnel. Supervisors have to look after the comfort of working personnel and also ensure proper sorting to avoid losses (such as good fruit being removed along with the culls) or any 'under-sorting' (culls being retained with good fruit). In advanced automatic electronic sorting systems, fruits can be sorted on the basis of color, size, shape, and blemishes by the cameras (machine-vision systems) that can detect culls and drop/remove them on cross belts running below. These operations are possible with the help of image analysis by a computer programmed for that purpose. Oranges, tangerines, and lemons can be sorted in different color categories as desired. Lemons are sorted in four color classes: dark green, light green, silver, and yellow by electronic sorters.

Chen et al. (1992) reported on an electro-optical sorter for color and size sorting of citrus fruits. This system includes five microprocessors to control feeding, size detection, color detection, discharge, and master control. Size grading is accomplished by a number of LED (light emitting diode) light beams blocked by the projected area of the fruit, while color grading is conducted by using a dual-wavelength IVRF quality index at 880 and 650 nm. System accuracy is reported to be 91–92 percent at a speed of 24–48 fruit/min.

3. Pre-sizing

The sorting and pre-sizing operations are done at the same time in advanced systems, where very large and very small fruits are also removed as culls along with blemished fruit. Sometimes size grading is done after sorting to remove very large or very small fruit that is not suitable for packing. This fruit, which is small and blemished but edible, can otherwise be sent to the processing unit for juice extraction and value addition. Electronic pre-sizers and pre-graders can reduce the cost of materials and labor considerably if used at the point of entry of fruit in the packinghouse.

4. Washing

Washers have a special type of brushes to remove dirt and natural wax partially from the fruit surface. Various types of bristles are used for washers; soft bristles usually cause no abrasions or bruises on the fruit. This is very important for limes, lemons, and mandarins, which have delicate skin. Excessive brushing can injure fruit, and dry, brown circular scars (indicating circular movement of idling fruit on brushes) develop on the fruit surface after few days. Unnecessary idling of fruit on the brush rollers causes severe injury to fruit, rendering it unfit for the market. The brushes have to be saturated with water so as to minimize the injury of fruit by the bristles (Fig. 9.3, see also Plate 9.2).

Tougher brushes cause bruises to fruits and soft-brush bristles with minimum washer brush speeds and brushing times (10–20 sec) are preferred for the minimization of rind staining and breakdown, especially in navel oranges. Brushing fruit in washers needs to be done at roller speeds lower than 100 rpm. Washing with detergent or disinfectant such as chlorine is followed by a fresh/plain water rinse to remove any residue of detergent or chlorine. Disinfectant or detergent is applied as foam or sprayed on the fruit. The nozzles through which disinfectant or detergent is discharged should be clean and deliver an accurate quantity. High-pressure brushing can also be used to remove scale insects, sooty mold, and debris. These are non-recovery sprays and are expected to remove maximum spore load and infection from the fruit surface. Fruits are then run on sponge rollers to remove excess water so that fruits are just damp at the time of wax coating.

5. Grading

Grading of fruit is done visually on the basis of surface blemishes and color or any other deformity as the fruit rolls by on the conveyer. Conventionally, grading is done manually by experienced graders. Graders have to be trained to recognize the percentage of blemished parts on the fruit surface, since fruit with 10–50 percent surface blemish may sometimes be allowed for various grades.

FIGURE 9.3 Washer Unit.

Most experienced graders should be at the last point in the conveyer so that no off-grade fruit is included. Fruits of second grade are removed and packed later and off-grade fruit is sent for processing/juice extraction. Grading can be done automatically also using high-speed electronic graders that recognize blemishes and color using photo-optical grading systems. Electromechanical graders used by the Sunkist Growers, Inc. facility in California uses an X-ray scanner (50 Roentgen/min) to scan the fruit for frost damage, granulation, and *Alternaria* core rot. Then fruit is graded for color blemishes and scars using electronic machine-vision graders. This information is combined (internal and external), a computer-operated software system decides the grade and marketability of fruit, and the fruit is dropped on the proper conveyer for further processing. The system is commonly used for lemons and oranges. The capacity is as high as 480 fruit/min in each line (Johnson, 1981).

Industrialized countries are trying to develop and introduce newer technologies to reduce labor costs so as to be competitive with developing countries where labor is cheaper. Sorting and grading are the two most labor-intensive activities. These operations engage 40–50 percent of the workforce in the packinghouse and even then results are not satisfactory. There is scientific as well as economic interest in developing automatic sorting and grading machines. The Videograding TM is a commercial automatic grading system in Italy. Reliability and cost are the major factors in the adoption of these technologies. New sensors are being developed to evaluate internal quality along with color vision and multiple cameras used to better locate external defects so that quality in totality can be given to consumers (Blandini et al., 1992).

6. Fungicide Application

Fungicide application is a major operation in the packinghouse. There are several methods of fungicide application. Cost, efficiency of operation, and effectiveness of treatment determines the application procedure that is selected. The postharvest fungicide application is done in soak tanks, where fruit is immersed in water containing fungicide and in the same tank fruit is transported by water current. This practice was found to be ineffective – it promoted disease – and hence discontinued for reasons of contamination. In India, wet dumping was carried out in the case of Kinnow mandarins but later discontinued for the same reasons. The drenching refers to the treatment of fruit held in bins or in containers by application of fungicide solution from nozzles above as the fruit passes on the conveyer. The volume of delivery of solution is high in this method. Fungicides can also be applied as a fine spray from nozzles to fruit passing on conveyer.

Sometimes fungicide, such as sodium ortho-phenyl phenol (SOPP), is added in a liquid cleaning-soap solution. The fungicide is also applied in high concentration as non-recovery spray before wax coating. When fungicide is applied separately, coating does not include fungicide again. Alternatively, fungicide can also be included in wax coating.

7. Waxing

Wax coating is a special kind of operation in citrus fruit packinghouses since it accomplishes a triple objective: (1) Providing the required gloss on which aesthetic value or cosmetic appearance of fruit depends, (2) Protecting from water loss as coating replaces natural wax which is removed to some extent during washing operation, (3) Acting as carrier for fungicide or any bio-gent and/or PGRs such as 2,4-D. When bio-agents are used to control disease pathogens, coating should be compatible with the bio-agent to sustain its growth. Wax coating reduces respiration; excessive reduction results in an off-flavor. Effective wax coating should reduce weight loss by about 30 percent.

Wax coatings are made of different chemicals that may be non-edible but can be evaporated during the drying operation. Thus only edible wax (bee wax, carnauba wax, candelila wax, sugarcane wax) remains on the fruit. Generally these coatings must be approved as 'Food Grade' by competent authorities of the respective countries and certified for use on fruits. For overseas marketing of fruit, such coatings should meet the standards of importing countries.

The coating is applied with traversing hydraulic nozzle mounted about 1 foot above the rollers. The nozzle moves constantly across the width of the horse-hair roller-brush bed of the waxer (Fig. 9.4, see also Plate 9.3).

The nozzle moves once every 1–1.5 s across the roller bed. The rollers are saturated before running the fruit. The liquid wax is metered through the pump, and spray delivery is atomized with compressed air to provide a fine spray over the initial 1 or 2 rollers. Wax may be dripped on the first roller, as with aqueous waxes. The applied coating is then uniformly coated with brushing action on subsequent rollers. The surface drying is done before the wax coating, particularly

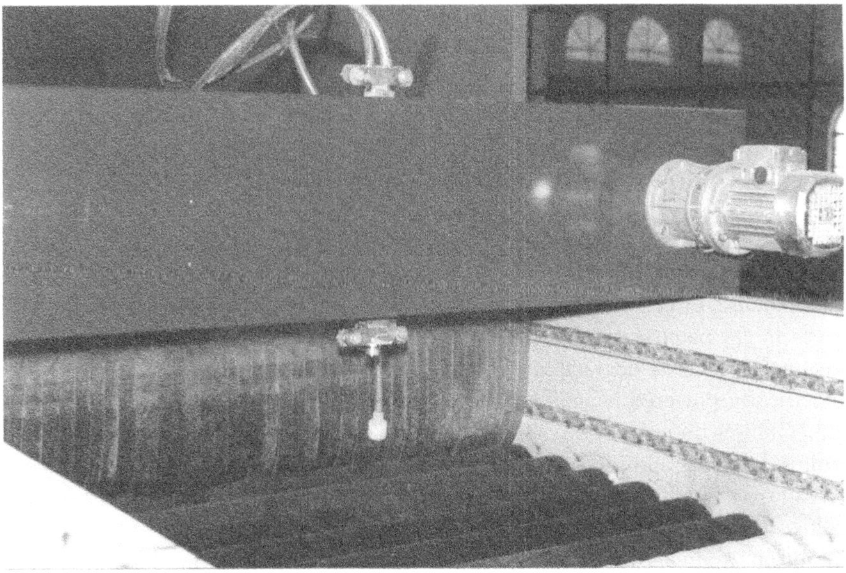

FIGURE 9.4 Traversing Nozzle of Waxer and Horse-Hair Brushes.

if solvent-based wax is applied. In the case of aqueous waxes, fruit surface is slightly damp, as excess moisture is removed by sponge rollers before wax application. Solvent waxes need a dry fruit surface for better application and a good shine. The guidelines of manufacturers are generally followed with respect to dilutions made while applying coatings. Generally, one gallon of wax formulation is sufficient for 4–4.5 tons of fruit if applied in a very fine spray. This will depend on the wax delivery/discharge mechanism and fruit flow on the conveyer. Fungicides are often added to coating to ensure their uniform application over the entire surface. Sometimes fungicides may be applied separately to achieve better decay control. In those cases, immediately after washing, fungicide or a mixture of fungicides is applied to the brush rollers so that the fruit rind may absorb some of the fungicide and have required contact time before the fruit is waxed with non-fungicidal coating. The coatings and fungicides are covered elsewhere (see later in this chapter and Chapter 16).

It is convenient to add fungicide to wax emulsion so as to avoid double operations such as first application of aqueous fungicide and then waxing. However, the incorporation of fungicide in wax reduces substantially its antifungal activity as compared to separate aqueous applications. The residue is also often greater when fungicide is applied in wax. The separate use of fungicide has given encouraging results commercially (Gutter, 1981).

Oranges and tangelos receive color addition in Florida but not in California. The color Citrus Red (the food dye citrus red No. 2, [2,5- dimethoxyphenylazo]-2-naphthol) is added to the wax coating or applied separately before waxing.

After grading, color is generally applied as a liquid solution and then rinsed before fungicide application. It is applied in emulsified form since it is dissolved in citrus oil component (i.e. *d*-limonene). The federal law has set its residue limit as 2 ppm and it is not allowed unless oranges achieve minimum maturity index of 9:1 Brix:acid ratio (Ting and Rouseff, 1986). The color-adding process is also followed in Mexico for oranges and the same dye is used.

8. Surface Drying

Surface drying is as important as coating because drying temperature has an immense effect on fruit peel. If temperature is excessively high, it may result in dry patches on fruit skin. This can be the cumulative effect of previous treatments such as degreening, brushing, waxing, and the hot air of the dryer. Electrically powered heaters and blowers are common features in drying units; temperatures do not exceed 50–55°C. A blast of hot air is often diverted onto the fruit as it passes and rotates on the roller conveyer under the blower. Air may be forced over the heaters on the crop or it may be sucked from the heaters placed below the conveyer. Maximum heat is preferably applied when wet fruit enters the drying tunnel. Among all packing line operations, it is the dryer that fruit passes through for the longest time so as to dry the coating thoroughly.

9. Quality Grading and Sizing

Sizing of citrus fruit on the basis of dimensions is better than weight-sizing because the problem of puffiness (particularly in mandarins) may be encountered, which may lead to erroneous sizing. Microprocessor-based (computer-aided) modern sizers can weigh individual fruit in specially designed cups. These cups weigh the fruit and accordingly drop it on cross belts below the cups. If fruit is of higher weight it moves farther, to next cup. Fruit with the highest weight falls last. Weight sizers may be useful with oranges and grapefruits.

A dimension sizer has to be designed to measure and segregate fruit on the basis of two or four contact points. Measurement on the basis of four contact points is most accurate since diameter is usually greater or smaller in citrus fruit as compared to height or length. When fruit is in contact at four points it is called a volumetric sizer. These are best suited for lemons, which are oblong to ovate in shape with greater length and relatively smaller diameter. Sizing of fruits should be fairly accurate for it may affect volume and lead to problems in uniformity of packaging. Packers will face problems maintaining uniformity of fruit in the packages, which can slow their speed in searching for the properly sized fruit.

The two-point contact sizers can be of two types. In roller sizers, the singulated fruit is dropped on the divergent rollers, which are positioned with increasing spacing. As the fruit moves from the beginning to the end of the roller, small fruit falls onto the pull-out belt below the roller while larger fruit continues along the rollers until appropriate spacing between the rollers allows the fruit to drop. The rollers can be adjusted for different sizes of citrus. In another method, fruit is conveyed to different-sized openings: a small opening first, increasing gradually to separate small, medium, big, and large sizes. These sizers have been fairly appropriate for oranges and mandarins with their near-spherical shape. With flat-shaped fruit in some cultivars of mandarins/tangerines with greater diameter and smaller height (the length between stem and stylar) sizing could be erroneous and would solely depend on the orientation of the fruit at the contact points. The divergent roller-type sizers or rotating disc–type dimension sizers (with two contact points) have given fairly good results in case of Kinnow and 'Nagpur' mandarins respectively, provided that contact takes place at the equatorial diameter. The rotation speed of the disc has to be adjusted according to the feeding of fruit to avoid congestion and mixing of grades. At the edge of the rotating disc is an adjustable opening for different sizes of fruit. Small fruit is segregated first followed by medium-sized and then large fruit. Reducing the speed of the sizer's rotary disc from 10 to 8.5 rpm resulted in satisfactory sizing at the feeding rate of 1 ton/h with 'Nagpur' mandarins (Ladaniya et al., 1994). This type of sizer could segregate Mosambi orange fruit on a dimension basis at two points with more than 90 percent accuracy in sizing. There was no injury to fruit in mechanical sizing (Ladaniya, 2001). A divergent roller grader with the capacity to grade 400 kg of acid lime (*C. aurantifolia* Swingle) fruit per hour can also be operated manually for small-scale operations (Pandiarajan et al., 1993). A similar grader with a capacity of 200–250 kg/h is reported to segregate Kinnow mandarin fruits on the basis of diameter (Anonymous, 1995). Graders with a capacity of 5000 fruit/h and 92–97 percent accuracy for various sizes (as per AGMARK, India) in *Pola* (Puffy) and *Battidar* (round and firm fruit) grades of 'Nagpur' mandarins has also been reported (Anonymous, 1992).

More accurate electronic sizers using optical technology are now available. They segregate on the basis of dimension and shape. These sizers use light beams and on the basis of light blockage determine shape and size. The singulation and orientation of fruit when light is focused on it is very important for sizing accuracy. In Japan, fully automated opto-electric sizing units have been used for more than 25 years. These units use the optical properties of the fruit: reflection, transmittance, or delayed emittance. With these sizers dropping impact is reduced. About 20–25 tons of fruit are sorted per hour in large and efficient packinghouses. In opto-electrical sizers, light-source and detector combination is used. The horizontal sizing is done by relating fruit size to a time length required for a fruit to continue intersecting a light beam that traverses a conveyer belt (Iwamoto and Chuma, 1981).

The most important requirement of good sizer is its ability to adjust the sizes and flow of fruit even when in operation. After sizing, fruit of only the required grade can be washed, waxed, dried, and packed. The other grade can be waxed and packed later, or small sizes can be sent for juice processing.

10. Packing

Packing is the last operation on the packing line and place-packing (manually placing fruits in the containers and arranging them in layers) is most common for mandarins and oranges. For hand packing, fruits are collected in tubs, chutes, bins, or long sliding tables. Fruits are arranged in the box to sit with desired orientation. The padding or paperboard separators or a paper pulp tray can be used

FIGURE 9.5 Place-Packing of Mandarin Specialty Fruit.

between fruit layers. Like sorting, packing is another operation that requires trained personnel, since this is the operation that has a direct impact on buyers when the box/container is opened in the market (Fig. 9.5, see also Plate 9.4). The convenience of the packer is of utmost importance in designing the packing table and a sitting/standing facility for easy reach of fruit and the container/tray. The position of the box should be as near as possible to the body of the packer. This allows packer to look at the bottom of the box and place the fruit properly and conveniently. The filled boxes are lifted and put on conveyers/rails traveling sideways. The empty boxes and other materials, such as trays and separators, should be within arms' reach of the packers or at the most within a one- or two-step movement. Automatic place-packing units are also operational in many packinghouses in the U.S. and elsewhere. The desired number of carefully sized and sorted fruit is lifted and placed automatically in the box in specific orientation. The box is then inspected and closed. In orange-packing machines used in some facilities of Sunkist growers in California, fruits are metered into pattern formers, where they are picked up by spring-loaded suction cups. The fruit is then moved over the carton and placed automatically. Machines with capacity as high as 315 cartons of 18 kg capacity per hour (Johnson, 1981) are operational. The actual speed is up to 200 cartons/h. The machine reduces labor requirements by 75 percent.

Limes and lemons are sometimes jumble-packed/volume-filled. Jumble-packing can be automatic or semi-automatic. Fruits for local markets are usually jumble-packed where prices are to be kept low. The automatic volume-fill or jumble-pack machine fills the box with pre-sized fruit to the desired weight and shakes it to immobilize the fruit. The vibration of the box helps to settle the fruit below the desired top level of the box so that any fruit protruding above the top edge is not pressed while closing. The box is weighed on the scale manually and fruit is added or removed so that the box is filled to the desired weight. Then the box is closed and sealed automatically. The fully automatic large machines have facilities as high as 400–600 boxes per hour.

Whatever the method of packing, fruit must be immobilized and fit into the container space without pressing.

11. Marking and Labeling

Labeling with individual labels/stickers on individual oranges, grapefruits, lemons, or tangelos is done just before packing. The singulated fruit in cups are synchronized to pass beneath a cassette of pre-gummed labels. A label is applied to each fruit with gentle pressure. Inks made of food-quality dyes have also been developed to stamp on the fruit; however, their acceptance depends on the rules and regulations of the country where the fruit is to be marketed. Recently, new laser technology has been developed to tattoo each fruit individually. A product look-up (PLU) code, country of origin (COOL) label, grower lot number, and any other requested security information can be marked on the fruit directly. This is a part of the produce industry's latest effort to develop technology to ensure tracking and tracing. The technology was designed by Durand-Wayland Inc., U.S.A (Anonymous, 2005). Most grocers require PLU labeling, as the number can be punched in to get the price.

Pre-gummed labels with all required information are also stuck to the package box. The marking stamps of the date of packing or any special treatment to fruit, and the packinghouse name and address are also used.

12. Packing in Net Bags/Polyethylene Film Bags

Oranges, mandarins, and grapefruits are also packed for retailing in net bags or plastic film bags using automatic machines. These machines can be used by big retail chain stores for packing in their own warehouses, since these bags are difficult to transport if packing is done in production areas. These bags can be placed in master cartons that are then transported, but box space would not be judiciously utilized in that case.

The machines used for this purpose are fully automatic and capable of packing 2–6 kg fruit in a bag and up to 12 bags in a minute. Larger throughput is achieved in automatic machines than in semi-automatic machines. A polyethylene tube is filled and sealed rapidly – thus it is a form, fill, and seal operation. The sealing is accomplished with a heat sealer. Most of the information about produce and packer can be printed on the polyethylene bag; however, labels/tags with bar codes, fruit weight, and price can be attached easily to net and film bags.

Over-filling is usual problem in net or plastic bags. Automatic bagging machines that minimize weight overfill are used to pack fresh citrus in Florida in commercial packinghouses for bagging early tangelos (large fruit, low count/bag) and Robinson tangerines (small fruit, large count/bag). A non-parametric statistical computer program has been developed to analyze automatic bagging machinery performance. In the weight-fill mode of operation, mixing of dimensional sizes reduces variations (Miller et al., 1986). In netted fruit, in order to meet the net weight requirement as depicted on the label, fruit with excess weight has to be shipped in each bag (Albrigo et al., 1991).

13. Palletization of Boxes

Palletization can be automatic or manual. Fruit is unitized for easy mechanical handling with forklift trucks. Fresh fruits must be shipped unitized on 120 × 100 cm pallets. This unit size is recommended by the Organization of Economic Cooperation and Development (OECD). Wood is the preferred material for pallets,

FIGURE 9.6 Pallet of Packed Boxes with Corner Posts and Strapping.

and usually pallets designed for four-way access are preferred. Reusable pallets can be used only if they are made to requirements of EEC standards for use in the European pallet pool system. Nonmetal straps are used for securing and stabilizing the pallet load. Corner boards are also used to immobilize the boxes in the pallet (Fig. 9.6, see also Plate 9.5). The load is generally 1metric ton or a little more per pallet.

The above information is based on the general sequence of operations in citrus packinghouses. The sequence of operations varies with the type of citrus fruit. Even treatments are different and operations can be changed to suit particular handling/marketing requirements. Following are some examples of packinghouse operations given in the form of schematic diagrams.

B. Commodity-specific Operations

Lemons – Usually lemons are cured after harvest to minimize oil spots on the rind during subsequent handling. Treatment with 2,4-D and fungicide before degreening is needed in order to maintain greenness of the pedicel. After degreening, the fruit is waxed or stored for later waxing as the case may be. A common handling sequence is as follows:

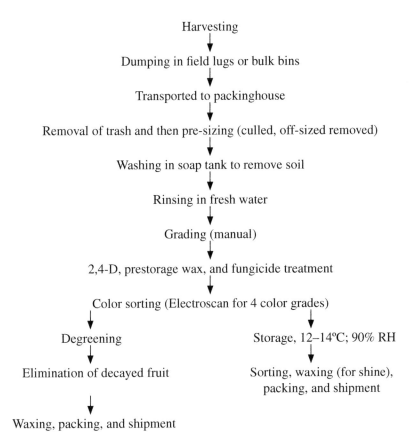

Harvesting
↓
Dumping in field lugs or bulk bins
↓
Transported to packinghouse
↓
Removal of trash and then pre-sizing (culled, off-sized removed)
↓
Washing in soap tank to remove soil
↓
Rinsing in fresh water
↓
Grading (manual)
↓
2,4-D, prestorage wax, and fungicide treatment
↓
Color sorting (Electroscan for 4 color grades)

↓	↓
Degreening	Storage, 12–14°C; 90% RH
↓	↓
Elimination of decayed fruit	Sorting, waxing (for shine), packing, and shipment
↓	
Waxing, packing, and shipment	

The sequential fungicide and wax coating treatment for lemons as practiced in California (Eckert and Brown, 1986) is given below. It is integrated with the normal handling sequence for lemons as mentioned earlier.

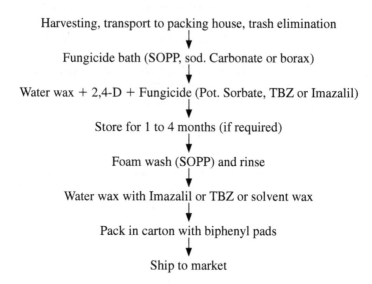

Harvesting, transport to packing house, trash elimination

↓

Fungicide bath (SOPP, sod. Carbonate or borax)

↓

Water wax + 2,4-D + Fungicide (Pot. Sorbate, TBZ or Imazalil)

↓

Store for 1 to 4 months (if required)

↓

Foam wash (SOPP) and rinse

↓

Water wax with Imazalil or TBZ or solvent wax

↓

Pack in carton with biphenyl pads

↓

Ship to market

Sweet orange- Sweet oranges are handled directly on the packing line after harvest if their natural color is acceptable. Otherwise fruit is sent for degreening.

In the U.S., sweet oranges and tangerines are handled in the following sequence

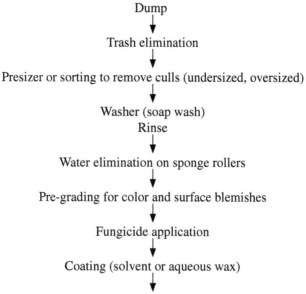

Dump

↓

Trash elimination

↓

Presizer or sorting to remove culls (undersized, oversized)

↓

Washer (soap wash)
Rinse

↓

Water elimination on sponge rollers

↓

Pre-grading for color and surface blemishes

↓

Fungicide application

↓

Coating (solvent or aqueous wax)

↓

Drying and careful grading
↓
Stamping/labeling
↓
Sizing
↓
Packing and palletizing
↓
Precooling
↓
Storage

If fruit is to be degreened, then after trash elimination a fungicide drench is given, followed by degreening for 24–72 h. Degreening is for early fruit only that has to be sent to market. From here, fruit is sent for pre-sizing. Groves are selected for fresh fruit packing so that high packout is achieved. Fruit from just any grove cannot be run on the packing line as there could be very low packout which would be uneconomical. Fruit has to be accumulated for certain minimum hours of operations because grove selection, harvesting, and transport are dependent on the efficiency of the harvesting crew and also the availability of suitable groves. At the accumulation site, fruit is stored at 10–20°C. The off-sized, off-grade, and culls are sent for processing/juicing. The fruit is generally sold from an inventory of packed and refrigerated fruit (Grierson and Wardowski, 1977).

Mandarins/Tangerines- Satsumas are handled in a different way in Japan. For high-quality fruit to be sold in the market, manual grading is done again if necessary after optical color and size grading.

Schematic diagram for Satsuma mandarin in Japan.

Receiving and degreening (2–3 days in ethylene) of early mandarin (June–October)
↓
Receiving in packing house
↓
Weighing and presorting
↓
Washing, fungicide treatment, and waxing
↓
Grading (manual)
↓

Optical color grading and sizing
↓
Manual grading if necessary
↓
Packing in 15 kg cartons
↓
Marketing

After harvesting, farmers cure the Satsuma fruit for 3–10 days. Fruit is also sometimes stored at the farmer's house for 1–3 months before forwarding to the packinghouse. Packinghouse operations usually take 2–3 days. Satsumas are sensitive to mechanical damage during handling on machines; fruit should not drop more than 30 cm (Yamashita and Kitano, 1981).

'Nagpur' mandarin is a major citrus fruit in India. Its mechanized handling is done in following way.

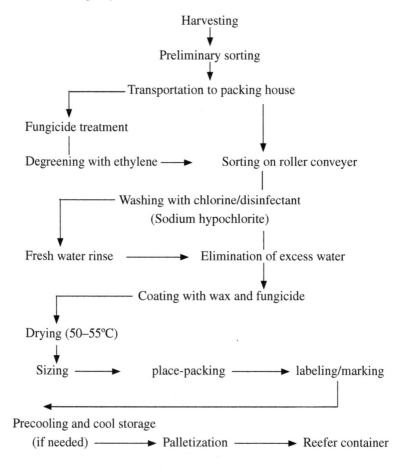

Harvesting
↓
Preliminary sorting
↓
Transportation to packing house

Fungicide treatment
Degreening with ethylene → Sorting on roller conveyer

Washing with chlorine/disinfectant (Sodium hypochlorite)

Fresh water rinse → Elimination of excess water
↓
Coating with wax and fungicide

Drying (50–55°C)
↓
Sizing → place-packing → labeling/marking

Precooling and cool storage (if needed) → Palletization → Reefer container

C. Care During Handling of Fruit on Packing Line

The most important point to bear in mind while designing the mechanical handling systems is to minimize drops, impacts, compressions, and vibrations to fruit. A system that injures or presses the fruit or softens it is self-defeating. The padding of abrasive surfaces, lining, sponge-cushioning, decelerating curtains of rubber or some other soft material should be used. Mandarins/tangerines, limes, and lemons are most sensitive to bruises. The optimum speeds of various conveyers and sizers, proper illumination of packing area and sanitary conditions inside the packinghouse are musts for optimum efficiency of working personnel and machines.

Belts, carton fillers, counters, shakers, and other equipment should be checked regularly against the possibility of excessive injury to the fruit (Eaks, 1956). Hasegawa et al. (1989) observed that causes of deterioration in different packinghouses included damage inflicted at receipt of boxes, mixing, brushing, sorting, transport within packing house, weighing, and packing. Partial improvements in these critical steps improved quality and reduced losses.

D. Manual versus Mechanical Operations

There are always doubts about the efficiency of mechanical systems to handle fruit without injuries, particularly in the case of easily peelable mandarin fruit. Attempts to assess handling injuries in 'Nagpur' mandarins indicated that the collection of harvested fruit from picker bags to vented plastic crates in the orchard and then transportation directly to packinghouse resulted in 0.28 major cuts and 0.65 minor bruises per fruit as compared with 0.69 major cuts and 6.1 minor bruises per fruit in conventional handling, which occurs 10–12 times including sorting (Ladaniya et al., 1994). Comparison of manual and mechanical operation revealed that sorting culled fruits on a conveyer was a faster and easier operation compared to manual sorting on the floor (Ladaniya and Dass, 1994). Manual washing is slow and laborious compared to mechanical washing. On a packing line of 1 ton/h capacity, sufficient cleaning is achieved with 105 rpm of nylon roller brushes. The 185 ml wax requirement in manual dip method for 150 fruits compared to 90 ml for the same number of fruit in a mechanized process indicated a higher cost for the manual operation. Manually dip-treated (for wax coating) fruit need a large space for surface drying while in the drying tunnel (55°C ± 5°C temperature) for satisfactory drying to be achieved within 1.5 min. Manual sizing of fruit was not as accurate as in rotary-disc type sizer. Fungicidal wax coating on a mechanized packing line reduced decay effectively as compared with a manual operation (Table 9.4). This difference was attributed to non-recovery sprays of washer and waxer. In the dip method of waxing there is always the possibility of contamination of treatment solution and thus spores get lodged in the bruises and wounds of healthy fruit.

TABLE 9.4 Effect of Manual and Mechanical Waxing on Decay of 'Nagpur' Mandarin in Storage

	Decay (%)			
	Ambient condition (1 week)		Refrigerated condition (1 month)	
Treatment	Manual	Mechanical	Manual	Mechanical
Stayfresh storage wax (6%)	6.78	2.38	8.97	1.17
Stayfresh storage wax (4%)	4.32	3.95	6.26	1.48
Stayfresh packout wax (6%)	7.98	7.53	2.74	1.29
Stayfresh highshine wax (6%)	8.88	8.97	7.68	4.00
Polyethylene wax (6%)	12.32	3.06	5.26	4.19
Polyethylene wax (4%)	7.35	7.08	5.80	1.56
Control (washed)	15.48	14.90	10.18	7.39

Ladaniya and Dass (1994).

E. Packinghouse Environment, Worker Conditions, and Safety Rules

The noise level in the packinghouse environment should be minimal. The entire area should be clean, amply ventilated, dust-free, and cool, with a relative humidity of 50–70 percent. Efforts should be made to minimize the distraction of workers. The floor should not be slippery and workers must wear shoes with right type of sole to have a good grip of the floor. The elevated platforms should have minimal vibrations when belts, conveyers, and sizers are in operation and good-quality matting should be placed on platforms. The stairs to elevated platforms must have railings on both sides. Principles of ergonomics apply to all workers in the packinghouse; comfort in the workplace can increase the efficiency many times over.

The safety of workers is of utmost importance in the packinghouse. Even the starting of machines should be in sequential order. For example, if heaters in the drying tunnel are switched on without switching on the conveyer, the conveyer may be damaged by the heat. Similarly, washer and waxer conveyers and spray pumps need to be started before fruit is fed to the machine. Usually the last machine is switched on first and the feeding conveyer and dumping should be started last so that machine operations are optimized by the time fruit starts moving on the machines.

From a quality-assurance point of view, certain rules such as dress code for workers and use of gloves must be followed as per international norms (see Chapter 19).

F. Sanitation

Sanitation is the most important aspect to keep in mind for running a packinghouse successfully, since the results of fungicide treatments and losses from decay depend on sanitary conditions, which affect spore load in the air, on machinery, and on handling container surfaces. Machinery, bins, handling containers (plastic and wooden), and flooring have to be cleaned with chlorinated water at 200 ppm with pH 7 (quaternary ammonia compounds or Sodium hypochlorite) or phenyl solution. After presorting, decayed fruit should not be left nearby because the spores can be carried by wind or insects into the packinghouse. These practices are invariably useful in reducing incidence of decay without use of fungicides or to minimize use of fungicides.

G. Wastewater Disposal

Considerable quantities of wastewater come from both large and small packinghouses. This water contains detergents, disinfectants, fungicides, and aqueous waxes. This wastewater can be easily treated (physical and chemical treatments) and recycled. About 94 percent of recycled water is considered fit for washing and rinsing use in packinghouses again (Ismail, 1981). The sludge coming out of the treatment plant can be used in orchards. For treatment, ferric chloride 80 mg/lit is used as a coagulant. Subsequently solids are flocculated, clarified, and filtered. This treatment results in removal of 98 percent of suspended solids. This water has 84 percent chemical oxygen demand and 74 percent biological oxygen demand.

H. Economics and Energy Needs of Packinghouse Operations

Packinghouse costs per carton vary depending on the scale of the operation or the size of the business. Trends indicate that for each increase of 100000 boxes packed, the cost of packing and selling decreased by 2.1 percent (Brooke and Spurlock, 1977). The common size of packinghouse in Florida during the late 1970s had a capacity of 200000–300000 packed boxes per season. There are much bigger packinghouses currently. Some of the large packinghouses in Florida pack up to 2.5 million boxes in a year. Large-scale operations reduce cost and thus fruit price to be competitive. Management is another important factor that can influence cost. The itemized cost of packing California oranges as reported by the Economic Research Service of the USDA in 1974 provides some insight about various costs in packinghouse operation during that time. Total cost of operations was 1.092US$ per carton. Packaging material and labor (about 0.30 and 0.25US$/carton, respectively) accounted for the main cost of packing and selling packed fresh citrus. The cost of transportation to packinghouses and profits were not included in this figure. The average charges for fruit hauling, packing elimination, and drenching (not the costs) as reported by Muraro (2006) were 0.58–0.53$, 0.54–0.55$ and 0.181–0.189$ per box in Florida for grapefruit, oranges and tangerines respectively. The average packing charges for domestic grapefruit, export grapefruit, oranges, and tangerines were 4.016$, 4.395$, 4.347$, and 5.469$, respectively per 4/5 bushel carton. These charges included plastic bags, labels, price codes, supervisory and labor charges, fruit treatment, power, water, maintenance, insurance, taxes, depreciation, rent, selling expenses, and administrative charges.

The total operational cost, including washing, chlorine treatment, fungicide application, and wax coating ranged between 6–6.5US$ per ton of fruit in California, whereas the same operations cost 5–5.5US$ in Florida (Johnson, 1991). In India, the operational cost of sorting, washing, waxing, drying, sizing, and carton filling and packing (excluding packing material) during 2004–2005 varied as Rs. 350–400 per ton of fruit (that comes to about 7–8US$ per ton of fruit at the exchange rate of Rs. 50 per US$). This included the interest on loans but does not include cost of packing material, depreciation, rent, and advertisement.

Mechanization and complete automation is the need of the hour in some countries considering costs and unavailability of labor. Miller and Drouillard (2000) analyzed key economic factors affecting the selection of automatic grading equipment for citrus in Florida for replacing manual graders and for saving in fungicides and waxes. For a $400 000 expenditure, an 8-grader reduction yielded an equivalent 5-year cost projection. A 14-grader reduction would be required for a high-level $800 000 capital outlay. Mechanization has the major drawback of job loss for many graders and packers.

Idling machines in a packinghouse consume energy without productivity. This situation can be avoided by meticulous planning of operations. Among various units on the packing line, the warm air drier is the largest consumer of electricity. In the study conducted by Schillacci (1988), packinghouse work productivity per operator was 0.11 ton/h and for sorting, it was 0.42 ton/h per operator. Sorting and packing were responsible for 31 percent and 27 percent of total work time since these are the most labor-intensive operations in the packinghouse.

I. Conventional Operations in Developing Countries and Recent Developments

Conventionally, handling operations have been manual in India and neighboring South Asian and Southeast Asian countries for most fruits including citrus. Handling fruit in bamboo baskets and transporting loose fruit from orchard to packinghouse cause considerable invisible bruises that can potentially lead to rotting during subsequent handling. In India, for the domestic market, mandarins, oranges and small-fruited acid limes are sorted and graded mostly manually with the exception of Kinnow mandarins.

In Central India, 'Nagpur' mandarin fruits are handled manually several times for sorting in orchard and before and after auction in market. Paddy straw is used as cushioning material at all stages of handling on the floor. Almost similar is the practice for Kinnow mandarins in conventional mode of handling, although it has been discontinued to a large extent. With increasing production and challenges of export in the early 1990s, the need for mechanization was clear.

During the last 10 years, considerable developments have taken place in mechanized handling of citrus fruits in India. Kinnow mandarins grown in Punjab (India) are increasingly being mechanically washed, waxed, and packed in corrugated cartons. With the initiative of the state government there are five operational waxing and packing units in Punjab and many more are in the works. These units have been using Citrashine wax (wood-resin and shellac-based) imported from Italy. In these packing units, as the fruits are brought by growers in ventilated plastic crates, they are delivered as dry dump on the packing line conveyer. Sorting, washing with plain water, removal of excess water with sponge rollers, coating, drying, and sizing on divergent rollers are carried out in that sequence. Fruits are sized in six sizes: very small, small, medium, big, large, and extra large. In Rajasthan, especially in Ganganagar district, such modern postharvest handling facilities have been built for Kinnows in four locations; their capacities are 5–6 ton/hour. About 60–70 tons of fruit is handled, washed, waxed, and graded in each of these packinghouses daily and thus about 100 000 tons of fruit is being handled in a season of 3–4 months in these two states. Facilities such as mechanized loading bays and precooling and cold storage facilities have also been built. In Andhra Pradesh, a packing line has been established by the cooperative of the farmers for oranges and small-fruited acid limes in Anantpur district.

During the 1960s through the 1980s, washing and waxing 'Nagpur' mandarins was first practiced manually by a few progressive growers in Central India. A mechanized process was adopted during 1990s for 'Nagpur' mandarins in Maharashtra state in Vidarbha region. Three units with 2–5 ton/h packing capacity have been established so far. Vented plastic crates of 19–20 kg capacity (55 × 35 × 30 cm) are being increasingly used as bulk handling containers in fields and packinghouses, replacing traditional bamboo baskets. This container has a smooth internal surface that results in minimum bruises. Mandarins placed in crates are being handled with forklift trucks to minimize fruit injuries in some of the large fruit-handling facilities in the country. The Maharashtra Agro-industries Development Corporation, Government of Maharashtra, has set up two processing units in the production area – one at Morshi in Amravati district and another – at Katol in Nagpur district. At morshi (Amravati), a fresh-fruit packing unit with a capacity of 5 ton/h can handle a minimum of 50–60 tons of mandarins a day. Precooling, degreening, and cold-storage infrastructure – which is a must for export – are also available and indigenously manufactured.

Acid lime fruits are handled manually. However, manual sizing is a time-consuming and labor-intensive operation. It is observed that the mechanical sorting, washing, and grading system suitable for mandarins (Ladaniya and Dass, 1994) can also be used for acid limes and Mosambi sweet oranges with modifications in treatments (Ladaniya, 2001; Ladaniya, 2002).

J. Some Tips for Export of Citrus Fruits

Mechanized handling is a must for export as it ensures hygienic conditions and avoids contamination. Citrus exporting units must follow the general guidelines for efficient export and achieving the targets.

(1) Identify the market, its demand, and the season of import.
(2) Target the production to fulfill the demand quantitatively and qualitatively. Maintain internal quality and uniformity of fruit shape,

size, appearance, and color as per quality standards of the importing country.

(3) Minimize time between harvesting and packing followed by precooling.
(4) Make sure that fruit is handled promptly at 20–25°C and 60–70% RH and packed quickly after harvest (except lemons).
(5) Two-piece full telescopic boxes are preferable to open trays.
(6) Mark the cartons with all the required information on at least two opposite sides or ends so that it is visible on the pallet.
(7) Maintain required temperature, RH, and ventilation during shipment.
(8) Phytosanitary certificate and certificate of origin must be obtained in time.
(9) Arrange transport route and shipment in such a way that consignment reaches importing country within shortest possible time.
(10) Complete the documentation in time to avoid delays and deterioration in quality. Be sure about fungicides and residues in fruit considering regulations of importing countries.

III. POSTHARVEST TREATMENTS

A. Curing

Water content of fruit peel is an important factor, determining turgidity of rind tissues that in turn determines the extent of injuries and disorders after harvest. When these fruits are harvested early in the morning or in wet weather, they tend to develop oleocellosis or brown spots on the rind with slight pressure or bruises. Curing is a pre-treatment and a conditioning of the fruit before running it on the packing line. Curing for few hours or days depending on variety or species removes slight moisture from the peel so that it becomes suitable for mechanical handling. This curing is achieved by just holding the fruit under shade in ambient conditions. Lemons are generally cured at 15–17°C for 1–2 days in many countries so that rind physiological disorders are minimized. In Israel, fruit is cured at 36°C for 72h in vented plastic crates covered on top with non-vented HDPE. The fruit is then coated with water wax containing TBZ (2000ppm). Temperature, relative humidity, and wrapping/lining in the box are the critical factors as far as water loss from citrus fruit is concerned.

B. Disinfection and Cleaning

Disinfection of water used in washers or for general cleaning purpose can be achieved by chlorination, ozone, and/or UV treatment. The most commonly used chemicals are sodium or calcium hypochloride for chlorination. Liquid soaps and detergents are also added to the wash water used for general sanitation of the packing house and for cleaning fruit. Fruit cleaners of various companies are available in the market. For example, FMC corporation, USA supplies fruit cleaner detergent (FMC fruit cleaner detergent 395) used as a foam application. The SOPP is commonly added to the water along with detergent before fruits are coated. Fruit cleaner, a biodegradable detergent, and FOMER (detergent + 20 percent SOPP) are commercial products from the Fomesa Company of Spain.

C. Surface Coatings

Waxes and coatings for polishing and improving sheen and reducing water loss of citrus fruits are available in various forms such as solvent waxes, aqueous emulsions, and resin solutions. 'Wax' has become a generic term, and any type of coating – whether it contains wax or not – is called wax. This is because the earliest coatings contained waxes and in fact in China, citrus fruits were preserved by applying molten waxes during ancient times. Common waxes are carnauba, paraffin, and oxidized polyethylene. Carnauba is extracted from leaves of the *Copernica prunifera* palm, which is native to Brazil and found in tropical forests. Oxidized polyethylene is a ethylene polymer with acidic groups for emulsification. Paraffin is hard white crystalline wax, mostly containing saturated hydrocarbons, and it is obtained from the petroleum industry. Edible coating can be developed from protein, polysaccharides, lipids, or from a blend of these compounds. The ability of these coatings to extend shelf life is due to differential permeability to CO_2, O_2, and water vapor, which reduce metabolic rate and water loss. Permeability of citrus coatings should be high to O_2 and low to water vapor to reduce transpiration as much as possible and not overly restrict respiration. From the fruit physiology point of view, coatings that excessively restrict respiration are harmful in that they result in off-flavor. Effective protective coating reduces weight loss by 30–40 percent (Johnson, 1991) only so as to avoid anaerobic respiration. In case of 'Nagpur' mandarins, respiration is reduced to about 30 mg/kg/h in aqueous wax emulsion-coated fruit compared to 42 mg/kg/h in non-coated fruit.

1. Wax Coatings

Initially (early twentieth century) waxing and polishing of citrus fruit was done using solid wax slabs. Slab waxing, though simple, was not considered satisfactory. The hot fog method is also no longer used, although it gives excellent results when carried out correctly. Solvent waxing was widely used in Australia during the 1950s and 1960s (Long and Leggo, 1959). Applying a water-wax emulsion by dipping was the most common method in the days before mechanization. In India, wax emulsions were tried in the 1960s and 1970s to extend the shelf life of citrus fruits; the application method was manual dipping of fruit placed in perforated buckets into drums containing a wax emulsion. Fruits were then surface-dried on bamboo mats, which took several hours.

There could be hundreds of permutations and combinations in developing a suitable composition for a coating. Most of these compositions are patented or trade secrets. The most commonly used waxes are emulsion waxes, which contain wax in soap or detergent. In emulsion-based water waxes (aqueous), melted waxes (sugarcane wax, carnauba wax, bee wax, candelila wax), oxidized polyethylene (synthetic) and microcrystalline (paraffin and acrylic compounds) are emulsified and then boiling water is added. Fungicide is also added for decay control. Soft, cold water is used to adjust total solids. A shellac or resin solution provides shine on the fruit surface. Shellac is prepared from raw lac after refining and bleaching. The storage wax does not contain additives for shine. In many countries the materials used in formulation of waxes must meet the regulations to qualify as food grade materials. Generally wax compositions are mixtures of waxes from plant and petroleum sources. Combinations of paraffin or polyethylene wax and carnauba wax are also used. Paraffin wax offers good control of water loss while carnauba provides luster. Water-type waxes have 4 percent polyethylene, 2 percent paraffin, up to 4 percent shellac, and 4 percent colophony, or wood resin. Carnauba wax level is usually 4–6 percent. Wax can be a carrier of fungicide to reduce rot or a carrier of anti-senescent PGR to slow aging. Waxes are brushed, sprayed, fogged, or foamed onto produce. Usually solids are of food-grade material in these waxes and the liquid component evaporates on drying, leaving only the wax. Water waxes give good results on clean fruit. Fruit should be at least damp (not completely dry) because a good bonding of water wax with fruit surface is necessary. Fruit should be clean to avoid 'sweating'. Water dissolves the wax or forms bubbles and leaves white spots when drying. Water waxes may chalk by re-emulsification when they get wet from sweating.

Solvent-type waxes include coumarone-indene resin with a varying range of 10–12 percent depending on composition. This resin was developed as a base for chewing gum. The so-called 'solvent waxes' are prepared by dissolving one or more types of resin in a blend of petroleum solvents. Resins are synthetic or from natural sources such as wood resin. These formulations of solvent waxes also contain plasticizing and/or leveling agents to assist in forming a shiny and flexible film on the fruit surface. In solvent waxes there is no water component. Fruit should be completely dry before the application of these waxes. Another type of wax is resin-solution wax or resin solutions that contain alkali-soluble resins, gums, shellac, and so on. Resins are sometimes modified with organic or mineral acids. The organic acids, wetting agents, and oils are also added because they act as plasticizers. These waxes also contain water as a component.

Even today, in most underdeveloped and developing countries, the most commonly used methods of application include dipping the fruit in an emulsion. Mechanical spraying, foaming, and dripping of emulsion onto roller brushes is gaining popularity slowly. In developed countries, the mechanized process of coating fruits has been common for the last 50–60 years.

Different citrus fruits respond differently to these wax formulations. Overwaxed fruits have an off-flavor, while underwaxed fruits shrivel from water loss. Eaks and Ludi (1960) reported that temperature, washing, and the method of waxing also affect the composition of the internal atmosphere of orange fruits.

Higher storage temperatures resulted in higher carbon dioxide and a lower oxygen content of the internal atmosphere. Washing and waxing tend to increase the carbon dioxide content and decrease the oxygen content; the degree of change depends upon the wax used. Washing increases weight loss in fruit and waxing reduces weight loss compared with that of unwashed fruit. Under the conditions of tests conducted by these workers, washing and waxing had little influence on the respiratory rate of the fruit at 20°C.

Oranges coated with commercial solvent-type wax lost less weight than those with comparable amounts of water-wax polyethylene coatings. Ethanol buildup in juice and off-flavors occurred after multiple coatings of water wax or solvent-type wax. Less ethanol accumulation and no off-flavors were noted in polyethylene wax-coated fruits (Davis and Hofman, 1973). Generally volatiles are analyzed by head-space gas chromatography and the amount is presented as mg per 100ml of juice. Ethanol and acetaldehyde are major volatiles. Clementines waxed with a formulation consisting of polyethylene (6 percent), lac (4 percent), and colophony (4 percent), and stored at 4–5°C had ethanol levels of 119mg/100ml and 362mg/100ml juice after 8 and 13 weeks respectively. An off-flavor was detected after 13 weeks, only. In sweet oranges (Valencia and navel), off-flavor was detected at the juice ethanol level of 140mg/100ml. The highest ethanol level was recorded in water-wax (Cuquerella et al., 1981).

In Valencia oranges, waxing increases internal fruit CO_2 concentration and juice ethanol content to some extent, but no off-flavor develops if the right type of wax is used. Waxing with a 15 percent resin formulation resulted in the highest internal CO_2 concentration and ethanol content (6.7 percent and 60.2mg/100ml respectively). After 120 days of storage at 3°C with 85 percent RH, followed by seven days at 20°C with 70 percent RH, weight loss was lower in coated fruits (8.6–12.3 percent). Waxing combined with sealed packaging increased internal CO_2 concentration and juice ethanol content (Martinez-Javega et al., 1991). Silverhill Satsuma mandarins grown in New Zealand, dipped in Citrus Gleam or carnauba wax and stored at 6°C had reduced O_2 and/or elevated CO_2 in their internal atmospheres (Lawes et al., 1999). O_2 and CO_2 levels changed little during 6 weeks in cold storage, but changed rapidly when transferred to ambient temperature. Citrus Gleam also increased the concentrations of ethanol and acetaldehyde in the headspace gas of juice extracted from fruits. Coating fruits with Citrus Gleam reduced weight loss and permeability of the peel by water vapor, but only by a maximum of 30 percent in 3 weeks of cold storage.

Florida-grown Valencia oranges coated with the polyethylene wax with Imazalil lost the least weight over 22 weeks (0.075 percent per day) at 3–4°C, while maintaining low anaerobic respiratory products and good internal quality (Peeples et al., 2000). Washing and lining treatment resulted in the lowest total weight loss of 2.38 percent over 20 weeks. Shellac-based coatings were not optimal for extended storage because of their relatively poor control of water loss and low permeability to O_2 and CO_2, which led to off-flavor development within 5–10 weeks of storage and, with long-term storage, fruit physiological breakdown.

Hagenmaier (2000) evaluated polyethylene-candelilla-wax coating formulation in comparison to a high-gloss shellac-and-wood-resin citrus coating for storage of Valencia oranges at 15–25°C. Oranges with the wax coating had relatively high flavor scores (8.9–10.4) even after 9–16 days in storage at 15–25°C. By contrast, the high-gloss coating resulted in flavor scores as low as 3.7–4.1 after 9–16 days at 25°C or 16 days at 21°C. Flavor was especially low for fruits with internal O_2 <1 percent. Flavor tended to decrease almost linearly with increasing ethanol content, which in turn was highly dependent on internal CO_2. Oranges with the shellac-and-wood-resin coating temporarily had better gloss than fruits with the wax coatings, but this advantage was lost after eight days in storage at 15–25°C.

Waxing *per se* and various wax compositions may also affect susceptibility of fruit to rind breakdown and pitting. Marsh white grapefruits coated with shellac-based waxes develop pitting when stored at 12°C and 93 percent relative humidity for 12 days. Pitting developed to a much lesser extent on fruits coated with carnauba or polyethylene-based wax. The application of shellac-based waxes significantly reduces internal O_2 levels and increase CO_2, ethanol, and acetaldehyde levels within 1 day of application (Petracek et al., 1998). The effects of individual wax components such as shellac and resin, and commercially available carnauba and shellac waxes on pitting incidence of white Marsh grapefruits and Fallglo tangerines were examined by Dou et al. (2000). Postharvest peel-pitting incidence was significantly reduced when grapefruits were coated with shellac solutions (pitting incidence <2 percent) or resin (<3 percent) rather than the commercial shellac waxes (20–46 percent). With the diluted shellac or resin solutions, pitting was also low (<3 percent) but fruit shine was reduced. Commercial shellac waxes or higher-concentration shellac/resin solutions increased the shine. Carnauba wax reduced pitting to a minimum (<1 percent). In Fallglo, pitting incidence was higher than in grapefruit.

Waxing sharply arrests the color change from ethylene degreening of Valencia oranges and Marsh grapefruits, but has comparatively little effect on ethephon-dipped fruits. Sta-fresh-451 resin-based aqueous wax emulsion applied mechanically on the packing line significantly retarded color development in Mosambi sweet oranges stored at ambient conditions (25–30°C) for 30 days in vented polyethylene-lined boxes. Non-waxed fruits inside vented polyethylene liners gradually turned from green to yellow during that period (Ladaniya, 2001).

Wax coatings also reduced susceptibility to chilling injury in grapefruit and Temple mandarins but enhanced it in limes (Pantastico et al., 1968). In wax-coated 'Nagpur' mandarin fruit stored at 3.5°C (constant), chilling was found to be less than in non-coated fruit (Ladaniya and Shyam Singh, 2005).

There is a voluminous literature on evaluation of waxes and coatings applied by manual dip method to extend the storage ability of various citrus fruits grown in India. The non-waxed + non-packed (control), waxed and polyethylene-wrapped 'Coorg' mandarins had a shelf life of 7, 12, and 20 days respectively at 26–30°C and 30–60 percent RH (Subbarao et al., 1967). Waxed

TABLE 9.5 Effect of Various Coatings Combined with Growth Regulators, and Lining/Wrapping on Storage Life of Mandarin Fruits

Mandarin cultivar	Effective treatment	Storage Temperature (°C) and RH (%)	Storage (Days)
Nagpur	Wax (12.5%) + 2,4-D (1000 ppm)	4.44	90
Nagpur	Wax (6%) + 2,4-D (100 ppm)	25	24
Coorg (Main)	Wax(4%)	7°C; 85–90% RH	75
Coorg	Vented PE liner + diphenyl wrapper	4–5; 85–90% RH	135
Coorg	Wax (3%) + 2,4-D (100 ppm) + PE	20–26°C	21
Kinnow	Wax 6% + Polyethylene	1–3°C; 85–90%	90
Kinnow	Wax (6%) + diphenyl paper wrapping	19±8	90

Lodh et al. (1963), Subba rao et al. (1967), Soni and Gupta, (1983), Dalal et al. (1974), Gopalkrishna rao and Krishnamurthy (1983), Josan et al. (1983), Singhrot et al. (1987c).

Kinnows could be stored for 30 days at ambient conditions and for 85 days at 2.4–4.4°C (Mann and Randhawa, 1978; Mann and Sandhawalia, 1983). Various postharvest treatments with and without wax coatings were reported to be effective in extending storage life of mandarin fruits grown in India (Table 9.5). Edible oils such as sesame oil (2 percent) and mustard oil (12 percent) were also effective in addition to waxol (a wax emulsion) in reducing weight loss of Kinnow fruits (Singh et al., 1978a; Singh et al., 1978b; Ran Singh et al., 1989; Sharma et al., 1991). Waxol (12 percent) with and without Captan (0.1 percent) and Topsin-M (0.1 percent) and packing in polyethylene liners also extended storage life of Baramasi lemons (Sharma et al., 1989; Sharma et al., 1986; Singhrot et al., 1987a; Singhrot et al., 1987b). For yellow limes, a mixture of carnauba wax, sugarcane wax and lac emulsion (8 percent solids) with 1 percent SOPP gave better control of loss in weight and rotting for up to 2 weeks at room temperature than polyethylene wax (Thomas et al., 1964). When packed in perforated polyethylene, a lower concentration (3 percent) of wax was also effective in reducing weight loss of lime fruits (Bhullar, 1983). 'Nagpur' mandarins could be kept in acceptable condition for 21 and 60 days at ambient and refrigerated condition respectively with fungicidal wax coating applied mechanically on the packing line followed by packing in vented polyethylene-lined boxes (Ladaniya and Sonkar, 1997; Ladaniya and Sonkar, 1999).

Washington Navel oranges treated with wax (3 percent) + sodium ortho phenyl phenate (500 ppm) could be stored up to 21 days under ambient conditions. Untreated fruits lost marketability after 2 weeks (Sharma et al., 1965). Different types of coatings and other postharvest treatments are reported to be effective in extending storage life of sweet oranges (Table 9.6). The Sta-fresh (451) wax–coated Mosambi sweet orange fruits packed in perforated polyethylene could be

TABLE 9.6 Effect of Coatings Combined with Growth Regulators and Lining Material on Shelf Life of Sweet Orange Fruits

Sweet orange cultivar	Effective treatment	Storage Temperature (°C) and RH (%)	Storage (Days)
Mosambi	Perforated polyethylene + Wax (6%)	10°C 34–40°C	75 28
Mosambi	Wax (8%) + 2,4-D (100 ppm) + PE	Room temperature	40
Mosambi	Wax (9%) + 2,4-D (500 ppm) + Bavistin (0.1%) + non perforated PE	Ambient condition	63
Mosambi	Sta-fresh 451 (1:1 dilution) + PE	25± 5°C 5–6°C; 90–95% RH	30 90
Sathgudi	SOPP (2%) + hexamine (2:1) + wax (6%)	5.5–7°C	127
Sathgudi	Vitamin K_3 (Menadine)	28–30°C	16
Sathgudi	Wax (12%) + vented polyethylene bag	3.3°C 5.6°C 31°C	125 75 25
Blood Red, Valencia	2,4-D (100 ppm) + Polyethylene bag	19–32	30–60
Blood Red	Wax (12%) + 2,4,-D (50 ppm)	Room temperature	45
Pineapple	Wax (6%)	Room temperature	45

Kohli and Bhambota (1965), Dalal and Subramanyam (1970), Sadashivam et al. (1972), Bhullar (1981), Bhullar (1982), Ghosh and Sen (1984), Rao et al. (1987), Dashora and Shafaat (1988), Tarkase and Desai (1989), Ladaniya (2001).

satisfactorily stored up to thirty days at 25 ± 5°C (Ladaniya, 2001). The respiration rate was reduced by 30 percent in waxed fruit as compared with a control.

In China, Rinrei's wax has been tried by Chen et al. (2000) for coating citrus fruits. The best concentrations for coating shaddock, Ponkan mandarins, Liucheng oranges, Tankan tangors, grapefruits, and Haili tangor fruits were 50, 40, 60, 30, 70, and 100 percent respectively. After storage at 25–28°C for 8 weeks, the weight loss of shaddock treated with 50 percent Rinrei's wax was reduced from 26.1 percent (control) to 10.3 percent. The weight loss of waxed Ponkan, Liucheng, Tankan, grapefruits, and Haili fruits stored at 15°C for twelve weeks were 2–5 percent, compared with 4–11 percent in controls and 1 percent for fruits in plastic bags. The chilling injury of Ponkan and Tankan at 10°C was reduced when waxed before storage.

a. Examples of Some Wax-coating Formulations

Different formulations of coatings are available commercially and some examples of wax coatings are as follows: (1) Tropical Fruit Coating 213 (EcoScience, Orlando, FL, USA). It contains 5 percent carnauba wax and fatty acid soaps for a total solids content of 10.8 percent, (2) FMC coatings such as StaFresh 451, StaFresh

921 (California, USA), (3) Gloss Guard-31(Stay-fresh, India), (4) Brogdex waxes, USA, (5) Citrashine (Elf Atochem Agri Itali s.r.l.). Citrashine is a water wax containing natural resin and shellac-based wood resin E 904-E914. The StaFresh 451 of FMC, Corporation (USA) is a resin-based, high-gloss aqueous wax with 15.9 percent solids containing a fungicide 2-(4 thiazolyl benzimidazole) TBZ 0.1 percent. Sta-Fresh 705 is a water-based coating for lemons to be stored. Sta-Fresh 320 is for imparting high shine for packing and distribution. Sta-fresh 360 is a coating for very high luster as required in export. Flavor Seal- 93' is a solvent wax of FMC Corporation. Water wax TTT/21UE is a 18 percent wax with 0.2 percent Imazalil and 0.5 percent TBZ from FOMESA, Spain, for use in Europe. These coatings are generally diluted before use as they are manufactured with 18–24 percent solids. The dilution can be in the proportion of 1:1, 1:2, 1:3 and so on. Addition of 1 part soft water to 1 part wax emulsion gives one-half of the initial solid concentration. If it is 1 part wax and 2 parts soft water, the concentration is one-third of the original. If it is 1 part wax and 3 parts water, the dilution is one-fourth of the original concentration. The Stayfresh gloss Guard-31, which is available with 18 percent solids is diluted to 3–4 percent solids for 'Nagpur' mandarins and 6 percent for acid limes.

Various agencies regulate the use of coating materials in different countries. In the U.S., the agency that regulates use of coatings is the Food and Drug Administration (FDA). This agency approves ingredients as direct food additives or as 'GRAS' (Generally regarded as safe) based on compliance with extensive performance and safety evaluations. Carnauba wax and wood resin are allowed as nonsynthetic ingredients in coatings for organic citrus. In Japan, the Ministry of Health and Welfare publishes the Japanese standards of food additives in which approved ingredients are mentioned for preparation of coatings and addition to food. In EEC countries, per-country standards in addition to codex standards (FAO) are followed.

2. Edible Coatings

Edible coatings are developed from edible ingredients (carbohydrates, fats, and proteins) and therefore they have recently generated considerable interest. These coatings can be classified in five categories: lipid-based, polysaccharide-based, protein-based, composite, and bilayer coatings (Baldwin et al., 1995). Functional requirements of the ideal edible coating are (1) Neutral organoleptic properties, (2) Water-vapor tightness to prevent desiccation, (3) Predetermined permeability of O_2 and CO_2, (4) Ability to enhance surface appearance (brilliance) with minimum stickiness. Edible coatings generally stabilize the product, thereby increasing shelf life. Specifically, the coatings have the potential to reduce the moisture loss, restrict oxygen entrance, lower respiration, retard ethylene production, and carry additives that retard microbial growth. Coatings may change flavor as a result of anaerobic respiration and increase ethanol concentration. Coatings can also trap aroma compounds, thereby increasing their concentration (Baldwin et al., 1995).

Semi-permeable coatings create a modified atmosphere (MA). However, this change in atmosphere is in response to the surrounding environment (temperature and humidity), fruit respiration, and coating permeability. Permeance is the permeability expression without accounting for thickness and is used for evaluation of coating performance under equilibrium conditions of gases on two sides of coatings (Donhowe and Fennema, 1994). The RH has significant effect on permeance as the edible coatings may have hygroscopic properties. With an increase or decrease in RH, the permeance of cellulose-based coating changes. Polysaccharide-based coatings are also more permeable to water vapor and less permeable to oxygen and CO_2 compared to carnauba wax. Under high-humidity conditions, however, the difference in permeability is reduced (Hagenmaier and Shaw, 1992; Baldwin et al., 1999). The combination of reduced O_2 content sufficient to retard respiration without allowing anaerobic conditions to occur and increase in CO_2 to inhibit microbial growth can be beneficial. Myrna et al. (1991) developed an experimental edible coating (US Patent # 07/679,849) and compared it with commercial coatings. Acetaldehyde increased in coated fruit; however, commercially coated fruit had significantly higher ethyl acetate and ethanol than that with the edible coating. Ethyl butyrate was also higher in commercial coatings.

Several edible and eco-friendly (biodegradable) materials such as cellulose-based coatings, sucrose esters of fatty acids, and chitosan, have been used to extend the shelf life of citrus fruits. N′O-carboxymethyl chitosan is a water-soluble derivative of chitin and reduces respiration to a large extent. Oranges treated with chitosan solution can be stored for 2 months without adverse effects on fruit appearance. Chitosan also reduces fungal growth (Li and Cao, 1997). A range of sucrose ester coatings are available in the market. These are based on one or more esters, a carrier (sodium carboxy methyl cellulose) and anti foam (mono and diglycerides of fatty acids). Semperfresh is an example of this type of material; it is used at a 0.25–2 percent concentration. Traditional wax coatings adversely affect the taste of the fruit but edible coatings based on sugar esters are reported to enhance taste (Curtis, 1988).

Baldwin et al. (1995) compared conventional aqueous waxes and edible coatings. Shellac-based aqueous wax-coated Valencia orange fruit showed higher concentrations of ethanol, ethylbutanoate, ethyl acetate, and α-pinene with lower levels of valencene, α-terpineol, and hexanol. Shellac-coated fruits had the lowest and highest amounts of O_2 and CO_2 respectively, and showed significantly less weight loss than fruits subjected to polysaccharide coating.

Sucrose ester (4–4.5 percent) and rhamnolipid (0.5 percent) coatings have been reported to effectively reduce the percentage of decay, water loss, and respiration rate in Fortune mandarin fruits at room temperature (28–33°C). Ascorbic acid content and firmness were greater in coated fruits stored at 1°C. Coating with rhamnolipid gave better results than coating with sucrose esters during storage studies (Li et al., 1998). Fruit coatings not only reduce water loss but also slow down change of color from green to yellow. Full-strength Nature Seal 2020, a cellulose-based edible coating, significantly delayed ripening of lemons (light-green and green fruits) as indicated by color ('L','a', and 'b') values during storage at 13°C for 3 weeks (Chen and Grant, 1995).

Carbohydrate and lipid emulsions and carboxymethylcellulose emulsions rendered a glossy appearance on oranges while sucrose ester (0.5–2 percent) did not provide gloss. Carbohydrate and lipid emulsions also provided better weight-loss control. The storage life of orange fruit ranged from 3–5 weeks for control and 4–8 weeks for fruit coated with sucrose ester (2 percent) and carbohydrate and lipid emulsion (Yicheng and Tingfu, 1990). Sucrose esters favored the development of yeast populations (support survival of the yeast *Candida oleophila*, a bioagent that controls postharvest decay) to a greater extent than those based upon shellac. Grapefruits with these coatings were slower to decay during 6-month storage (McGuire and Dimitroglou, 1999).

a. Examples of Some Edible Coatings:

(1) 'Nature-Seal'-2020 (Hydroxy propylcellulose at 5 percent and includes preservatives, acidulant, and emulsifiers with total solids 7 percent). (2) 'Tal Pro-Long'- sucrose esters of fatty acids and carboxy methyl cellulose (Banks, 1984). (3) 'Semperfresh,' 'Nu–Coat' and 'Brilloshine' (Surface systems International, UK) for citrus fruits to protect shine and extend shelf life. (4) Palm-oil emulsions and anti-transpirants (5) N, O-carboxymethyl chitosan. This coating is selectively permeable to CO_2, O_2 and ethylene (Davies et al., 1988). (6) Trehalose sugar disaccharide of glucose linked by reducing sugar carbon (stable reducing sugar i.e. inert and non-toxic) can also be used in coating material (Roser and Colaco, 1993).

D. Plant-Growth Regulators

Plant-growth regulators (PGRs) control physiological processes at extremely low concentrations. Most of these compounds occur naturally and hence their use in postharvest citrus treatments is expected to receive consumer acceptance. Auxins, gibberellins (GA), cytokinins, abscissic acid (ABA), and ethylene are five important types of PGRs that occur naturally in fruits. The first three types of compounds are used to extend the vitality of fruit tissues while last two are known to promote aging and senescence processes. Synthetic compounds with auxin-like action (2,4-D) are also used. The jasmonates are also naturally occurring PGRs and known to regulate various aspects of plant development and responses to biotic and abiotic stresses.

Gibberellic acid (GA_3) and 2,4-dichloro phenoxy acetic acid (2,4-D) are the most widely and commercially used PGRs in citrus. Both these chemicals have pre-harvest and postharvest applications. As a preharvest application 2,4-D delays and reduces abscission of mature fruit and increases fruit size. As a postharvest application in lemons, 2,4-D delays button abscission by maintaining its vitality and thus reduces *Alternaria* rot. GA_3 is primarily used as preharvest spray on navels and Minneola tangelos to delay peel senescence and fruit maturity in lemons while postharvest application in lemons is aimed at delaying coloration and reducing sour rot (Coggins, 1991). Gibberellin A_3 is observed to be highly persistent in orange peel (half life = 80 days). Ethylene causes a slight enhancement of GA_3 metabolism in orange fruit (Shechter et al., 1989). Gibberellin A_1 is metabolized by orange peel at a relatively high rate (half life < 24h) and ethylene slightly reduces this rate. Ethylene-induced enhancement of senescence does not involve major effects on the deactivation of applied gibberellins.

The effect of GA_3 treatment in retarding color change during the storage of lemon fruits is associated with the lowest levels of ABA (Valero et al., 1998a). 2,4-D treatment lowers the ABA concentration and prolongs the storage life of Valencia oranges (Liu and Xu, 1998).

In California, GA_3 application is recommended for lemons since it delays ripening and improves storage quality. Postharvest application of gibberellic acid (50ppm) in storage wax reduces the incidence of *Geotrichum* decay, probably by delaying senescence and thus retaining resistance of fruit to the decay pathogen (Coggins et al., 1992).

The quality of GA_3 dip–treated (100ppm) mature green fruits of the Mahaley orange was better than that of the controls throughout the storage period at 4 or 7°C (Al-Doori et al., 1990). GA_3 (0 or 250ppm) treatment to *C. latifolia* fruits prevented degreening more at low temperatures (8°C or 10°C for 28°C days) than at ambient temperature (Mizobutsi et al., 2000).

Postharvest use of synthetic growth retardants such as CCC and Alar (500–4000ppm) as a dip treatment for Valencia oranges is reported to hasten color development while 2,4-D, 2,4,5-T (500–1000ppm) and gibberellic acid (500–2000ppm) retarded color development. Gibberellic acid reduced weight loss and the percentage of discarded fruit during long storage at 3.3°C (El- Nabawy et al., 1981).

Considerable work has been done in India with respect to evaluation of growth regulators to extend the storage life of citrus fruits. Application of 2,4-D delayed ripening and color development (Lodh et al., 1963) and retained greenness of buttons (Sonkar et al., 1999) in mandarins. Maleic hydrazide and 2,4,5-T have not been so effective. The Kinetin and Benzyl adenine (40μM) treatments appreciably reduced chlorophyll degradation (Nagar, 1993) and polymethylesterase, polygalacturonase, and cellulase enzyme activities in stored Kinnow mandarin fruits (Nagar, 1994). Plant-growth regulators, along with wax and polyethylene lining have been very effective in extending storage life of various mandarin cultivars (Table 9.5).

The postharvest application of 2,4-D (500ppm) reduced decay in small-fruited acid limes (Kohli and Bhambota, 1966). GA_3 + waxol (6 percent) + Bavistin (0.1 percent) extended storage life of limes up to 3 months at 10°C (Ghosh and Sen 1985). At ambient conditions, indole butyric acid (200ppm) + waxol (12 percent) also extended storage life of limes up to seventeen days (Rodrigues et al., 1963). IBA was effective in retaining color and ascorbic acid content in limes (Ram et al., 1970). GA_3 (200–500ppm) and Cytokinin (10–25ppm) alone (Jitendar Kumar et al., 1987) and along with Topsin-M (1000ppm) reduced weight loss and decay in Baramasi lemons (Sindhu and Singhrot, 1996). GA_3 (20ppm) + polyethylene liners effectively minimized storage losses in Marsh Seedless grapefruits (Sandhu et al., 1982).

Wax emulsion with and without PGRs has been effective extending the shelf life of sweet orange fruits also (Table 9.6). Mosambi fruits could be kept for

20–24 days after wax (3 percent) + 2,4-D (100 ppm) dip treatment at 25–30°C (Das and Dash, 1967). Jaffa oranges treated with GA_3 (50 ppm) followed by wrapping in perforated polyethylene could be stored for 4 weeks at 15–26°C (Chattopadhyay et al., 1992). 2.4-D along with fungicides benomyl (Benlate) and carbendazim was most effective in controlling rots in Mosambi sweet oranges (Kumbhare and Choudhari, 1979).

Ethylene accelerates senescence of fruits and chemicals have been developed to mitigate its effect (Sisler and Serek, 2003). Plant bioregulators such as 1-methylecyclopropene (1-MCP) and Aminocthoxyvinylglycine (AVG) have inhibitory activity against ethylene production. 1-MCP is an ethylene action inhibitor that apparently binds to the cellular ethylene receptor proteins and effectively inhibits ethylene responses by making plant tissue insensitive to ethylene. 1-MCP is non-toxic and odorless. In Shamouti oranges, 1-MCP is effective in inhibiting negative effects of ethylene and delays degreening. However, it is shown to be ineffective against ethylene effects such as enhancing chilling injury and increasing decay (Porat et al., 1999). This compound can be used in commodities marketed green, such as limes and lemons. AVG is an analog of rhizobiotoxine. The root-nodule bacterium *Rhizobium japonicum* of soybean produce this phytotoxin. This phytotoxin competitively inhibits the conversion of S-adenosylmethionine (SAM) to 1-aminocyclopropane-1-carboxylic acid (ACC) in the synthesis of ethylene (Boller et al., 1979). AVG is (also available as commercial product named ReTain) reported to inhibit ACS (ACC Synthase) enzyme activity during production of ACC. Methyl jasmonate is also known to retard ethylene action.

E. Other Chemicals

Polyamines have been found to extend storage life of fruits by delaying senescence. Vacuum infiltration with 1 mM putrescine significantly increased fruit firmness in color break of lemons, and delayed color change associated with low ABA concentrations (Valero et al., 1998b). Putrescine-treated fruits showed higher levels of firmness and lower weight loss than calcium chloride-treated or non-treated fruits during storage. The concentrations of putrescine, spermidine, and spermine were higher in color break than in yellow fruits; the opposite was found for ABA. Putrescine was the most effective at maintaining higher levels of endogenous putrescine and spermidine, but only for yellow fruits.

Lemon cultivar Verna fruits treated with putrescine and calcium (1 mM) maintained higher firmness values (fruit deformation force) and more resistance to peel rupture during storage at 15°C (Martinez et al., 1999). Treated lemons also showed less deformation when the compression force (50N) to induce mechanical damage was applied. In damaged control fruits, an increase of spermine and ABA levels was observed.

High concentrations of endogenous polyamines and salicylic acid during Ponkan mandarin fruit maturation are considered good for storage. Fruits treated with 100 ppm putrescine, spermidine, or spermine, or 400 ppm salicylic acid exhibited one-sixth to one-half of the decay and weight loss when compared with control fruits during storage for 3 months. Salicylic acid and spermine gave best results, followed by spermidine and putrescine (Zhang et al., 2000).

The above mentioned literature on postharvest treatments of citrus fruits reveals that many chemicals have been experimented with and quite a number of them are effective. For commercially important citrus fruits, the cost/benefit ratio, effectiveness of treatment, tolerance limits, and safety issues are some of the important points.

REFERENCES

Ahrens, M.J., and Barmore, C.R. (1987). Interactive effects of temperature and ethylene concentration on postharvest colour development in citrus. *Acta Hort.* 201, 21–27.

Albrigo, I.G., Burns, J.K. and Hunt, F.M. (1991). Weight loss consideration in preparing and marketing weight-fill bagged citrus. *Proc. Fla. Sta. Hort. Soc.* 104, 74–77.

Al-Doori, A., Hanna, K.R., Daoud, D.A., and Shakir, I.A. (1990). The effect of gibberellic acid and storage temperature on the quality of cv. Mahaley orange. *Mesopotamia J. Agric.* 22(2), 45–47.

Anonymous (1992). Twenty-five years of research on citrus in Vidarbha (1967–1992). Director of Extension, Dr. PDKV, Akola, 54 pp.

Anonymous (1995). Annual Report for 1993–94 and 1994–95. Central Institute on Post-harvest Engineering and Technology (CIPHET), Ludhiana, 31 pp.

Anonymous (2005). Tattooed fruit is on the way. *Processed Food Industry*, New Delhi **8**(10), 40–41.

Arora, J.S., Pal, R.N., and Jawanda, J.S. (1973). Effect of ethrel on degreening of Valencia sweet orange (*C. sinensis* Osbeck). *Punjab Hort. J.* 13, 13–17.

Baldwin, E.A., Nisperos, M.O., Shaw, P.E., and Burns, J.K. (1995). Effects of coatings and prolonged storage conditions on fresh orange flavour volatiles, degree Brix and ascorbic acid levels. *J. Agric. Fd. Chem.* 43, 1321–1331.

Baldwin, E.A., Burns, J.K., Kazokas, W., Brecht, J.K., Hagenmaier, R.D., Bender, R.J., and Pesis, E. (1999). Effect of two edible coatings with different permeability characteristics on mango (*Mangifera indica* L.) ripening during storage. *Postharvest Biol. Technol.* 17, 215–226.

Banks, N.H. (1984). Some effects of Tal-Prolong coating on ripening of bananas. *J. Expt. Bot.* 35, 127–137.

Barmore, C.R. (1975). Effect of ethylene on chlorophyllase activity and chlorophyll content in Calamondin rind tissue. *HortScience* 10, 595–596.

Basiouny, F.M., and Biggs, R.H. (1975). Effects of UV-B radiation on pigment changes and quality of citrus fruits. *Proc. Fla. Sta. Hort. Soc.*, 87, 281–289.

Baudoin, A.B., and Eckert, J.W. (1985). Development of resistance against *Geotrichum candidum* in lemon peel injuries. *Phytopathology* 75, 174–179.

Ben-Yehoshua, S. (1991). New developments in seal-packaging of fruits and vegetables. *Proc. Int. Citrus Symp.*, China, pp. 757–771.

Ben-Yehoshua, S., Shapiro, B., and Moran, R. (1987). Individual seal packaging enables the use of curing at high temperatures to reduce decay and heal injury of citrus fruits. *HortScience* 22, 777–783.

Ben-Yehoshua, S., Shapiro, B., Kim, J.J., Sharoni, J., Carmeli, S., and Kashman, Y. (1988). Resistance of citrus fruits to pathogens and its enhancement by curing. *Proc. 6th Int. Citrus Congress*, Israel, pp. 1371–1379.

Ben-Yehoshua, S., Shapiro, B., and Shomer, M. (1990). Ethylene enhanced heat damage to flavedo tissues of cured citrus fruits. *HortScience* 25, 122–124.

Ben-Yehoshua, S., Kim, J.J., Rodov, V., Shapiro, B., and Carmeli, S. (1992). Reducing decay of citrus fruits by induction of endogenous resistance against pathogen. *Proc. Int. Soc. Citric.*, Italy, pp. 1053–1056.

Bhullar, J.S. (1981). Effect of wax emulsion, Benlate and some growth regulators on storage life of Pineapple sweet oranges. *Haryana J. Hort. Sci.* 10, 147–150.

Bhullar, J.S. (1982). Effect of various treatments on extending the post-harvest life of blood red fruits. *Indian Fd. Packer* 36, 44–48.

Bhullar, J.S. (1983). Storage behaviour of Kagzi lime fruits. *Haryana J. Hort. Sci.* 12, 52–55.

Blandini, G., Levi, P., and Pappalardo, R. (1992). Problem analysis of an automatic citrus grading machine. *Proc. Int. Soc. Citric.*, Italy, pp. 1033–1035.

Boller, T., Herner, R.C., and Kende, H. (1979). Assay for and enzymatic formation of an ethylene precursor, 1-aminocyclopropane-1-carboxylic acid. *Planta* 145, 301–303.

Brooke, D.L., and Spurlock, A.H. (1977). Production costs – fruit production, harvesting, packinghouse, processing. In *Citrus science and technology* (S. Nagy, P.E. Shaw, and M.K. Veldhuis, eds.), AVI Publishing Company, Connecticut, USA, Vol. 2, pp. 141–170.

Brown, G.E., and Craig, J.O. (1989). Effectiveness of aerosol fungicide application in the degreening room for control of citrus fruit decay. *Proc. Fla. State Hort. Soc.* 102, 181–185.

Burg, S.P., and Burg, E.A. (1967). Molecular requirements for the biological activity of the ethylene. *Pl. Physiol.* 42, 144–152.

Chattopadhyay, N., Hore, J.K., and Sen, S.K. (1992). Extension of storage life of sweet orange (*C. sinensis* Osbeck) cv. Jaffa. *Indian J. Pl. Physiol.* 35, 245–251.

Chauhan, K.S., and Rana, R.S. (1974). Effect of ethrel on degreening of Washington Navel orange. *Indian J. Hort.* 31, 154–156.

Chauhan, K.S., and Parmar, C. (1981). Degreening of Mosambi with ethrel. *Proc. Third Int. Symp. Subtrop. and Trop. Hortic.*, Bangalore, 8–14 February 1972. Horticultural Society of India, pp. 153–156.

Chen, S.M., Fon, D.S., Hong, S.T., Wu, C.J., Len, K.C., Tien, B.T., and Chang, W.H. (1992). Electro-optical citrus sorter. American Society of Agricultural Engineering, No. 92-3520, 19 pp.

Chen, X.H., and Grant, L.A. (1995). Nature-Seal delays yellowing of lemons. *Proc. Fla. Sta. Hort. Soc.* 108, 285–288.

Chen, R.Y., Tsai, M.J., and Liu, M.S. (2000). Effect of plastic bag packaging and fruit coatings on the storage quality of citrus fruits and melon. *J. Chinese Soc. Hort. Sci.* 46(1), 35–44.

Chundawat, B.S., Gupta, O.P., and Singh, J.P. (1973). Effect of 2-chloroethyl phosphonic acid (Ethrel) on degreening and quality of Ganganagar Red grapefruit (*C. paradisi* Macf). *Haryana J. Hort. Sci.* 2, 45–49.

Coggins, C.W., and Jones, W.W. (1977). Growth regulators and colouring of citrus fruits. *Proc. Int. Soc. Citric., Int. Citrus Congress*, Florida, Vol. 2, pp. 686–688.

Coggins Jr., C.W. (1991). Present research trends and the accomplishments in the USA. *Proc. Int. Citrus Symp.*, Guangzhou, China, 1990.

Coggins Jr., W., Anthony, M.F., and Fritts Jr., R. (1992). The postharvest use of gibberellic acid on lemons. *Proc. Int. Soc. Citric.*, Italy, Vol. 1, pp. 478–481.

Cohen, E. (1977). Some physiological aspects of citrus fruit degreening. *Proc. Int. Soc. Citric.*, Vol. I, pp. 247–249.

Cohen, E. (1978). The effect of temperature and RH during degreening on the colouring of shamouti orange fruit. *J. Hort. Sci.* 53, 143–146.

Cohen, E. (1981). Methods of degreening Satsuma mandarin. *Proc. Int. Soc. Citric.*, Vol. 2, 748–750.

Cohen, E. (1991). Investigations on post-harvest treatments of citrus fruits in Israel. *Proc. Int. Symp.*, Guangzou, China, 9–11 November 1990. International Academic Publication, pp. 32–36.

Craft, C.C. (1970). Respiratory response of lemons to ethylene. *J. Am. Soc. Hort. Sci.* 95, 689–692.

Cuquerella, J., Martinez-Javega, J.M., Jimenez-Cuesta, M. (1981). Some physiological effects of different wax treatments on Spanish citrus fruits during cold storage. *Proc. Int. Soc. Citric.*, Japan, Vol. 2, pp. 734–737.

Curtis, G.J. (1988). Some experiments with edible coatings on the long term storage of citrus fruits. *Proc. 6th Int. Citrus Congress*, Israel, Vol. 3, pp. 1515–1520.

Dalal, V.B., and Subramanyam, H. (1970). Refrigerated storage of fresh fruits and vegetables. *Clim. Contr.* 3, 37.

Dalal, V.B., Krishnaprakash, M.S., and Nagaraja, N. (1974). Effects of coatings on fruit and vegetables during refrigerated storage. Paper presented in *3rd Nat. Symp. on Refrigeration and Air Conditioning*. CFTRI Mysore, July 1974.

Das, R.C., and Dash, J. (1967). The effect of wax emulsion 2,4,-D and 2,4,5-T on the storage behaviour of Mosambi fruits. *Proc. Int. Symp. Subtrop. Trop. Hort.*, New Delhi, pp. 104–107.

Dashora, L.K., and Shafaat, M. (1988). Effect of 2,4-D, wax emulsion and their combination on the shelf life of sweet orange (*C sinensis*) cv. Mosambi. *South Indian Hort.* 36, 172–176.

Davies, D.H., Elson, C.M., and Hayes, E.R. (1988). N, O-carboxy methyl chitosan, a new water soluble chitin derivative. *Fourth Int. Conf. Chitin and Chitosan*, 22–24 August 1988, Trontheim, Norway, 6 pp.

Davis, P.L., and Hofman, R.C. (1973). Effects of coatings on weight loss and ethanol buildup in juice of oranges. *J. Agri. Fd. Chem.* 21, 755–458.

Davis, P.L., Roe, B., and Bruemmer, J.H. (1974). Biochemical changes in grapefruit stored in air containing ethylene. *Proc. Fla. Sta. Hort. Soc.*, 87, 222–227.

Deason, D.L., and Grierson, W., (1972). Heating of citrus fruits during degreening and associated temperature gradients within the typical horizontal airflow degreening room. *Proc. Fl. St. Hort. Soc.* 84, 259–264.

Del-Rio, M.A., Cuquerella, J., and Ragone, M.L. (1992). Effect of post harvest curing at high temperatures on decay and quality of Marsh seedless grapefruit and navel oranges. *Proc. Int. Soc. Citric.*, Italy, pp. 1081–1083.

Donhowe, G., and Fennema, O. (1994). Edible films and coatings: characteristics, formation, definitions, and testing methods. In *Edible coatings and films to improve food quality* (J.M. Krochta, E.A. Baldwin, and M.O. Nisperos-Carido, eds.). Technomic Publishing, Lancaster/Basel, pp. 1–24.

Dou, H.T, Ismail, M.A., and Petracek, P.D. (2000). Reduction of postharvest pitting of citrus by changing wax components and their concentrations. *Proc. Fla. Sta. Hort. Soc.* 112, 159–163.

Dutt, S.C., Sarkar, K.P., and Bose, A.N. (1960). Storage of mandarin orange. *Indian J Hort.* 17, 60–68.

Eaks, I.L. (1956). Brush effects on Oranges. *California Citrog.* 45, 67–68.

Eaks, I.L. (1977). Physiology of degreening – summary and discussion of related topics. *Proc. Int. Soc. Citric.*, Vol. 1, pp. 223–235.

Eaks, L., and Dawson, A.J. (1979). The effect of vegetative ground cover and ethylene degreening on 'Valencia' rind pigments. *J. Am. Soc. Hort. Sci.* 104, 105–109.

Eaks, I.L., and Ludi, W.A. (1960). Effects of temperature, washing, and waxing on the composition of the Internal Atmosphere. *Proc. Am. Soc. Hort. Sci.* 76, 220.

Eckert, J.W., and Brown, G.E. (1986). Post-harvest citrus diseases and their control. In 'Fresh Citrus Fruits' (E.F. Wardowski, S. Nagy, and W. Grierson, eds.). AVI Publishing Co., West Port, CT, USA, pp. 315–360.

Eckert, J.W., and Eaks, I.L. (1989). Post-harvest disorder and diseases of citrus fruits. In *The citrus industry*, (W. Reuther, E.C. Calavan, and G.E. Carman, eds.),Vol. V. Division of Agricultural and Natural Resources, UC, pp. 179–260.

El-Nabawy, S.N., El-Hammady, A.M., El–Hammady, M.M., and Nasim, W.H. (1981). Effect of post harvest application of 2,4,-D, 2,4,5-T, GA, CCC and Alar on keeping quality of Valencia orange. *Proc. Int. Soc. Citric.*, Japan, Vol. 2, pp. 725–728.

Gautam, D.R., Dhar, R.P., Bhutani, V.P., and Dhuria, H.S. (1977). Ethephon for postharvest degreening of Kagzi lime. *Indian J. Agric. Sci.* 47, 282–284.

Ghosh, S.K., and Sen, S.K. (1984). Extension of storage life of sweet orange (*Citrus sinensis* Osbeck) cv. Mosambi. *South Indian Hort.* 32, 16–22.

Ghosh, S.K., and Sen, S.K. (1985). Extension of storage life of lime. *Punjab Hort. J.* 25, 46–52.

Godara, S.L., and Pathak, V.N. (1995). Effect of plant extracts on Post-harvest rotting of sweet orange fruit. *Global Conference on Advances in Research on Plant Diseases and Their Management.* Rajasthan College of Agriculture, Udaipur, 12–17 February, p. 172.

Gopalkrishna, R.P., and Krishanmurthy, S. (1983). Studies on shelf life of Coorg mandarin (*C. reticulata* Blanco). *South Indian Hort.* 31, 55–61.

Grierson, W., and Newhall, W.F. (1960). Degreening of Florida citrus fruits. *Fla. Agric. Exp. Sta. Bull.* 620.

Grierson, W., and Wardowski, W.F. (1977). Packinghouse procedures relating to citrus processing. In *Citrus science and technology* (S. Nagy, P.E. Shaw, and M.K.Veldhuise, eds.), Vol. 2. AVI Publishing Co., Westport, CT, pp. 128–140.

Grierson, W., Cohen, E., and Kitagawa, F. (1986). Degreening. In *Fresh citrus fruits* (W. Wardowski, W. Grierson, and S. Nagy, eds.). AVI Publishing Co., Westport, CT, USA, pp. 253–274.

Gupta, O.P., Chauhan, K.S., and Daulta, B.S. (1983). Effect of ethrel on the storage life of citrus fruits. *J. Res., HAU* 13, 458–463.

Gutter, Y. (1981). Investigation on new postharvest fungicide in Israel. *Proc. Int. Soc. Citric.*, Japan, Vol. 2, pp. 810–811.

Hagenmaier, R.D. (2000). Evaluation of a polyethylene-candelilla coating for 'Valencia' oranges. *Postharvest Biol. Technol.*, 19, 147–154.

Hagenmaier, R.D., and Shaw, P.E. (1992). Gas permeability of fruit coating waxes. *J. Am. Soc. Hort. Sci.* 117, 105–109.

Hale, P.W., and Risse, L.A. (1974). Exporting grapefruits in tray pack cartons. *Citrus Veg. Mag.* 37(12), 12–13.

Hasegawa, Y., Yano, M., and Iba Y. (1990). Effect of high temperature pretreatment on storage of citrus fruits. *Proc. Int. Symp. Citrus*, China, 1990, pp. 736–739.

Hasegawa, Y., Yamo, M., Iba, Y. Makitay, and Kona-Kahara, M. (1989). Factors deteriorating the quality of Satsuma mandarin (*C. unshiu*) fruit on the packing house line and their improvement. *Bull. Fruit Tree Res. Stn.* (Okitsu) *Japan*, 16, 29–40.

Hatton, T., Hearn, J., and Smoot, J. (1987). Degreening and storage of 'Sunburst' citrus hybrid fruit. *Proc. Fl. Sta. Hort. Soc.* 99, 127–128.

Ismail, M.A. (1981). Treatment and recycling of waste water in a Florida citrus packing house. *Proc. Int. Soc. Citric.*, Japan, Vol. 2, pp. 835–837.

Iwamoto, M., and Chuma, Y. (1981). Recent studies on development in automated citrus packinghouse facility in Japan. *Proc. Int. Soc. Citric.*, Vol. 2, pp. 831–834.

Jahn, O.L. (1973). Degreening citrus fruit with post-harvest applications of (2-chloroethyl) phosphoric acid (ethephon). *J. Am. Soc. Hort. Sci.* 98, 230–233.

Jahn, O.L. (1976). Degreening of waxed citrus fruits with ethephon and temperature. *J. Am. Soc. Hort. Sci.* 101, 597–599.

Jahn, O.L., and Young, R. (1975). Effects of maturity stages and ethylene on the induction of carotenoid synthesis in citrus fruits by 2(4-chlorophenyl thio)-triethylamine (CPTA). *J. Am. Soc. Hort. Sci.* 100, 244–246.

Jahn, O.L., and Young, R. (1976). Changes in chlorophyll a, b and a/b ratio during color development in citrus fruit. *J. Am. Soc. Hort. Sci.* 101, 416–418.

Jahn, O.L., Chase Jr., W.G., and Cubbedge, R.H. (1969). Degreening of citrus fruits in response to varying levels of O2 and C2H4. *J. Am. Soc. Hort. Sci.* 94, 123–125.

Jahn, O.L., Chace Jr., W.G., and Cubbedge, R.H. (1973). Degreening response of Hamlin oranges in veluation to temperature, ethylene concentration and fruit maturity. *J. Am. Soc. Hort. Sci.* 98, 177–181.

Jitender Kumar, Ramkrishan, R., Yamadagni and Ran Singh (1987). Effect of growth regulators on shelf life of lemon cv. Baramasi. *Res. Develop. Rep.* 4, 21–25.

Jorgensen, K.R. (1969). New process cuts citrus degreening time. *Aust. Citrus News* 45(11), 12–15.

Jorgensen, K.R. (1977). Trickle degreening of citrus fruits. *Queensland Agric. J.* 103, 85–91.

Johnson, T.M. (1981). Electronic sorting and automated packaging for fresh citrus fruit. *Proc. Int. Soc. Citric.*, Japan, Vol. 2, p. 826.

Johnson, T.M. (1991). Citrus postharvest technology to control losses. *Proc. Int. Citrus Symp.*, China, pp. 704–708.

Josan, J.S., Sharma, J.N., and Chohan, G.S. (1981). Effect of post-harvest application of ethrel on degreening of Italian Round lemon. *Indian J. Hort.* 38, 178–183.

Josan, J.S., Sharma, J.N., and Chohan, G.S. (1983). Effect of different lining materials and wax emulsion on post-harvest life of Kinnow fruits. *Indian J. Hort.* 40(3–4), 183–187.

Kim, J.J., Ben-Yehoshua, S., Shapiro, B., Henis, Y., and Carmeli, S. (1991). Accumulation of scoparone in heat-treated lemon fruit inoculated with *P. digitatum* Sacc. *Pl. Physiol.* 97, 880–885.

Kitagawa, H., Adachi, S., and Tarutani, T. (1971). Studies on the colouring of Satsumas II. A practical and convenient method of colouring or degreening with ethylene using plastic film. *J. Japanese Soc., Hort. Sci.* 40, 195–199.

Kitagawa, H., Kawada, K., and Tarutani, T. (1977). Degreening of Satsuma mandarin. *Proc. Int. Soc. Citric.* 1, 219–223.

Knee, M., Tsantili, E., and Hatfield, S.G.S. (1988). Promotion and inhibition by ethylene of chlorophyll degradation in orange fruits. *Ann. Appl. Biol.* 113, 129–135.

Kohli, R.R., and Bhambota, J.R. (1965). Storage of oranges. *Indian J. Hort.* 22, 167–174.

Kohli, R.R., and Bhambota. J.R. (1966). Storage of limes (*C. aurantifolia* Swingle). *Indian J. Hort.* 23, 140–146.

Kohli, R.R., and Ladaniya, M.S. (1988). Degreening of 'Nagpur' mandarin. Paper presented in *6th Workshop of AICRP on Post-harvest Tech. of Hort. Crops*. KKV, Dapoli.

Kuc, J. (1982). Induced immunity to plant diseases. *Bioscience* 32, 854–860.

Kumbhare, G.B., and Choudhari, K.G. (1979). Storage decay of sweet orange and control measures. *J. Maharashtra Agric. Univ.* 4, 230–231.

Ladaniya, M.S. (1998). Intermittent ethylene treatment technique for degreening of fruits with special reference to Nagpur mandarin. *Indian Fd. Packer* 52, 5–10.

Ladaniya, M.S. (2000). Response of 'Nagpur' mandarin to degreening by ethylene trickle method, mechanical wax coating and storage at ambient condition. *J. Fd. Sci. Technol.* 37, 103–110.

Ladaniya, M.S. (2001). Response of 'Mosambi' sweet orange (*Citrus sinensis*) to degreening, mechanical waxing, packaging and ambient storage conditions. *Indian J. Agric. Sci.* 71, 234–239.

Ladaniya, M.S. (2002). Standardization of harvesting, handling and storage techniques for acid lime and Mosambi sweet orange. *Annual Report, NRCC, 2002–2003*, 130 pp.

Ladaniya, M.S., and Dass, H.C. (1994). Mannual v/s mechanical sorting, washing, waxing and size grading of Nagpur mandarin. *J. Maharashtra Agric. Univ.* 19, 159–160.

Ladaniya, M.S., and Singh, S. (2001). Use of ethylene gas for degreening of sweet orange (*Citrus sinensis* Osbeck) cv. Mosambi. *J. Sci. Ind. Res.* 60, 662–667.

Ladaniya, M.S., and Sonkar, R.K. (1999). Effect of pre and post-harvest treatments and packaging containers on 'Nagpur' mandarin fruit in storage. *Proc. Nat. Symp. Citric.*, 17–19 November 1997, NRCC, Nagpur, pp. 454–460.

Ladaniya, M.S., Naqvi, S.A.M.H., and Dass, H.C. (1994). Packing line operations and storage of Nagpur mandarin. *Indian J. Hort.* 51, 215–221.

Ladaniya, M.S., Singh, S., and Mahalle, B. (2005). Sub-optimum low temperature storage of 'Nagpur' mandarin as influenced by wax coating and intermittent warming. *Indian J. Hort.* 62, 1–7.

Lancaster, J.E., Lister, C.E., Reay, P.F., and Triggs, C.M. (1997). Influence of pigment composition on skin colour in a wide range of fruit and vegetables. *J. Am. Soc. Hort. Sci.* 122, 594–598.

Lawes, G.S., Prasad, L., and Michalczuk, L. (1999). Peel permeance and storage changes in internal atmosphere composition of surface-coated mandarin. *Acta Hort.* 485, 249–254.

Li, H.Y., and Cao, B. (1997). Effect of chitosan for maintaining freshness of fruits and vegetables and its mechanism of action. *J. Fruit Sci.* 14(Suppl), 92–95.

Liu, X.Z., and Xu, M.X. (1998). Study on the relationship between the content of endogenous ABA and the storeability of Valencia orange. *South China Fruits* 27(2), 15.

Li, J.S., Li, Z.Y., and Yuan, C.Q. (1998). Effects of sugar esters and rhamnolipid on fresh keeping of fruits. *Pl. Physiol. Commun.* 34, 115–117.

Lodh, S.B., De, S., and Bose, A.N. (1963). Storage behaviour of mandarin oranges treated with growth regulators. *Sci. Cult.* 29, 48–49.

Long, J.K., and Leggo, D. (1959). Waxing citrus fruits. *Fd. Pres. Quart.* 19, 32–37.

Ma-De-Los, A. Torres, and Pividal, F. (1981). Degreening and colour problems of Cuban 'Valencia' oranges. *Proc. Int. Citrus Congress*, Tokyo (Japan), Vol. 2, pp. 761–764.

Mann, S.S., and Randhawa, J.S. (1978). Preliminary studies on the effect of wax emulsion and growth regulators on the storage behaviour of Kinnow mandarin at ambient storage conditions. *Progressive Hort.* 10, 35–41.

Mann, S.S., and Sandhawalia, H.S. (1983). Effect of waxing and polyethylene packs on the low temperature storage of Kinnow mandarin. *Proc. Int. Citrus Symp.*, Bangalore, 17–22 December 1977. Horticultural Society of India, pp. 246–251.

Martinez-Javega, J.M., Cuquerella, Rio, J., Del, M.A.H., and Mateos, M. (1991). Coating treatments in post-harvest behavior of oranges. In *Technological innovations in freezing and refrigeration of fruit and vegetables, Paris, France.* International Institute of Refrigeration, pp. 79–83.

Martinez, R.D.,Valero, D., Serrano, M., Martinez, S.F., and Riquelme, F. (1999). Effects of postharvest putrescine and calcium treatments on reducing mechanical damage and polyamines and abscisic acid levels during lemon storage. *J. Sci. Fd. Agric.* 79, 1589–1595.

Martinez-Javega, M., Cuquerella, J., Del-Rio, M.A., and Mateos, M. (1989). Coating treatments in post harvest behaviour of oranges. IIR. Commissions C2/D1, D2/3. Technical innovations in freezing and refrigeration of fruit and vegetables. University of California Publication, pp. 51–55.

Mazumdar, B.C., and Bhatt, D.N.V. (1976). Effects of pre-harvest application of GA and Ethrel on sweet orange. *Progressive Hort.* 8, 89–91.

McCornack, A.A. (1971). Effect of ethylene degreening on decay of Florida citrus fruit. *Proc. Fla. Sta. Hort. Soc.* 84, 270–272.

McCornack, A.A., and Wardowski, W. (1977). Degreening Florida citrus: procedure and physiology. *Proc. Int. Soc. Citric.*, Vol. 1, pp. 211–215.

McGuire, R.G., and Dimitroglou, D.A. (1999). Evaluation of shellac and sucrose ester fruit coating formulations that support biological control of post-harvest grapefruit decay. *Biocontrol Sci. Technol.* 9, 53–65.

Miller, W.M., and Drouillard, G.P. (2000). Engineering economic analysis for automatic grading. *Trans. ASAE* 43, 220–223.

Miller, W.M., and Ismail, M.A. (1995). Computer-based monitoring and control system for quality control in Florida citrus packinghouses. *Thirty-fourth annual citrus packinghouse day*, September 7, 1995. Cooperative Extension Service, Institute of Food and Agricultural Sciences, University of Florida, Gainsville.

Miller, W.M., Muraro, R.P., and Wardowski, W.F. (1986). Analysis of automatic weight fill bagging machinery for fresh Citrus. *Appl. Eng. Agric.* 2, 252–256.

Mizobutsi, G.P., Borges, C.A.M., Siqueira D.L., and de Siqueira, D.L. (2000). Postharvest conservation of Tahiti limes (*Citrus latifolia* Tanaka) treated with gibberellic acid and stored at three temperatures. *Rev. Brasileira de Fruticultura*, 22 Special Issue, pp. 42–47.

Muraro, R.P. (2006). Average packing charges for Florida fresh citrus – 2005–06 season. University of Florida, IFAS Extension Service, CREC, Florida. http://www.crec.ifas.ufl.edu/extension/economics.

Myrna, O., Carriedo, N.S., Baldwin, E.A., and Shaw, P.E. (1991). Development of an edible coating for extending post-harvest life of selected fruits and vegetables. *Proc. Fla. Hort. Soc.* 104, 122–125.

Nagar, P.K. (1993). Effect of plant growth regulators on the natural and ethylene induced pigmentation in Kinnow mandarin peel. *Biol. Plant.* 35, 633–636.

Nagar, P.K. (1994). Effect of some ripening retardants on fruit softening enzymes of Kinnow mandarin fruits. *Indian J. Pl. Physiol.* 37, 122–124.

Naik, K.C. (1948). *South Indian fruits and their culture.* P. Varadachari Co., Madras, 335 pp.

Oh, S.D., Kim, Y.Y., Hong, S.B. and Chvng, S.K. (1979). Effect of post-harvest application of ethephon, ethylene and methionine on colour and quality of Satsuma. *J. Korean Soc. Hort. Sci.* 20(2), 142–147.

Ohishi, H., Watanabe, J., and Kadoya, K. (1996). Effect of red light irradiation on skin coloration and carotenoid composition of stored 'Miyauchi' iyo (Citrus iyo hort. ex Tanaka) tangor fruit. *Bull. Exp.-Farm Coll. Agric, Ehime-University*, No. 17, pp. 33–37.

Pandiarajan, T., Pugalendhi, S., Devadas, C.T. and Gothandapani, L. (1993). Development and testing of lime fruit grader. In *Abstracts*, National Seminar on optimization of production and productivity of acid lime. Horticulture College and Research Institute, Periyakulam, Tamil Nadu.

Pantastico, Er. B., Soule, J., and Grierson, W. (1968). Chilling injury in tropical and subtropical fruits: II Limes and grapefruit. *Proc. Trop. Reg. Am. Soc. Hort. Sci.*, 12:171–186.

Peeples, W.W., Albrigo, L.G., Pao, S., and Petracek, P.D. (2000). Effects of coatings on quality of Florida Valencia oranges stored for summer sale. *Proc. Fla. Sta. Hort. Soc.* 112, 126–130.

Petracek, P.D., and Montalvo, L. (1997). The degreening of fallglo tangerine. *J. Am. Soc. Hort. Sci.* 122, 547–552.

Petracek, P.D., Dou, H., and Pao, S. (1998). The influence of applied waxes on postharvest physiological behavior and pitting of grapefruit. *Postharvest Biol. Technol.* 14, 99–106.

Porat, R., Weiss, B., Cohen, L., Daus, A., Goren, R., and Droby, S. (1999). Effects of ethylene and 1-MCP on postharvest qualities of Shamouti oranges. *Postharvest Biol. Technol.* 15, 155–163.

Purandhare, N.D., Khedkar, D.M., and Sontakke, M.B. (1992). Physico-chemical changes during degreening in sweet orange. *South Indian Hort.* 40, 128–132.

Qing, S., and Petracek, P.D. (1999). Grapefruit gland oil composition is affected by wax application, storage temperature and storage life. *J. Agric. Fd. Chem.* 47, 2067–2069.

Ram, H.B., Srivastava, R.K., Singh, S.D., and Singh, L. (1970). Effect of plant growth regulators on the storage behaviour and post-harvest physiology of Kagzi lime (*C. aurantifolia* Swingle). *Progressive Hort.* 2, 51–56.

Ramana, K.V.R., Saroja, S., Setty, G.R., Nanjundaswamy, A.M. and Moorthy, N.V.N. (1973). Preliminary studies to improve the quality of Coorg monsoon mandarin. *Indian Fd. Packer* 27, 5–9.

Ramana, K.V.R., Setty, G.R., Moorthy, N.V.N., Saroja, S., and Nanjundaswamy, A.M. (1979). Effect of ethephon, benomyl, TBZ and wax on the colour and shelf life of Coorg mandarin. *Trop. Sci.* 21, 265–272.

Ran Singh, Sharma, R.K., and Kumar, J. (1989). Effect of some chemicals on shelf life of Kinnow mandarin. *Res. Develop. Rep.* 6(1), 87–91.

Rana, R.S., and Chauhan, K.S. (1976). The effect of Postharvest ethrel treatment on degreening of fruits of Kagzi lime. *Punjab Hort. J.* 16, 30–32.

Rao, D.V.R., Reddy, M.L.V., and Murthi, V.D. (1987). Effect of new ripening retardants on shelf life of sweet oranges. *Indian J. Pl. Physiol.* 30, 208–211.

Reid, M.S. (1985). Ethylene in post-harvest technology. In *Postharvest technology of horticultural crops*. University of California Publ. 3311, pp. 68–74.

Rodov, V., Peretz, J., Ben-Yehoshua, S., Agar, T., and D' hallowin, G. (1997). Heat application as complete or partial substitute to postharvest fungicide treatments of grapefruit and Oroblanco fruits. *Proc. Int. Soc. Citric.*, Sun city, South Africa, pp. 1153–1157.

Rodrigues, J., Dalal, V.B., Moorthy, N.V.N., and Srivastava, H.C. (1963). Effect of post-harvest treatment with plant growth regulators in wax emulsion on storage behaviour of limes. *Indian Fd. Packer* 17(5), 9–11.

Roser, B., and Colaco, C. (1993). A sweeter way to fresher food. *New Scientist*, 15, May 25–28.

Sadasivam, R., Muthuswamy, S., Sundaraj, S.S. and Vasudevan, V. (1972). Storage studies with Sathgudi fruit. *South Indian Hort.* 20(1–4), 37–40.

Sandhu, S.S., Dhillon, B.S., and Sohansingh (1982). Effect of Post-harvest application of gibberellic acid and wrappers on the storage behaviour of Marsh seedless grapefruit. *J. Research, PAU* 19(3) 198–202.

Sanghvi, K.U., and Patil, A.V. (1983). Effect of Ethrel on nucellar Mosambi fruits (*C. sinensis* Osbeck). *Proc. Int. Citrus Symp.*, 17–22 December 1977, Bangalore, Horticultural Society of India, pp. 322–325.

Schillacci, G. (1988). Energy consumption and work productivity in citrus handling. *Rivista di Ingegneria Agraria Quaderno No.* 10, 1001–1006.

Sharma, G.D., Pillai, K.C.R., and Kapur, N.S. (1965). Role of skin coating on the transportation of perishable fruits. Part III. Studies on transportation of oranges. *Indian Fd. Packer* 19(5), 14–16.

Sharma, R.K., Singh, R., Kumar, J. and Sharma, S.S. (1989). Shelf life of Baramasi lemon as affected by some chemicals. *Res. Develop. Rep.* 6(1), 78–82.

Sharma, R.K., Sandooja, J.K., and Singhrot, R.S. (1991). A note on enhanced shelf life of Kinnow by some chemicals. *Haryana J. Hort. Sci.* 20(3–4), 216–217.

Sharma, S., Mishra, B.P. and Sharma, R.K. (1986). Effect of some antifungal compounds to combat the *Aspergillus* rot of Baramasi lemon. *Progressive Hort.* 18(1–2), 71–72.

Shechter, S., Goldschmidt, E.E., and Galili, D. (1989). Persistence of [14C]gibberellin A3 and [3H]gibberellin A1 in senescing, ethylene treated citrus and tomato fruit. *Plant Growth Reg.* 8, 243–253.

Shimokawa, K., Sakanoshita, A., and Horiba, K. (1978). Ethylene induced changes of chloroplast structure in Satsuma mandarin. *Pl. Cell Physiol.* 19, 229–236.

Sindhu, S.S., and Singhrot, R.S. (1996). Effect of oil emulsion and chemicals on shelf life of Baramasi lemon (*C. limon* Burm). *Haryana J. Hort. Sci.* 25, 67–73.

Singh, B.P., Chundawat, B.S., and Gupta, A.K. (1978a). Effect of various treatments on storage of mature green Kinnow fruits. *Udyanika* 2(1/2), 33–38.

Singh, B.P., Gupta, A.K., and Chundawat, B.S. (1978b). Effect of various treatments on storage of Kinnow fruits. *Punjab Hort. J.* 18(3–4), 161–165.

Singh, H.P., Srivastava, K.C., Ganapathy, K., Muthappa, D.P., and Randhawa, G.S. (1978c). Effect of ethephon (2-chloroethyl phosphonic acid) on post-harvest degreening of mandarin and sweet orange. *Vatika* 1, 56–63.

Singh, K. (1971). Storage behaviour of sweet orange and mandarins. *Technical Bulletin* (Agric. Series), ICAR, No. 35, 106 pp.

Singhrot, R.S., Sharma, R.K., Sandooja, J.K., and Singh, J.P. (1987a). Effect of some chemicals to enhance shelf life of Baramasi lemon. *Haryana J. Hort. Sci.* 16, 25–30.

Singhrot, R.S., Sharma, R.K., Sandooja, J.K., and Singh, J.P. (1987b). Studies on shelf life of Baramasi lemon as affected by some chemicals. *J. Res., HAU* 17, 234–239.

Singhrot, R.S., Singh, J.P., Sharma, R.K., and Sandooja, J.K. (1987c). Use of diphenyl fumigant in wax coating with different cushionings to increase the storage life of Kinnow fruits. *Haryana J. Hort. Sci.* 16, 31–39.

Sisler, E.C., and Serek, M. (2003). Compounds interacting with the ethylene receptor in plants. *Pl. Biol.* 5, 473–480.

Soni, S.L., and Ameta, S.L. (1983). Effect of pre-harvest application of ethrel on colouration and physico-chemical composition of grapefruit. *Proc. Int. Citrus Symp.*, 17–22 December 1977, Bangalore, Horticultural Society of India, pp. 309–315.

Soni, S.L., and Gupta, M.S. (1983). Effect of wax emulsion, plant growth regulators and their combinations on the physico-chemical changes and storage behaviour of mandarin. *Proc. Int. Citrus Symp.*, 17–22 December 1977, Bangalore, Horticultural Society of India, pp. 252–257.

Sonkar, R.K., Ladaniya, M.S., and Singh, S. (1999). Effect of harvesting methods and post –harvest treatments on storage behaviour of 'Nagpur' mandarin (*Citrus reticulata* Blanco) fruit. *Indian J. Agric. Sci.* 69, 434–437.

Stewart, I. (1977). Citrus color-A review. *Proc. Int. Citrus Congress*, Florida, Vol. 1, pp. 308–311.

Stewart, I., and Wheaton, T.A. (1972). Carotenoids in citrus: their accumulation induced by ethylene. *J. Agric. Fd. Chem.* 20, 448–449.

Subbarao, K.R., Narasimham, P., Anandswamy, B., and Iyengar, N.V.R. (1967). Studies on the storage of mandarin oranges treated with wax or wrapped in diphenyl treated papers. *J. Fd. Sci. Technol.* 4, 165–169.

Subramanyam, C., and Cama, H.R. (1965). Carotenoids in Nagpur orange pulp and peel. *Indian J. Chem.* 3, 463–464.

Subramanyam, H., Lakshminarayan, S., Moorthy, N.V.N., and Subhadra, N.V. (1970). The effect of iso-propyl *N*-phenyl carbamate and fungicidal wax coatings on Coorg mandarins to control spoilage. *Trop. Sci.* 12, 307–313.

Sulikeri, G.S., Bhandari, K.R., and Badakar, K.N. (1980). Effect of ethrel (2 chloroethyl phosponic acid) on degreening of Tahiti lime fruits. *Curr. Res.* 9, 42–43.

Tarkase, B.G., and Desai, U.T. (1989). Effects of packaging and chemicals on storage of orange cv. Mosambi. *J. Maharashtra Agric. Univ.* 14, 10–13.

Tesson, C. (1970). Citrus degreening trials with Ethrel. *Fruits d' Outre Mes*, 25, 210.

Thomas, P, Dalal, V.B., and Srivastava, H.C. (1964). A comparison between the effect of wax and Polyethylene coating on the storage behaviour of yellow limes (*C. autrauti folia* Swingle). *Indian Fd. Packer* 18, 7–10.

Ting, S.V., and Rouseff, R.L. (1986). *Citrus fruits and their products: analysis and technology*. Mercel Dekker, Inc., New York, 293 pp.

Tuset, J.J., Garcia, J., and Hinarejos, C. (1988). Effect of intermittent degreening method on decay of Satsuma mandarin. *Proc. 6th Int. Citrus Congress*, Vol. 3, pp. 1461–1465.

Vakis, N.J. (1976). Effect of degreening of Cyprus-grown lemons and grapefruit. *J. Hort. Sci.* 50, 311–319.

Valero, D., Martinez, R.D., Serrano, M., and Riquelme, F. (1998a). Post-harvest gibberellin and heat treatment effects on polyamines, abscisic acid and firmness in lemons. *J. Fd. Sci.* 63, 611–615.

Valero, D., Martinez-Romero, D., Serrano, M., and Riquelme, F. (1998b). Influence of postharvest treatment with putrescine and calcium on endogenous polyamines, firmness, and abscisic acid in lemon (*Citrus lemon* L. Burm Cv. Verna). *J. Agric. Fd. Chem.* 46, 2102–2109.

Wardowski, W.F. (1981). Packing house operations and shipping conditions of citrus for export. *Proc. Fla. Sta. Hort. Soc.* 94, 254–256.

Warner, H.L., and Leopold, A.C. (1969). Ethylene evolution from 2-chloroethyle phosphonic acid. *Pl. Physiol.* 44, 156–158.

Wheaton, T.A., and Stewart, I. (1973). Optimum temperature and ethylene concentrations for postharvest development of carotenoid pigments in citrus. *J. Am. Soc. Hort. Sci.* 98, 337.

Yamashita, S., and Kitano, Y. (1981). Factors causing deterioration in the quality of Satsuma mandarin during sorting and packing. *Proc. Int. Soc. Citric.*, Japan. Vol. 2, pp. 827–839.

Yicheng, X., and Tingfu, S. (1990). Some effects of various skin coatings on orange fruits. *Proc. Int. Citrus Symp.*, China, 5–8 November, pp. 745–749.

Young, R., and Jahn, O.L. (1972). Preharvest sprays of 2-chloroethyl-phosphonic acid for colouring Robinson tangerines. *Proc. Fla. Sta. Hort. Soc.* 85, 33–37.

Zhang, Q.M., Zheng, Y.S., Wei, Y.R., Liu, K.Y., and Xie, S.X. (2000). Studies on polyamine metabolism and its regulation of growth and fruit set in citrus. II. Changes of polyamines and salicylic acid during citrus fruit maturation and their effects on fruit storage. *J. Hunan Agric. Univ.* 26, 271–273.

10

PACKAGING

Water loss and microbial decay are the two most important factors that render fresh citrus fruit unfit for sale within few days after harvest. Fruits cannot be stored or shipped unpacked. Besides serving as an efficient handling unit in a specified volume, packaging also protects the fruit from the hazards of transportation and storage. While keeping the fruit clean and hygienic, it promotes sales because it is attractive to the customers. Packages should be economical. The container for transport/shipment must have required stacking strength. It should provide adequate ventilation to the commodity to avoid anaerobic respiration during storage and transportation. The interior surface of the package should have a smooth surface to avoid bruising the fruit.

Packaging of the produce is a necessary component in its marketing. Packaging is done to delay physiological and pathological deteriorative changes. In order to develop a suitable package for particular citrus fruit, the understanding of the biology of the fruit is essential. Because fruit is a living entity, the biological factors that are involved in the deteriorative changes are: (1) Physiological factors: During respiration, CO_2 and several other volatiles are released. The commodity also emits ethylene. Water loss takes place by transpiration. (2) Pathological factors: Disease pathogens, mostly fungi that infect fruit. (3) Biochemical/metabolic changes: Compositional changes take place in sugars, acids, ascorbic acid, pigments, and volatiles. Sensory or organoleptic qualities such as flavor, texture, color, and firmness change as the activities of degradative enzymes increase after harvest (Ladaniya et al., 2000).

Packages can be of two main types depending on purpose for which they are required: (1) Container for shipment/transport and (2) Consumer/retail package. There is a third type of container for bulk handling only, such as pallet bins/boxes, plastic crates/boxes, and baskets that are used in the field and packing houses.

Because the temperature, relative humidity, and aeration have a profound role to play in determining postharvest life of a commodity, the package needs to be developed accordingly. High humidity (HH) packages sometimes referred in the literature could be a shipment/transport container lined from the inside with perforated or non-perforated plastic film (polyethylene or any film resistant to water transmission), or it could be a consumer package in the form of a micro-perforated or non-perforated plastic film bag or a heat-shrinkable film wrapping of individual fruit or six or twelve pieces of fruit placed in a tray.

I. CONTAINERS AND PACKAGING MATERIALS

A. Bulk Handling Containers

Containers are required for handling harvested fruit in the field and packing houses or for storage before it is finally packed for market. These containers are made of wood, metal, and plastic as bulk boxes or pallet bins used in the U.S.A. and many other countries. The capacity of these containers is 275–300 kg or 400–499 kg fruit. These are to be handled with forklift trucks. Small plastic crates made of high density polyethylene (HDPE) measuring $50 \times 30 \times 30$ cm with an empty weight of about 1.7–2.0 kg can hold 18–19 kg fruit (gross weight 20–21 kg). These plastic crates with net or perforation for ventilation are commonly used in India. There are different styles of these crates, and nesting types, which require less space in the return trip, are also available. The 36 lit plastic returnable crate is considered suitable in Australia because it is the nesting type and also ideal for retail display and self service by the consumers in stores (Tugwell, 1981). Increasing cost of conventional packaging material is forcing the industry for the reusable, returnable (multi-trip) alternatives. Fruits are also packed in 20-carton equivalent bins in Florida for display in supermarkets. These are common with many brands of citrus. They serve the purpose of easy access for the shopper, advertisement, and economy of packing.

B. Containers for Transport/Shipment

These containers are designed for long distance transportation in units varying from 4–5 kg (for limes) to 20–25 kg (for oranges, mandarins, grapefruits, and pummelos) in capacity. The most important requirements of the design of shipping/transport containers are their compatibility with the handling system, specifically, (1) Package should be amenable to the equipment at the packing facility (mechanization), and the unitization/palletization process. These packages must withstand impacts, compression, and vibration during transport and have sufficient resistance to high relative humidity. These containers are not expected to withstand direct contact of water for prolonged periods because they are made from corrugated paper board. Therefore, such conditions must be avoided during handling of these containers. Containers removed from refrigerated transport vehicles or storage rooms condense water from the atmosphere on the cold surface. Waterproofing treatment is essential for these packages.

Wooden boxes, corrugated paper board boxes, and trays of molded pulp and foamed polystyrene are commonly used as shipping/transport packages. Sacks of jute or synthetic material and bamboo baskets are also used as shipping packages in developing countries. They are commonly used for sweet oranges, limes, and Kinnows transported in India. Corrugated fiberboard (CFB) containers have replaced the wooden boxes worldwide. In Florida, the 56 lit capacity wirebound boxes were used in the 1950s for oranges, but by the late 1960s, these were almost gone. Thereafter, 28 lit wire-bound boxes were in vogue. Use of nailed wooden boxes was almost negligible by the early 1970s in the U.S. Since then, 28 lit fiberboard boxes have been in use.

Cuban oranges and grapefruits are packed in fiberboard telescopic boxes of 16–20 kg capacity. Tangerines and limes are packed in 9–10 kg capacity boxes. Mexican citrus, mainly oranges and grapefruits, are packed in wirebound boxes (41.3 × 26 × 26 cm internal) with net weight of 18.5 kg for the domestic market. For export to Japan, fruits are packed in fiberboard cartons measuring 42 × 27.3 × 24.1 cm.

1. Regular slotted corrugated board box: Usually made of 150–200 g/sqm craft paper and to be erected as single piece box; closure is by stapling and taping (Fig. 10.1, see also Plate 10.1). These boxes can also be printed and vented.

2. Telescopic corrugated board box: These boxes can be folded and erected in two pieces. These are stapled, and the lid or upper part must be placed manually. Plastic tape is used. These boxes can be stacked in columns or placed in an interlock arrangement. Transport and handling in palletized form or loose/single is possible. Printing directly on the outside of boxes in single or multicolor ink is possible (Fig. 10.2, see also Plate 10.2).

3. Wooden boxes and crates: The wooden logs of babool, neem mango, or locally grown species are cut in saw mills, and boards of various sizes and thickness are assembled by nailing. Even wood of senile rubber trees are used in India. These boxes are relatively rigid and have good stacking strength, but

FIGURE 10.1 One-Piece Regular Slotted Type Box Filled with Acid Lime.

FIGURE 10.2 Two-Piece Telescopic Box Filled with Mandarin Fruit.

they are not as thick as other kinds of boxes, so they sometimes give way during transport. The unfinished wood can cause abrasion to the fruit. In central India, it is a normal practice to tie a coir rope all around the box as a reinforcement after packing and nailing. Dry paddy straw is used as cushioning, and newspaper is placed as lining so that damage to fruit by the rough surfaces of the wood is minimized. These boxes are likely to absorb moisture during transport and in cold storage. Moisture could be from storage air or fruit. The trademark or other information is stamped on the sides of board in magenta and green ink. Boxes are handled singly by laborers. Boxes are built at the site by nailing because delivery of assembled boxes would be uneconomical for transportation. Size cut boards and battens are transported to the site where boxes are erected. These boxes provide sufficient ventilation – 0.5–1 cm gaps are left between two adjacent boards during assembly. The sides are made of 3–4 mm thick solid wood board of desired length and 5–10 cm width. Battens at the end corners are 1–1.2 cm thick and 4–5 cm wide. Sizes vary from 45 × 32 × 32 cm to 49 × 35 × 35 cm. These boxes are heavy – empty boxes weigh 4–5 kg. These boxes are relatively sturdy but do not have dimensional uniformity because they are made manually, and space remains in the truck leading to movement and damage to boxes and produce.

Wooden crates are assembled with wider gaps of 3–4 cm between the boards on all six sides. Thus crates provide more ventilation as compared to boxes, and with crates, the product is visible from the outside.

4. Plastic crates: Due to increasing cost of packaging materials, the plastic crates with holding capacity of 18–20 kg are also being used for transportation of fruit to distant places. Because these crates are reused, the cost of packaging comes out to be much less. These crates are emptied at the destination and returned, so the commodity must be suitably handled further at the destination. HDPE and Polypropylene (PP) injection molded crates/boxes are common. PE has higher impact strength and higher density than PP.

5. Plastic cartons: Collapsible plastic cartons are very economical because they can be returned and reused after cleaning. These cartons occupy about one fourth of the space after collapsing, and fruit can be exported in them.

6. Bamboo baskets: Bamboos are split open, and strips of about 2–3 cm width and 2–4 mm thickness are woven to form conical baskets. These baskets are reused because they are quite sturdy (Fig. 10.3, see also Plate 10.3). These type of long conical baskets (height 3–4 feet) holding 600–700 Khasi mandarins are common in Meghalaya and neighboring states of the North-Eastern-Hill (NEH) region of India. In central India, for short distance transportation, fruits are packed in small conical baskets (height 1–1.5 ft. with flat base and made of thinner strips of bamboo) holding 100–200 mandarins. Fruits are usually packed with newspaper lining and paddy straw as cushioning to avoid abrasion of the fruit's surface. These baskets are not reused. In this way, local employment is generated, and bamboo, an eco-friendly packaging material, is utilized. Due to the thinness of the bamboo strips, these baskets cannot withstand the weight if

FIGURE 10.3 Bamboo Basket for Packing and Transport of 'Khasi' Mandarin.

stacked, and they do not have dimensional stability and uniformity. Therefore, these baskets are not suitable for long distance transport in trucks.

C. Commodity-specific Containers

1. Mandarin/Tangerine

In India, mandarin fruits are packed in wooden boxes for long distance shipments. It is estimated that 8–9 million wooden boxes are required annually to pack Nagpur mandarin fruit produced in central India alone. The common lining material is newspaper in the box with paddy straw as a cushioning. With unavailability of wood, over a period of time, wooden boards and battens have become thinner (thickness reduced from 5–6 to 2–4 mm) resulting in damage to 5–10 percent of boxes during transport. As a substitute for wooden boxes, CFB boxes from commercially available kraft paper and the kraft paper from cotton plant stalks (CPS) have been designed and developed (Ladaniya et al., 1999; Shaikh et al., 2003). To a limited extent, cotton stalks are utilized as domestic fuel by the rural masses in central India, while the bulk of the stalks are disposed by burning in the field itself. The material is rich in cellulose and is very similar to hard wood in fiber dimensions. It has been found to be a suitable raw material for the preparation of pulp and paper. The stacking strength of the box can be achieved with kraft paper of desired weight (grammage).

TABLE 10.1 Fruit Sizes and Count of Nagpur Mandarin in a Conventional Wooden Box and a CFB Carton

Wooden box with paddy straw cushion (45.5 × 32.5 × 32.5 cm)			Telescopic CFB box with paper board dividers (50 × 30 × 30 cm)		
Fruit size	**Fruit diameter average (cm)**	**Count**	**Fruit size**	**Fruit diameter (cm)**	**Count**
Extra large	7.55	80–120	Extra large	7.31–8.00 (Avg. 7.65)	96–104
Large	7.00	120–140	Large	6.71–7.30 (Avg. 7.00)	112–120
Medium	6.40	150–180	Medium	6.11–6.70 (Avg. 6.40)	140–175
Small	5.80	170–200	Small	5.50–6.10 (Avg. 5.80)	200
Very small	5.00	225–270	Very small	4.70–5.49 (Avg. 5.00)	225–250

Source: Ladaniya et al. (1999).

Since water absorption is a major problem in CFB boxes under high humidity conditions of refrigerated storage, water vapor barrier options such as (1) lamination with PP film, (2) coating with wax, (3) use of bitumen paper, and (4) resin coating can be utilized. The use of bitumenized paper as one of the plies inside the board depends on acceptance by regulatory authority and is subject to rules and regulations of the different countries.

The cost of the CFB box is a major constraint in its wide acceptance by the mandarin industry in India, so boxes with a cost comparable to that of conventional wooden boxes have been developed. The one-piece regular slotted type cartons cost less. The cost of CPS craft paper boxes is also on par with wooden boxes. CFB cartons of the following specifications have been suitable for packing, transport, and storage of Nagpur mandarin fruits: two piece, 5 ply, fully telescopic, laminated with polypropylene film from the outside, B flute vertical, 48.5 × 28.5 × 28.5 cm inner dimensions, 50 × 30 × 30 cm outer dimensions, bursting strength 10 kg/cm^2 (5 ply), compression strength 460 kg, weight of box (at 25–30°C; 50–60 percent RH) about 1.10 kg with cobb value nil. These boxes are well-suited to shipping fruit because of good fruit count capacity and utilization of space in transport vehicles. The number of fruits that can be packed in the box varies with the size of the fruit but is similar to that of a conventional wooden box (Table 10.1). Boxes with 19.25 kg/cmsq bursting strength of the board are desirable for export use.

Aeration vents taking up 4–5 percent of the side areas punched as four long slits (9.5 cm × 1.75 cm) and 1.65 percent of end area punched as one slit (8 × 1.7 cm) as a handling slot are sufficient. Ventilation up to 6 percent of the side areas of the box provided better aeration during precooling in shipping containers (Ladaniya and Shyam Singh, 2002).

TABLE 10.2 Sizes and Count of Kinnow in a 10-kg Box

Kinnow size (mm)	Number of fruits
60–64	84
65–69	72
70–72	60
72–74	54
75–79	51
80–85	48

TABLE 10.3 Nagpur Mandarin Fruit Size, Weight, and Count in 10-kg Box

Fruit size (mm)	Approximate fruit weight (g)	Number of fruits
50–54	80–85	117–125
55–59	95–100	100–105
60–64	110–120	83–90
65–69	140–148	68–71
70–74	150–165	61–65
75–79	170–190	53–59

Nearly 40 to 50 percent of Nagpur mandarin fruit produced in central India is packed in conventional wooden boxes (48 × 35 × 35 cm) for dispatch to distant markets either from orchard or assembly markets. Kinnows produced in Himachal Pradesh, Punjab, and Rajasthan are transported by truck after packing in gunny bags or bamboo baskets with paddy straw (Chauhan et al., 1987; Chopra and Dhar, 1990). Since 1995, CFB cartons are increasingly being used for packing of Kinnow fruit because the fruit is waxed in modern packing houses. Kinnows from Rajasthan and Punjab states, India are packed in 10 kg-capacity boxes (3 layers per box; 48–84 per carton count; carton size 45 × 23 × 18 cm) for export (Table 10.2). 10-kg capacity cartons were also used by some exporters for Nagpur mandarins (Table 10.3) Anandswamy and Venkatsubbiah (1976) improved wooden cases (42 × 32 × 29 cm internal dimensions) to hold 110–120 mandarins and also developed ventilated CFB boxes (46.5 × 27.5 × 27.5 cm internal dimensions) with 18 kg gross weight for export of Coorg mandarins. As far as cost is concerned, gunny bags and bamboo baskets are cheaper than CFB and wooden boxes for transporting Kinnows, but losses are higher in bags and baskets (Jain and Chauhan, 1994).

The CFB cartons used for mandarins in California are two-piece, fully telescopic, half-slotted with side vents. The dimensions are 41.6 cm (L) × 27.2 cm (W) × 19.0–21.6 cm (H) and 41.6 cm (L) × 27.2 cm (W) × 26 cm (H) with

TABLE 10.4 Containers Generally Used for Fresh Citrus in the U.S.A.

		Net weight	
Fruit	Container	Kg	lb
Oranges	Fiberboard carton	20	45
	Fiberboard box	39	86
	Carton for place packing	17–18	37–40
Tangerines/mandarins/hybrids	Fiberboard carton	20	45
	Fiberboard carton	11–12	25
	Master carton with 10 bags of 1.3 to 1.4 kg	13–13.5	30
	Master carton with 16 bags of 1.4 kg	22	48
Grapefruit	Fiberboard carton (28.9 lit)	19	42.5
	Fiberboard carton	17–18	40
	Master carton with 6 bags of 3.6 kg	22	48
	Master carton with 10 bags of 2.4 kg	24	52
Limes	Small carton	4.5	10
	Fiberboard carton (28.9 lit)	17–18	38–40
Lemons	Fiberboard carton (28.9 lit)	17	38

volume of nearly 28 lit. The fruit count capacity of this carton is 100 for jumbo (6.35–6.98 cm diameter), 135 for large (5.72–6.35 cm diameter), 180 for medium (5.08–5.72 cm diameter), and 230 for small (4.44–5.08 cm). The standard container used for mandarins measures 41.6 × 27.2 cm with height of 19 or 21 cm.

In Japan, 15-kg cartons are used for Satsuma mandarins (Yamashita and Kitano, 1981). Because fruits cannot bear more weight, the static load should not be more than 0.5 kg per fruit. The container height is suggested to be a maximum of 35 cm with stacking of not more than 5 high.

2. Sweet Orange, Lemon, and Lime

Fiberboard boxes/cartons with 15–18 kg fruit capacity are used in many citrus growing countries for domestic trade of oranges and lemons and also for export. The standard carton size varies from region to region. The variety of shipping containers generally used for citrus fruits in the U.S. are given in Table 10.4. In Florida, standard 28.2 lit (4/5 bushel) cartons are packed with 64 to 163 oranges depending on fruit diameter. The net weight per carton for oranges in Arizona and California is 17.19 kg, and in Florida, it is 20.4 kg. Texas oranges (8–9 cm diameter) are packed in 49.3 lit cartons and count about 125. In 4/5 bushel cartons with 15 kg fruit weight, usually 80 Hamlin fruits are packed. For lemons packed in California and Arizona, 17.3 kg is the net weight. Lemons are also jumble-packed in cartons by mechanical devices to the required weight. In Australia, oranges are packed in 30-liter CFB boxes (Tugwell, 1981).

Persian limes (*C. latifolia*) are packed in fiberboard cartons containing 4–5 kg of fruit for export. Full or partial telescopic CFB cartons are common. The uniformity of fruit size is important, and count/box is 48, 54, or 63 with 5-kg net weight packing. In India, acid limes (small fruits) are packed in jute bags for long distance transportation. For export of kagzi acid limes, small fully telescopic boxes of 4 to 5 kg capacity (30.5 × 24.4 × 12.7 mm inside) are preferred in European markets. Uniformity of fruit size is very important in packing.

3. Grapefruit

Conventional fiberboard cartons used for packing of grapefruits in the U.S. have a 28.2 lit capacity. Forty grapefruits of size code 4 (as per codex standards) can be packed in this box. The net weight of a carton of grapefruit in Florida is 19.3 kg, whereas in California, it is 14.2–18.2 kg (Brooke and Spurlock, 1977).

For successful export, two-piece, fully telescopic fiberboard containers should have a minimum strength of 160 kg for the body (40–15–40 kg paper board weight/ 92 m^2) and 90 kg test fiberboard for the cover (19–15–19 kg). Waterproof adhesive should be used throughout in both the body and cover of the shipping container (Hale et al., 1981). Different types and sizes of containers, but mostly CFB cartons, can be seen in the international market. Grapefruit containers with dimension of 38.5–47 cm (L) × 25–32 cm (W) × 25–32 cm (H) are common. Palletization, unitization with webbing (stretch type film to secure), and use of 1200 × 1000 mm pallets are the export requirements.

Florida grapefruit in tray pack cartons were observed to be in better general condition and appearance. Tray packing reduces even slight fruit deformation in fruit exported to Europe and Japan (Hale and Risse, 1974). Folded or Telefil-type cartons are preferred over glued boxes because customers have to open and check for cull fruits. The size and grade for the European market are decided by EEC; these requirements must be followed while filling the box and packaging.

Western European preferences for packaging and unitizing of fresh fruit containers are to be followed while exporting fruit to these countries. Type, style, closure method, and accessory material for shipping containers are considered while deciding which box to use. For best printing results, shipping containers should have a base color other than craft brown. Usually white color is preferred as the base color for bright color printing on it.

D. Filling and Packing of Boxes

Fruits are usually place packed or jumble packed in case of oranges. Mandarins must be place packed systematically in layers with dividers between the layers. Mandarins (*C. reticulata*) lose firmness and therefore their shape changes due to weight of other fruit if jumble packed. Overfilling of boxes leads to deformation particularly in mandarins and grapefruits. Fully telescopic CFB boxes have better strength than regular slotted types. The use of a honeycomb cell pack arrangement in the containers provides maximum protection to ripe fruits (Hale et al., 1981). Molded PVC trays, PVC film with entrapped air bubbles, corrugated pads, and vertical separation reduce bruising. Corrugated pads and separators are the most cost effective. During transportation on unpaved roads fruit gets more bruises if not properly cushioned and immobilized in the box.

Results have shown that carton liners such as ICI Lifespan® can retain freshness of oranges on long voyages. Plastic liners in Valencia orange cartons shipped from Australia to Singapore reduced weight loss from 9 percent in non-lined package to 1 percent in plastic lined cartons. This also led to retention of the round shape of fruit and minimized moisture absorption by corrugated cartons that caused bulging of the package and possible collapse of box bottoms (Tugwell and Chvyl, 1997). Packaging in non-ventilated polyethylene-lined cartons leads to high humidity and water condensation that results in greater decay (Ladaniya, 2001). The liner must be ventilated, and fruit must be treated with a suitable fungicide. With mandarins and sweet oranges, ventilation up to 0.5 to 1 percent of the area of the PE bag gives satisfactory results. Using plastic liners in cartons can be risky at temperatures of 15–20°C and above because most pathogens grow rapidly at higher temperatures and lead to unacceptable decay (Grierson, 1969; Tugwell and Chvyl, 1997). The normal overfill of 20 mm by selecting the appropriate size range of fruit is the common practice for packaging oranges in Australia.

Nagpur mandarin fruits are packed in layers using 3-ply CFB vented separators/dividers which act as cushions between the layers. CFB board is also placed at the bottom of the box. The vented polyethylene liner or bag is very useful as a moisture barrier to minimize weight loss of fruit. After loading the fruits, the polyethylene liner is closed. For taping of flaps, plastic synthetic tape (BOPP 50 mm wide) is used. The general filling patterns for Nagpur mandarins are as follows:

Regular/linear fruit arrangement (Nagpur mandarin)

Box externally measuring 50 × 30 × 30 cm, full telescopic type with 48 × 29 × 29 cm internal dimensions

(1) Fruit with 5.55–6.1 cm diameter (5.8 cm average diameter)

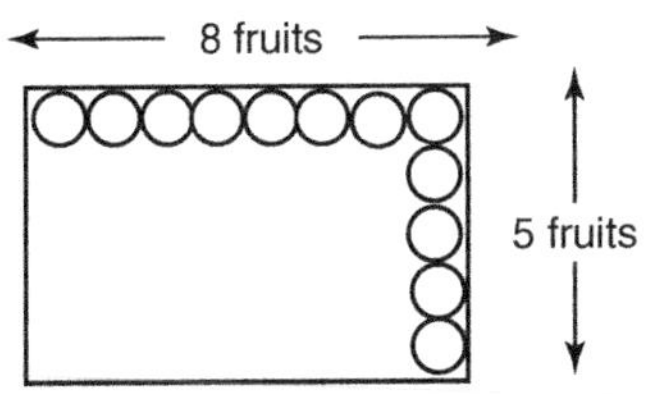

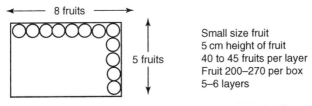

Small size fruit
5 cm height of fruit
40 to 45 fruits per layer
Fruit 200–270 per box
5–6 layers

(2) Fruit size 6.11–6.7 cm diameter (6.4 cm average diameter)

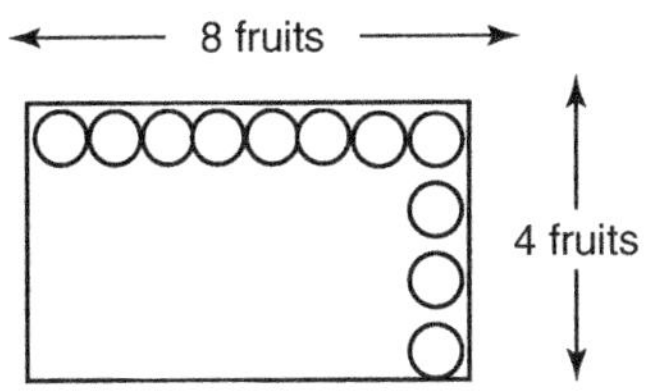

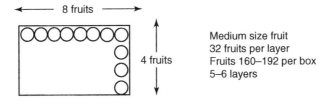

Medium size fruit
32 fruits per layer
Fruits 160–192 per box
5–6 layers

(3) Fruit size 6.71–7.30 cm diameter (7.00 cm average diameter)

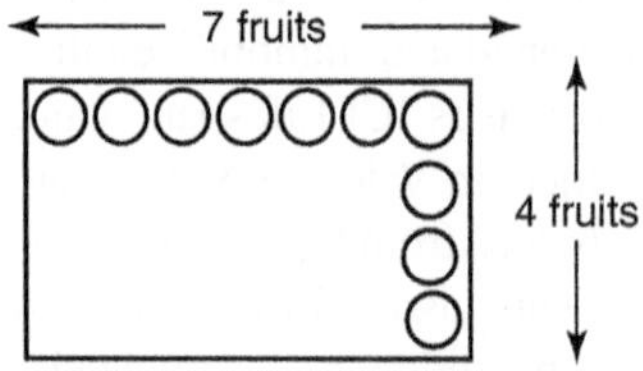

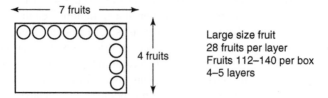

Large size fruit
28 fruits per layer
Fruits 112–140 per box
4–5 layers

(4) Fruit size 7.31–8 cm diameter (7.65 cm average diameter)

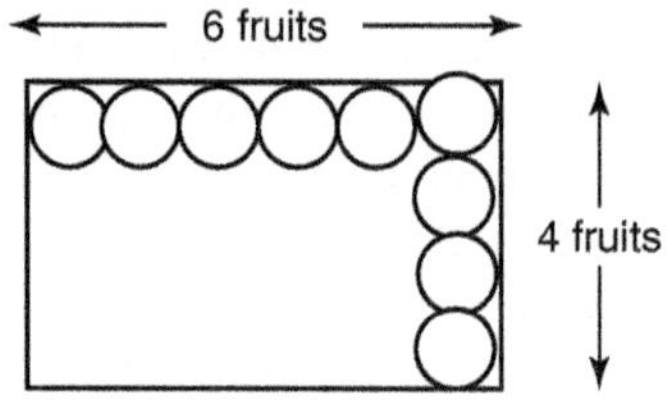

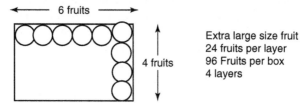

Extra large size fruit
24 fruits per layer
96 Fruits per box
4 layers

Offset type fruit arrangement
(1) Medium-sized fruit (6.4 cm average diameter)

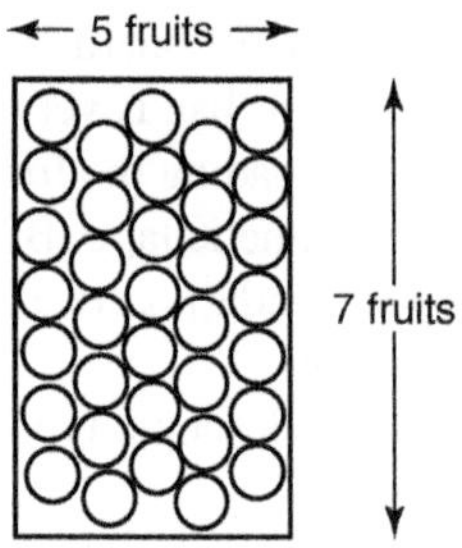

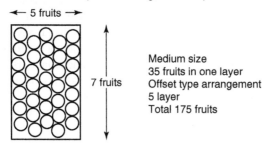

Medium size
35 fruits in one layer
Offset type arrangement
5 layer
Total 175 fruits

(2) Large-sized fruit (7.0 cm average diameter)

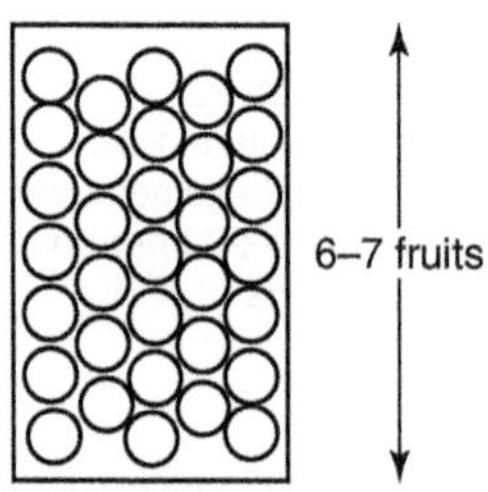

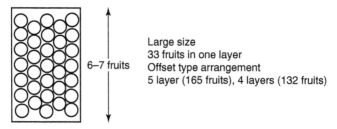

Large size
33 fruits in one layer
Offset type arrangement
5 layer (165 fruits), 4 layers (132 fruits)

E. Transportworthiness and Strength of the Boxes

CFB and carton tests: The carton developed for packaging of citrus fruits must be tested for its transportworthiness and stacking strength. The compression test of the carton indicates the strength of the box to sustain weight during stacking in trucks or cold storage. The box is placed between two steel plates and increasing force measured in kg is applied until it collapses. Drop tests and inclined impact tests of the cartons are conducted to assess their strength to sustain impacts without any damage. Cobb value indicates the water absorption property of the board. Bursting tests are conducted by puncturing the board; bursting strength is measured in kg/sq cm.

In order to evaluate the transport-worthiness of the containers filled with Nagpur mandarin fruits, they are subjected to an inclined impact test (speed 8 km/hr, 6 impacts, one on each face with a trolley distance 1.5 m), vibration test (frequency of 180 cpm and amplitude of 2.54 cms for 1 h) and a drop test of the filled boxes (drop height of 90 cm, total 4–5 drops on corners, edges, bottom, long side, and ends). The drop height taken for the simulated test is 90 cm, but in actual handling and transport, boxes are never dropped from such a height. Although dropping 5 times from a 90 cm height is considered to be too rigorous a test to judge transport-worthiness of CFB boxes, this test could be decisive in determining strength of the box. The lids of telescopic boxes must be secured with plastic tape before undergoing the drop test. CFB boxes made from cotton plant stalk kraft paper and commercial kraft paper should be able to withstand these tests (Ladaniya et al., 1999; Shaikh et al., 2003). The CFB cartons should be able to withstand impacts, vibrations, and bouncing, which occur during actual transport at rear, middle, and front side of the truck. There should not be load shifting and damage to boxes. There should be no gaps between the box stacks and also between the box stacks and sides of the truck body. Compactness of the load can be ensured due to uniformity in box size and compatibility with the truck dimensions. Due to their light weight, more CFB boxes can be transported as compared with conventional wooden boxes thus reducing transportation costs per box. The number of wooden boxes in the truck need to be decreased in case dimensions increase slightly during manual assembling/fabrication. Such possibility does not arise with CFB boxes because they are machine-made and uniform in size.

CFB boxes with the above specifications stacked to the height of 210 cm in a refrigerated chamber with 85–95 percent. RH should be able to withstand the load, and there should not be collapse/breakdown or compression damage. Boxes should have a waterproof coating or lamination, and there should not be bulging at the bottom of the stack. If the outside of a box is waxed, BOPP tape cannot be used for sealing of flaps, so the flaps are stapled when the boxes are formed.

The sides of the bottom box usually bulge as the box is pressed from the top. Very slight or no bulging occurs in top boxes. Two-piece telescopic boxes with resin coating and bursting strength of 16–19 kg/cm^2 can withstand 90 percent RH in a chamber for 45 days without collapsing or damage to fruit. Care needs to be taken that moisture or water vapor from humidifiers or air ducts doesn't come in direct contact with the box. The polyethylene liner in the box helps to avoid contact of moisture on fruit to the box and thus helps in maintaining strength of the box.

F. Stacking and Handling

Fiberboard boxes should not be stacked beyond their stacking limit. Generally, boxes with combination board-weight strength of 160 kg for the body and 90–100 kg for the cover should not be stacked more than seven layers high. To avoid physical damage to the shipping containers during unitizing and handling, the following practices should be observed: (1) When unitizing on wooden pallets, bases, or elevated pallet bases, avoid stacking patterns that allow containers to overhang the pallet. (2) Containers must interface with the top deck board on the wooden pallet for the particular unitizing pattern being used. (3) When loading break bulk refrigerated ships, all boxes should be stacked in register and in direct vertical alignment.

G. Packaging for Retail Market

Retail packages are small in size designed to hold 1 or 2 dozen pieces or 2–3 kg of fruit. The materials used for retail or small unit packaging vary in different locations depending on demand, availability, and economics. However, they can be categorized as films, boxes, trays, and mesh bags. Films are made of different materials such as cellophane, polyethylene (HDPE, LDPE), polyvinyl chloride (PVC), polypropylene (PP), cellulose acetate, and polystyrene. Ventilation is necessary in plastic film packaging or else very high humidity and water accumulation can lead to decay. Important characteristics of these materials are as follows:

Polyethylene (PE): It is usually low density and easily heat-sealed. It is the most widely used film for bagging applications and excellent for firm products such as oranges and grapefruits. For retail marketing, mesh or ventilated polyethylene bags of 2.3 and 3.6 kg capacity are used for grapefruits and oranges. These bags are carried to markets in master cartons holding 5–8 bags. The master box measures 51 × 33 × 28 cm. In Florida, PE bags are made of 38–51 micron film, and they are suitably printed and ventilated.

Polyvinyl chloride (PVC): This stretch film is in wide use. It is non-fogging and does not rip or tear unless punctured. It is an excellent film for consumer packaging.

Cellulose acetate film: It is highly transparent and sparkling in appearance, making an attractive package. It is relatively high in permeability to O_2 and CO_2.

Polystyrene: This material is similar to cellulose acetate film. It is extensively used for packaging.

Trays or Backings: Trays that are used to hold produce in conjunction with a film over-wrap or sleeve are called backings.

Bags: Bags made of plastic film, fiber net bags, and plastic net bags are used for packaging of oranges for retail marketing.

Closures: For closing fruit bags, draw-strings and straps are widely used.

1. Plastic film and mesh bags: These packages are designed mainly for providing a handy consumer package to carry and also for sales promotions.

The vented plastic film bags control water loss and delay shriveling and senescence. These are called high humidity (HH) packages as mentioned in the beginning of this chapter.

Packaging of fruit in plastic film can substitute for wax coating on the fruit. Wax coating is reported to result in off-flavor development in Nova and Murcott easy peeler mandarin type fruits and injured the peels (Cohen et al., 1989a; 1989b). A major drawback of these HH packages is that they promote decay due to high humidity or water condensation inside. These packages are mostly combined with fungicide treatment or any other treatment of fruit that would reduce decay. Perforations are necessary to regulate water vapor transmission to the desired level, which would be a compromise between reduction of shriveling and reduction of relative humidity. Mesh bags and perforated polyethylene film results in less decay (Grierson, 1967). Different degrees of ventilation in consumer packages results in different levels of decay, although these packages are enclosed in bags in master carton for as long as 4 weeks. The most successful combination was prefumigation with 2-amino butane (2-AB) followed by packing in a Vexar net bag (Grierson and Hayward, 1968). PE bags are superior in terms of restricting weight loss, undesirable change in peel color and gloss, and stem-end rot. Weight loss is linearly correlated with susceptibility of citrus fruit to deformation (Kawada and Albrigo, 1979).

This type of packing of fruits will be preferred by the buyers in supermarkets and department stores which are gaining acceptance in India and will also be helpful to promote prepackaging of citrus fruits in the country on a large scale. Polyethylene films have been extensively tried for prepacking oranges. Films of 100, 150, and 200 gauge contained water loss in Sathgudi oranges; the decay increased with decreasing percentage of ventilation (0.6, 0.4, and 0.2 percent) although water loss remained unaffected (Sadasivam et al., 1973).

In many countries including Australia for retail marketing, 4.5-kg net bags are used, which are then packed in master cartons (Tugwell, 1981). The product is visible in these bags, but weight loss is not controlled.

II. FILM-WRAPPING AND SEAL-PACKAGING

A. Seal-packaging

Seal-packaging of citrus fruit with non-perforated plastic film extends the shelf life and reduces shrinkage, weight loss, and the occurrence of various blemishes. This is achieved with special equipment and heat-shrinkable film that stretches over the individual commodity or a tray holding it according to the shape. The films used are partially permeable to gases and water vapor and can modify atmosphere if hermetically sealed. Commercially available films are very transparent and add to the sheen of the fruit. The heat-sealing machines are semi-automatic or fully automatic with varying speeds depending on packaging requirements. Weight loss reduction is up to 10 times or more in seal-packaged fruit than in non-packaged fruit. It can be an alternative to short-term refrigerated storage. In China and Japan, seal-packaging is a commercial practice, and several hundred metric tons of oranges are being wrapped and stored well until April or May of the following year. More than 0.1 million metric tons of Hassaku and Amanatsu fruits were commercially stored during late 1970s and early 1980s in Japan in LDPE film (Kawada et al., 1981). Fruits are harvested at the correct stage of maturity, and no irrigation is applied 2 weeks before harvest. After harvest, the fruits are dipped in 2, 4-D + Topsin (Thiophenate methyl) solution for 30 s and then wrapped in polyethylene film. The fruit stem remains green and fruit skin glossy with good eating quality. Fruit weight loss was only 7 percent with a total of 13–15 percent of decay loss during storage (Liu, 1998). In Japan, Natsudaidai and Hassaku fruits (natural hybrids of pummelo) are stored for several months by individual seal-packaging with 10–20 micrometer (μm) LDPE film at ambient temperature in winter. Individual seal-packaging also has the advantage of preventing secondary fungal infection. Hassaku fruit packaged individually within 20–40 days after harvest shows a low rate of the incidence of Kohansho, a common physiological disorder in this fruit. Individual seal-packaging with thin LDPE film is also useful in storing green Sudachi and Kabosu fruit because this type of packaging alleviates chilling injuries of the fruit at low temperature (Murata, 1997). Enclosure of an ethylene and acetaldehyde absorbent in LDPE packaging can inhibit degreening and alleviate the physiological disorder of green Kabosu.

At present citrus fruits are not packed in polyolefin or other types of permeable films during wholesale or retail marketing in India and many other developing countries, but considering hygiene in handling of fruit and the moisture barrier and selective gas permeation properties of these materials, the scope is tremendous.

B. Commodity-specific Response to Seal-Packaging

1. Sweet Orange

Individual fruit seal-packing with low-density polyethylene film effectively reduces fungal rot in sweet oranges by preventing spread. Mosambi sweet orange fruits individually and tray over-wrapped with heat shrinkable polyethylene (LDPE) and Cryovac (polyolefin) films can be satisfactorily stored up to 40 days under ambient conditions (20–30° C). Loss in weight varies from 1–3 percent in film-wrapped fruit whereas decay is reduced to 4–6 percent by pre-harvest spray with carbendazim (Ladaniya and Shyam Singh, 2001; Ladaniya, 2003). In general, the incidence of decay caused by *Diplodia natulensis* and *Alterneria citri* was reported to be high in film-wrapped navel oranges particularly at ambient temperatures and indicated the necessity of pre and/or postharvest fungicide treatments (Gilfillan, 1985).

Seal-packaging also preserves fruit quality. Seal-packaging reduces/prevents desiccation, which adversely affects peel thickness and weight of the non-wrapped orange fruit but not juice volume (El-Mughrabi, 1999). Juice flavor is adversely affected although juice volume remains the same in non-wrapped fruit. Seal-packaging preserves the natural flavor and prevents loss of juice in Mosambi orange (Ladaniya, 2003). After 3 months' storage at 21–33° C, fruit quality (firmness, texture, and organoleptic properties) of Valencia oranges was the best in fruit individually wrapped in polyethylene film (Alfaro et al., 1986).

2. Mandarin/Tangerine

Heat-shrinkable films (polyolefin) extend storage life of unipacked and tray over-wrapped Nagpur mandarins (*C. reticulata* Blanco) up to 3 weeks at 30–35° C and 25–35 percent relative humidity (Ladaniya et al., 1997). Film wrapping extends the storage ability of Nagpur mandarin fruits after refrigerated storage also. The CO_2 and ethylene level increases inside the film over-wrapped tray of Nagpur mandarins (Sonkar and Ladaniya, 1998; Ladaniya et al., 2000). Considering higher ambient temperatures and higher humidity inside polyethylene films, disinfection of the fruit and suitable fungicide treatment are necessary (Ladaniya et al., 1997; Dhatt et al., 1997; Randhawa et al., 1999).

Sealing of individual fruit in HDPE (10 μm thick) film increased storage life of Kinnow mandarins under ambient conditions up to 8 weeks (Dhatt and Randhawa, 1994). Thiabendazole (TBZ) and imazalil (1000 ppm dip for 2 min) successfully minimizes decay in seal-packaged Nova and Ora mandarin fruits. Weight loss was reduced while firmness was maintained in the sealed non-perforated package (Peretz et al., 1998). Juice characteristics of Minneola tangelos were retained by seal-packaging with Cryovac MR 15 film during storage at 20°C. Imazalil (500 ppm) was more effective than sodium hypochlorite (500 ppm) in reducing decay caused by green mold (D'Aquino et al., 1998). The shelf life of Dancy mandarin fruits was extended up to 8 weeks by wrapping individual fruits in polyolefin film followed by cold storage at 5°C and 85–90 percent RH. Wrapping significantly lowered percentage of CO_2 and ethanol levels in the juice. No off-flavor or internal aromas were detected (Saucede et al., 1997).

Use of ethylene absorbents in seal-packaging can reduce the concentration of dimethyl sulfide that is one of the causal compounds of off-flavor in Satsuma mandarins during storage (Kwak et al., 1992).

Sealing improves the taste of easy peeler Hadas mandarins by reducing its high acid content and thus improving the TSS:acid ratio (Ben-Yehoshua, 1990). Similar are the results with sealed Ellendale mandarins after 15 weeks of storage as compared with the control (Tugwell and Gillespie, 1981). The panelists of organoleptic tests have found that wrapped (Cryovac MD and MY; 19 and 20 micron thick) Okitsu Satsuma fruits were more acceptable than non-wrapped fruits (D'Aquino et al., 1998). In polyethylene-film–seal-packed Satsuma and Ponkan mandarins stored for 90 and 110 days, the weight loss was 3.6–6.5 percent and 5.53 percent, which was nearly 1/3rd and 1/4th of the non-wrapped fruit, respectively, under room temperature conditions (winter season). After storage, fruit color changed completely with natural luster, and the fruit had its own pure flavor, appearance, quality, and vitamin C contents (Biying Zhou, 1990).

Kawada et al. (1981) reported that film wrapping of mandarins is not beneficial because it causes peel puffing and thus increases decay. However, such effect was not observed in individually sealed and tray-wrapped Nagpur mandarin fruit (Ladaniya et al., 2000). The increase in *Alternaria* rot may occur due to high humidity (Kawada et al., 1981), and similar observations were recorded in Nagpur mandarins. *Alternaria* rot incidence is dependent on field infection as in the case of *Diplodia* and *Phomopsis* rots. Tray-wrapped (in Cryovac-Impact/D-955; 15 and 25 micron) Nagpur mandarin fruit did not rot even after 8 weeks at ambient conditions (30–35°C and 25–35 percent RH) indicating that if there is no field infection and subsequent injury at or after harvest, no decay would occur. Although fruit is fresh in appearance and as such acceptable from a marketing point of view, the flavor loss occurs slowly due to loss of titratable acidity if fruits are stored beyond certain limit (4 weeks or so) at such high temperatures. Here the role of selectively permeable film is important so that natural fruit flavor is retained. Similar results with respect to loss of titratable acids and its effects on taste were reported by Poretz et al. (1998).

3. Grapefruit

The resistance of grapefruit to deformation can be maintained by minimizing weight loss through individual seal-packing of fruit. RH reduction would also reduce moisture absorption by the fiberboard boxes and permit the maintenance of box strength. Retention of box strength would allow boxes in lower layers in a stack to better withstand the pressure exerted on them by boxes in the stack and thus allow less fruit deformation (Kawada and Hale, 1980). If grapefruit are individually film wrapped, they can be shipped without refrigeration or humidity control, thereby saving energy. However, for successful shipment without refrigeration, proper harvesting, handling, and decay control will be required.

Seal-packaging of individual grapefruits not only control the spread of decay but also contain depletion of the O_2 level in the carton as rotting fruit consumes O_2 faster and releases volatile chemicals that adversely affect the flavor of healthy fruits. The flavor of sound grapefruit that was stored in a polyethylene film-sealed container along with decaying fruit remains similar to that of sound fruit stored in the absence of decaying fruit (Barmore et al., 1983).

When weight loss during shipment exceeded 5 percent of original weight, the gloss or sheen was not maintained with a single waxing (Albrigo et al., 1980). Double waxing reduced weight loss more than the weight loss in single waxed fruit. Unipacking reduced weight loss to <2 percent of the original weight of Florida grapefruit. Control of weight loss results in maintenance of fruit gloss and firmness.

Unipacking (seal-packing of individual fruit) enabled export of Florida grapefruit to Rotterdam (The Netherlands), in non-refrigerated van containers (in dry-freight van containers). Thiabendazole and Imazalil spray reduced stem-end rots. Imazalil coated film had no advantage over spray applications of fungicides directly on the fruit to prevent decay in seal-packed grapefruit. Realized savings from lower freight rates, greater usable cargo space, lower equipment investment cost, lower energy uses, etc. could be benefits to users of this equipment (Miller et al., 1986).

4. Lemon

Very good results can be obtained by shrink wrapping lemons with 15 micron high-density polyethylene film. Because of marked reduction in weight loss, the wrapped lemons can be kept at ambient temperatures up to 75–90 days which is similar to the storage life of refrigerated non-wrapped lemons. Wrapped lemons kept in a cold room maintain good commercial quality for 120 days. Un-refrigerated non-wrapped lemon becomes unsaleble in less than 30 days (Testoni and Grassi, 1983). In Australia, lemons dip-treated in a mixture of imazalil, Guazatine, and 2, 4-D (500 ppm) followed by shrink-wrapping (Cryovac 925 and Cryovac MD) could be stored up to 6 months at 7–8°C (Wild, 1987).

Seal-packaging of individual Baramasi lemon fruit in HDPE (10 μm thick) film bags increases storage life up to 4 weeks under ambient conditions (Dhillon et al., 1993). Individual seal-packaging with HDPE film also extends the storage life of Eureka lemons stored at 13°C. Ethephon (10 ppm) treatment combined with seal-packaging (HDPE films) inhibits the development of blemishes in lemons (Ben Yehoshua et al., 1982). In individually sealed lemons, quality retention is better after 6 months at 13°C than non-sealed fruit. Juice and acid content were not significantly different in sealed and non-sealed fruit, but fruit firmness was better in sealed fruit (Cohen, 1991).

5. *Citrus Iyo*

In Iyo (*Citrus iyo*) fruit, weight loss was 5 percent for individually sealed (high-density 0.03 mm polyethylene film) fruits and 10–20 percent for waxed fruits at 10 or 20°C up to 6 weeks. Juice sugar content was low in sealed fruits especially at 20°C. Juice ethanol concentration increased with increasing temperature and at 20°C it was high, particularly with seal-packaging or oil emulsion (Hino et al., 1990).

C. Advantages of Seal-packaging

Fruit softening is highly correlated with declining water potential. Seal-packaging reduces water loss and thus maintains fruit turgor. Sealing inhibited softening and changes in cell wall pectins and delayed membrane disintegration as shown by inhibited leakage of amino acids in particular and electrolytes in general. The effects of sealing were prevented by including hygroscopic $CaCl_2$ in a sealed enclosure, which reduces ambient humidity (Ben-Yehoshua et al., 1983).

Sealing reduces the chilling injury in grapefruit and lemons. Differentially permeable transparent film, notably pilofilm (rubber hydrochloride), is reported to establish an in-package atmosphere that reduces chilling injuries and extends the shelf life of citrus fruits (Grierson, 1971). Grapefruit chilling injuries were prevented for 1 month at 4.5°C by sealing in PVC, polyethylene, and cellophane films (Wardowski et al., 1973).

Wrapped grapefruits conditioned at 34°C (30 or 95 percent RH) for 72 h and stored at 1–4°C for 4 weeks + 1 week at 21°C had less chilling injury (Miller and Risse 1988). Individual fruit sealing also reduces the commodity's respiration rate. Effect of temperature on fruit condition is, of course, greater than that of film wrapping of fruit. Fruits stored at higher temperatures lost acidity faster than those stored at lower temperatures (Purvis, 1983). Individual-fruit sealing also affects the internal composition, and it is dependent on film thickness and the nature of the polymer. Ethanol levels were higher in sealed grapefruits and tangerines wrapped in those plastic films that gave maximum reduction in weight loss (Albrigo and Fellers, 1983). Increased ethanol is associated with lower taste ratings. The adverse effect of sealing on lemon, however, is less significant than with wax coatings (Hale et al., 1983). As a rule of thumb, any excessive restriction in respiration causes anaerobic conditions and ethanol production leading to reduced taste ratings, and therefore selection of film of optimum permeance for gases is necessary. The sealing has a significant advantage with respect to the control of spread of decay and restricting off-flavor contaminating other healthy fruit. In sealed grapefruit packed in cartons, decay has little effect on remaining sound fruits. This can be attributed to lower CO_2 build-up and O_2 depletion. The flavor of remaining healthy fruit is not adversely affected because there is no depletion of O_2 and volatile chemicals emanated by rotting fruit are contained within the film-sealed package (Barmore et al., 1983). Another important advantage of sealing is that there is no soilage due to sporulating fruit, and discarding of rotten fruit becomes easy. Seal-packing of fruit in trays or as individual fruit delays color development in grapefruit (Kawada and Albrigo, 1979) and lemons (Hale et al., 1983), and this could be an advantage in fruit that are preferred green.

Fruit can be cured or conditioned by wrapping with plastic film and storing it for a short time at higher temperatures. Curing of lemons, pummelos, and oranges at 34–36°C under higher RH conditions for 48–72 h (within 2 days from harvest) is effective in reducing decay (Ben-Yehoshua et al., 1987). With individually wrapped fruit, curing can be accomplished without controlling RH because the fruit is surrounded by a saturated atmosphere inside the film (Ben-Yehoshua et al., 1990). This process should be cautiously done because it may result in heat damage to fruit and can increase ethanol content (Martinez et al., 1989).

The seal-packing also results in less green mold rot as compared to similarly treated waxed fruit. Curing has been found to reduce decay caused by *P. digitatum*; however, other organisms resistant to higher temperatures may cause decay. Curing by itself cannot control decay during long-term storage. Individual fruit wrapping of navel oranges and Marsh grapefruit and curing at 35°C for about

3 days (within a day after harvest) is effective in minimizing decay. This treatment can reduce chilling injury in oranges stored at 2° C (Del Rio et al., 1992).

Wrapping of Miho Satsuma fruits in Cryovac MD 15 followed by curing (38°C and 100 percent RH for 72 h) can maintain freshness of fruits for 24 days at 20°C and 75 percent RH. Curing combined with wrapping greatly decreases weight loss in addition to decay (D'Aquino et al., 1996).

Seal-packaging of grapefruit also helps in curing. Curing without seal-packaging leads to weight loss, shrinkage, and softening of fruit. Curing (at 25–42°C for 1–3 days) of sealed Goliath pummelos inhibited postharvest decay without deleterious effects on quality and prevented development of *P. digitatum* (Ben-Yehoshua et al., 1987). Curing sealed grapefruit at 30°C reduces damage caused by mechanical harvest, probably because of the water saturation in the micro-atmosphere of the sealed fruit (Golomb et al., 1984).

III. MODIFIED ATMOSPHERE PACKAGING (MAP)

Fresh citrus fruit continue to respire after harvest, and spoilage is aggravated when fruit is shipped to distant markets. One way to reduce decay and extend the shelf life is by promptly cooling and maintaining the produce at low temperatures. Modification of the atmosphere surrounding the product to create a new atmosphere that usually has a lower level of O_2 and a higher level of CO_2 could be a complement to temperature management techniques. This technique can be used where refrigeration is not available or not economical because the shelf life can be easily extended for a few weeks to months particularly when ambient temperatures are low. At these levels of O_2 and CO_2, the respiration rate of fruit will decrease, and their shelf life will increase. Unlike climacteric fruits, citrus fruits do not respond much to controlled atmosphere (CA) and modified atmosphere (MA). The principal benefit is reduction of metabolic activity by reduction in respiration and almost inhibition of water loss. Fruits can be packed individually or in a tray using heat-shrinkable polymeric selectively permeable films. The most important part is air-tightness of the seal to get MA inside the package. If the package is not sealed properly (not air-tight) or the seal is broken later on, the MA will not be created. The other advantages of sealing, including prevention of cross infection, prevention of wetting of the paper board box due to decayed fruit, and retention of fruit firmness and freshness by maintaining saturated microclimate will not be achieved fully.

Gas exchange (O_2, CO_2, and to some extent ethylene) take place from fruit to surrounding atmosphere and viseversa. If the commodity is in equilibrium with its environment, the rate of gas exchange is same in both directions. If by using permeable film for wrapping, the concentration of gases (O_2 and CO_2) is changed outside the commodity, the concentration of these gases inside the commodity would also change. This concentration of gases inside the commodity would change depending on permeability of the film, which in turn would depend on the films' chemical properties and thickness plus the ambient temperature, and relative humidity. When the O_2 and CO_2 concentration inside the film package (tray containing six or twelve fruit over-wrapped with semipermeable plastic) are stable without wide fluctuation, it can be termed as an equilibrium condition. Usually as CO_2 increases and O_2 depletes inside the package, the respiration rate of the commodity decreases and solely depends on the permeability of plastic film used for wrapping. Initially, the commodity respires at a normal rate consuming O_2 inside the package leading to accumulation of CO_2. This condition has an inhibitory effect on the respiration rate of the commodity. Excessive suppression of respiration leads to accumulation of CO_2 inside the commodity (orange or mandarin core and tissues) that can lead to anaerobic conditions and off-flavor development.

In MAP, fruits are enclosed in a sealed pack constructed with a selectively gas permeable plastic film (Fig. 10.4, see also Plate 10.4). Inside the MA package, a steady state of O_2 and CO_2 concentration is achieved when the rates of O_2 uptake and CO_2 production by the product are equal to the rates of O_2 and CO_2 flow through the film (Beaudry et al., 1992). In properly designed MA packages, the steady-state gas concentrations can correspond to the optimal storage for the given commodity (Beudry, 1999).

Factors such as type of fruit, temperature, optimum O_2 and CO_2 partial pressures, respiration rate, product weight, permeability of the film, and atmosphere outside the package are considered when designing the package (Chinnan, 1989). Surrounding gas concentrations, type and maturity of fruit, and temperature of the fruit determine the respiration rate of the commodity (Kader et al., 1989; Talasial et al., 1992). When the optimization of different influencing factors is achieved for a particular type of fruit, the shelf life can be reasonably extended without quality loss.

MA packages should be designed keeping in mind surrounding temperatures during handling, transportation, storage, and marketing, because the respiration rate of the commodity would change with temperature and this would be different from that of the permeability of the film. Packages are normally designed for specific constant surrounding temperatures. A new type of film has been introduced that automatically adjusts the permeability in response to temperature changes by a phase change in the polymer structure (Clarke and De Moor, 1997).

Two approaches are followed: First is testing of a large number of films under required temperature regimes to find out the most suitable film for the given fruit. This is called the empirical approach which is somewhat effective, but the best film for the commodity may not be obtained. As mentioned earlier, if a fruit is in equilibrium with its environment, the rate of gas exchange is same in both the directions. To achieve this, the oxygen transmission rate (OTR) of the film should be same as that of respiration rate of the commodity.

Another approach could be mathematical and computer modeling in developing equations to predict permeability of film to identify suitable film so as to obtain the desired modified atmosphere in a particular temperature range. By

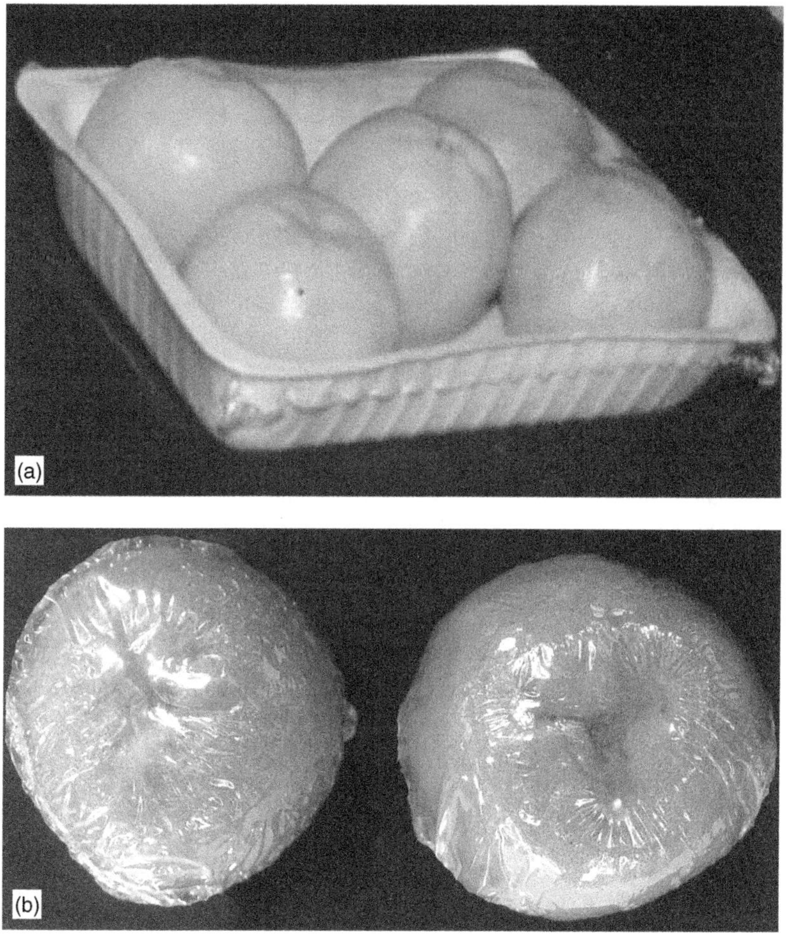

FIGURE 10.4 (a) Modified Atmosphere Packaging of 'Nagpur' Mandarin in Trays and (b) Individual Fruit Seal-Packing (MAP of Single Fruit).

using a mathematical model, it can be predicted that the particular film will be suitable for particular commodity. These models needs to be validated with data obtained by experimentation. These models must be based on dynamic handling conditions considering changes in respiration of type of fruit and permeation of film with changing temperature and relative humidity. The optimum gas concentration for the commodity, gas diffusion coefficients of the commodity, respiration rate, ethylene evolution rate, transpiration rate, etc. need to be determined in order to develop and test mathematical models. Some researchers have developed mathematical models (Lopez–Briones et al., 1993) that can help predict the atmosphere around the commodity sealed in plastic film. Modified atmosphere packaging is reviewed by Church (1994) who covered ethylene absorbers, ethanol vapor generators, MAPs in trays and bulk MAPs, and leak and gas detectors in detail.

A. Plastic Films and Their Characteristics

The film used for MAP should be sufficiently permeable to water vapor to avoid accumulation of water. Besides extending shelf life and improving appearance, this type of packaging can be a hygienic and convenient unit for retail marketing. There is a wide range of plastic films used for MAP. These include polyethylene (HDPE, LDPE, MDPE), polypropylene, polyvinyl chloride (PVC), and linear low density polyethylene (LLDPE), which combines properties of LDPE and HDPE. The thickness of plastic film is measured in gauge, millimeters, or micrometers (100 gauge = 25 micrometers = 0.025 mm). One micrometer is approximately 1/4th of the gauge unit. The thickness of plastic film may vary slightly across the sheet as the polymer is co-extruded or laminated. Several films are multi-layered and their thickness would also vary. The thickness of plastic film is measured for its permeability to gases with a Mitutoyo film thickness gauge, which can measure thickness of the film from 1 micron to 1 mm. It is a very handy and sensitive instrument. A square piece of plastic film is to be placed on the flat sample plate and gently pressed under the measuring probe. The indicator needle shows the thickness on the dial with graduation of 1 μm micro-meter. Instruments are also available to measure gas permeability of films.

The physical properties of some of the plastic films are given in Table 10.5 and should be considered while designing a MAP. HDPE has greater tensile strength and stiffness and hardness. LLDPE and LDPE have greater transmission rate for O_2 and CO_2 than HDPE (Schlimme and Rooney, 1994)

Keeping in mind properties of films, respiration rate, and weight (quantity) of produce, the modeling of MAP can be done. Gas flushing with the desired mixture of gases can be done to achieve MA faster i.e. active MAP.

Film permeability of gases is by active diffusion where gas molecules pass through the film matrix and diffuse through. The diffusion takes place through the film depending on the concentration gradient.

The permeability of the film can be calculated by the formula given by Crank (1975):

$$\text{Permeability (P)} = Jx/A\,(P_1 - P_2)$$

Where J = the volumetric rate of gas flow through the film at a steady state (ml), X = the thickness of film (m), A = the area of permeable surface (m^2), P_1 = gas partial pressure on side 1 of the film, and P_2 = gas partial pressure on side 2 of the film ($P_1 > P_2$) (Pressure in kPa).

Out of so many parameters that can affect gas transmission through film, the fruit weight, temperature, RH, surface area for gas transmission and O_2 and CO_2

TABLE 10.5 Some Physical Characteristics of Plastic Films Used in MAP

Films	Thickness (μm)	WVTR (g/m²/day)	Permeability (cc/m²/day) O_2	CO_2	Density (g/cc)
Cryovac BDF-2001 (multi-layered co-extruded polyolefin)	30	12.66	3	12	0.96
Cryovac D-955 or Impact (cross linked polyolefin)	25	13.66	4941	19764	0.92
'Cryovac'D-955 or Impact (cross linked poly-olefin)	15	23.00	8548	34192	0.92
LLDPE	25	16–35	7000–9300	30000	
RD 106 (WR Grace)			10200	23200	
LDPE	25	6–23	3900–13000	7700–77000	0.91–0.93
HDPE	25	4–10	520–4000	3900–10000	0.97–0.99

Note: Based on information supplied by the manufacturers before shrinkage of the film and at 20–23° C and 60–65% relative humidity. WVTR: Water vapor transmission rate.

transmission rates of film are the most important. With lower commodity weight, equilibrium gas concentration can be achieved with higher O_2 and lower CO_2 keeping the film unchanged and all other variables constant (Zagory, 1990). The relative humidity can also alter gas transmission rate across the film (Roberts, 1990).

In modified atmosphere packages, potassium permanganate is used to oxidize ethylene at room temperature. Ethysorb and Purafil are commercial proprietary products available in the market, and they can be placed in the box or a tray. The activated alumina carrier (Al_2O_3) is used to impregnate potassium permanganate. The oxidation of ethylene is irreversible because H_2O and CO_2 are formed. Other materials such as vermiculite, perlite, charcoal, and zeolite (hydrated aluminum silicate [($Al_2Sl_{30}O_{72}$) $24H_2O$] can be used as carrier of potassium permanganate.

B. Citrus Fruit Responses to Modified Atmosphere Packaging

Among citrus fruits, limes desiccate rapidly, and their color and texture changes rapidly. In seal-packed limes, degreening take place and weight loss is prevented. Degreening decreased with use of ethylene absorbents in the bag (Thompson et al., 1974). According to Ben-Yehoshua et al. (1983), sealing effects in citrus fruit could not be related to possible modified atmosphere mechanisms of O_2,

CO_2, and ethylene. These workers suggested that the mode of action of sealing in PE relates mostly to the alleviation of water stress in the fruit because some of the effects of sealing could be observed by maintaining non-sealed fruit in a water saturated atmosphere.

Polymeric films (D955 or Impact of 25 and 15 μm thickness) can be heat-sealed over plastic trays holding six fruit to produce MA around Nagpur mandarin (*C. reticulata* Blanco) fruit (Ladaniya, 2005). Proper atmosphere can be obtained by selecting a film with the right gas transmission rate and having the appropriate surface area for the package. In the MAP created by Cryovac films as mentioned earlier, at 27 ± 3°C with 25 ± 5 percent RH, O_2 level dropped and CO_2 level increased rapidly during the initial stage (first two days) inside both the films. From the 5th day onward, steady state was recorded up to 28th day with CO_2 concentration ranging from 6.1–8.4 percent and O_2 concentration ranging from 11–13.5 percent in D955–15 μm film (Fig. 10.5). In D955–25 μm film, steady state was recorded between the 7th and 28th day with CO_2 and O_2 levels 7.8–9.7 percent and 9.4–11.2 percent, respectively (Fig. 10.6). D955–15 μm film was considered better than D955–25 μm film because it had greater O_2 and CO_2 permeance under varying ambient temperatures during March and April. Fruit quality was very much acceptable up to 4 weeks in both the films but better in D-955–15 μm film. The inhibitory concentration of CO_2 that can retard

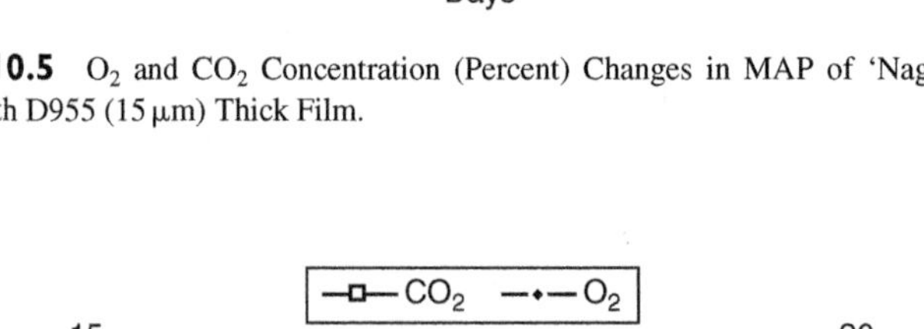

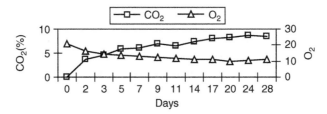

FIGURE 10.5 O_2 and CO_2 Concentration (Percent) Changes in MAP of 'Nagpur' Mandarin Wrapped With D955 (15 μm) Thick Film.

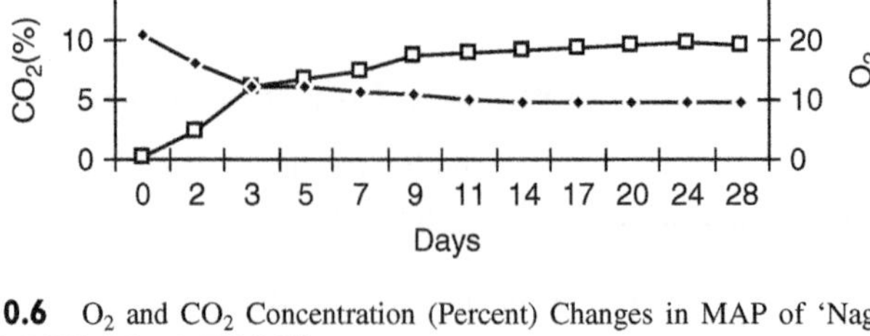

FIGURE 10.6 O_2 and CO_2 Concentration (Percent) Changes in MAP of 'Nagpur' Mandarin Wrapped With D955 (25 μm) Thick Film.

fungal growth without off-flavor development in mandarin fruits in MA packages needs to be determined. The reduced pressure of CO_2 (10–12 percent) can retard fungal growth and spore germination (Brooks et al., 1932), and this effect forms the basis for MA and CA in some berries and cherries (Beaudry, 1999). Nagpur mandarin fruits appear to tolerate O_2 and CO_2 level up to 9 percent, and this needs to be investigated further (Ladaniya, 2007). Acidic citrus fruits are more tolerant to higher CO_2 levels and therefore MA with high CO_2 levels may be more promising in these fruits.

Packaging of Nagpur mandarins in Cryovac BDF-2001 film of 30 micron thickness resulted in CO_2 levels as high as 12–14 percent inside the package at 20–24°C and 40–50 percent RH. The O_2 level reduced to about 9–10 percent. There was a slight adverse effect on fruit flavor. The color development was adversely affected as green spots remained unchanged up to 3 weeks, although ethylene level increased to 30 ppm inside (Ladaniya, 2003). This could be attributed to a strong presence of CO_2 which had an inhibitory effect on ethylene activity. Zamba (1986) reported that packing of Navel oranges in 1.5–2 kg capacity polyethylene packets (in standard boxes) with the inside atmosphere consisting of 3–4 percent CO_2 and 4.8–7 percent O_2 increased storage life of fruit up to 150 days at 5–7°C.

Plastic films are mostly made from by-products of the mineral oil refining industry. These films are non-biodegradable and hence can cause environmental pollution. Recently, biodegradable plastic films have been developed combining starch and low density polyethylene (LDPE). Granules made of 10–40 percent starch and LDPE and a binding agent are used to make a bio-degradable sheet. The starch component, being organic in nature, degrades in the soil, and once the molecule of the compound broke, its vulnerability to bacterial attack increases resulting in its disintegration within 60 days.

REFERENCES

Albrigo, L.G., Kawada, K., Hale, P.W., Smoot, J.J., and Hatton, Jr. T.T. (1980). Effect of harvest date and pre-harvest and postharvest treatments on Florida grapefruit condition in export to Japan. *Proc. Fla. Sta. Hort. Soc.* 93, 323–327.

Albrigo, L.G., and Fellers, P.J. (1983). Weight loss, ethanol, CO_2 and O_2 of citrus fruits wrapped in different plastic films. *HortScience*, 18, 615.

Albrigo, L.G., and Ismail, M.A. (1981). Shipment and storage of Florida grapefruit using unipack film barriers. *Proc. Int. Soc. Citric.*, 714–717.

Alfaro, D., Casamayor, R., Toledo, J.L., and Lopez, B. (1986). Postharvest storage of Valencia late orange. *Centro Agricola*, 13, 42–48, 57.

Anandswamy, B., and Venkatsubbaiah, G. (1976). Wooden and corrugated shipping containers for the export of Coorg oranges. *Indian Fd. Packer*, 30(3), 44–49.

Barmore, C.R., Purvis, A.C., and Fellers, P.J. (1983). Polyethylene film packaging of citrus fruit: Containment of decaying fruit. *J. Fd Sci.* 48, 1558–1559.

Beaudry, R.M., Cameron, A.C., Shirazi, A., and Dostal-Lange, D.L. (1992). Modified atmosphere packaging of blueberry fruit: effect of temperature on package O_2 and CO_2. *J. Am. Soc. Hort. Sci.* 117, 436–441.

Beaudry, R.M. (1999). Effect of O_2 and CO_2 partial pressures on selected phenomenon affecting fruit and vegetable quality. *Postharvest Biol. Technol.* 15, 293–304.

Ben-Yehoshua, S. (1990). Individual seal-packaging of fruits and vegetables in plastic films. In *Controlled/modified atmosphere/vacuum packaging of foods* (A. Brody, ed.) Foods and Nutrition Press, Inc., pp. 101–107.

Ben-Yehoshua, S., Barak, E., and Shapiro, B. (1987). Post-harvest curing at high temp reduces decay of individually sealed lemons, pummelos and other citrus fruits. *J. Am. Soc. Hort. Sci.* 112, 658–663.

Ben-Yehoshua, S., Shapiro, B., and Chem, Z.E. (1983). Mode of action of plastifilm in extending life of lemon and bell paper fruit by alleviation of water stress. *Pl. Physiol.* 73, 87–93.

Ben-Yehoshua, S., Shapiro, B., and Kobiler, I. (1982). New method of degreening lemons by a combined treatment of ethylene-releasing agents and seal-packaging in high density polyethylene films. *J. Am. Soc. Hort. Sci.* 107, 365–368.

Ben-Yehoshua, S., Shapiro, S.B., and Moran, R. (1987). Individual seal-packaging enables the use of curing at high temps to reduce decay and heat injury of citrus fruits. *HortScience*, 22, 230–231.

Biying, Zhou, Yuanyn, W., and Bojun, L.I. (1990). Investigation on storage of loose-skin mandarin fruits sealed individually in plastic film. *Proc. Int. Citrus symp.* China 5–8 November. 791–795.

Brooks, C., Miller, E.V., Bratley, C.O., Cooley, J.S., Mook, P.V., and Johnson, H.B. (1932). Effects of solid and gaseous carbon dioxide upon transit disease of certain fruits and vegetables. *USDA Tech. Bull.* 318, 1–59.

Chauhan, K.S., Sandooja, J.K., Sharma, R.K., and Singhrot, R.S. (1987). A note on assessment of certain prevailing practices for marketing of commercial fruits. *Haryana J. Hort. Sci.* 16, 229–232.

Chinnan, M.S. (1989). Modeling gaseous environment and physiological changes of fresh fruits and vegetables in modified atmosphere storage. In *Quality factors of fruits and vegetables* (J.J. Jen, ed.), Am. Chem. Soc. Washington DC. pp. 189–202.

Chopra, S.K., and Dhar, R.P. (1990). Packing and storage of citrus with special reference to Kinnow. *Proceedings of the Seminar on Problems and Prospects of Kinnow cultivation.* PAU, Ludhiana, pp. 205–209.

Church, N. (1994). Developments in modified atmosphere packaging and related technologies. *Trends in Fd Sci. Technol.* 5, 345–352.

Clarke, R., and DeMoor, C.P. (1997). The future in film technology: a tunable packaging system for fresh produce. *Proc. 7th Controlled At. Res. Conf.* Vol. 5, 13–18 July 1997. Univ. California. Davis, pp. 68–75.

Cohen, E., (1991). Investigations on postharvest treatments of citrus fruits in Israel. *Proc. Int. Citrus symp*, Guangzou, China, pp. 32–35.

Cohen, E., Shalom, Y., and Rosenberger, I. (1989a). Keeping the quality of Nova mandarin (Suntina) in storage and in shelf life. *Alon Hanoteia*, 43, pp. 1354–1356.

Cohen, E., Shalom, Y., and Rosenberger, I. (1989b).The effect of waxing on the flavour of Murcott easy peeler. *Alon Hanoteia*, 43, 1361–1364.

Crank, J. (1975). *The mathematics of diffusion.* 2nd edition, Oxford, Clamden Press.

D'Aquino, S, Piga, A., Agabbio, M., and McCollum, T.G. (1998). Film wrapping delays ageing of Minneola tangelos under shelf-life conditions. *Postharvest Biol. Technol* 14, 104–107.

D'Aquino, S., Piga, A., Agabbio, M., and Tribulato, E. (1996). Improvement of the post-harvest keeping quality of 'Miho' Satsuma fruits by heat, Semperfresh and film wrapping. *Advances in Hort. Sci.* 10, pp. 15–19.

D'Aquino, S., Piga, A., Petretto, A., Agabbio, M., and Ben-Yehoshua, S. (1998). Respiration rate and in-package gas evolution of 'Okitsu' Satsuma fruits held in shelf-life condition. Proc. 14th International congress on plastics in agriculture, Tel Aviv, Israel, March 1997, pp. 626–632.

Del-Rio, M.A., Cuquerella, J., and Ragone, M.L. (1992). Effect of postharvest curing at high temperatures on decay and quality of Marsh seedless grapefruit and navel oranges. *Proc Intn. Soc. Citric.* Italy, pp. 1081–1083.

Dhatt, A.S., and Randhawa, J.S. (1994). Extending the postharvest life of unrefrigerated Kinnow fruit by individual seal packaging. *Punjab Hort.* J. 34, 27–29.

Dhatt, A.S., Sidhu, A.S., and Randhawa, J.S. (1997). Effect of polyethylene liners on physiological parameters of individually sealed Kinnow fruit. *Nat. Symp. Citric.*, Nagpur, pp. 90–91.

Dhillon, B.S., Sandhu, A.S., and Grewal, G.P.S. (1993). Effect of high density polyethylene wrapping on postharvest life of uni-packed Baramasi lemon. *Symp. Hort. Res- Challenging scenario*, HIS, Bangalore 1993.

El-Mughrabi, M.A. (1999). Effect of bagging, individual wrapping and temperature regimes on quality attributes of Baladi oranges. *Arab Univ. J. Agric. Sci.* 7, 145–158.

Gilfillan, I.M. (1985). Preliminary trials on polyethylene film wrap for South African citrus export fruit. *Proc. of the 15th annual congress of the South African society of crop production.* Pietermaritzburg (1985) Outspan citrus centre, Nelsfruit, South Africa, pp. 186–94.

Golomb, A., Ben-Yehoshua, S., and Sarig, Y. (1984). High-density polyethylene wrap enhance wound healing and lengthens shelf life of grapefruit. *J. Am. Soc.Hort.Sci.* 109, pp. 155–159.

Grierson, W. (1967). Consumer packages for Florida citrus fruits. *Proc. Fla. Sta. Hort. Soc.* 79, 274–280.

Grierson, W. (1969). Consumer packaging of citrus fruits. *Proc. Int. Soc. Citric.*, California, Riverside, Vol. 3, pp. 1389–1401.

Grierson, W. (1971). Chilling injury in tropical and sub-tropical fruits: IV. The role of packaging and waxing in minimizing chilling injury in grapefruit. *Proc. Trop. Region. Am. Soc. Hort. Sci.* 15, 76–88.

Grierson, W., and Hayward, F.W. (1968). Simulated marketing tests with pre packaged citrus. *Proc. Fla. Sta. Hort. Soc.* 80, 237–241.

Hale, P.W., and Risse, L.A. (1974). Exporting grapefruits in tray pack cartons. *Citrus and Veg. Mag.* 37, 12–13.

Hale, P.W., Smoot, J.J., and Hatton, Jr. T.T. (1981) Factors to be considered for exporting grapefruit to distant markets. *Proc. Fla. State Hort. Soc.* 94, 256–258.

Hino, A., Li, S.Y., Kawahara, S., and Kadoya, K. (1990). The effect of seal packaging in high density polyethylene waxing and temperature on iyo fruit during storage. *Memoirs of the college of Agri. Ehime Uni,* 34(2), 327–336, 757.

Jain, P.K., and Chauhan, K.S. (1994). Influence of packaging and modes of transportation on marketability of Kinnow. *Indian J. Hort.* 51, 251–254.

Kader, A.A., Zagory, D., and Kerbel, E.L. (1989). Modified atmosphere packaging of fruits and vegetables. CRC critical reviews, Food science and nutrition, 28, 1–30.

Kawada, K., and Kitagawa. (1988). Plastic film package storage of citrus and some other fruits in Japan. In *Proc. 6th Int. Citrus Cong.* Israel, Vol. 3, pp. 1555–1556.

Kawada, K., Wardowski, W.F., Grierson, W., and Albrigo, L.G. (1981). Unipack: individually packed storage of citrus. *Fruits. Proc. Int. Soc. Citrus.* Japan, Vol. 2, p. 725.

Kawada, K., and Albrigo, L.G. (1979). Effects of film packaging, in-carton air filters and storage temperatures on the keeping quality of Florida grapefruit. *Proc. Fla. State Hort. Soc.* 92, 209–212.

Kawada, K., and Hale, P.W. (1980). Effect of Individual film wrapping and relative humidity on quality of Florida grapefruit and condition of fibreboard boxes in simulated export tests. *Proc. Fla. State Hort. Soc.* 93, 319–323.

Kwak, S., Ueda, Y., Kurooka, H., and Yamanaka, H. (1992). Effect of gas condition of polyethylene package on the occurrence of off-flavour of stored Satsuma mandarin (*C. unshiu* Marc.) fruit. *J. Japanese Soc. Hort. Sci.* 61, 453–459.

Ladaniya, M.S. (2003). Shelf life of seal-packed 'Mosambi' sweet orange fruits in heat shrinkable and stretchable films. *Haryana J. Hort. Sci.* 32(1/2), 50–53.

Ladaniya, M.S.(2004). Reduction in postharvest losses of fruits and vegetables. NATP Project Final Report of NRCC, Nagpur, 135.

Ladaniya, M.S. (2007). Quality and carbendazim residue of 'Nagpur' mandarin fruit in modified atmosphere package. *J. Fd Sci. Technol.* 44, 85–89.

Ladaniya, M.S., and Shyam Singh (2001). Tray over-wrapping of Mosambi sweet orange J. *Food Sci. Technol.* 38, 362–365.

Ladaniya, M.S. and Shyam Singh (2002). Packaging of horticultural produce with special reference to citrus fruits. Packaging India (IIP, Mumbai) December 2001–Jananuary 2002, Vol. 34(5), 9–19.

Ladaniya, M.S., Sonkar R.K., and Shyam Singh (2000). Shelf life of 'Nagpur' mandarin fruits wrapped with heat-shrinkable and stretchable cling films under ambient conditions. *Proc. Int. Symp. Citric.*, NRC for Citrus, Nagpur, 23–27 November 1999, pp. 1155–1167.

Ladaniya, M.S., Sonkar R.K., Shaikh A.J., and Varadarajan P.V. (1999). Corrugated fibre board containers for packaging, transport and storage of 'Nagpur' mandarin in domestic market. *Indian Fd. Packer* 53,(4), 5–15.

Ladaniya, M.S., Sonkar, R.K., and Dass, H.C. (1997). Evaluation of heat-shrinkable film wrapping of Nagpur mandarin (*C. reticulata* Blanco) for storage. *J. Fd. Sci. Technol.* 34, 324–327.

Ladaniya, M.S., and Shyam Singh (2000). Influence of ventilation and stacking pattern of corrugated fibre board containers on forced-air pre-cooling of 'Nagpur' mandarins. *J. Fd. Sci. Technol.* 37, 233–237.

Liu, H.F. (1998). Experience of storing orange wrapped per fruit with film. *South China fruits* 27(5) 22.

Lopez–Briones, G., Varoquaux, P., Bareau, G., and Pascat, B. (1993). Modified atmosphere packaging of common mushroom. *Int. J. Fd. Sci. Technol.* 28, 57–68.

Martinez-Javega, M., Cuquerella, J., Del-Rio, M.A., and Mateos, M. (1989). Coating treatments in post harvest behaviour of oranges. IIR. Commissions C2/D1, D2/3. Technical innovations in freezing and refrigeration of fruit and veg., Univ. Calif. Publication. pp. 51–55.

Miller, W., and Risse, L.A. (1988). Recent research of film wrapping of fresh produce in Florida. *Proc. 6th Int. citrus Congr.*, Israel, pp. 1521–1530.

Miller, W., Hatton, T.T., Hale, P., Rasmussen, G., and Hoogendoorn, H. (1986). Overseas tests of wrapped grapefruit in non-refrigerated van containers. *Proc. Fla. State Hort. Soc.* 99, 114–117.

Murata, T. (1997). Citrus. In *Postharvest physiology and storage of tropical and subtropical fruits.* (S.K. Mitra, ed.) CAB International. U.K., pp. 21–47.

Peretz, J., Rodov, V., Ben-Yehoshua, S., and Ben-Yehoshua, S. (1998). High humidity packaging extends life of easily peeled citrus cultivars (*C. reticulata*). *14th Int. congress on plastics in agriculture, Tel Aviv*, Israel, pp. 617–625.

Purvis, A.C. (1983). Effects of film thickness and storage temperature on water loss and internal quality of seal-packed grapefruit. *J. Am. Soc. Hort. Sci.*, 562–566.

Randhawa, J.S., et al. (1999). Studies on prolongation of storage life of Kinnow fruits with individual seal-packaging. *Proc. Nat. Symp. Citric,* Nagpur, pp. 461–464.

Roberts, R. (1990). An overview of packaging material for MAP. International conference on modified atmosphere packaging. Part I. Campden, U.K.

Sadasivam, S. et al. (1973). Note on the effect of polyethylene film packs on post-harvest life of sweet oranges in storage. *Ind. J. Agric. Sci.* 43, 211–212.

Saucede, V.C., and others (1997). Effect of individual film wrapping on quality and storage life of mandarin fruit Dancy. *Proc. 7th Int. CA Res. Conference*. Vol. 3, pp. 230–233.

Schlimme, D.V., and Rooney, M.L. (1994). Packaging of minimally processed fruits and vegetables. In *Minimally processed refrigerated fruit and vegetables* (R.C. Wiley ed.) Chapman and Hall New York, pp. 135–182.

Shaikh, A.J., Ladaniya, M.S., Varadarajan, P.V., and Shyam Singh (2003). Paper and corrugated boxes from cotton plant stalks for effective packaging of oranges. *J. Sci Industrial Res.* 62, 311–318.

Sonkar, R.K., and Ladaniya, M.S. (1999). Individual film wrapping of 'Nagpur' mandarin with heat-shrinkable and stretch cling film for refrigerated storage. *J. Food Sci. Technol.* 36, 273–276.

Sonkar, R.K. and Ladaniya, M.S. (1998). Effect of tray over-wrapping by heat shrinkable and stretchable films on Nagpur mandarin fruits. *Indian Fd Packer*, 52, 22–26.

Sonkar, R.K., and Ladaniya, M.S. (1999). Unipacking of Nagpur mandarin fruits with heat-shrinkable and stretchable films on Nagpur mandarin fruits. *Proc. Nat. Symposium on Citric.*, 17–19 November. 1997 NRCC, Nagpur, pp. 465–469.

Talasila, N., Chau, K.V., and Brecht, J.K. (1992). Effects of gas concentrations and temperature on O_2 consumption of strawberries. *Trans. ASAE.* 35, 221–224.

Testoni, A., and Grassi, M. (1983). Individual shrink wrapping with plastic film: a new technology for citrus fruits. Annali dell Istituto Sperimentale per le Valorissazione Technologies des prodolti Agricori 14, 49.

Thompson, A.K., Magzoub, Y., and Silvis, H. (1974). Preliminary investigations in to desiccation and degreening of limes for export. Sudan. *J. Fd. Sci. Technol.* 6, 1–6.

Tugwell, B.L. and Gillespie, K. (1981). Australian experience with citrus fruits wrapped in high density polyethylene film. *Proc. Int. Soc. Citric.* Vol. 2, pp. 710–714.

Tugwell, B.L., and Chvyl, W.L. (1997). Modified atmosphere packaging for citrus. *Proc. Int. citrus congress*, Sun City S. Africa, 1150–1152.

Tugwell, B.L. (1981). Marketing of citrus in plastic returnable crates. *Proc. Int. Soc. Citric.* Japan. Vol. 2, pp. 845–847.

Wardowski, W.F., Grierson, W., and Edwards, G.J. (1973). Chilling injury of stored limes and grapefruit as affected by differentially permeable packaging films. *HortScience.* 8, 173–175.

Wild, B.L. (1987). Report on plastic shrink-wrap treatment for lemons. Rural Newsletter 102, 30–36.

Yamashita, S., and Kitano, Y. (1981). Factors causing deterioration in the quality of Satsuma mandarin during sorting and packing. *Proc. Int. Soc. Citric.*, Japan. Vol. 2, pp. 827–839.

Zagory, D. (1990). Application of computers in the design of modified atmosphere packaging to fresh produce. International conference on modified atmosphere packaging. Part I. Campden, U.K.

Zamba, A.I. (1986). Orange storage in gaseous medium. Sodovotstavo i Vinogradortvo, Moldavii, 834–835.

PLATE 2.1 Valencia Orange – Widely Grown Late Maturing Orange World Over.

PLATE 2.2 Mosambi Orange-Widely Grown Early Maturing Orange in India.

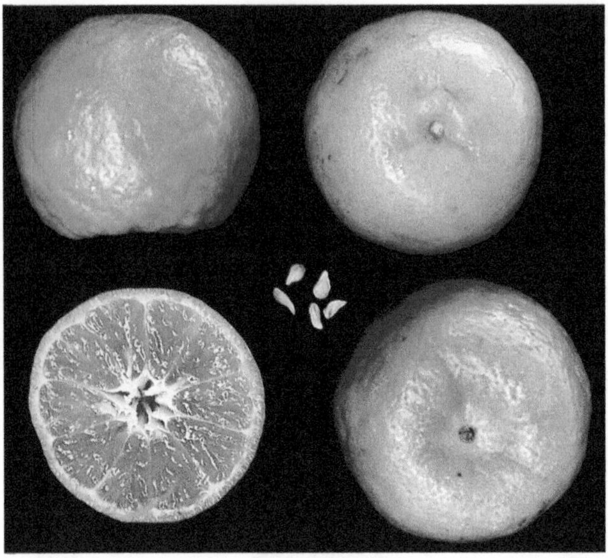

PLATE 2.3 'Nagpur' Mandarin – the Leading Mandarin Cultivar of Very Fine Taste.

PLATE 2.4 'Khasi' Mandarin – Widely Grown Mandarin in North-Eastern India.

PLATE 2.5 Nepali Oblong (Assam Lemon) – Common Lemon Cultivar in North-Eastern India.

PLATE 2.6 Galgal or Hill Lemon – Popular Lemon Cultivar in North and North-Western India (Left), Possible Lemon × Citron Hybrid (Right).

PLATE 2.7 Seedless Lemon – A Promising Seedless Lemon Variety of India.

PLATE 2.8 Key or 'Kagzi' Lime – Common Small Fruited Acid Lime of the World.

PLATE 2.9 Calamondin-Small Mandarin – like Fruit with Edible Peel.

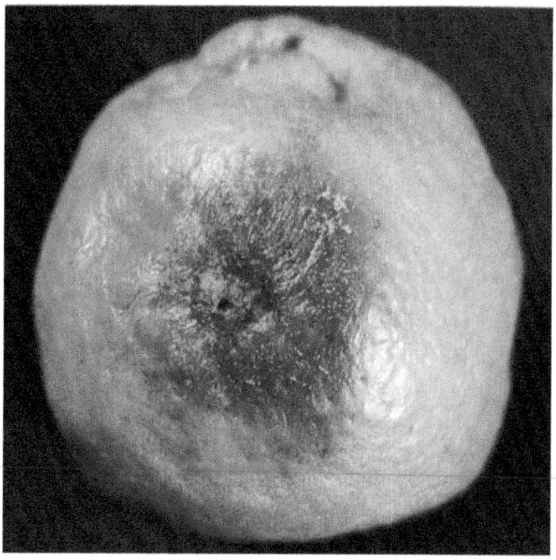

PLATE 4.1 Damage to Nagpur Mandarin by Fruit Sucking Moth.

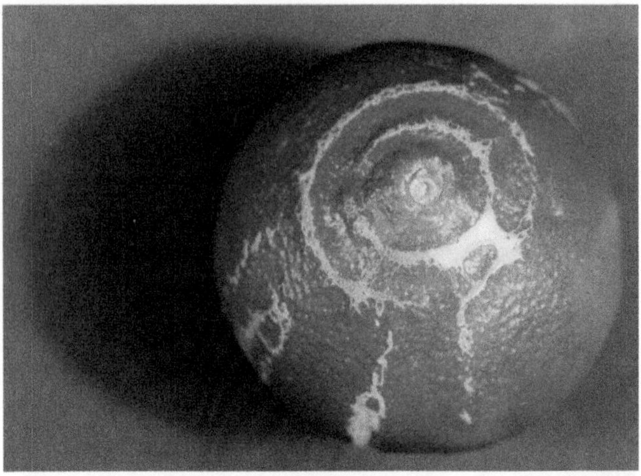

PLATE 4.2 Very Common and Peculiar Damage by Citrus Thrips in Ring Form Around the Fruit Stem of Nagpur Mandarin.

PLATE 4.3 Gibberellic Acid Sprays Delay Fruit Color Development and Facilitate On-Tree Storage of Nagpur Mandarin.

PLATE 8.1 Harvesting of Acid Lime with a Hook as a Practice in Central India. Note the Plain Field Inside the Canopy Covered with Cloth to Avoid Fruit Injury and Contact with Soil.

PLATE 8.2 Injury to 'Nagpur' Mandarin Caused Due to Long Stems of Other Fruit during Handling.

PLATE 9.1 Manual Sorting of Lemons.

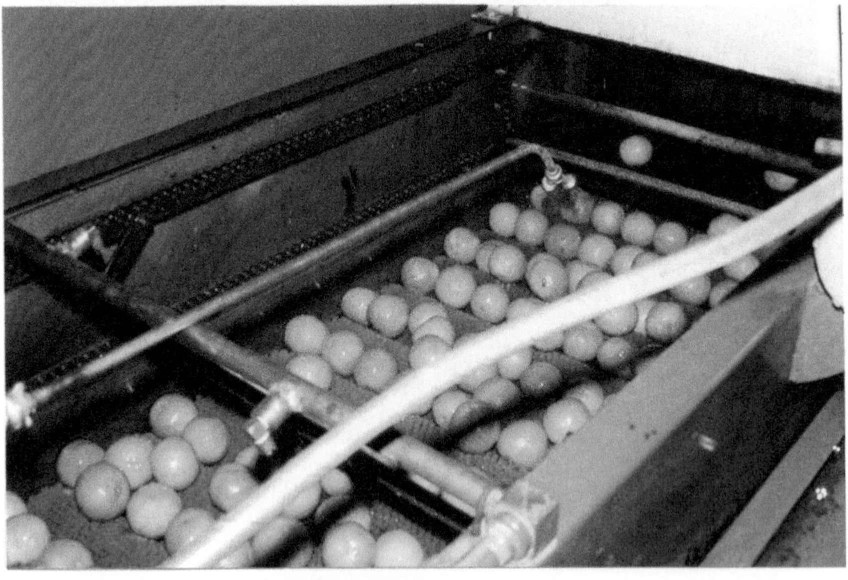

PLATE 9.2 Washer Unit.

PLATE 9.3 Traversing Nozzle of Waxer and Horse-Hair Brushes.

PLATE 9.4 Place-Packing of Mandarin Specialty Fruit.

PLATE 9.5 Pallet of Packed Boxes with Corner Posts and Strapping.

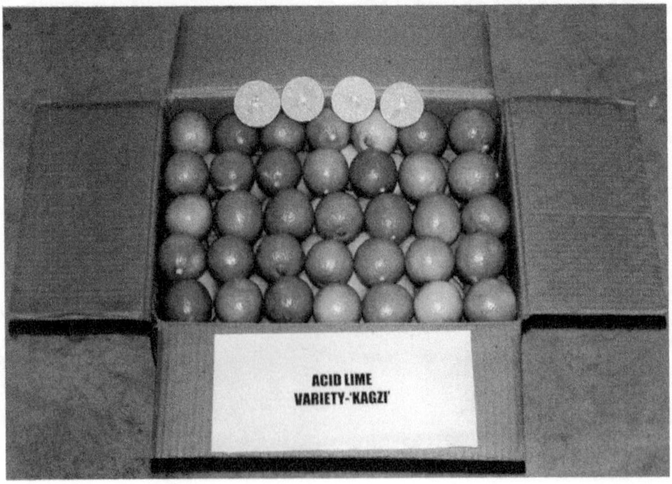

PLATE 10.1 One-Piece Regular Slotted Type Box Filled with Acid Lime.

PLATE 10.2 Two-Piece Telescopic Box Filled with Mandarin Fruit.

PLATE 10.3 Bamboo Basket for Packing and Transport of 'Khasi' Mandarin.

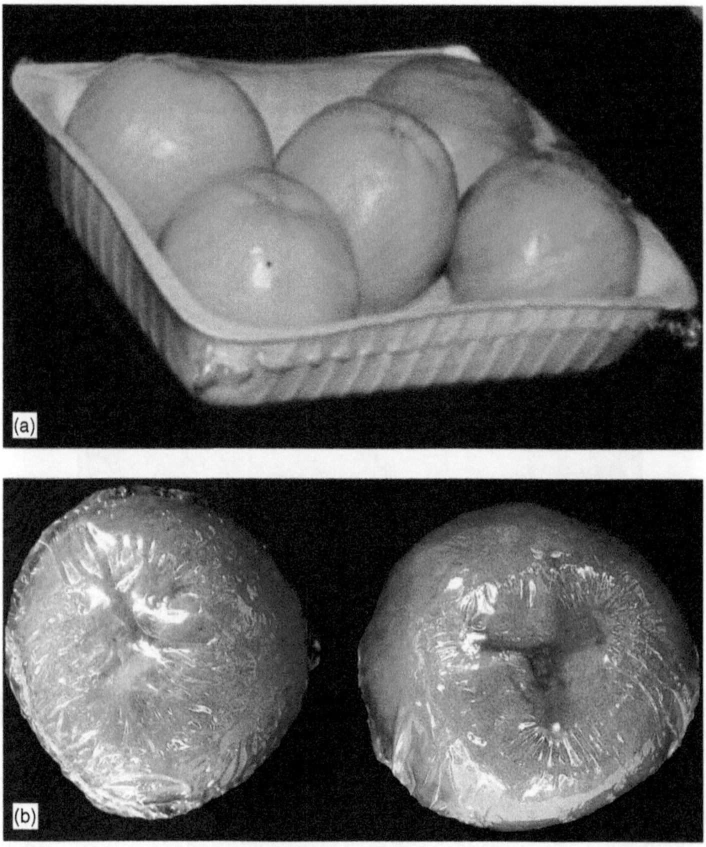

PLATE 10.4 (a) Modified Atmosphere Packaging of 'Nagpur' Mandarin in Trays and (b) Individual Fruit Seal-Packing (MAP of Single Fruit).

PLATE 12.1 Chilling Injury Symptoms on 'Nagpur' Mandarin Fruit.

PLATE 13.1 Reefer Container with Cooling Unit.

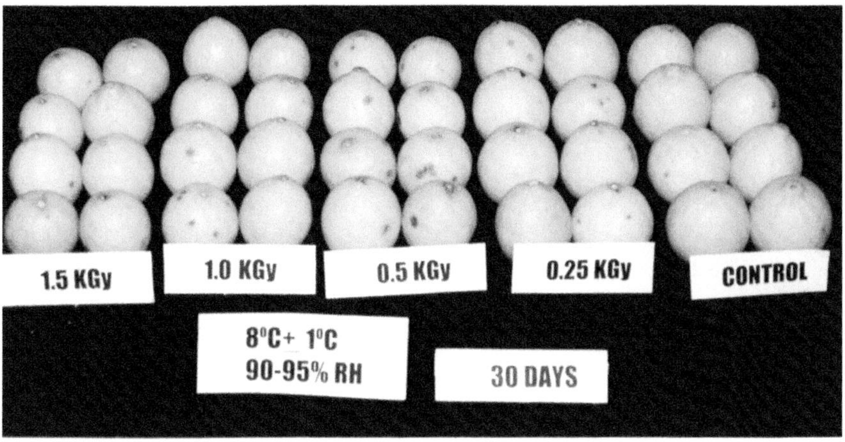

PLATE 15.1 Peel Damage Due to Irradiation in Acid Lime.

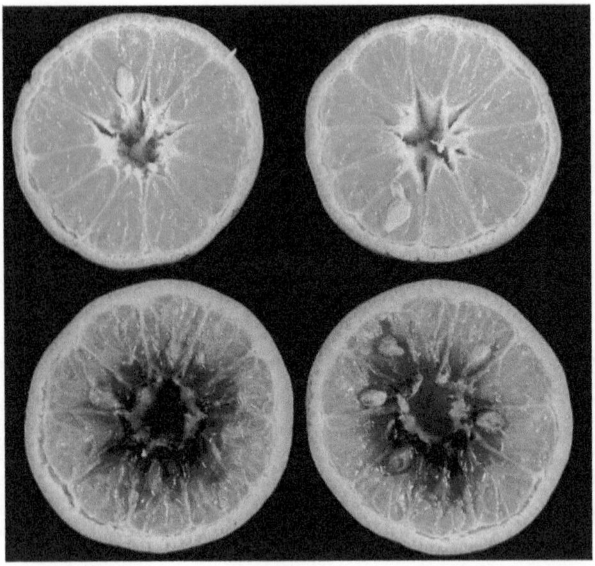

PLATE 16.1 *Alternaria* Rot of 'Nagpur' Mandarin.

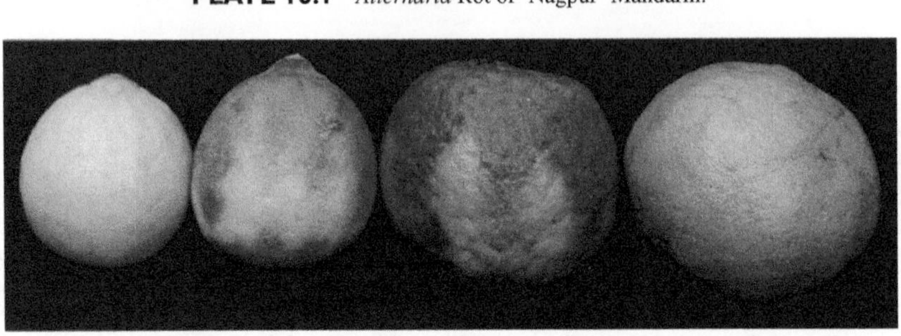

PLATE 16.2 *Botryodiplodia* (*Physalospora rhodina* Berk & Curt) Stem-End Rot.

PLATE 16.3 Blue Mold Rot.

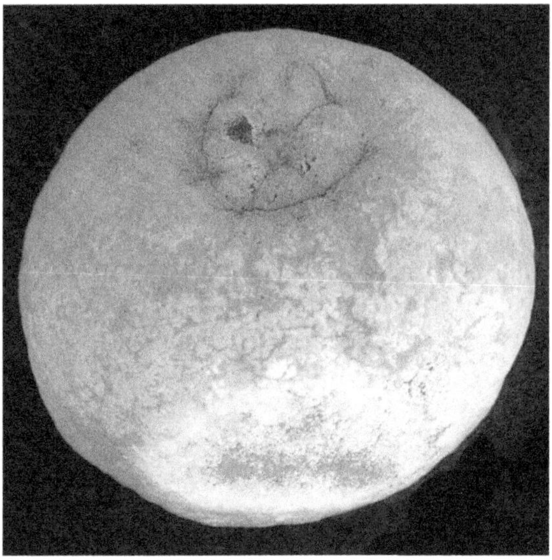

PLATE 16.4 Green Mold Rot.

PLATE 17.1 Granulation of Nagpur Mandarin Fruit.

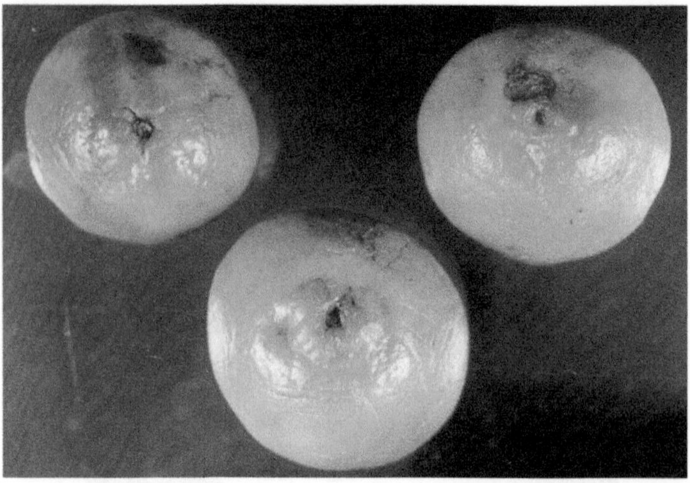

PLATE 17.2 Sunburn or Sunscald on Peel of Nagpur Mandarin Fruit.

PLATE 18.1 *Bactrocera dorsalis* (Oriental Fruit Fly) Adult Male (Without Ovipositor) on Citrus Fruit.

11

PRECOOLING AND REFRIGERATION

Rapid cooling after harvest is generally referred to as 'precooling'. Special facilities for fast cooling of the produce after harvest are essential because land vehicles, ship holds, and shipping containers (reefers) are not designed for rapid removal of field heat. Adequate cold air circulation within and around individual boxes may not take place to cool the warm fruit if loaded directly into the shipping/transport vehicle. Transport units are designed to maintain the temperature of precooled produce. Maximum acceptable loading temperatures are commonly and closely controlled in order to help ensure that fruit will reach the destination in good condition following refrigerated transport. Even the cool storages do not have a higher refrigeration capacity for faster removal of heat as required for precooling. Hence, fruit to be stored should also be preferably precooled. Precooling involves precise management of temperature and relative humidity for optimum results.

The temperature of citrus fruits at harvest is close to that of ambient air and can be as high as 40°C or more for the produce held in direct sunlight on summer days. The quicker the temperature is reduced to the optimum storage temperature, the longer is the storage life of the produce. In case of citrus fruits, precooling is done after postharvest treatments and packaging in to the containers. Precooling essentially refers to rapid heat transfer from commodity to cooling medium. Prompt precooling inhibits the growth of decay caused by microorganisms, restricts enzymatic and respiratory activities, inhibits water loss, and reduces ethylene production.

I. TEMPERATURE, RH, AND VAPOR PRESSURE DEFICIT

The temperature of the surrounding air and hence that of the commodity is the most important variable because it influences metabolic rate and thus an array of chemical and biophysical changes. Lower temperatures (within reasonable limits) are beneficial. Temperature also has a pronounced effect on the behavior of water vapor in the air. Dry- and wet-bulb temperatures, relative humidity (RH), and dew point are common psychrometric variables. These variables are related – a change in one variable is bound to affect another. Understanding these variables is necessary for the postharvest management of citrus fruits. As the temperature decreases, the vapor pressure of the air also decreases, although RH is 100 percent with the difference in dry- and wet-bulb temperatures being zero. This also indicates that at lower temperatures, air has a lower water-holding capacity. That is why air at higher temperatures (warmer air) loses water when cooled. When cooled fruit is brought out of storage, the air in immediate contact with the fruit loses water as it cools. Water condenses in the form of droplets on the fruit surface. At lower temperatures, if the difference in dry- and wet-bulb temperature increases, the RH decreases and vapor pressure decreases. With a decrease in the vapor pressure of the air, the deficit increases, leading to moisture loss in the fruit.

The amount of water vapor (WV) in air – that is, the psychrometric state of the atmosphere – can be specified either by the water content or by its vapor pressure (VP) and either in absolute or relative terms. RH is the ratio of water vapor pressure in the air to saturation vapor pressure at the same temperature, expressed as a percentage. RH can be compared only at the same temperature and barometric pressure. The absolute or (specific) humidity is the measure of the weight of water vapor contained in a known weight of dry air.

$$RH = (P/P_0)\, T \times 100 \text{ percent}$$

P = Actual WV pressure of air at temperature T.

P_0 = Saturation vapor pressure at the same temperature T.

Saturation VP is the VP of water in equilibrium with a free water surface. Alternatively it can be defined as the pressure exerted by the maximum amount of water that can be contained in air at a given temperature. Dew point is the temperature at which saturation occurs when air is cooled without change in water content. Change in air temperature above the dew point does not affect water content, but cooling below dew point removes moisture from the air through condensation on cooler surfaces. That is why we find dew on the cooled grass in the early morning.

The direct cause of evaporation is called the Vapor Pressure Deficit (VPD). VPD represents the difference in the vapor pressure on the product's surface and that of the air around it. The higher the VPD, the greater the amount of evaporation and thus the greater the water loss from fruit. At a fixed temperature, VPD can be decreased by raising the RH in the storage chamber. If RH is kept constant, VPD may be reduced by lowering the temperature. VPD is a difference in the saturated and non-saturated air condition. Hence RH has to be increased in a storage environment to reduce the deficit.

VPD indicates difference in VP of air at saturated condition (100 percent RH) and at lower saturation (such as 95 percent, 85 percent, or 50 percent RH). For example, at 20°C VP of air at 100 percent RH is 23 millibar. (It is at saturated condition.) At 30 percent RH VP is 7.1 millibar at the same temperature. Hence the difference, or deficit (between 23 and 7.1 VP), at different RH indicates VPD. This behavior and property of water vapor in the air with respect to temperature needs to be kept in mind during the storage of fruit. (The study of this variable is called psychrometrics.) Since the fruit is detached from the mother plant at harvest, its supply of water stops at that moment. Thereafter its water level (moisture content) depends on the surrounding temperature and relative humidity of air. The prime concern in postharvest handling is to retain the fruit's water level, since with loss of water fruit also loses marketability.

Water is in vapor form in the air and in saturated air (100 percent RH) the VPD is zero. The vapor pressure is at a maximum in saturated air (23 millibar at 100 percent RH at 20°C). The difference in dry- and wet-bulb thermometer is zero at 100 percent RH. As this difference increases, RH in the air decreases and the vapor pressure in the air also decreases. (We say that the air is getting dryer.) As the vapor pressure in the air decreases, fruit starts losing water because surface cells and tissues are at maximum vapor pressure (saturated condition) because of the high water content of fresh fruit. As water in fruit evaporates, weight decreases and shriveling and softening occur, with a resultant drop in quality.

The relative humidity of the air in the precooler should be 90–95 percent so that there is no drying effect of the air, which is flowing with high velocity at a low temperature. Fiberboard and wood absorb water and may decrease RH in a room. A fiberboard box held at 50 percent RH has a moisture content of 7 percent (dry-mass basis). At 90 percent RH, the moisture content would be 16 percent (Soroka, 1995). High RH will not prevent moisture loss if the product temperature is not near the air temperature. Therefore, efforts are made during precooling to minimize the temperature difference in commodity and coolant as rapidly as possible.

For precooling, mechanical refrigeration is the most common means to achieve lower temperatures. Before the invention of mechanical refrigeration, natural ice was used in ice bunkers to cool the fruit. Cool air drawn from ice is used for precooling. Ice cannot be applied to citrus fruit directly as it causes chilling injury.

II. MECHANICAL REFRIGERATION

In mechanical refrigeration, heat is transferred by convection, conduction, and radiation. All of these heat-transfer principles play a role in cooling fruit and maintaining storage area cool. Liquefied refrigerant flowing in closed tubes absorbs heat as it changes its physical state from liquid to vapor and again to liquid with the help of a compressor, thus cooling the product kept in the storage area.

A. Refrigeration Equipment and Cooling Plant

The major parts of any mechanical refrigeration system or cooling system are: (1) compressor, (2) condenser, (3) evaporator, and (4) valves (Fig. 11.1). Cooling systems are custom-built according to capacity requirement using civil construction work (reinforced cement concrete, or RCC, structure and brick walls) or prefabricated panels of polyurethane foam sandwiched between steel sheets. Compact cooling units can be fitted to these cold rooms and detached as and when required.

Compressors are generally rated on the basis of their capacity of cooling, or tons of refrigeration. (If the capacity is 1 ton of refrigeration, as in 12000 BTU/hr, the compressor can remove 12000 BTU of heat/hr from the storage area.) One ton of refrigeration equals 3024 kcal per hr or 72576 kcal or 288000 BTU in 24 h. This is enough to melt 907 kg (one ton) of ice. The capacity of the compressor is based on piston displacement (volume) and speed of operation. Variable speed drive compressors can be used to adjust the refrigeration requirements. Spare compressors are used in cold-storage plants to take care of emergencies and heavy cooling loads.

Water cooling-type condensers are common in most ammonia-based refrigeration plants. Water is recirculated; lost water (from evaporation) is replenished automatically. In small cold rooms holding 2–3 tons of fruit, air-cooled condensers and compact sealed compressors with Freon refrigerants are easily installed.

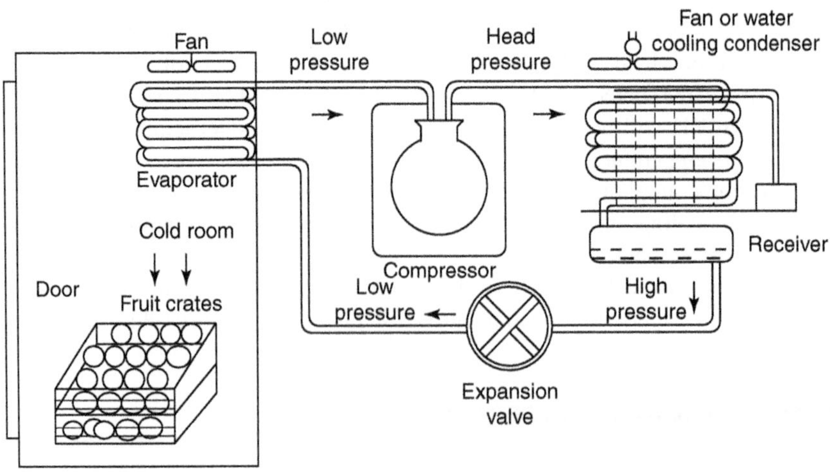

FIGURE 11.1 Schematic Diagram of Mechanical Refrigeration and Cool Store.

Ammonia is a common refrigerant used in big cold stores as it is inexpensive and has reasonably good heat-transfer capacity. Compressors are located in cold-storage engine rooms. Reciprocating-type compressors are used commonly and can be used for ammonia and Freon 12 and 22 (ASHRAE, 1972).

Evaporators consist of coils of pipe with fins to increase heat transfer and fans of required capacity to blow air through the coils or suck the air through coils. In large storage facilities, air is circulated through the coils with the help of ducts. The movement of air past the coils facilitates heat transfer.

Evaporators should be large enough and the temperature difference in room air and refrigerant in the coil should be minimal at the time of fruit loading. It is also advisable to bring down the commodity temperature in stages if it is loaded directly into the storage room. This is done using a thermostat of the cooling unit so that temperature difference between cooling coils and room/commodity temperature is minimized. The storage room is also cooled before the commodity is stored. When the difference between the temperature of the cooling coil and the room air is greater, the commodity may lose moisture and dehydration takes place. The amount of air circulated through the evaporator is generally about 0.007–0.008 cum/min/kcal. The load should be stacked to provide aisles for air movement.

If the evaporator coil temperature is lower than the air temperature in the chamber, water condenses on the evaporator coil and fins of the refrigeration system, leading to frosting and reduced air delivery. This water is taken from the air of the cool chamber and also from the commodity.

The pressure of the refrigerant is controlled in the system. A thermal expansion valve operates under high pressure. A hot gaseous refrigerant is condensed in the condenser. (The condenser is air- or water-cooled.) Before the liquid enters the evaporator coil the cut-off valve operates by thermostat control. The evaporator coil operates under low pressure and refrigerant gets vaporized by absorbing heat. Gas is compressed by the compressor and again circulated to the condenser. The entire system is gas-tight; any leakage results in inefficient cooling or no cooling at all. In large cold-storage plants, the back pressure valve and head pressure valve can be adjusted to improve cooling capacity/efficiency.

B. Construction of Cold Storage Rooms

The flooring and walls are constructed on reinforced concrete foundations to facilitate the use of heavy palletized load of forklifts. Insulating materials such as expanded polyurethane and expanded extruded polystyrene are used for low thermal conductivity (0.020–0.057 W/m-k). The best quality insulation is required in the roof/ceiling, as the maximum heat is likely to enter through the top. Insulation requirements are greater in hot, tropical areas than in regions with relatively cool climates. Vapor barriers such as resinous coatings, metal foils, and plastic barriers are also used to minimize the penetration of moisture. Water-impervious insulating materials are also used, serving both purposes. Foamed-in-place polyurethane

TABLE 11.1 Thermal Properties of Insulating and Construction Material Used in Cold Storage

Material	Thermal conductivity W/m-K	Specific heat KJ/kg-K	Density kg/m^3
Cork	0.040	2.0	140–170
Cork granules	0.176	2.03	220–260
Cork plate	0.042–0.058	1.76	148–198
Glass wool	0.042–0.059	–	154–206
Ply wood	0.15	2.51	600–700
Expanded polystyrene	0.020	0.0326	10–40
Polyurethane foam	0.025	–	20–40
Hard rubber	0.157	1.38	1200
Saw dust	0.25–0.26	–	210
Sand	1.16	0.80	1400–1900
Concrete	1.55	0.84	2200
Brick work with mortar	0.81	0.88	1700

Prasad (1989), Pereira (2007).

is used in many cold-storage facilities, as it is versatile and can meet any type of insulation requirement. Sawdust was once used widely in cold storage; thereafter it was replaced by thermocole and then polyurethane foam. The thermal conductivity, specific heat, and density of insulating material vary; their use is primarily determined by cost and the type of insulation required (Table 11.1). Air ducts and blowers are provided for air circulation in large storages. The joints must be effectively sealed with a waterproof silicon sealant so that air leakages and the entry of water is avoided. Care needs to be taken for defrosting when temperatures are kept at 1–2°C.

Prefabricated insulated stainless steel panels are assembled at the site and interlocked to erect the cold-storage structure in a very short time. These storage structures have single floors as opposed to the multi-floored cold storages constructed with civil RCC and brickwork.

The refrigerated containers, or reefers, are of an integral type that has a refrigeration unit operating on diesel engine or electricity. The porthole-type reefers have to be supplied with cool air from a central unit on land or on a ship since they are not fitted with their own units.

Air circulation and air velocity in storage areas are very important since water loss depends mostly on RH, temperature, and air velocity. Spatial variation of no more than ±1°C is required for good temperature control. Particularly at lower temperatures (below 5°C), a variation of 1°C can have significant effect on fruit quality – more so than at higher temperatures (10–15°C) during long-term storage. Humidifiers (spinning-disc or sonic type) are sometimes used to add water in the form of a very fine mist in the stream of air. However, if the dew point of the air is reached, the moisture is likely to condense on the evaporator, necessitating defrosting. Very sensitive humidistats are required to regulate humidifiers, particularly at a high RH range (85–95 percent).

Ventilation of the storage area is a must to remove accumulated ethylene and carbon dioxide. The inspection of stored fruit boxes is necessary in citrus storage because most of the mold pathogens grow slowly at low temperatures. Lemon storage rooms (14–15°C) are usually ventilated but this may increase the refrigeration load. A ventilation facility is created during the construction of large storage rooms so that fresh air is let in at least during the night when ambient temperatures are low.

The recording and display of temperature and RH (preferably digitally) is necessary for monitoring and troubleshooting during the operation of cooling/refrigeration units. The measurement of dry- and wet-bulb temperatures also provides RH values from the psychrometric charts. Modern humidity-control systems provide a digital display of RH outside the storage room.

Usually the room is cooled to the desired temperature before loading the fruit, and doors are not opened frequently. It is advisable to cool fruit before it is placed in large storage rooms. While loading the fruit, a required space of 8–10 cm between walls and pallets as well as between pallet rows is necessary.

C. Heat Load and Thermal Properties

Heat to be removed from the fruit is measured in British Thermal Units (BTUs in the British system) or calories or joules (in the metric system). (1 K calories = 1000 calories). In order to calculate the heat load of the produce and other packaging material for mechanical refrigeration, the following factors should be known: specific heat, the weight of the commodity, and the temperature difference. The number of people working, the number of forklift trucks operating, the number of electric lamps, and how often doors open and close are all considered when calculating heat load. The specific heat of commonly used material for packing, such as wood and paper, is 0.30 BTU/Ib/°F. The thermodynamic properties of various citrus fruits are given in Table 11.2. Before the advent of mechanically refrigerated transport vehicles, ice bunkers were used to cool the fruit during transport. In these specially designed vehicles, air is forced through ice and then over the produce and then recycled over ice again. One ton (2000 lbs) of ice absorbs 288 000 BTU of heat to melt completely.

For example, under ideal conditions, the heat transfer coefficient of water to apples is about 120 for water as compared with 6–10 for air (air to apple). This means that if the temperature difference between the fruit surface and water or air is 1°F, the heat transferred per square foot of fruit surface is 120 BTU in water and 6–10 BTU h in air. This shows that coefficient of heat transfer is low in air-cooling systems compared to water-cooling systems and thus heat transfer or rate of cooling is slower in air cooling. If air at a low temperature is used with high velocity, the coefficient of heat transfer can be increased considerably. With

TABLE 11.2 Water Content and Thermal Properties of Various Citrus Fruits

Fruit	Water (%)	Sp. Heat (KJ.kg.°C)	Specific gravity	Thermal diffusivity α(m²/s)	Thermal conductivity W.m-K	Latent heat kJ/kg
Grapefruit	89.1	3.82	0.88	0.90×10^{-7}	0.457	–
Lemon	87.4	3.76	0.95	1.25×10^{-7}	0.431	295.40
Lime	89.0	3.81	–	–	–	–
Mandarin	87.3	3.76	–	–	–	–
Orange	86.4	3.73	0.96	0.95×10^{-7}	0.559	–

Grierson (1976), Grierson (2002), Turrell and Perry (1957), USDA (1986), Sp. Heat (kJ.kg. °C) = 0.0335 × (water content) + 0.8374.

increased air velocity, more fruit surface area is brought into contact with the cooling medium, thus increasing the cooling rate. This is the main principle of pressure cooling, or forced-air precooling. Here again air has to be at a high RH to avoid fruit desiccation.

III. PRECOOLING

Methods of precooling include room cooling, forced-air or pressure cooling, hydrocooling, evaporative cooling, and vacuum cooling. Evaporative cooling has limitations as the temperature can be lowered only a certain amount. Citrus fruits are amenable to forced-air cooling because of the nature of the commodity. The cooling rate in hydrocooling is actually faster than forced-air cooling, but because of the risk of decay it is not used for citrus. Moreover, packed fruits cannot be subjected to hydrocooling. Vacuum cooling is also not suitable for citrus fruits because the surface area is less when compared to weight or volume. Cooling of fruit depends on the following factors:

1. Initial and final temperature of fruit.
2. Temperature of cooling medium.
3. Accessibility of fruit to cooling medium.
4. Surface-to-mass ratio of fruit.
5. Thermal properties of fruit and cooling medium.
6. Volume and velocity of cooling medium.

Citrus fruits are normally cooled with cold air in forced-air precoolers, either as a batch or continuous-flow system. It is technically and economically feasible and being practiced commercially worldwide. The rate of cooling with cold air may be significantly increased if the surface area available for heat transfer is enlarged by forcing air through packages and thus around each fruit rather than only over the surfaces of packages. Such 'pressure cooling' can cool fruit in about 10–25 percent the time required for room cooling. Pressure cooling involves passing cold air along induced pressure gradients through warm produce in specially vented containers. The pressure differential is created by a blower that pulls the cold air through the packed fruit (creating negative pressure around the fruit). The pressure differential between opposite faces of packages ranges from barely measurable to 20–25 mm of water head. Airflow rates can vary from 1 to 2 lit/s/kg. The speed of cooling can be adjusted by varying the rate of airflow. Immediate or subsequent physiological injury to citrus fruit from rapid forced-air precooling is negligible provided that the surface temperature does not go below −2.0°C (Soul et al., 1969). Desiccation of fruit is insignificant within limits. For maximum performance, the precooler should always be loaded up to the capacity and all the air should pass through the product voids.

During forced-air cooling, the following points need to be recorded.

1. Air velocity – Meter per minute or seconds.
2. Volume of air – Cubic meter per minute
3. Pressure difference (mm of water) in air plenum and storage area.
4. Temperature of air reaching the fruit.
5. Weight (kg) of the commodity being cooled.
6. Temperature of the commodity during cooling and at beginning, half cooling time, and 7/8th cooling time.

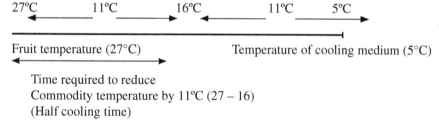

27°C ← 11°C → 16°C ← 11°C → 5°C

Fruit temperature (27°C) Temperature of cooling medium (5°C)

Time required to reduce
Commodity temperature by 11°C (27 – 16)
(Half cooling time)

Cooling rate decreases as cooling progresses, as the temperature of the commodity approaches that of the cooling medium. Cooling time is recorded as half cooling (time required to remove one-half the difference between the initial fruit pulp temperature and the cooling medium temperature) and 7/8th cooling (Fig. 11.2). Seven-eighths cooling time is a practical estimate of cooling in commercial operations, since 87 percent of the temperature difference between commodity and coolant is removed. From graphs for a particular system cooling time can be predicted.

A. Effect of Precooling on Various Citrus Fruits

Among various methods of precooling that have been tried, forced-air cooling at 6–7°C and 90–95 percent RH with airflow of minimum 0.1 cum/min/kg has been found to be most suitable for 'Nagpur' mandarins. A forced-air precooler

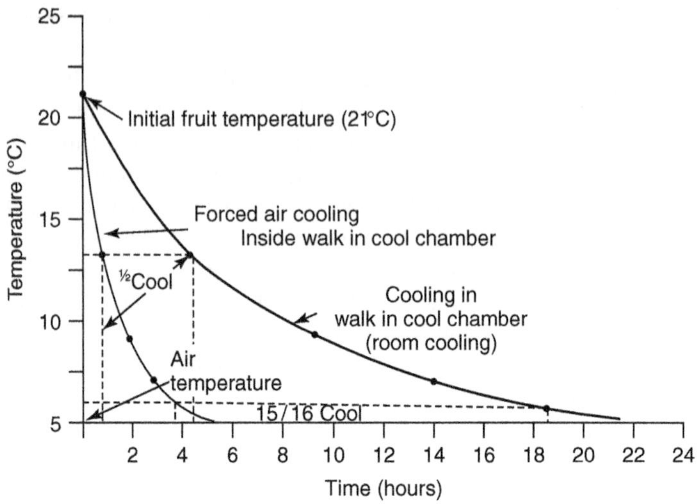

FIGURE 11.2 Cooling Rate of 'Nagpur' Mandarin Fruit Subjected to Forced-air Cooling and Room Cooling.

with more than 1/2 tons of fruit-holding capacity at a time has been designed and developed for the precooling of citrus fruits (Ladaniya, 1995). Once the precooling is done, the cold-chain must follow until retail marketing. Transporting precooled fruit in non-refrigerated trucks serves little purpose because rewarming takes place within hours. Precooling fruit before storage helps to minimize weight loss and spoilage and to retain better quality. Precooling is detrimental if fruit is to be fumigated. Water on the surface causes peel injury during fumigation. Mandarins have been shown to derive more benefit from precooling than oranges and grapefruits.

B. Packaging Containers and Their Alignment during Precooling

Container size and ventilation usually varies, which may affect cooling rates. The acceptability of the container depends on the extent of venting for rapid cooling. Roughly, venting 4 percent of the side areas of CFB containers results in moderately fast cooling, while containers with less than 2 percent vents do not cool much faster than non-vented containers (Mitchell et al., 1972). Citrus fruit packed in wire-bound boxes or ventilated fiberboard cartons cools at about one-half the rate of fruit in ventilated bulk bins (Soul et al., 1969). Fruits in polyethylene bags or other wrappers cool slowly. Talbot and Baird (1990) observed that a major disruption of air flow occurs if box vents do not match. A few large holes provide more sufficient air passage than many small holes.

Studies with 'Nagpur' mandarin using 50 × 30 × 30cm corrugated fiberboard container indicated that a ventilation level of 5–6 percent on side panels reduces cooling time by 35–50 percent as compared with cooling in boxes having 2 percent vents during forced-air cooling (Ladaniya and Shyam Singh, 2000) in

TABLE 11.3 Air Velocity Inside the Air Plenum during Precooling

Ventilation on sides of container (%)	Air velocity (m/min) Containers and stacking pattern 8*	16#	16**	32+	Loss in weight (%) 8*	16#	16**	32+
2.0 (CFB box)	66–72	60–72	78–84	48–54	0.29	0.35	0.21	0.77
3.0 (CFB box)	77–90	60–72	90–96	60–66	0.19	0.41	0.33	0.88
6.0 (CFB box)	90–108	72–84	90–96	84–90	0.15	0.30	0.16	0.38
22.56 (Plastic crate)	150–180	180–192	192–210	90–120				

Ladaniya and Shyam Singh (2000),
* Stack width 30 cm; height 120 cm; length 50 cm,
# Stack width 60 cm; height 120 cm; length 50 cm,
** Stack width 30 cm; height 120 cm; length 100 cm,
+ Stack width 60 cm; height 120 cm; length 100 cm,
Note: Air velocity values in the centre of the plenum tunnel.

different stacking patterns without changing refrigeration capacity. The upstream fruit in the stack cools earlier than downstream fruit near the air tunnel. Increasing ventilation from 2 percent to 6 percent on side panels increases air flow (m^3/min/kg fruit) in CFB boxes and thus reduces cooling time. Very high air velocity and air flow in plastic crates from a higher percentage of vented areas on side panels (22.56 percent) do not necessarily increase cooling because the air may bypass the fruit in open crates (Table 11.3).

Besides ventilation, stacking patterns also greatly influence the air movement, cooling time, and water loss. Longer cooling times also increase the cost of precooling. Therefore, the proper stacking of boxes for the alignment of vents and the reducing of air leaks are the prime requirements of precooling (Fig. 11.3). Sealing air leaks to force the air through fruit increases the efficiency of the cooling system (Talbot and Baird, 1990; Ladaniya and Shyam Singh, 2000).

Depth of fruit parallel to the direction of airflow should be from 12 to 24 in (Soul et al., 1969). During forced-air cooling of 'Nagpur' mandarins, a stack width of 30 cm resulted in higher air velocity as compared with 60 cm stack width (Table 11.3). Higher velocity resulted in better cooling. More depth of fruit means an increase in fan capacity. Mitchell et al. (1972) reported that with higher stack width, static pressure requirements (fan capacity) are higher for efficient cooling of downstream fruit, although floor space can be used efficiently. Performance of the precooler is found to be optimum when loaded up to its maximum capacity and air pass through the product voids.

Desiccation of fruit is insignificant within limits if RH maintained and cooling time is not long. Higher RH (about 90–92 percent) maintained during precooling minimizes fruit weight loss (Table 11.3) in spite of high air velocity. Some weight loss of fruit does occur, which can be due to high water-vapor pressure of the fruit as compared to the refrigerated air (Talbot and Baird, 1991).

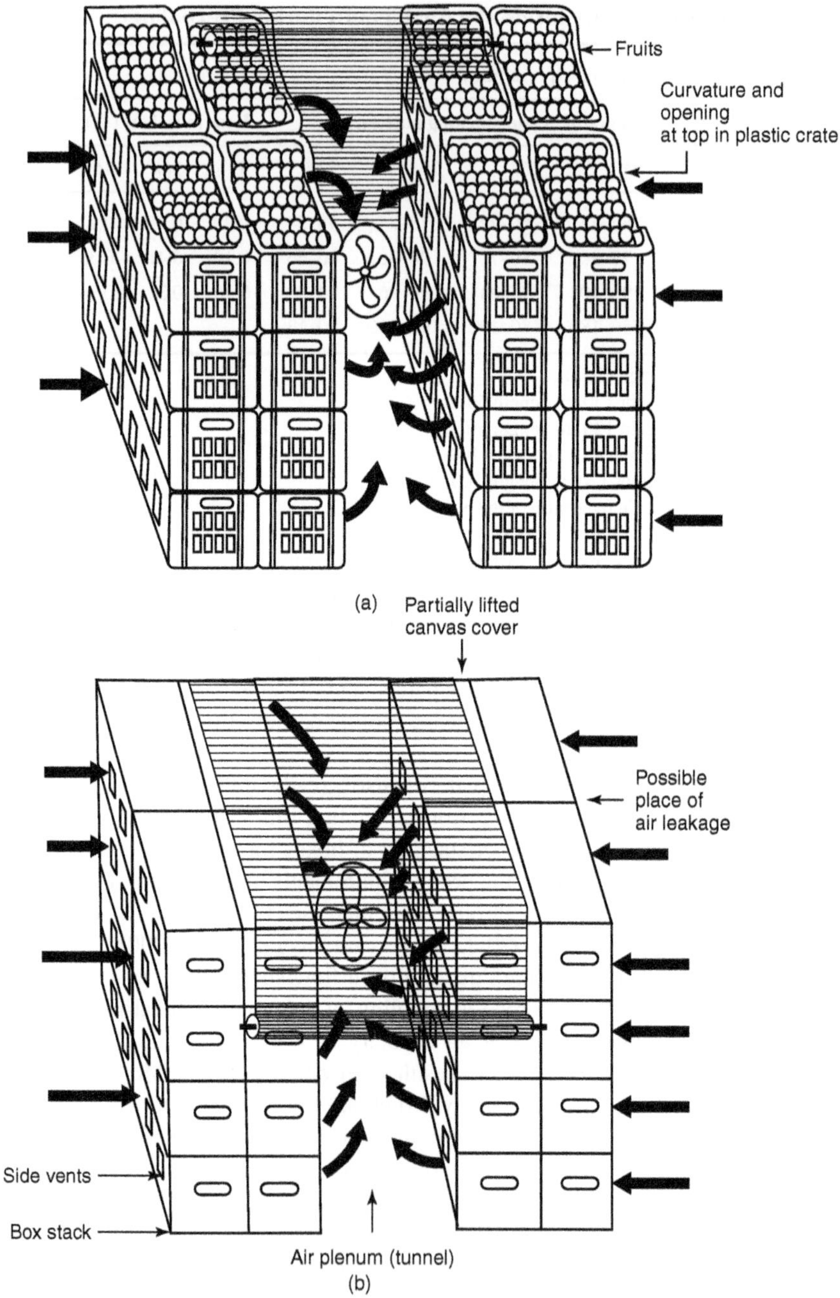

FIGURE 11.3 Air Flow Pattern in Forced-air Cooling of 'Nagpur' Mandarin Fruit: (a) Air Flow Through Side Vents and Top Opening of Plastic Crates (b) Air Flow Through Side Vents of Corrugated Cartons. (Ladaniya and Singh, 2000).

REFERENCES

ASHRAE (1972). Compressors (equipment). *ASHRAE guide and data book.* Chapter 2, pp. 101–126.

Grierson, W. (2002). Fruit development, maturation, and ripening. In *Handbook of plant and crop physiology.* (M. Pessarakli, ed.). Marcel Dekker, New York, Basel, pp. 143–159.

Grierson, W. (1976). Preservation of citrus fruit. *Refrig. Serv. Engineers Soc. Serv. Appln. Manual.* Section 7, 10 p.

Ladaniya, M.S. (1995). Studies on precooling of 'Nagpur' mandarin fruit. *Haryana J. Hort. Sci.* 24(3–4), 212–218.

Ladaniya, M.S., and Singh S. (2000). Influence of ventilation and stacking pattern of corrugated fibreboard containers on forced-air precooling of 'Nagpur' mandarin. *J. Fd. Sci. Technol.* 37, 233–237.

Mitchell, F.G., Guillou, R., and Parsons, R.A. (1972). Commercial cooling of fruits and vegetables. University of California, Expt. Stn. Extn. Service Manual 43, pp. 1–44.

Pereira, H. (2007). Cork: Biology, Production and Uses. Elsevier, Amsterdam, The Netherlands, pp. 336.

Prasad, M. (1989). *Refrigeration and air conditioning data book.* New Age International (P) Ltd., New Delhi, 304 pp.

Soroka, W. (1995). *Fundamentals of packaging technology.* Institute of Packaging Professionals, Herndon, VA.

Soul, J., Yost, G.E., and Bennett, A.H. (1969). Experimental forced air precooling of Florida citrus fruit. *Mktg. Res. Rep.* US Department of Agriculture, 845, 27 pp.

Talbot, M.T., and Baird, C.D. (1990). Evaluating commercial forced-air precooling. *Proc. Fla. Sta. Hort. Soc.* 103, 218–221.

Talbot, M.T., and Baird, C.D. (1991). Psychrometrics and postharvest operations. *Proc. Fla. Sta. Hort. Soc.* 104, 94–99.

Turrell, F.M., and Perry, R.L. (1957). Specific heat and heat conductivity of citrus. *Proc. Am. Soc. Hort. Sci.* 70, 261–263.

USDA (1986). The commercial storage of fruits, vegetables and florist and nursery stock. *Agric. Handbook.* No. 66, 130 pp.

12

STORAGE SYSTEMS AND RESPONSE OF CITRUS FRUITS

Nearly two-thirds of citrus fruits produced in the world are consumed as fresh produce. This trend clearly indicates the importance of preserving the natural qualities of fresh citrus after harvest, either for domestic market or for export. Storage is the most important operation during the marketing of fruit. All operations, including harvesting, pre- or postharvest treatments, packaging, transportation, and temperature and humidity management during handling influence the storage life of fruits. For reasonable storage life, fruit has to be properly protected from decay losses, and the moisture loss needs to be slowed to an extent that facilitates a fresh appearance. The stress of water deficit results in desiccation, leading to wilted rind and a shriveled appearance. Peel wilting has an adverse effect on gaseous exchange and metabolic alterations, because enzymatic activity and ion flux are also altered by the loss of cell turgor. In general about 5–6 percent loss of water is considered acceptable before the citrus is unmarketable. Fruit loses aroma and flavor during prolonged storage even if decay and moisture loss are checked and commodity appears acceptable. The fruit is no longer palatable, although its external appearance is acceptable. Research has been going on to understand the basic metabolism of the fruit and thereby delay or slow down biochemical reactions that lead to unacceptable flavor development during prolonged storage.

Citrus fruits mature and ripen on the tree. Thus the stage of maturity not only affects fruit quality at harvest but also the storage life. Fruits harvested too early are more sensitive to chilling than late-harvested fruit. Late harvesting causes deformation of mandarins and grapefruits in storage as the fruit becomes soft. Long-term storage under high-humidity conditions leads to the puffing of Satsuma mandarins. In sweet oranges and mandarins, delayed harvesting causes the granulation of juice vesicles, particularly in larger fruits. Late harvesting also causes senescent disorders of the rind. Late-harvested yellow-colored acid limes (*Citrus aurantifolia* Swingle) are more prone to aging and rind breakdown than green fruits during storage. Late-harvested mandarins do not store well. Green peel or peel with color break is considered 'young', and such fruit has better storage life. This is true in the case of limes and lemons.

Compared with many subtropical fruits, citrus fruits have a relatively long postharvest life. Several storage systems have been tried and many of them are being commercially used. In most developing countries, citrus fruits are handled, marketed, and stored under ambient conditions with much less commercial storage in refrigerated conditions. In developed countries, after harvesting and postharvest treatments, fruits enter to the cool-chain and remain at lower temperatures until they are consumed. Handling and storage facilities and marketing structures vary greatly among the citrus-producing countries. Storage techniques are being improved to extend the storage life of these fruits. Efforts are made in this chapter to present an overview of storage systems, fruit storage, and various qualitative changes that occur.

I. STORAGE SYSTEMS

The increasing demand for and higher production of citrus fruits has resulted in a greater emphasis on the design, construction, and management of storage houses. In Japan and China, since ancient times, simple and open storage houses were constructed for short-term storage. Simple-framed structures of storage houses were inexpensive and normally constructed within or around the orchard for temporary storage during harvest. The common storage houses currently in use are an improved and expanded design of simple-framed storage houses. They contain 2–4 rooms with ventilation (Sangey et al., 1999). Different types of evaporative cool structures, stone quarries and abandoned caves, are still used in some parts of Asia. Refrigerated, low-temperature storage houses are relatively modern storage facilities. The improved design and installation of sophisticated equipment in refrigerated storage houses have enabled the long-term storage of large quantities of citrus fruits. Controlled-atmosphere (CA) storage is the latest innovation, but prohibitive costs deter their use. Different types of citrus fruits respond in different ways to high-tech CA storage. The technology has thus enabled growers to meet the high demand for quality citrus fruits year-round.

At present, refrigerated storage is the storage system being used extensively commercially to extend marketing time. Alternatives to refrigeration systems are evaporative cool storage, cellar storage, underground tunnels, and quarries and caverns. Fruit is also being stored under ambient conditions for short durations wherever ambient temperatures are low. Several other storage systems being tried for storage of citrus include (1) controlled-atmosphere storage, (2) hypobaric storage, (3) storage at chilling temperatures with intermittent warming and other treatments. A modified atmosphere can be achieved by wrapping fruit with permeable films. This method is known as modified atmosphere packaging (MAP). MAP and seal packaging are already being used in many countries on large scale to extend the shelf life of citrus fruits (see Chapter 10).

Each variety and cultivar has different optimum conditions for storage depending on its tolerance to low temperature, high humidity, low O_2, high CO_2, ethylene level, and mechanical injury.

A. Refrigerated Storage

It has become possible to market fruit long after harvest only through mechanical refrigeration systems. Fruits such as lemons are harvested in winter month but held until summer – that is, for almost 6 months – when demand is high. In earlier days, ice was used to prolong the shelf life of fruit. In the U.S., commercial cold storages started in the 1880s, while in Britain, refrigerated storages started even before. In India, cold storage started on a commercial basis in cities such as Bombay, Calcutta, and New Delhi in the early twentieth century. With growing population and urbanization, large refrigerated warehouses became a necessity.

The climate during cultivation of fruit affects its response to low temperature during storage. In addition to temperature and relative humidity during storage, pre- and postharvest factors, are often critical to success or failure during storage. The quality of the fruit to be stored, its maturity, and careful handling are important. With oranges, a curing at 30°C with relative humidity of 90–95 percent before degreening promotes the healing of minor injuries by lignification. The use of the approved fungicides as postharvest dips or in-package volatiles is desirable. The storage life of citrus fruits of different species varies, and different varieties of same species also respond differently to storage conditions.

Specific conditions are necessary for the storage of each cultivar, as fresh fruit is likely to develop off-flavor, lose fresh appearance, and marketability. Longer storage duration with more than 10–15 percent spoilage is not economical. Fruit quality and shelf life after removal from refrigerated storage are the key focuses of storage studies. Recommended storage conditions and storage life of important citrus fruits are shown in Table 12.1. Cool storage for fresh citrus is required to operate within the range of ± 1°C. The set relative humidity and the temperature

TABLE 12.1 Recommended Temperatures, Relative Humidity, and Storage Life of Citrus Fruits

Citrus fruit	Temperature (°C)	Relative humidity (%)	Storage life (Weeks)
(1) Sweet oranges			
Baladi orange (Egypt)	2–3	85–90	8
Blood Red	5–7	85–90	8–10
Valencia (California and Arizona)	3–9	85–90	4–8
Valencia (Florida and Texas)	1–2	85–90	8–12
Valencia, Navel (Spain)	2–3	85–90	4–8
Washington Navel (California)	5–7	85–90	6
Shamouti	5–6	85–90	8–12
Malta, Mosambi	5–7	90–95	12
Sathgudi	5–7	90–95	12
(2) Mandarins, Tangerines and hybrids			
Ponkan	4–6	85–90	4–5
Satsuma	2–3	80–85	12–18
Clementine	3–4	85–90	4–6
Clementine (Spain), Lee	4–5	85–90	4
Ellendale, Murcott			
'Nagpur'	6–7	85–90	6–8
'Coorg' (main crop)	5–6	85–90	6–8
Temple, Orlando, Dancy	4–6	90–95	2–5
Kinnow mandarin	3–4	85–90	8–12
(3) Lemons			
Dark green	13–14	85–90	16–24
Light green	13–14	85–90	8–16
Bright Yellow	10–12	85–90	3–4
Sicilian lemon	7–8	85–90	8
Australian lemon	10	85–90	24
California lemon	12.5–15	85–90	8–12
(4) Limes			
Tahiti	9–10	90–95	6–8
'Kagzi', Mexican	9–10	90–92	10–12
Dark green			
Yellow	8–9	90–92	8–12
(5) Grapefruit	10–14	90	6–8
Texas and Florida grapefruit	10	90–92	4
California and Arizona grapefruit	14–15	90	5–6
Foster, Ruby, Saharanpur Special	9–10	90	16–20
(6) Pummelo	8–9	85–90	10–12

Chase et al. (1966), Wild and Rippon (1973), Ryall and Pentzer (1974), El-Ashwah et al. (1975), Bleinroth et al. (1977), Hardenburg et al. (1986), Ladaniya and Sonkar (1996), Ladaniya (2004a,b).
Note: Various supplementary treatments as mentioned in Chapters 9, 10, and 16 are to be used during storage at low temperatures. Fruit held for short term in pallet boxes can be kept at 90–98 percent RH, while 85–90 percent RH is recommended for fruit in CFB boxes.

have to be uniform throughout the storage room and over the period of storage time. At lower relative humidity (66–65 percent) fruits exhibit higher transpiration rates, earlier senescence, and greater deterioration in visual appearance than fruits stored at higher humidity (90–95 percent). Internal CO_2 increases and O_2 decreases rapidly under low RH conditions. D-Aquino et al. (2003) observed that if elevated levels of humidity are applied and fruits are adequately protected against decay with an effective fungicide, fruits can be stored at relatively higher temperatures (5–10°C higher than normally used) with the advantage of saving energy and reducing the risk of physiological changes caused by prolonged exposure to low temperatures. Air ventilation and air movement are other important parameters of storage. Fresh air intake for ventilation is considered important; fresh air equal to one volume of empty storage area per hour is essential.

1. Sweet Oranges

In most countries with subtropical climates, where fresh oranges of different maturity periods are available for 6–8 months, long storage is not common except for the international trade. Sweet oranges, in general, can be stored at 2–7°C for 8–12 weeks, depending on the cultivar and the producing area. These fruits are relatively sensitive to chilling and subject to pitting at temperatures below 2–3°C. Green and non-ethylene-treated fruits also develop color at 15–25°C. Fairly acceptable color can be obtained if oranges are held at 15–25°C for longer periods. However, weight loss and decay are higher at these temperatures. Hence, a suitable storage temperature schedule is necessary depending on the duration of storage and the response of the cultivar to those temperatures.

Sathgudi oranges grown in Andhra Pradesh, India, can be stored up to 16 weeks at 5.5–7.2°C (Singh, 1971). Malta and Mosambi oranges grown at Saharanpur (India) can be more satisfactorily stored at 4.4–6.1°C than at 2.2–3.9°C (Mukherjee and Singh, 1983). Mosambi sweet oranges grown in central India (Nagpur) cannot be stored at 3–5°C for long periods because chilling injury develops. The optimum temperature for storage of Mosambi is 5.5–7°C, and fruits can be stored up to 90 days in acceptable condition (Ladaniya, 2004a).

California Navel orange are sometimes stored for 2–3 weeks during winter and spring to balance supply with demand. These oranges are susceptible to chilling injury, so temperatures above 5°C are often employed. Oranges tend to have a shorter storage life as the harvest season progresses. Decay and off-flavors develop quickly in storage because the fruit remains on the tree longer after reaching commercial maturity, but the fruit harvested earlier is somewhat more susceptible to chilling injury. Valencias grown in the desert regions of the Arizona and California and harvested in March develop rind pitting when stored at 3°C but remained in good condition for 20 weeks when stored at 9°C. On the other hand, Valencias from the same areas but harvested in June developed decay when stored at 9°C but remained in good condition when held at 3°C (Khalifah and Kuykendall, 1965). Florida- and Texas-grown Valencia oranges can be stored with minimum decay and rind pitting for 8–12 weeks at 1°C with 85–90 percent relative humidity (Hardenburg et al., 1986).

Navel and Valencia oranges grown in Australia develop surface pitting after a few weeks at temperatures below 4.5°C. Storage at 7°C for Valencias and early-picked Navels and 4.5°C for mid- or late-season Navels has been suggested. Careful harvesting and handling is necessary to minimize oleocellosis rind blemishes in Washington Navel oranges; a carriage temperature of 10°C for 2 weeks followed by 5°C is recommended to minimize mold wastage, weight loss, and rind blemishes caused by chilling during shipment to U.S.A. (Tugwell and Chvyll, 1995).

Valencia late oranges grown in Sardinia, Italy, can be stored well for 2 months at a temperature of 6°C with 85 percent RH (Agabbio et al., 1982). Imazalil (1500 ppm) in wax provided best protection against *Penicillium*, *Botrytis*, and *Alternaria* rots.

Israeli-grown Shamouti oranges develop chilling injury (CI) at 6°C and 90 percent RH during storage up to 12 weeks followed by 2 weeks at 17°C. At 12°C, the fruit do not develop chilling. Shamouti is more sensitive to CI than Valencia.

The Tarocco blood-red oranges grown in Italy are also sensitive to low temperatures, and chilling injury develops between 2 and 8°C, while rotting is higher at 12°C during prolonged storage of 12 weeks. A very high accumulation of volatiles (ethanol and acetaldehyde in juice) has been reported in fruit stored below 6°C (Table 12.2). The endogenous ethylene levels were higher at higher temperatures. When the fruit was removed from cold storage, higher biochemical changes, respiratory rate and volatile evolution were noticed (Schirra and Chessa, 1988). Chilling injury resulted in a greater development of volatiles.

Ethylene accumulation induces stem-end decay and off-flavor in stored Valencia oranges. Blue-mold and green-mold rots are extensive in fruit stored at relatively higher temperatures for long periods. Export shipments of Valencia fruits taken from cold storage is not recommended, especially with fruits packed at the end of the harvest season (Cruz, 1995).

TABLE 12.2 Respiration Rate, Ethylene, Ethanol, and Acetaldehyde Levels in 'Tarocco' Orange Fruit Stored at Low Temperatures and After Shelf Life (5 Days at 20°C; 60–70% RH)

	Respiration ($mgCO_2/kg/h$)		Ethylene (ppm)		Acetaldehyde (mg 100 ml)		Ethanol (mg/100 ml)	
Temp. (°C)	At low temperature	After shelf life	At low temperature	After shelf life	At low temperature	After shelf life	At low temperature	After shelf life
2	4.63	17.48	0.27	0.43	0.48	1.17	49.63	221.61
4	5.59	20.92	0.23	0.43	0.51	1.22	80.74	309.56
6	6.43	19.82	0.35	0.84	0.45	0.96	45.07	174.36
8	7.41	13.73	0.27	0.84	0.42	0.84	48.07	141.93
12	7.04	16.41	0.90	1.83	0.39	0.60	42.85	135.42

Schirra and Chessa (1988).

Ventilation rates in the storage room influences the internal atmosphere and thus the incidence of decay. The development of decay, loss of typical orange flavor and occurrence of off-flavor are all associated with the presence of ethylene during storage of Valencia oranges for 12 weeks at 10°C (McGlasson and Eaks, 1972).

Rates of ventilation affected the CO_2 concentrations more than the O_2 levels of both the external and internal atmospheres of the orange cultivars Shamouti and Valencia. In small-scale tests, ventilation rates as low as 10 percent per hour of the empty volume of the storage space did not cause major changes in gas composition, nor did they affect fruit quality adversely. In commercial tests, however, ventilation up to 70–100 percent per hour was needed to achieve similar results. A balance in ventilation rates and cooling temperatures is necessary to lower the costs of refrigeration (higher ventilation can cause an increase in cooling requirement) while still maintaining good fruit quality. A reduction in the ventilation rate in commercial citrus storage rooms from 150 to 200 percent per hour to 100 percent per hour is recommended (Waks et al., 1985).

2. Mandarins/Tangerines

Tangerines/mandarins and their hybrids (tangelos, tangors, Temples) are specialty fruits and generally not suited to long-term storage when compared to sweet oranges, grapefruits, limes, and lemons. These fruits are very susceptible to stem-end rots and blue-mold decay. Storage of mandarins is not generally recommended, but if some storage is needed for orderly marketing it should be limited to 2–3 weeks or at the most 4 weeks (Hardenburg et al., 1986). These fruits can be stored best between 4°C and 6°C in general with slight variations in temperature range depending on the tolerance of the specific cultivar (Table 12.1). In Japan, pretreatment of Satsuma mandarins before 3–4 months of storage is a common practice. This pretreatment is given at 5–7°C at natural air circulation and low RH (60–70 percent) for 2–3 weeks. Hasegawa et al. (1990) reported that high-temperature pretreatment at 20°C and 60–70 percent RH for 2–3 weeks also resulted in the development of a deep orange color and a reduction in *Penicillium* rot. Satsuma fruits are usually reduced by 3–5 percent of their weight before storage. Pre-storage curing also reduces the incidence of puffing and chilling injury during storage. The storage life of mid-season and late-season Satsuma mandarins is relatively long. Ponkan (*C. reticulata* Blanco) and Satsumas puff up and lose quality during long-term storage at high relative humidity.

Mandarins should be marketed promptly after storage. Temples, Orlando tangelos, and tangerines have shown susceptibility to chilling injury at 0–1°C and hence are stored at 4°C. Mandarins develop surface pitting at low temperatures. Akamine (1967) recommended 2°C with 93 percent relative humidity for Dancy tangerines grown in Hawaii. Careful handling and pre-storage application of an approved decay-control treatment is necessary for tangerines intended for storage.

A ventilation facility is essential in mandarin storage – air movement reduces the level of volatiles from the fruits and removes excess water vapor in the storage room. Satsuma mandarins grown in tropical areas can be stored up to 45 days at 4°C and RH >85 percent with minimum decay losses or loss of green color (Guerra et al., 1988). Ponkan fruits grown at Miaoli, Taiwan and harvested at the optimum maturity could be stored at the optimum temperature of 12.5–15°C and had a reasonable storage-life span of 3 months. Fruits stored at 5°C for 3 months developed severe chilling injury. Fruits harvested early were of poorer quality, while those harvested late had higher decay and weight-loss rates than fruits harvested at optimum maturity (Liu et al., 1998a). The Tankan (hybrid of *C. reticulata* and *C. sinensis*) could be stored best up to 4 months at 15°C. Tankan stored at 0°C and 5°C had chilling injuries. The fruits stored at 10°C, 12.5°C, and 15°C had similar decay and water-loss rates, but some fruits stored at 10°C and 12.5°C developed off-flavors. The fruits stored at 20°C had higher rates of decay and weight loss and poorer orange-color development on the rind when compared to similar fruits stored at 10–15°C (Liu et al., 1998b).

Late-harvested Kinnow mandarins grown in Punjab (India) do not store well (Mann, 1978), whereas fruits harvested in mid-January suffer the least spoilage and weight loss (Dhillon and Randhawa, 1983). Temperatures of 4.4–6.1°C are better than 2.2–3.9°C for Kinnow mandarins grown at Saharanpur (Mukherjee and Singh, 1983). Completely orange-colored Kinnows can be stored longer and retain higher TSS and vitamin C content than green and yellow-green fruits (Chundawat et al., 1978). The activity of polymethylesterase, polygalacturonase, and cellulase enzymes increase with delays in harvesting Kinnows, and late-harvested fruits deteriorate faster in storage (Nagar, 1995).

Nagpur mandarins grown near Nagpur (central India) develop chilling injury even at 3.5–5.0°C temperature after 30 days, while fruit grown in northern India can be stored at 2.2–3.9°C without chilling injury (Fig. 12.1, see also Plate 12.1). Injury increases with extension of storage up to 90 days (Mukherjee and Singh, 1983; Ladaniya and Sonkar, 1996). Yellow-green fruits are more susceptible to chilling than orange-colored ones. Temperatures of 5.25–6.85°C and 90–95

FIGURE 12.1 Chilling Injury Symptoms on 'Nagpur' Mandarin Fruit.

percent relative humidity were optimum storage conditions without any chilling injury to fruit, but after 60 days, juice content declined and taste turned insipid even though fruit appeared acceptable from the outside. Longer storage of orange-colored fruit result in higher losses from *Alternaria* black-core rot; hence storage beyond 60 days is not recommended for Nagpur mandarins. Fruit should not be ripe orange-colored at the beginning of storage. Firm fruits with orange color on one-third to one-half of the surface have better storage life. Darjeeling mandarins store well at 4–6°C, whereas chilling occurs at 1.5–3°C (Trivedy and Mazumdar, 1982). Temperatures of 3.88–5.55°C are optimum for Coorg mandarins with a storage life of 12 weeks (Singh, 1971). Coorg mandarin fruits of main- and monsoon-season crops can be stored for 56 and 42 days at 5.5°C and 7.2°C respectively (Dalal and Subramanyam, 1970).

For Satsuma, Clementina, Lee, Clemendor, Ellendale, Malvasio, Ortanique, Murcott, Montenegrina, and Malaquina mandarins grown in Uruguay, the best cold-storage temperature was 1 ± 0.2°C. Fruit remained in good condition for 20 days at this temperature followed by ambient conditions (18–20°C) for 6 days (Puppo et al., 1988).

Malvasio is a late-maturing mandarin in Sardinia, Italy. It has good quality and excellent cold-storage ability, which makes it available for a period of 8 months on the market, when the production of mandarins in Italy is almost over. Physico-chemical characteristics change very little from January to May during harvest. Thereafter fruits can be stored for 3 months at 4°C and 90 percent RH with 1 week of marketing at 20°C and 70–75 percent RH. Fruits are resistant to chilling injury and not affected by the physiological disorders common to most mandarins (Agabbio et al., 1999).

3. Grapefruits and Pummelos

Preharvest factors critical to success in grapefruit storage include rootstock, weather during growth, tree conditions, and orchard treatments (Greirson and Hatton, 1977). Fruit maturity at harvest, harvest and packinghouse handling, and postharvest treatment all affect the success of storage. Two major problems in grapefruits storage are (1) chilling susceptibility of the fruit, which results in severe rind damage when stored below 10°C and sometimes even at higher temperatures, (2) development of stem-end and green-mold decay at temperatures favorable to spore germination and growth of rot organisms. In grapefruits (cultivar Ruby) from the exterior of canopy, juice acidity and rates of weight loss and fruit decay were lower, while the content of total soluble solids, total sugar, and ascorbic acid were higher than in interior fruits during storage at 10°C and RH 85–90 percent (Abed-el-Wahab, 1990).

If harvested and handled gently, grapefruits can be stored for 6–10 weeks under refrigerated conditions without serious spoilage. Besides fungal rot, moisture loss and deformation are the major problems. Wax emulsion with fungicides extends the storage life of fruit by reducing decay and minimizing moisture loss during transportation and storage. Rotting often increases ethylene and carbon dioxide levels inside the package, and hence such fruits need to be promptly removed. Low O_2 causes an increase in the ethanol concentration in the juice of sound fruit that results in the development of off-flavor. Grapefruit is very susceptible to chilling injury at temperatures below 10°C. Treatment with ethylene induces rind stickiness after storage for 8–12 weeks at 10°C.

Ethylene treatment for degreening grapefruits is not recommended for fruit meant for storage, because such treatments increase susceptibility to decay. Natural degreening occurs during storage at 10°C or above, so fruit that is not completely degreened at harvest will usually attain satisfactory color during several weeks in storage. Waxing the fruit before storage interferes with degreening during storage. Grapefruit harvested in October can be degreened with the good yellow color in 3 weeks if they are washed but not waxed before storage at 10°C. However, washed and waxed fruit did not degreen sufficiently during 3 weeks at 10°C or 15.5°C or after an additional holding period of 2 weeks at 21°C. A 36–38h degreening period with ethylene was necessary for development of satisfactory color if the fruit was waxed before storage (Chace et al., 1966).

Temperatures for storage have varied, depending primarily on the relative prevalence of rind disorders and decay during storage life. Temperatures in the range of 10–15°C for storage of more than 3 or 4 weeks is desirable (Table 12.1). The recommendations for California and Arizona fruit are 14.5–15.5°C and for Texas and Florida grapefruit 10°C. Grapefruit harvested in October or December develop serious rind pitting within 3 weeks at 10°C, but relatively little at 15.5°C (Chace et al., 1966). Fruit harvested later in the season develop little pitting at 10°C, so lower temperature is recommended for grapefruit harvested late (January–March). Late-harvested fruits (January and May) in Florida and Texas are not suitable for extended storage because of increased decay susceptibility. Generally, early fruit is stored at 15°C and mid-season fruit at 12–13°C. Because of the seasonal pattern of grapefruit production in major growing areas of the U.S. such as Florida and Texas, fruits are stored to extend the marketing season through the summer months when there is no harvest.

Under conditions in northern India, early harvest of grapefruit results in less loss during storage than harvesting in late December. Foster Pink grapefruits can be stored longer than Ganganagar Red grapefruits (Gupta et al., 1981a). Grapefruit cultivars Ruby, Saharanpur Special, and Marsh Seedless can be stored up to 135 days at 9–10°C (Singh, 1975).

Cuba exports Marsh Seedless grapefruit, and the optimum storage temperature was found to be 14°C for degreened fruit for a maximum period of 12 weeks. For non-degreened fruit, the optimal temperature was 12°C. Total losses were highest for fruit picked at the end of the harvest season. Changes in juice composition were generally not significant, even for extended storage periods (Rosa, 1981).

4. Lemons

Dark-green lemons store best – at this stage they remain in marketable condition for 4–6 months (Table 12.1). During storage, color changes from green to yellow, fruit becomes juicier, and flavor improves. Less-ripe fruit is resistant to decay, and by shipping greener fruit losses can be reduced considerably (Gilfillan and Saunt, 1991). Eureka lemons grown in South Africa (harvested in March–July) can be marketed in the Northern Hemisphere up to September–October. Fruits picked after color break in early April and treated with fungicides, graded, sized, and waxed in the packinghouse are stored at 10°C until August–September. These fruits are repacked in 8 kg export cartons after the removal of blemished and decaying fruits, and shipped at 11°C to overseas markets for sale during September–October. Fruit condition and color were excellent and there were low levels of soft decay of fruit 2 months after discharge from the ship. The main antifungal compound in green lemon flavedo has been identified as monoterpene aldehyde citral. During long-term storage, citral content decreases in parallel with the decline of antifungal activity in the peel and with an increase of decay incidence. The level of citral in the flavedo is reported to be related to the resistance of lemon fruit to postharvest decay (Rodov et al., 1995b).

Lemons are prone to chilling injury during prolonged storage; they must be held at 14.5–15.5°C. Lemons that are fully colored when picked and intended for immediate marketing can be held at lower temperatures (0–4.5°C) for a few weeks. During prolonged storage, temperatures below 14.5°C usually produce surface pitting, membrane staining, and red blotch. Storage temperatures above 15.5°C favor the development of decay and increase weight loss.

Lemons grown in different areas have different optimal temperatures. Australian lemons are stored at 9–12°C depending on the duration of storage. California lemons are stored at 12.5–15°C, while Cuban lemons are stored at 12°C. Relative humidity in storage should be maintained at 85–90 percent. At optimum storage temperature, humidity, and ventilation, fruit will lose weight at the rate of 2–3 percent a month. Some water loss is desirable because it leads to thinning and smoothing of the peel. Lemons need conditioning after harvest so that color turns yellow and rind damage is minimized. Conditioning is usually done at 13–15.5°C and 85–90 percent relative humidity. Ventilation is provided to remove ethylene. Benzimidazole and imidazole fungicides (thiabendazole and Imazalil) reduce decay. Femminelle, Teresa, Verna, and Limone di Messa lemons are better-adapted to cold storage than other cultivars (Arras et al., 1997).

The date of harvest is a principal factor in the occurrence of decay in lemons; *Alternaria* rot is the major decay problem. The increase in black buttons occurred after an average storage period of 23 weeks for lemons harvested in March, but occurred after 12 weeks in fruit harvested in April and after 10 weeks in fruit harvested in May (Harvey, 1946). In the lemons grown in the desert areas of Arizona and California, there was no button darkening closely related to fruit decay even after 7 months at 14.5–15.5°C (Rygg and Harvey, 1959). However, softening and bronzing of the lemons after 3 months affected commercial acceptability. Juice content increased from 48–52 percent (at harvest) to 50–53 percent (after 2 months' storage). Citric acid content decreased steadily after the first month in storage and rind weight decreased in proportion to total fruit weight.

In conventional lemon storage practice in Israel, fruits are kept at 13°C for 3 months. This leads to weight loss, color change from green to yellow, drying/wilting and decay (Cohen, 1991). The storage at chilling temperature with intermittent warming is a commercial practice in Israel as fruits can be stored in marketable conditions up to 6 months.

5. Acid Limes

Mature acid lime fruits of normal size with green color and smooth and shiny skin store well. Tahiti or Persian limes (*C. latifolia*) and Mexican or Key limes (*C. aurantifolia* Swingle) can be stored at 9–10°C at 85–90 percent RH. Carefully harvested and handled Persian and Key limes can be maintained in satisfactory condition for 8–12 weeks when stored at 9–10°C. However Persian limes show some undesirable yellow color after 3 or 4 weeks at this temperature, and after 8 weeks they may become completely yellow. Because Key limes are usually marketed when yellow, the color changes occurring during storage in these limes are more acceptable than in Persian limes.

Chilling injury develops in limes at temperatures below 7.5°C. Surface pitting occurs after 4 weeks at 4.5°C, but no pitting is seen after storage at 10°C (Eaks, 1955). Chilling injury was not observed until 8 weeks at 7.5–8°C (Eaks and Masias, 1965). Oil spotting at bruised areas appears in storage within 1 week and is more severe on fruit that is very turgid at harvest. Waxed limes can be safely marketed after 8 weeks of storage at 7.5–8°C. Soluble solids, total acids, and ascorbic acid content increase during storage.

Yellow-colored fruit suffer more losses; hence greener fruit are preferred for storage. Rind color remained greener for longer periods in storage at 5.5–7°C, but pitting occurred in limes grown in northern India (Mukherjee and Singh, 1983). Chilling injury symptoms appeared at 5.5–7°C after 30 days in yellow Kagzi acid limes (*C. aurantifolia* Swingle), whereas at 9.5–11°C, aging of the peel was faster (Ladaniya, 2004b). The optimum temperature for tree-ripe yellow limes is 7.5–9°C (Table 12.1) and the maximum storage life at this temperature is up to 90 days. A vented polyethylene liner is necessary in corrugated cartons to check water loss in addition to pre- and postharvest carbendazim treatment to control decay.

6. Other Citrus Fruits

Citrus fruits of local importance, such as Natsudaidai, Hassaku, and Amanatsu are usually stored in Japan. Hassaku and Amanatsu fruits placed in plastic bags are stored in refrigerated houses run by grower cooperatives near major cities. These fruits have good storage ability at temperatures around 5°C. Fruit is washed, waxed, and marketed when required (Kitagawa 1981). Treatment with thiabendazole (1000 ppm) + 2,4-D (200 ppm) 24 h after harvest, when combined with wrapping (individually in polyethylene film) and storage at 8.42 ± 3.10°C and 79.82 ± 5.11 percent RH, significantly reduced decay, controlled the occurrence of brown blotch, and reduced weight loss in Natsudaidai (*C. natsudaidai*) fruits. This treatment had no marked effect on eating quality (sugar content, titratable acidity, and the sugar/acid ratio). Low relative humidity is considered to be a factor in the development of brown blotch on Natsudaidai fruits (Zhang et al., 1994).

7. Compatibility of Commodities in Refrigerated Storage

Perishable commodities are stored together in the same chamber in commercial cold storages. Every commodity has its own flavor and imparts it to other fruits or vegetables. Citrus fruits transmit their odors to meat, eggs, and dairy products, or they may absorb the odors emitted by onions, cabbages, and garlic. Limes, lemons, and grapefruits can be stored together with tropical fruits such as papaya, mango, guava, and pineapple above 10ºC at 90 percent RH.

B. Storage at Suboptimal or Chilling Temperatures

Low temperatures slow down metabolic rate, retard or inhibit fungal growth and thus increase the keeping quality of fruit. The lower the temperature the better the result, provided that freezing does not occur. All citrus fruits have an optimal lowest safe temperature for long-term refrigerated storage without chilling injury. If fruits are stored below this safe temperature, chilling injury is bound to occur. Various techniques such as intermittent warming; conditioning of fruit at higher temperatures; and the use of coatings, wrappings, and chemical treatments have been tried to minimize chilling injury and thus harness the advantage of low temperatures that are just above freezing the fruit tissues. Very precise temperature management is the key to the success of this technology. Various citrus fruits have different tolerance levels for chilling; these levels also vary with stage of maturity.

Postharvest treatments can be applied prior to or during chilling temperature storage to reduce the development of chilling injury. Various treatments that have been tried with varying degrees of success in reducing chilling injury are:

1. Sealing/Coating Fruit

Sealing fruit in polyethylene film and coating fruit with wax, vegetable oils, and vegetable oil–water emulsions reduces chilling (McDonald, 1986; Cohen et al., 1990).

Wax coatings reduce susceptibility to chilling injury in grapefruit and Temple mandarins but enhance susceptibility to chilling injury in limes (Pantastico et al., 1968). Pitting develops to a much lesser extent on Marsh white grapefruit coated with carnauba wax or polyethylene-based wax when stored at 12ºC and 93 percent relative humidity for 12 days. The application of shellac-based waxes significantly reduces internal O_2 levels and increases CO_2, ethanol and acetaldehyde levels within 1 day after application and lead to development of chilling (Petracek et al., 1998).

2. Anaerobic Conditions

Short-term anaerobiosis during postharvest storage for subsequent adaptation to cold-storage chilling stress has also been found to be effective (Pesis et al., 1992).

3. Jasmonic Acid Treatment

Application of jasmonate and jasmonic acid to counteract stresses in general and chilling injury in particular could be useful (Sembdner and Parthier, 1993).

Methyl jasmonate treatment (dipping fruits for 30 s in 1–100 μM) reduced the severity of chilling injury symptoms and the percentage injury in Marsh Seedless grapefruit stored at 2°C for 4–10 weeks. The best results were observed in grapefruits treated with 10 μM. The application of methyl jasmonate by gassing for 24 h was equally effective. This indicates that methyl jasmonate mediated the plant's natural response to chilling stress, and thus might provide a simple means to reduce chilling injury (Meir et al., 1996).

4. Application of Fungicides

Fungicides, especially thiabendazoles, have been found effective in reducing chilling injury in grapefruit (Schiffmann-Nadel et al., 1975). In Star Ruby grapefruits, susceptibility to chilling injury is highest in fruits harvested in November and January, but is lower in April and negligible in June. Treatments with 200 ppm Imazalil at 50°C effectively controlled CI as well as treatment with water dips at 50°C for 3 min. Beneficial effects were also achieved after treatment with 1200 ppm TBZ at 20°C, although its efficacy in reducing CI was markedly improved with reduced doses (200 ppm) at 50°C (Schirra et al., 2000). Acid lime fruits dipped in Imazalil at 3000 ppm a.i. at 25°C for 5 min and TBZ at 1000 ppm a.i. at 25°C for 2 min showed the lowest incidence of chilling injury during cold storage at 3°C for 6 weeks. There was a positive correlation between CI and water loss. Fruits that showed the lowest incidence of chilling injury also had the lowest weight loss (Ganji and Rahemi, 1998).

5. Temperature Conditioning or Curing of Fruit

Conditioning fruit for 72 h at 34°C and 95 percent RH or at 34°C and 30 percent RH for waxed or film-wrapped fruit tended to reduce CI during storage at 4°C or 1°C followed by 1 week at 21°C. Compared to waxing, film-wrapping of fruit reduced weight loss, maintained fruit freshness, reduced pitting, and reduced the development of *Penicillium* rot (Miller et al., 1987).

Curing at 37°C has also been tried for 3 days in saturated atmosphere (90–95 percent RH) to reduce the sensitivity of fruit to chilling temperatures (2.5°C). Flavedo lipid composition is affected by chilling stress and also by high-temperature conditioning. CI appeared to enhance the peroxidation of unsaturated fatty acid (Raison and Orr, 1990). High-temperature conditioning is reported to induce an increase in fatty acid concentration in polar lipids. Linoleic acid plays an important role in this increase. Large amounts of unsaturated fatty acids are believed to prevent membrane alterations and thus prevent or reduce chilling injury. The conditioning of chilling-sensitive fruits at higher temperatures (3 days at 37°C and 90–95 percent RH) has been related to some physiological modification of flavedo tissue (Lafuente et al., 1994) and as a consequence, chilling stress resistance has been linked to the tissue's ability to modify the fatty-acid composition toward higher levels of unsaturation.

In fatty-acid composition of the Star Ruby grapefruit (flavedo), the degree of unsaturation measured by the unsaturated/saturated fatty-acid ratio (U:S) and the double-bond index (DBI) was higher in fruits stored at 4–8°C than in those stored at 12°C. A negative linear correlation was observed between palmitic and linoleic acids. Increased unsaturation of fatty acids at low temperatures was considered a response to the rigidifying effect of chilling in order to maintain the fluidity of the lipid membrane (Schirra, 1993).

The effect of a hot-water dip at 53°C for 2–3 min on chilling injury and decay in various citrus fruits has been compared with the effect of curing at 36°C for 2 h. Pre-storage hot-water dips significantly reduced the sensitivity of fruits to chilling. Polyamine content increased as a result of hot-water treatment that significantly decreased chilling injury. Hot-water dip also reduced decay. Hot-water dip has been suggested as a more practical and easier way to reduce chilling and decay compared with curing (Rodov et al., 1995a). Conditioning Fortune mandarins for 3 days at 35°C before storage at 2.5°C increased the concentration of putrescine, spermidine, and spermine 2,5-, 4-, and 5-fold respectively. Conditioning increased the chilling tolerance of fruit. Storage of the conditioned fruit resulted in slight reduction in putrescine concentration and a substantial reduction in spermidine and spermine concentrations (Gonzalez et al., 1994).

Catalase enzyme activity was also induced when fruits were conditioned for 3 days at 37°C and 90–95 percent RH. Catalase was considered a major antioxidant enzyme involved in the defense mechanism of Fortune mandarin fruits against chilling stress (Sale and Lafuente, 2000).

Porat et al. (2000) reported that conditioning and curing requiring longer exposure periods of 3–7 days at relatively high temperatures increase fruit weight loss, enhance peel color alteration, and increase juice TSS:acid ratios. Short postharvest heat treatments, including either hot water dip at 53°C for 2 min or a hot water brushing (HWB) at 60°C for 30 s are preferable for Star Ruby grapefruit because they effectively induce tolerance to cold temperatures. HWB was faster and could be used to clean and disinfect the fruit and, simultaneously, enhance its CI tolerance during 6 weeks of cold storage at 2°C and an additional week at 20°C.

6. Foliar Sprays

Preharvest proline sprays (0.5–1.5 percent, applied at full bloom, at fruit set, and 4 weeks before harvest) reduced chilling injury during cold storage (5°C at 80 percent RH) of Marsh grapefruits and Washington Navel oranges. Peel proline, free amino acids, and reducing sugars contents increased with foliar proline sprays (Ezz, 1999).

7. Intermittent Warming

The mechanism by which IW works is still not well understood. Warming treatment may allow tissues to metabolize toxic substances that accumulate during chilling, or may promote restoration of materials depleted during chilling (Lyons, 1973). Acetaldehyde and ethanol are the major metabolites that accumulate in citrus fruit during chilling (Davis, 1971; Davis et al., 1974). Changes in ethanol and acetaldehyde occurring in cold-stored fruit are different in lemons and grapefruits, indicating that these volatiles apparently do not play a role in causing pitting in the peel. These volatiles are, however, related to changes occurring in the fruit flesh that affect the internal quality such as taste and flavor (Cohen et al., 1988).

Intermittent warming has been used in commercial practice for storage of lemons. The technique is effective in reducing chilling injury and it may not be difficult to apply a weekly cycle of intermittent warming on a large scale. Developments in storage equipment and modifications to storage structures can make it possible to apply such treatments on a commercial scale. The standardization of the duration and temperature of warming is necessary in each commodity. The storage at chilling temperature with IW can offer a non-chemical solution to the problem of chilling injury.

The response of citrus fruits to the application of IW at chilling temperatures during storage is briefly discussed below.

a. Sweet oranges: In the sweet orange cultivar Olinda (*C. sinensis* Osbeck) intermittent warming (3 weeks at 3°C followed by 2 weeks at 15°C) delayed the onset of chilling injury by approximately 10 weeks and greatly enhanced resistance to CI development compared with storage at a constant 3°C. Decay percentage was low (6 percent) after 25 weeks of storage with negligible difference caused by storage conditions. Acetaldehyde and ethanol levels in the juice significantly increased in fruit stored at constant 3°C than in intermittently warmed fruits after 15–20 weeks of storage. The respiration rate in fruits stored at 3°C was significantly lower than the respiration rate in intermittently warmed fruits (Schirra and Cohen, 1999).

b. Mandarins: Intermittent warming during cold storage and pre-storage hot-water dip treatment can be combined to limit fungicide use in postharvest treatments (Schirra and Mulas, 1995). Hot-water treatment at 47°C is the most effective temperature for reducing chilling injury in Fortune mandarin fruit during storage at 2°C and 90–95 percent RH up to 4 weeks. Scald damage occurs at temperatures above 53°C. Gonzalez et al. (1998) reported that increasing periods of conditioning at 37°C delayed the development of CI and increased the concentration of polyamines (spermine, spermidine, and putrescine) in Fortune mandarin fruit, but changes in polyamine levels were dependent on the ripening stage of the fruits. Mulas et al. (1998) found no correlation between polyamine content and chilling injury.

c. Lemons: Intermittent warming of lemons at 13°C during storage at 2°C is a commercial practice in Israel (Cohen, 1999). Lemons are harvested at the light-green stage, washed, disinfected, coated with storage wax, and stored. Relative humidity is kept constant at 95 percent, with air ventilation changes of half the air every hour during warming. After 5–6 months of storage, the lemons are again disinfected, coated with polyethylene water-emulsion wax, transferred in open trucks to a refrigerated vessel, and shipped to Europe. Lemons retain good external appearance, reasonable firmness, marketable internal quality, and negligible color development and decay. No major external or internal quality changes are observed in these lemons during a further month of holding at 18°C.

C. Controlled Atmosphere (CA)

Atmosphere consists of gases and water vapor, and in totality 'CA storage' refers to control of temperature, gases (nitrogen, oxygen, carbon dioxide, and volatiles – mainly ethylene), water vapor, or RH. Of course there should be no phytotoxic gas in any CA storage area – it is injurious to fruit. CA storage has to be airtight and equipped with sensors to monitor temperature, RH, and gaseous composition. The refrigeration system should be efficient and gas controls should be reliable (Thompson, 1998).

A controlled atmosphere is more precise in composition than a modified atmosphere, and 'controlled atmosphere' is the term generally used for storage rooms with controllable gaseous atmospheres. A modified atmosphere can be achieved by packaging the commodity inside permeable films, while special equipment is required for controlled atmosphere, which achieve a very precise control of gases such as O_2, CO_2, and N_2. The major benefit of CA storage is derived from the reduction in respiration as a result of low O_2 content. High CO_2 competes for ethylene receptor sites (Burg and Burg, 1967). Ethylene production is reported to be suppressed in CA storage (Wang, 1990). It is possible that fruit becomes less susceptible to ethylene because of high CO_2 and low O_2. CA storage can also be a effective organic technology for storage when it uses only O_2, CO_2, and N_2 (Prange et al., 2006).

CA is a supplement to refrigerated storage to increase the shelf life of the fruit. CA storage technology has the potential to extend the marketable life of citrus fruits, particularly acid fruits such as limes, lemons, Natsudaidai, and Hassaku – provided that an economically feasible technique is developed. CA gases are measured at periodic levels and adjusted to predetermined levels as leakage through the wall and doors take place. The objective in this technique is to control the atmosphere surrounding the fruit and also inside the fruit so as to have the required effect on commodity respiration and its metabolic activity. In static CA systems, the required concentrations are generated in an airtight room by respiring fruit (O_2, is consumed and CO_2 is evolved) and intermittently monitored and adjusted. In continuous-flow systems, the atmosphere is constantly flushed with a required mixture of gases (mainly O_2, CO_2, and N_2). For commodities requiring high CO_2, the mixture is injected directly. In commercial CA storage, maintaining a flow of the appropriate mixture of gases through the system would be very expensive.

In static systems, scrubbers are required to absorb or adsorb CO_2. N_2 is flushed to remove O_2 to desired levels. During storage, CO_2 and O_2 are monitored and regulated. Usually levels of gases are maintained at ± 0.5 percent of the required level. Calcium hydroxide or soda lime is used for maintaining CO_2 level. Storage air is passed over the calcium hydroxide and then over the cooling coil before entering the CA storage. Theoretically, 1 kg of lime adsorbs 0.5 kg of CO_2. In commercial operations, 0.4 kg CO_2 is usually absorbed by 1 kg of CaOH (Bishop, 1996).

$$2CO_2 + 2CaOH \text{ Calcium hydroxide (lime)} — 2CaCO_3 \text{ Calcium carbonate} + H_2O + \text{heat}$$

Renewable scrubbers such as activated charcoal and molecular sieves (aluminum calcium silicate) are also used. When air is blown through the activated charcoal columns, the CO_2 in the air that is coming from CA storage is removed. Molecular sieves remove CO_2 as the air is blown over the heated sieve. More recently, selective-diffusion membrane scrubbers are being used. For ethylene, suitably activated alumina and some other patented scrubbers are also used.

Catalytic converters remove ethylene by chemical reaction as the air is passed over a device and is heated to 220°C in the presence of an appropriate catalyst, usually platinum. Ethylene is oxidized into water and CO_2 (Wojceichowski, 1989). These catalytic converters operate on electricity. Activated charcoal (also known as activated carbon), which is used for scrubbing CO_2, can also be used for removing ethylene.

Another way to achieve CA conditions is to store the fruit in airtight rooms and let it consume oxygen and emit carbon dioxide. Thereafter, required oxygen is introduced and the emitted carbon dioxide is absorbed to maintain the composition. Carbon dioxide and ethylene scrubbers are used. Generators are also used to flush out or consume oxygen in the room; the required carbon dioxide is then introduced (Metlitskii et al., 1985). Carbon dioxide absorbers are used to absorb excess carbon dioxide produced by generators and by fruit. Nitrogen generators are also used to flush out oxygen from the storage area. Hollow-fiber membrane systems or pressure-swing absorption systems produce nitrogen of high purity, thereby preventing any other gaseous product from being introduced in to the storage area. Once oxygen concentration is achieved by flushing, the carbon dioxide is controlled by scrubbers (activated charcoal or dry hydrated lime). Similar systems are used in ship holds and refrigerated containers.

The levels of CO_2 and O_2 are generally expressed in percentages and also in kPa (kilo Pascal) in CA literature. The kilo Pascal is a partial pressure of gas around fruit and is approximately equivalent to percentage since normal atmospheric pressure is 1 atmosphere, which equals 101 kPa. Hence approximately 21 kPa of $O_2 = 21\%$, as in air. (1 kPa= 1% Atmospheric Pressure).

1. Construction of CA Storage Systems

CA storages are constructed with gastight steel linings inside the storage. The inner walls are lined with galvanized steel or polyurethane foam (insulation) sandwiched between steel sheets and made airtight with a locking system. Sliding-type doors have rubber gaskets to avoid any gaps. Pressure-release valves are fitted to regulate the pressure, since doors opening and closing can change pressure and gas concentration. The air and gaseous mixtures pass through cooling coils to avoid temperature change. However, it is important to check the gas-tightness of the storage chamber. Refrigeration units are attached to CA storages. CA storages with capacities as high as 600 tons or more are commercially operational in different parts of the world. Insulation is placed behind the steel lining with some gaps where refrigerated air is circulated. The airtight steel lining removes the product heat by conduction because there is no direct air circulation in the storage area. The controlled gaseous concentration is easily maintained inside the steel walls of the storage. However, cooling is quite slow in this system and air ducts may be necessary. Fruit should be precooled before storage. There are several variations on this system. The polyurethane foam–sandwiched metallic panels provide interiors that are gastight and very easy to assemble.

For transport purposes, a CA system unit can also fitted to refrigerated reefer containers of 30–60 cum volume capacities. These reefers have their own mechanical refrigeration units and are powered by their own diesel engine electricity units. Membrane technology is used to regulate oxygen and carbon dioxide levels. These units are compact and efficient for containers and are also easily detachable. CO_2 and O_2 are monitored using flow through gas analyzers (CO_2 with IR gas analyzers and O_2 with paramagnetic O_2 analyzers). The utility and economic value of such reefers for shipment of high-value specialty citrus fruit needs to be investigated.

CA conditions (gaseous composition) may also vary with temperature (Izumi et al., 1990). Humidity interacts with CO_2 and O_2 in storage, and the effects of gaseous composition can be different at different humidity levels.

2. Response of Citrus Fruits to CA Storage

Citrus fruit is non-climacteric, and fruit is fully mature/ripe when harvested. Therefore there is no vital process that leads to ripening after harvest and that can be slowed to extend green life or storage life. The decay level cannot be controlled in CO_2 atmosphere because the atmosphere required (high CO_2) to control most citrus postharvest pathogens is not tolerated by fruit, resulting in an off-flavor or peel injury. Very high doses of CO_2 as pretreatment or a lower concentration throughout storage could be used to minimize chilling injury (Vakis et al., 1970). In general, citrus fruits are difficult to store in controlled atmospheres except for acid fruits. Failures were associated with high concentrations of CO_2, which elevated fruit acidity, giving fruits an off-flavor and unpleasant odor. Storage experiments indicated that fruits could be stored at low O_2 (5–6 percent) and higher CO_2 (2.5–4 percent) concentrations for >5 months (Sun, 1998).

TABLE 12.3 Proposed Controlled Atmosphere Conditions for Citrus Fruits

Commodity	CO_2 (%)	O_2 (%)	N_2 (%)	Temperature (°C)
Sweet orange	0–5	10–15	80–90	5–10
Mandarin 'Satsuma'	0–2	8–12	86–92	4
Mandarin 'Nagpur'	2–4	10–12	84–88	5–6
Kinnow	5	10–15	80–85	4–5
Grapefruit	5–10	3–10	80–92	10–14
Lemon	0–5	5	90–95	13–15
Acid limes	0–10	4–5	85–96	9–12

Note: Relative humidity 85–90 percent in all cases. Values are average of several studies, Chase (1969), Hatton and Reeder (1968), Kubo and Heginuma, (1980), Kader (1980), Sea Land (1991), Bishop (1996), Amarjit Singh and Rajinder Singh (1996), Ladaniya and Shyam Singh (2000).

CA storage can be useful in the storage of limes and lemons, where preservation of natural green color is the objective. Persian limes are marketed green in the U.S., and grade-1 green fruit fetch a very good price. The green color of lemons are harvested in winter and marketed in summer (in California) can be preserved with CA storage or by removing ethylene. The specific objective to preserve flavor in high-value fruits such as Temples may fetch high prices to compensate for the high cost of CA storage. Though results have been promising in some cases, CA storage has not been used commercially for citrus fruits, partly for economic reasons and partly because of the physiological characteristics of the fruits. Elevated CO_2 has no beneficial effect on respiratory changes in storage, and low O_2 concentrations stimulate the accumulation of ethanol and acetaldehyde. However, this system of storage can be useful for a particular commodity in a specific situation if it is economically viable.

Considerable research has been carried out in various countries on the CA storage of citrus fruit, including lemons, oranges, grapefruit, mandarins, and Kabosu. Recommended CA conditions for citrus fruits are given in Table 12.3. Thompson suggested a threshold level for CA storage of citrus fruit (Table 12.4) in general, but cultivar-specific requirements have to be worked out.

a. Sweet orange: The Valencia sweet oranges stored in a CA atmosphere of 15 percent O_2 + 0 percent CO_2 for 12 weeks at 1°C plus 1 week at 21°C had higher flavor retention and lower incidence of rind pitting. CO_2 levels of 2.5–5 percent, when combined with 5 or 10 percent O_2, adversely affected flavor retention. With increased CO_2 and decreased O_2, the flavor score decreased. Extremely high concentrations of ethyl alcohol were found in fruit with off-flavors. Low alcohol concentrations were found in fruit held at 15 percent O_2 + 0 percent CO_2 (Chase, 1969). Temple oranges held in 10 percent O_2 + 5 percent CO_2 atmosphere for 5 weeks at 1°C plus 1 week at 21°C had a higher flavor rating than those held in other controlled atmospheres or in air. McGlasson and Eaks (1972) observed that the development of off-flavors, decay, and flavor loss in Valencia oranges were primarily associated with the presence of ethylene

TABLE 12.4 The Limits of CO_2 and O_2 Levels and Injury to Citrus Fruits Under CA Conditions

Citrus fruit	Maximum CO_2 (%)	Injury symptoms beyond maximum CO_2 limit	Minimum O_2 (%)	Injury symptoms beyond minimum O_2 limit	Reference
Orange	5	Off-flavor	5	Off-flavor, skin injury	Chace (1969)
Lime	10	Increased decay	5	Scald injury, off-flavor	Hatton and Reeder (1968)
Lemon	10	Decreased acidity, increased decay	5	Off-flavor	Smoot (1969)
Grapefruit	10	Scald, off-flavor	3	Off-flavor higher ethanol and acetaldehyde content	Davis et al. (1973)
Satsuma	4	Skin injury	10	Off-flavor	Kubo and Heginuma (1980)

The levels of CO_2 and O_2 at or beyond given limits are injurious.

in the storage atmosphere. The higher carbon dioxide in CA is also reported to counteract the ethylene action. Moreover, the continuous flushing of gases is likely to remove volatiles and ethylene from the storage atmosphere. The removal of ethylene has been reported to maintain calyx green in Valencia Late oranges (Testoni et al., 1992). Zamba (1987) reported that Valencia and Washington Navel oranges stored in a controlled atmosphere at 5–7°C and 85–90 percent RH retained more moisture, dry matter, acids, sugars, vitamin C, and rind firmness than fruits stored in air.

Smoot (1969) showed that for Valencia fruit, the controlled atmosphere combinations of 0 percent CO_2 with 10–15 percent O_2 resulted in less decay than storage in air. It was recorded that *P. digitatum* and *P. italicum* rots were retarded but not stopped at a level of 2.5 percent O_2, while higher (5, 10, and 15 percent) levels of O_2 did not have any effect on rot development.

Ke and Kader (1990) studied the tolerance of Valencia oranges to very low O_2 and very high CO_2 concentrations. Fruits tolerated exposure to 0.5, 0.25, or 0.02 percent O_2 up to 20 days at 5–10°C followed by storage in air at 5°C for 7 days without any detrimental effects on external and internal appearance. Oranges stored in 0.5, 0.25, or 0.02 percent O_2 at 10°C had lower respiration rates but higher resistance to CO_2 diffusion and higher ethanol evolution rates than those stored in air at the same temperature. Similar but less pronounced effects of the low O_2 atmospheres were observed at 0°C and 5°C. After 26 days, fruit kept at 0.02 percent O_2 at 5°C had moderate to severe off-flavor. Oranges kept in 60 percent CO_2 at 5°C for 5–14 days followed by holding in air at 5°C for 7 days developed slight to severe skin browning. Juice-quality parameters such as TSS, acidity, and vitamin C content were not significantly influenced by either the low O_2 or the high CO_2 levels, but ethanol and acetaldehyde contents did increase, which correlated with a decrease in the flavor score.

Hamlin, Pineapple, and Valencia oranges pretreated with controlled atmospheres (N_2, CO_2, or 0.1–0.7 percent acetaldehyde in air for 8–24 h) had a two- to three-fold increase in acetaldehyde, ethyl acetate, ethyl butyrate, and ethanol within a day after treatment. Methanol, methyl butyrate, and hexanal were unchanged. Changes in juice flavor were not consistent up to 8 days after treatment. Treatment with acetaldehyde had little effect on flavor or composition except for Hamlin oranges, in which acetaldehyde and ethyl butyrate levels increased (Shaw et al., 1991).

Brackmann et al. (1999) reported that Valencia oranges had 47.36–63.31 percent decay in controlled atmosphere storage (air + 5 percent CO_2 or 10 percent O_2 + 5 percent CO_2) at 3.0°C and 90–95 percent RH after 84 days. Percentage decay was lowest (5.55 percent) in iprodione-treated fruits stored in air at 3.0°C. Weight loss was highest in fruits stored in air at 7.0°C after 84 days of storage.

Very high levels of oxygen (60–80 percent) have also been tried for the storage of sweet oranges. Navel oranges stored in 80 percent oxygen with 20 percent nitrogen at 15°C resulted in a darker orange color in both peel and juice. The color persisted even when fruits were removed from the storage atmosphere. Ethylene (20 ppm) in the storage atmosphere adversely affected juice color (Houck et al., 1978). In Hamlin, Parson Brown, and Pineapple oranges, 80 percent oxygen with 20 percent nitrogen at 15°C resulted in a pale orange color while endocarp and juice were deep orange in color. Quality parameters such as TSS, titratable acidity, pH, and juice flavor were not adversely affected (Aharoni and Houck, 1980). In the case of blood-red oranges (Ruby, Tarocco, and Sanguinello) similar high O_2 conditions at 15°C deepened the red color of flesh and juice, while ethylene (20 ppm) resulted in an off-flavor (Aharoni and Houck, 1982).

The recommended CA conditions for storage of oranges are 5–10°C temperature with 0–5 percent CO_2 and 5–10 percent O_2 (Kader, 1992). Sea Land (1991) recommended 5 percent CO_2 with 10 percent O_2.

b. Mandarin/Tangerine: Mandarin/tangerines have the most delicate flavor and shortest storage life among all citrus fruits. CA storage has shown some promising results in the storage of these fruits. The proposed optimum conditions for the CA storage of Satsuma mandarins are 8–12 percent O_2, 0–2 percent CO_2 at 1–4°C, and 83–90 percent relative humidity (Kubo and Heginuma, 1980). Under high CO_2 conditions (1–10 percent O_2 and 0–4 percent CO_2) rind pitting occurred in some fruits with more free acid present in the flesh during 140 days storage at 3°C with 85–90 percent RH. Low humidity induced browning and dryness in some fruits, and low humidity with high CO_2 resulted in higher total-acid and total-sugar contents in the peel. During storage, formic acid and α-ketoglutaric acid fractions increased but oxalic and malic acids decreased. Under high CO_2, malic acid decreased and citric acid increased markedly. Satsuma mandarins stored in 9 percent O_2 and 1 percent CO_2 atmosphere combined with 5 ± 1°C temperature and 73–79 percent RH had less weight loss and decay (Oogaki and Manago, 1977).

Elevated CO_2 levels (2, 3, 4 and 5 percent) at 5 ± 1°C increased the storage life of Kinnow mandarins up to 140 days. Higher levels of CO_2 have been more effective in suppressing the intensity of fungal rotting (Amarjit Singh and Rajinder Singh, 1996).

Rind-color development was delayed and no adverse effect on flavor and the fruits' external appearance was observed in Nagpur mandarins stored at 6–7°C under controlled atmosphere conditions (2 and 4 percent CO_2 with 7 and 12 percent O_2 and balance nitrogen). O_2 concentration appeared to be primarily responsible for a color change from green to orange; the lower the O_2 level the slower the change. It was observed that decreased levels of O_2 (7 and 12 percent) and increased levels of CO_2 (2 and 4 percent) had no effect on *Alternaria* rot. After 9 weeks, losses from rotting, softening and off flavor were greater in fruit packed in polyethylene-lined cartons than fruit stored in continuous-flow CA system (Ladaniya and Shyam Singh, 2000c).

c. Lemon: In lemons, the disappearance of the chlorophyll in the rind was markedly affected by lower oxygen levels (5–10 percent), and the time required for green lemons to turn yellow was 16 weeks in 10 percent O_2 compared with 4 weeks in air at 15°C (Biale, 1953). Color change is delayed with high CO_2 and low O_2 in the storage atmosphere, but 10 percent CO_2 impairs flavor (Pantastico, 1975). The storage life of lemons can be increased considerably in CA storage. At 10°C with 0.1–0.2 percent CO_2 and 3–5 percent O_2, lemons can be stored up to 27 weeks provided that an ethylene absorbent is used. CA storage improved green-color retention (Wild et al., 1976). A CA atmosphere of 10 percent O_2 with no CO_2 and the continuous removal of ethylene was also useful for the storage of lemons for 6 months (Wild et al., 1977).

Storage in 0–10 percent CO_2 and 5–10 percent O_2 at 10–15°C had a good effect but was not used commercially (Kader, 1992; Bishop, 1996). According to Sea Land (1991), CA storage recommendations for lemons are 5–10 percent CO_2 with 5 percent O_2.

d. Acid lime: Pantastico (1975) recommended storing Tahiti limes in 7 percent O_2, which reduced the symptoms of chilling injury; however, CA storage increased decay and reduced juice content. The satisfactory storage of Persian limes was reported at 5 percent O_2 + 7 percent CO_2 levels (Hatton and Reeder, 1968). Rind remained green up to 60 days; however, scald-like injury was reported on rind during gas storage; limes from air storage were not affected. Spalding and Reeder (1974) reported that Tahiti limes can be stored in 7 percent CO_2 and 5 percent O_2 atmosphere at 10°C with acceptable flavor for 6 weeks.

Storage of limes at 10–15°C with 0–10 percent CO_2 and 5 percent O_2 has been recommended (Kader, 1992; Bishop, 1996). Acid lime fruits could be stored at low concentrations (3–6 percent O_2 and 2.5–4 percent CO_2)for more than 5 months (Sun, 1998). Sea Land (1991) reported that 0–10 percent CO_2 with 5 percent O_2 as optimum levels for limes.

e. Grapefruit: Under high CO_2 concentration, pyruvic acid, acetaldehyde, and ethanol contents increased, but citric acid and total solids decreased in grapefruit

(Davis and Bruemmer, 1973). The accumulation of ethanol and acetaldehyde, and the stimulation of pyruvic dehydrogenase and ADH under high CO_2 conditions indicated anaerobic conditions, and as a result off-flavor developed.

f. *Citrus natsudaidai*: At 98–100 percent RH, high CO_2 (20 percent) increased the water content of the peel of *C. natsudaidai* fruit and the ethanol content of the juice. Internal O_2 decreased and off-flavor developed. At 85–95 percent RH no injury occurred and high CO_2 produced beneficial effects (Kajiura, 1973).

D. Hypobaric Storage

Hypo means below and *baric* means atmospheric pressure. Hypobaric storage systems operate below normal atmospheric pressure. Normal atmospheric pressure (1 Atm) is 760 mm Hg as measured by barometer. One normal atmosphere equals 101 325 Newtons of pressure per sqm. In hypobaric storage systems, a partial vacuum is created and this pressure is reduced. Benefits are achieved by a reduction in O_2 and volatiles, including ethylene. The major problems with this system are higher costs of construction, since storage walls must be able to withstand lower pressure inside compared to outside. In addition, under vacuum, water evaporates from the fruit surface, and thus saturated air has to be introduced to avoid excessive water (weight loss) from the fruit.

The use of hypobaric storage depends on its economic viability, since the major varieties of citrus are cultivated in the countries of both the Northern and Southern Hemispheres and are made available in the off-season. Promising results have been achieved in the hypobaric storage of acid limes, but not Satsuma mandarins. Tahiti limes retained their green color, juice content, and flavor during hypobaric storage (170 mm Hg, 10–15.6°C, and 98–100 percent RH) for 6 weeks (Spalding and Reeder, 1976). Presently hypobaric storage is not being used commercially for storage of citrus fruit.

E. Evaporative Cool Storage and Other Natural Systems

Cellars and caves have been used in China, Japan, Europe, and many other regions to preserve freshness of perishables, including citrus. These structures are relatively cool in summer because they are below ground or constructed on the slopes of hills, as in Nepal. The structures are warmer when outside temperatures are below 0°C. These structures are provided with a lining of cement or stones and also have drainage so that water does not spill over the fruit. These structures have been used for storage of fruits since ancient times. In relatively more recent times, ice was used to cool the air that was circulated over the fruit and then back over the ice. Air carries the heat from fruit to ice. While it melts, 1 kg of ice absorbs 325 kJ of heat; this is how the hot surrounding air is cooled. In ice bunkers, large quantities of ice are required and the room/chamber has to be insulated. With the invention of mechanical refrigeration systems, the use of ice bunkers has been discontinued in most countries. Natural systems of storage are still being used by growers of citrus in many countries for small-scale, on-farm storage of fruits, as they cost less, are naturally available, or are easily constructed with locally available material.

1. Low-cost Evaporative Cooling Systems (Passive or Active)

Low-cost storage chambers developed on the principle of evaporative cooling are promising for small-scale storage of citrus fruits at the farm itself. This technique of storage can be very useful in developing and underdeveloped countries that have scarce mechanical-refrigeration facilities in remote areas. Evaporative cooling is achieved by blowing dry air over a wet surface so that reduction in temperature is achieved (per wet-bulb temperature). In the process of water evaporation, the latent heat of evaporation (2260 kJ) heat energy is absorbed (from the surrounding air) per kg of water. This system is very useful for citrus fruit storage in arid and semiarid areas. In many parts of Asia, America, Africa, and Australia this technique could be very useful.

These chambers, which can operate without use of electrical energy, are called zero-energy cool chambers (Roy and Khurdiya, 1986) or passive evaporative cooling systems. In a zero-energy cool chamber with one quintal capacity, the storage life of grapefruits, acid limes, Nagpur mandarins, Kinnows, and Darjeeling mandarins was extended up to 77, 35, 42, 60, and 42 days, as compared with 27, 11, 14, 14, and 14 days at ambient conditions in winter months in New Delhi (Anonymous, 1993). The 1.5-ton storage capacity of an evaporative cool chamber has been reported by Ladaniya (1997b). This above-ground, walk-in type of evaporative cool storage structure with 8.5 cum volume (inside dimensions of 225 cm length × 180 cm width × 210 cm height) can be constructed using kiln-baked bricks (22 cm L × 10.5 cm W × 6 cm H), sand, bamboo, wood-wool panels, and jute cloth. A drip system (PVC pipe of 16 mm outer diameter with emitters fitted 30 cm apart) and an overhead water tank (200-lit capacity) of plastic material with a control valve are installed for the continuous trickling of water. For air movement, a small exhaust fan (maximum capacity $3\,m^3$/min air flow) with a regulator is provided on the back of the structure. On two sides, double brick walls with a cavity (4–5 cm) are made with masonry work in mud or cement mortar. The brick surface is not covered with mud or plaster. The walls are supported from inside and outside by bamboo or PVC pipes fixed in four columns (35 L × 35 W × 210 H cm) erected at 4 corners using bricks, cement, and sand. Alternatively, for better stability, brick walls on both sides are also constructed by placing alternate layers of bricks in a crossed manner (Fig. 12.2b). The use of cement mortar should be kept to a bare minimum. The floor is made by arranging bricks in a single layer. In the front, a cement sheet or plastic door with a wooden frame is provided. The wood–wool and jute cloth panels (90 cm L × 45 cm W) of 4 cm thickness supported with fine iron netting in a wooden frame are fitted on front and back. The ceiling is made of wooden frame and cement sheets. This structure accommodates 70 plastic crates (55 cm L × 35 cm W × 30 cm H external dimensions); each crate holds 21–22 kg fruit – roughly 1.5 tons of fruit storage capacity. In order to provide shade and avoid direct sunlight on the structure, a hut-shaped roof cover made of bamboo, locally available grass panels, jute cloth, and plastic sheets can be fixed with the support of four columns over the ceiling. The structure can be constructed under a large tree on the citrus farm (Fig. 12.2). As an alternative, if a large tree is not available for shade on the farm, a shed of iron angle structure having trusses with cladding of cement sheets can be provided. In this case, a hut-shaped roof made of bamboo as stated above is not necessary. The 25–26°C and 11–14°C maximum and minimum average annual temperatures and nearly 90–95 percent relative humidity were achieved inside this cool chamber. During November–December, the storage life of Nagpur mandarins was extended up to 3 weeks compared with 6–7 days under ambient conditions. Outside temperatures were 10–15°C higher than temperatures inside the structure (Ladaniya, 2004c).

Evaporative cool storage structures with brick-batts (brick pieces) and sand in the wall cavities also reduce the temperature by 14–18°C and extend the storage life of oranges (Umbarkar et al., 1991). The evaporative cool chamber with 8 tons of fruit storage capacity has also been reported (Pal et al., 1997). Kinnow mandarins were stored up to 40 days with 15 percent weight loss, compared with 15 days' storage life at room temperature. The cost of the 8-ton fruit storage capacity structure would be quite high as it requires a cement concrete frame and brick walls in cement and sand mortar.

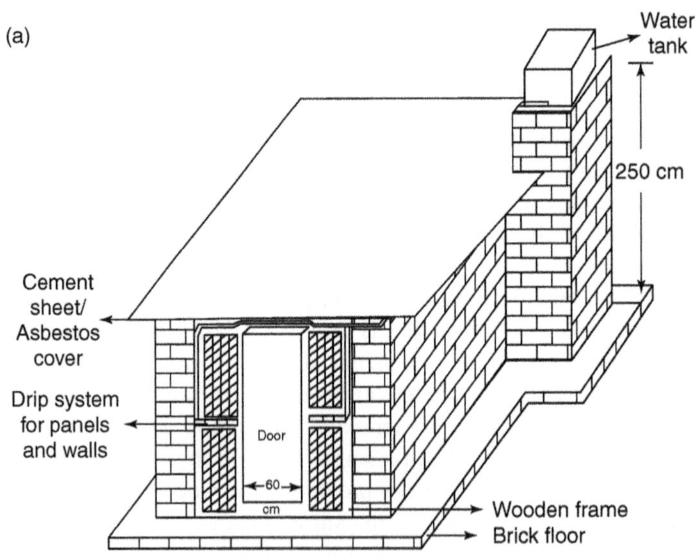

FIGURE 12.2 Evaporative Cool Chamber: (a) Structure for Construction Under the Shed for Protection from Rain and Direct Sun light (Note that Hut-shaped Roof is not Provided).

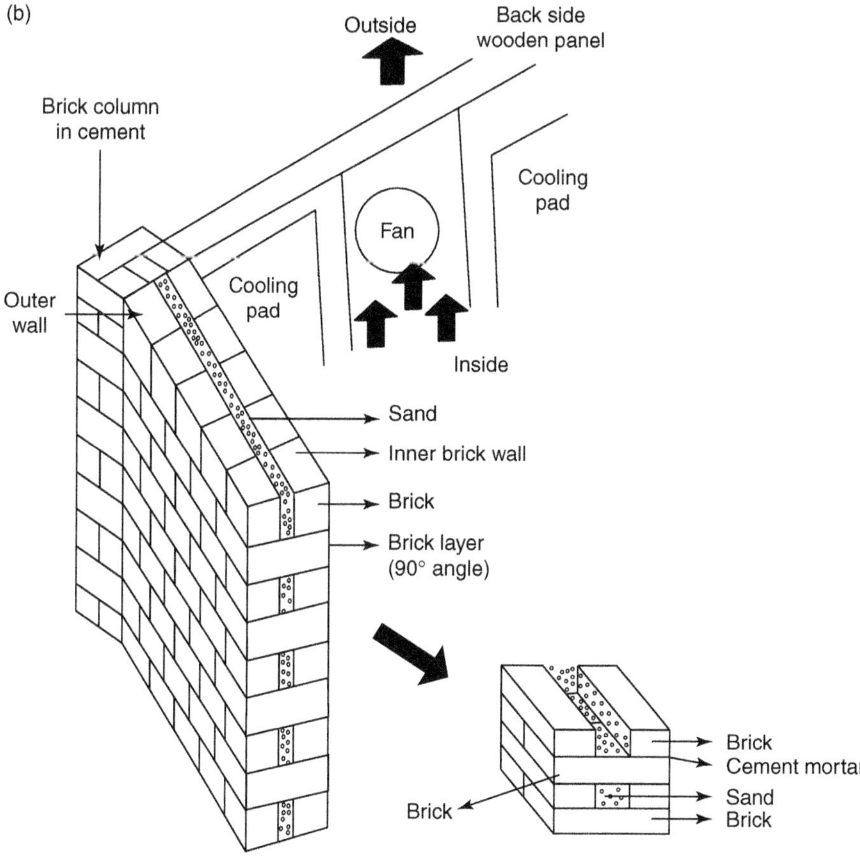

FIGURE 12.2 (Continued) (b) Construction Details for Brick wall.

In the local practice of storing fruits in underground pits of 60 × 60 cm size, Blood Red sweet oranges were stored up to 45 days (Singh et al., 1987). Considering the high humidity, treatments with disinfectants and fungicides are necessary for storage of fruits in evaporative cool structures. Treatment with Captaf (0.2 percent) and Bavistin (0.1 percent) reduced losses in lemons, sweet oranges, and Kinnows stored in evaporative cool chambers (Jitender Kumar et al., 1990; Jain and Chauhan, 1995; Kaushal and Thakur, 1996; Waskar et al., 1999). Pre-treatment with wax (4 percent) + 2,4-D (100 ppm) extended the storage life of acid limes (Mexican lime) up to 35 days in an evaporative cool chamber (Thangaraj et al., 1993). A modified atmosphere created by covering fruits with polyethylene followed by storage in zero-energy cool chambers increased storage life of Baramasi lemons up to 42 days (Sindhu and Singhrot, 1994) and of Kinnow mandarins up to 77 days (Jain and Chauhan, 1995).

Valencia oranges stored for 30 days in a brick-wall evaporative cooler lost only 4.2 percent in weight (Babarinsa, 2000). Control fruit left unpacked at ambient conditions lost 31.4 percent in weight. The weight loss in the cooler was further reduced to 3 and 1.2 percent for fruits in perforated and non-perforated polyethylene bags, respectively. Fruits left in ambient conditions shriveled and were unacceptable.

2. Coolers/Desert Coolers

Common desert coolers can also be used for cooling produce in hot, arid areas. These coolers are very common in India. They work on the principal of evaporative cooling, and water requirement varies depending on the size of the cooler, which in turn depends on room volume. These coolers can achieve temperature ranges of 15–20°C with 80–90 percent RH when outside temperatures are 30–35°C with 20–30 percent RH. With increasing air discharge (flow), water tank size also varies. A cooler that has a blower with air flow of 2000 cum/h would need tank of 60-lit capacity. For 8000 cum/h air flow, a 260 lit capacity water tank would be required (Prasad, 1989). Coolers with air velocity of 1.2–1.5 m/s across the cooling pad and 4–8 m/per second across the grill of the cooler are considered sufficient. The humidification should be 70–80 percent. The pad thickness (wood–wool) with 4.5 cm and a water flow rate of 0.133 lit/s across the pad gives sufficient cooling. The density of wood-wool in the pad should be around 60 kg per/cum.

3. Cellar Store

Underground cellars are used in Nepal for the storage of citrus fruits, mainly mandarins. Storage losses are higher at the lowest altitude and decrease with increasing altitude, indicating the effects of ambient temperatures. Dipping the fruits in paraffin wax or bee wax reduced fruit weight loss but not percentage decay. Fruits with 25–50 percent yellow skin had a long storage ability in the cellars (Subedi, 1998). In the Republic of Korea, a considerable quantity of citrus fruit is stored in conventional cellars. Less than 10 percent of the crop is kept in cold storage. More than 25 percent of crop stored in cellars goes to market in 1 month, 60 percent in 2 months, and 100 percent in 3 months (Kim et al., 1988).

4. Caverns/Stone Quarries

Abandoned caverns and underground stone quarries are used to store citrus in Japan. At higher altitudes, temperatures remain below 10°C with more than 90 percent RH in these structures throughout the year. These energy-saving, naturally cooled storage structures can provide considerable cost savings on electricity for refrigeration (Hasegawa, 1988).

5. Orchard Fruit-storage System (China)

In Hubei province (Huangshi county), structures that are half underground and half aboveground are constructed. These are ventilated structures; ventilation is provided at night by opening ducts so that 3–5°C temperature and 85–95 percent RH is maintained. Fruits are treated with 500 ppm Imazalil + 200 ppm

2.4-D dip and kept for wound healing before storage. Single fruits packed in polyethylene and placed in vented cartons are also stored up to 4 months. Imazalil and Tecto (TBZ) are widely used as fungicides. About 550 tons of Satsuma mandarins were stored in this way in China during the 1989–1990 season (Wai-chin, 1990).

6. Wooden Rooms

In Japan ambient temperatures are quite low during winter. Satsuma (*unshu* mikan), Natsudaidai, Hassaku, and Iyokan are stored in wooden rooms for 3–4 months. Satsumas are cured before storage so that fruit loses 3–4 percent water. Fruit is sensitive to high RH; therefore RH is usually adjusted to 85–90 percent. Fruits are inspected every 2 weeks for decay.

F. Storage Under Ambient Conditions

Ambient temperature and relative humidity vary considerably in subtropical and tropical regions when citrus fruits are harvested. In India, for example, the temperature range is 15–25°C with 60–70 percent RH in winter to 35–40°C with 20–30 percent RH in summer. Citrus fruits harvested during winter months in subtropical regions can be stored for longer periods under ambient conditions than fruits harvested in tropical regions. Higher temperatures and relative humidity in tropical areas increase the incident of postharvest decay loss. Since most developing countries are in tropical regions with inadequate postharvest infrastructures, the supplemental procedures of loss reduction can play an important role in extending the marketability of citrus fruit. After harvest, during the handling period of 10–15 days without a cool chain, a major part of the fresh fruit produced in these countries is distributed domestically and therefore requires proper postharvest management. The response of citrus fruits to various treatments during storage at ambient condition is briefly discussed as follows:

1. Sweet Orange

The response of sweet orange varieties varies under different ambient storage conditions. Pineapple oranges lose more weight than Washington Navel oranges during storage (Chattopadhayay and Ghosh, 1994). Packaging materials and containers play an important role in retaining the natural freshness of orange fruits during transport and storage. Polyethylene film bags of 100–200 gauge with 0.2–0.4 percent ventilation effectively controlled weight loss in Mosambi sweet oranges (Choudhari and Kumbhare, 1979).

2. Mandarin/Tangerine

Jawanda et al. (1978) reported that polyethylene film wrapping is more effective than waxing for reducing weight loss in Kinnows stored at room temperature. However, rotting appeared earlier if polyethylene wrapping was used without fungicide treatments. Combination of wax treatment with fungicide (Benlate) and polyethylene wrapping extended the storage life of Khasi mandarins up to 24 days at 11–19°C (Karibasappa and Gupta, 1988). Packaging Khasi mandarin fruits in wooden boxes or corrugated fiberboard boxes resulted in better retention of fruit firmness and ascorbic acid content during storage under ambient conditions (Barua and Yamadagni, 1996).

Kinnow mandarins stored under ambient condition (15–20°C and 65–70 percent RH) recorded lowest weight loss, highest juice content, and best overall quality after 40 days (Mahajan and Sharma, 1997).

3. Acid Lime

Polyethylene film wrapping has been found to be more effective than waxing for reducing weight loss in limes (Naik et al., 1993). Pre- or postharvest treatment with fungicides is a must to reduce decay losses in polyethylene-wrapped fruit. Treatment with GA_3 (50 or 100 ppm) or NAA (50 or 100 ppm) with packing in polyethylene increased percentage of marketable fruits after 20 and 30 days of storage respectively (Bandopadhyay and Sen, 1996).

II. REMOVAL OF ETHYLENE FROM STORAGE AREA

The ethylene concentration in fresh air is very low (0.005 μl/l) (Wills et al., 1998). This concentration is not effective to hasten physiological aging processes, but in storage it is most likely that over a period of time the ethylene level builds up considerably. The ethylene can be removed by the following methods:

1. Ventilation.
2. Inspection of boxes for removal of rotten fruit that produce high ethylene.
3. Scrubbers: (i) Potassium permanganate ($KMnO_4$) – Potassium permanganate is coated to material with large surface area such as alumina, expanded mica, and brick pieces. When ethylene comes in contact with potassium permanganate, it is oxidized to CO_2 and water. Potassium permanganate also reacts with water; in high-RH areas its efficiency is reduced. (ii) Other chemicals, such as tetrazine, which react with ethylene specifically can be used for ethylene scrubbing.
4. Oxidation with ozone: Ozone (O_3) can be produced from atmospheric oxygen using electric discharge. Ozone is a very active oxidizing agent and oxidizes ethylene. The reactive free radical is formed from ozone and oxidizes ethylene. Ozone is corrosive to metal parts and can be toxic to humans (Wills et al., 1999).

Ethylene-removal techniques are also discussed in Part I of this chapter.

III. CHANGES IN FRUIT COMPOSITION DURING STORAGE

Softening of fruit and development of insipid taste and off-flavor are the major deteriorative changes that take place during storage of citrus fruits. Changes are rapid if fruits are held under hot and dry ambient conditions, while under optimum refrigerated conditions with high relative humidity changes are gradual and at times seem insignificant. But at the end of long-term refrigerated storage, changes can be quite conspicuous. Compositional changes under CA and hypobaric storage conditions are determined by the gaseous composition. Postharvest treatments such as coatings, packaging, and various chemicals also influence these changes. Sometimes there are inconsistent and contradictory findings with respect to these changes; these can be attributed to experimental procedures and materials used.

A. Total Soluble Solids, Sugars, and Acidity

Changes in total soluble solids, sugars and titratable acidity (TA) contents of the juice depend on the conditions under which citrus fruits are stored. Higher ambient temperatures coupled with lower relative humidity result in rapid water loss and cause an increase in TSS and titratable acidity contents of mandarins (Dhillon et al., 1976; Chundawat et al., 1978). The trend is similar in acid lime (Kohli and Bhambota, 1966) and sweet orange fruits stored at room temperature. If relative humidity is kept higher (with polyethylene liner) acidity declines in various citrus fruits during ambient storage (Angadi and Shantha Krishnamurthy, 1992; Chattopadhyay et al., 1992). Rapid water loss resulted in alcoholic off-flavor development in Nagpur mandarins after 3 weeks, which was attributed to a tough and leathery peel that restricted natural gas exchange, leading to anaerobic conditions (Ladaniya et al., 1997). Kawada and Kitagawa (1986) reported that TA reduced in Satsuma, Seminole tangelo, and Natsumikan (*C. sinensis*) fruit held at 35°C in O_2 for 72 hours, at 35°C in air for 72 hours, and at 5°C in cold storage for 3 weeks respectively. Warming the fruits also reduced the ascorbic acid content but had little effect on sugar content. Reduction in acidity was mainly to the result of a decrease in citric acid. Both citric and malic acid contents declined during cold storage. No significant changes in TSS, acidity, or organoleptic properties of freshly harvested Dancy and Clementine tangerines, Minneola tangelos, and Olinda Valencia oranges were recorded at 23–25°C and 58–70 percent and at >95 percent RH up to 11 days (Akamine and Goo, 1979).

Seal-packaged mandarin and sweet orange fruits stored at 20–25°C lost a considerable amount of TA within 4–5 weeks, however, change in total soluble solids content were not significant. During prolonged storage (12 weeks) of Hadas mandarins, the sealing treatment reduced TA levels, whereas with wax treatment TA level increased. In sealed Nova fruits TA level decreased after 5 weeks during storage (Peretz et al., 1998).

In cellar storage, irrespective of the maturity stage of fruit and dipping treatments, total soluble solids content remained unchanged while total titratable acidity content decreased during storage (Subedi, 1998).

Under refrigerated storage conditions (about 4–6°C at 90 percent RH), juice content, TSS, and acidity contents declined gradually in Kinnow and Nagpur mandarins (Mukherjee and Singh, 1983; Ladaniya and Sonkar, 1996). Temple oranges (*C. reticulata* × *C. sinensis*) stored for 4 weeks at 14, 5, or 2°C, followed by 1 week at a shelf-life temperature of 17°C did not show any significant changes in the internal composition of the fruit. However, after 8 weeks, titratable acids and internal CO_2 were higher with lower O_2 content and unacceptable quality (Cohen et al., 1984). Losses in titratable acidity, sugars and vitamin C contents of Hayashi Satsuma were very slow and minimal at 5°C than at 15°C up to 3 months. Exterior canopy fruit recorded fewer losses than interior canopy fruit (Izumi et al., 1990). Sucrose and fructose contents decreased with increase in °Brix while citric acid content remained unchanged in Palestine limes during storage at 15°C and 95 percent RH. In grapefruits, also, total sugar content declined despite no change in °Brix value, indicating that the increase in °Brix value during storage in some citrus fruit cultivars is not always directly related to changes in the fruit's simple sugar content (Echeverria and Ismail, 1987; Echeverria and Ismail, 1990). Sweet orange fruits wrapped in polyethylene bags and stored at 4–6°C retained acidity up to 28 days (Falak-Naz et al., 1997). Juice content, soluble solids, and total acidity of lemons increased during storage for 4–12 weeks at 4, 13, and 24°C (Arras et al., 1997). A marked decline in soluble sugar content of Natsudaidai (*C. natsudaidai*) occurred up to 79 days in storage, but titratable acidity showed no changes. Thereafter, the trend reversed up until 160 days (Zhang et al., 1994). Ye et al. (2000) compared a highly storable variety of the mandarin Ougan and the weakly storable variety Hongju for compositional and physiological changes during storage. In Ougan fruits, soluble sugar content in pulp increased and then rapidly decreased, titratable acidity decreased, superoxide dismutase activity first rose and then decreased, and malonaldehyde content was lower. In Hongju, sugars gradually decreased within the whole period of storage, acidity decreased, superoxide dismutase activity continuously decreased from harvest to full granulation, and malonaldehyde content was relatively higher.

Satsuma mandarins stored under 9 percent O_2 and 1 percent CO_2 atmosphere with 5 ± 1°C temperature and 73–79 percent RH had higher levels of vitamin C than fruit stored in air (Oogaki and Manago, 1977). Juice-quality parameters of Valencia oranges, such as TSS, titratable acidity, and vitamin C content, were not significantly influenced by either the low O_2 (0.25, 0.5, and 0.02 percent at 5°C and 10°C) or the high CO_2 (60 percent at 5°C) levels (Ke and Kader, 1990). In CA storage with elevated CO_2 (2, 3, 4, and 5 percent) levels, the changes in acidity and sugars of Kinnow mandarins were very slow at 5 ± 1°C and 8–30°C (Amarjit Singh and Rajinder Singh, 1996). In Nagpur mandarins, ascorbic acid content declined and its retention was not significantly affected by controlled atmospheres (2 percent CO_2 + 12 percent O_2 and 4 percent CO_2 + 7 percent O_2). Loss of acidity was less under CA conditions. Changes in total soluble solids were inconsistent, and increased CO_2 levels (up to 2 percent and 4 percent) had no adverse effect on flavor and fruit appearance during 9 weeks of storage at 5°C and 90 percent RH (Ladaniya and Shyam Singh, 2000).

B. Volatiles

Volatiles, which give typical citrus odor, emanate from fresh citrus fruit during storage. They include essence oil compounds from peel. The higher the storage temperature, the more rapid the loss. Hamlin sweet oranges have been shown to emanate ethyl acetate, ethanol, ethyl butyrate, limonene, ethyl caproate, and ethyl caprilate. Production of ethanol and acetaldehyde increases with storage duration and temperature. The occurrence of ethanol is attributed to spoilage (Attaway and Oberbacher, 1968). Peel oil volatiles – limonene, linalool, myrcene, citronellal, and citral – of Nagpur, Kinnow, and Darjeeling mandarin fruits are lost rapidly at room temperature (14–22°C and 65–88 percent RH) than at lower temperatures (Sinha et al., 1990). Higher humidity and lower temperatures (5–6.5°C at 60–70 percent RH and 9.5–19°C at 88–98 percent RH) have been found to improve retention of aroma volatiles by preserving the freshness of fruits.

C. Enzymes

Activities of polyphenoloxidase and peroxidase (Chauhan et al., 1980) and polymethylesterase, polygalacturonase, and cellulase (Nagar, 1994) have been reported to be increased during ambient storage, indicating the senescence of citrus fruits. During CA storage (reduced O_2 and increased CO_2) of citrus fruits, activity of alcohol dehydrogenase increases, indicating anaerobic respiration (Davis et al., 1973). Superoxide dismutase activity is reported to decrease with an increase in malonaldehyde during storage of *C. reticulata* fruits at 5.4–14.6°C and 82–92 percent RH (Ye et al., 1997). Zheng et al. (1999) studied enzyme activity and the relationship with granulation in pummelos. The activity of superoxide dismutase in fruit rind and pulp increased to a peak around 60 days of storage and then decreased. In pulp, peroxidase activity was low initially, then increased gradually after 90 days of storage. Catalase in rind and pulp decreased after 60–90 days.

D. Ascorbic Acid

Ascorbic acid (vitamin C) content decreases during storage of citrus fruits under ambient and refrigerated conditions. The loss is more rapid at higher temperatures. There is a loss of 10–20 percent vitamin C in usual handling and marketing practices of fresh produce. The longer the storage duration, the greater the loss. Adisa (1986) reported that the ascorbic acid content of healthy oranges gradually decreased as the storage temperature (5–30°C) and period increased, More ascorbic acid was lost in infected oranges than in healthy fruits, and there was no total loss of ascorbic acid. The ascorbic acid content declines in modified atmospheres with elevated CO_2 (2, 3, 4, and 5 percent) during long-term storage of Kinnow mandarins (Amarjit Singh and Rajinder Singh, 1996). Sweet orange fruits wrapped in non-perforated polyethylene maintain higher ascorbic acid content during refrigerated storage (Falak-Naz et al., 1997).

E. Pectic Substances

Fruit firmness decreases from the degradation of insoluble protopectin into the more soluble pectic acid and pectin. In citrus fruits these changes are relatively slow and less pronounced compared with climacteric fruits.

F. Lipid Fraction

Lipid fraction in peel and pulp of citrus fruit changes with aging on the tree and also during storage. Fuh and Ichi (1987) studied the fatty acid composition in flavedo tissue of Naruto (*Citrus medioglobosa*) fruit. The linoleic acid (18:2) content in all lipid fractions increased from October to mid-winter as the fruits degreened. The proportion of linolenic acid (18:3) and the 18:3/18:2 ratio decreased. These changes were later reversed as the fruits began to regreen. Total fatty acids, particularly in neutral lipids and phospholipids, increased markedly in the winter. Gibberellic acid delayed the decline in linolenic acid and the 18:3/18:2 ratio that occurred in storage at 5°C. During storage at 18°C for 5 weeks, there was a rapid decrease in the total fatty acids and major glycolipids accompanied by loss of chlorophyll and linolenic acid. Free sterol content and the sterol/phospholipid ratio increased in the storage. GA treatment delayed the reduction in the linolenic acid contents of neutral lipids, phospholipids, and glycolipids at the expense of oleic and linoleic acids. Ethephon treatment enhanced the reduction in linolenic acid. Low temperature had a similar effect to GA in maintaining the linolenic acid content of all lipid fractions (Fuh et al., 1988). In Satsuma mandarins, the total lipid and phospholipid contents of pulp juice and the proportion of linoleic acid to total fatty acids increased at 5°C and 25°C during storage. Oleic acid increased while palmitic and linolenic acid remained constant at 5°C. At 25°C, linolenic acid decreased and palmitic and oleic acid remained constant (Kadota and Miura, 1982).

REFERENCES

Abed-el-Wahab, W.A. (1990). Effect of fruit size on some physical and chemical properties of grapefruit during cold storage. Bull. Faculty of Agriculture University of Cairo, 41, pp. 959–971.

Adisa, V.A. (1986). The influence of molds and some storage factors on the ascorbic acid content of orange and pineapple fruits. *Fd. Chem.* 22, 139–146.

Agabbio, M., Chessa, I., Schirra, M., and Arraas, G. (1982). Cold storage of Citrus. First experiment in Sardinia on the storage of oranges of the cultivar Washington Navel, Torocco and Valencia late. Studi Sassaresi, III, 29, 3–24.

Agabbio, M., D'-Aquino, S., Piga, A., and Molinu, M.G. (1999). Agronomic behaviour and postharvest response to cold storage of Malvasio mandarin fruits. *Fruits*, Paris 54(2), 103–114.

Akamine, E.K. (1967). Tangerine storage. *Hawaii Agric. Expt. Storage Bull.* 142, 15.

Akamine, E.K., and Goo, T. (1979). Effect of storage on compositional changes in ripe sweet citrus fruits. Res. Report, Agric. Expt.-Station, Hawaii University, No. 210, 8 pp.

Angadi, G.S., and Shathakrishnamurthy (1992). Studies on storage of Coorg mandarins (*C. reticulata* Blanco). *South Indian Hort.* 40(5), 289–292.

Anonymous (1993). *Final report of Indo-USAID subproject on post-harvest technology of horticultural crops*. ICAR, New Delhi, 271 pp.

Arras, G., Fronteddu, F., and Delegu, M. (1997). Evalution of cold storage capability of some lemon cultivars. *Rivista di Frutticoltura e di ortofloricoltura,* Italy 59(2), 67–70.

Attaway, J.A., and Oberbacher, M.F. (1968). Studies on the aroma of intact Hamlin oranges. *J. Fd. Sci.* 33, 287–290.

Babarinsa, F.A. (2000). Reduction of weight loss in Valencia oranges using a brick-wall cooler. *Trop. Sci.* 40, 92–94.

Bandopadhyay, A., and Sen, S.K. (1996). Studies on physico-chemical changes associated with storage of Kagzi lime fruits. *Indian Agricul.* 40(1), 65–69.

Barua, P.C., and Yamdagni, R. (1996). Post-harvest quality and storage life of Khasi mandarin as influenced by different packages and maturity stages. *J. Agric. Sci. Soc. North East India* 9(1), 12–17.

Biale, J.B. (1953). Storage of lemons in controlled atmospere. *Calif. Citrog.* 38, 427, 436–438.

Bishop, D. (1996). Controlled atmosphere storage. In '*Cold and chilled storage technology.*' (C.J. Dellin, ed.). Blackie, London.

Bleinroth, F.W., Hansen, H.A., Ferreira, V.L.P., and Angelucci, E. (1976). Preservation of Tahiti and Sicilian lemon at low temperature. Coletanca Institute Technol. Aliment. 7(2):343.

Brackmann, A., Lunardi, R., and Donazzolo, J. (1999). Cold storage and decay control in 'Valencia' orange. *Ciencia-Rural.* 29(2), 247–251.

Burg, S.P., and Burg, E.A. (1967). Molecular requirements for the biological activity of ethylene. *Pl. Physiol.* 42, 144–152.

Chace, Jr. W.G. (1969). Controlled atmosphere storage of Florida citrus fruits. *Proc. 1st Int. Citrus Symp.*, Riverside, California, Vol. 3, pp. 1365–1373.

Chace, Jr. W.G., Harding, P.L., Smoot, J.J., and Cubbedge, R.H. (1966). Factors affecting the quality of the grapefruit exported from Florida. United States Department of Agriculture, Marketing Research Report 739.

Chattopadhyay, N., Hore, J.K., and Sen, S.K. (1992). Extension of storage life of sweet orange (*C. sinensis* Osbeck) cv. Jaffa. *Indian J. Pl. Physiol.* 35, 245–251.

Chattopadhyay, N., and Ghosh, C.N. (1994). Studies on the storage life of some sweet orange cultivars. *Haryana J. Hort. Sci.* 23, 9–16.

Chauhan, K.S., Kainsa, R.L., and Gupta, O.P. (1980). Enzymatic activity in citrus fruit as affected by various packages and length of storage. *Haryana J. Hort. Sci.* 9, 34–37.

Choudhari, K.G., and Kumbhare, G.B. (1979). Post-harvest life of sweet orange (*C. sinensis* Osbeck) as influenced by various poethylene film packs. *J. Maharashtra Agric. Univ.* 4, 228–229.

Chundawat, B.S., Gupta, A.K., and Singh, A.P. (1978). Storage behaviour of different grades of Kinnow fruits. *Punjab Hort. J.* 18, 156–160.

Cohen, E. (1991). Investigations on postharvest treatments of citrus fruits in Israel. *Proc. Int. Citrus symp.* Guangzou, China, pp. 32–35.

Cohen, E. (1999). Current knowledge on storage of citrus fruit under intermittent warming In *Advances in postharvest diseases and disorders control of citrus fruit.* (M. Schirra, ed.). Research Signpost Publication, Trivandrum, India, pp. 93–105.

Cohen, E., Rosenberger, I., and Shalom, Y. (1984). Effects of prolonged storage at low temperatures on chemical composition, physiological behaviour and diseases of Temple orange fruit. *Hassadeh* 64, 2236–2240.

Cohen, E., Ben-Yehoshua, S., Rosenberger, I., Shalom, Y., and Shapiro, B. (1990). Quality of lemons sealed in high-density polyethylene films during long-term storage at different temperatures with intermittent warming. *J. Hort. Sci.* 65, 603–610.

Cohen, E., Rosenberger, I., and Shalom, Y. (1988). Effect of volatiles on the development of chilling injury in long term storage of citrus fruits at sub-optimal temperature. *Israel Agresearch* 2, 57–65.

Cruz, V.L. (1995). Evaluation of losses and damage in Valencia oranges. (*Citrus sinensis* L. Osbeck) during simulated transport under forced ventilation. *Alimentaria* 33(268), 69–75.

D'Aquino, S., Palma, A., Agabbio, M., Tijskens, L.M.M., and Vollebregt, H.M. (2003). Response of three citrus species to different hygrometric conditions. *Acta Hort.* 604, 631–635.

Dalal, V.B., and Subramanyam, H. (1970). Refrigerated storage of fresh fruits and vegetables. *Climate Control* 3(3), 37.

Davis, P.L. (1971). Further studies on ethanol and acetaldehyde in juice of citrus fruits during the growing season and during storage. *Proc. Fla. Sta. Hort. Soc.* 84, 217–222.

Davis, P.L., Roe, B., and Bruemmers, J.H. (1973). Biochemical changes in citrus fruits during controlled atmosphere storage. *J. Food. Sci.* 38, 225–229.

Davis, P.L., Hoffmann, R.C., and Hatton, T.T. (1974). Temperature and duration of storage on ethanol contents of citrus fruits. *HortScience* 9, 376–377.

Dhillon, B.S., and Randhawa, J.S. (1983). Harvesting maturity and cold storage of Kinnow oranges. *Proc. Int. Citrus Symp.*, Bangalore, 17–22 December 1977, pp. 235–238.

Dhillon, B.S., Bains, P.S., and Randhawa, J.S. (1976). Studies on storage of Kinnow mandarins. *J. Research*, PAU 14, 434–438.

Eaks, I.L. (1955). The physiological breakdown of the rind of lime fruits after harvest. *Proc. Am. Soc. Hort. Sci.* 66, 141–145.

Eaks, I.L., and Masias, E. (1965). Chemical and physiological changes in lime fruits during and after storage. *J. Food Sci.* 30, 509–515.

Echverria, E.D., and Ismail, M.I. (1987). Changes in sugars and acids of citrus fruits during storage. *Proc. Fla. Sta. Hort. Soc.* 100, 50–52.

El-Ashwah, F.A., Hussain, M.F., and Safwat, M.M. (1975). Studies on storage of Baladi orange. *Agric. Res. Rev.* 53, 41–42.

Ezz, T.M. (1999). Eliminating chilling injury of citrus fruits by preharvest proline foliar spray. *Alexandria J. Agric. Res.* 44, 213–225.

Falak, N., Jan, M., Manzoor, Khan, F.K., and Shah, T.H. (1997). Effect of storage temperature and packaging material on the extention of storage life of cv. Blood red orange. *Sarhad J. Agric.*, Pakistan 13, 299–302.

Fuh, B.S., and Ichi, T. (1987). Effects of aging and temperature on fatty acid composition of the flavedo tissue of Naruto (*Citrus medioglobosa*). *J. Japanese Soc. Hortic. Sci.* 56, 344–351.

Fuh, B.S., Ichii, T., Kawai, Y., and Nakanishi, T. (1988). Changes in lipid composition in the flavedo tissues of Naruto (Citrus medioglobosa) during storage, and the effects of growth regulators and storage temperature. *J. Japanese Soc. Hortic. Sci.* 57, 529–537.

Ganji, M.E., and Rahemi, M. (1998). Effect of warm emulsions of thiabendazole and imazalil on reducing chilling injury, decay and electrolyte leakage in lime fruits. *Iranian J. Pl. Path.* 34, 56–66.

Gilfillan, L.M., and Saunt, J.E. (1991). Long term storage of lemons. *J. Southern African Soc. Hort. Sci.* 1(2), 65–68.

Gonzalez, A.G., Lafuente, M.T., Zacarias, L., and Ait-Oubahou, A. (1994). Changes in polyamines in 'Fortune' mandarin with cold storage and temperature conditioning. *HortScience.*

Gonzalez, A., Zacarias, L., and Lafuente, M.T. (1998). Ripening affects high-temperature – induced polyamines and their changes during cold storage of hybrid Fortune mandarins. *J. Agric. Fd Chem.* 46, 3503–3508.

Grierson, W., and Hatton, T.T. (1977). Factors involved in storage of citrus fruits – a New evaluation. *Proc. Int. Soc. Citric.* Vol. 2, pp. 227–231.

Guerra, F., Izquierdo, I., Otero, O., and Mazaira, M. (1988). Cold storage of tropical Satsuma mandarin. *Cincia Y. Tecnica en la Agricultura, Citricos of Ros Frutales* 11(3), 45–51.

Gupta, O.P., Chauhan, K.S., and Singh, J.P. (1981a). Evaluation of harvesting time of grapefruit cultivars for storage under cold storage conditions. Paper presented in *Symposium on Recent Advances in Fruit Development*, PAU, Ludhiana, 14–16 December 1981.

Hardenburg, R.E., Watada, A.C., and Wang, C.Y. (1986). The commercial storage of fruits, vegetables, and florist and nursery stocks. USDA, *ARS Agric. Handbook No.* 66, ARS. 128 pp.

Harvey, E.M. (1946). Changes in lemons during storage as affected by air circulation and ventilation. *U.S. Dept. Agr. Tech. Bull.* 908.

Hasegawa, Y. (1988). *Energy-saving storage of citrus fruit in Japan. Post-harvest handling of tropical and sub-tropical fruit crop*. Published by FFTC for the Asian and Pacific region. FFTC book Series No. 37 pp. 92–103.

Hasegawa, Y., Yano, M., and Iba, Y. (1990). Effect of high temperature pretreatment on storage of citrus fruits. Proc. Int. Symp. Citrus, China, pp. 736–739.

Hatton, Jr. T.T. and Reeder, W.F. (1968). Quality of Persian limes after different packinghouse treatments and storage in various controlled atmosphere, *Proc. Trop. Reg. Am. Soc. Hort. Sci.* 11, 23–32.

Izumi, H., Ito, T. and Yoshida, Y. (1990). Changes in fruit quality of satsuma mandarin during storage, after harvest from exterior and interior canopy of trees. *J. Japanese Soc. Hort. Sci.*5, 885–893.

Jain, P.K., and Chauhan, K.S. (1995). Extending shelf life of Kinnow mandarin with different storage conditions. *Indian J. Hort.* 52(4), 267–271.

Jawanda, J.S., Singh R., and Vij, V.K. (1978). Studies on extending postharvest life of Kinnow mandarin.*Punjab Hort. J.* 18, 147–153.

Jitender Kumar, Sharma, R.K. Singh R., and Godara, R.K. (1990). Increased shelf life of Kinnow mandarin (*C. reticulata*) by different storage conditions and chemicals. *Indian J. Agric. Sci.* 60(2), 151–154.

Kader, A.A. (1980). Prevention of ripening in fruits by use of controlled atmosphere. *Food Technol.* 34(3), 51–55.

Kader, A.A. (1992). *Post-harvest technology of horticultural crops*. Division of Agriculture and Natural Resources, 2nd edition Univiversity of California, Oakland, California, USA, ANR Publication. 3311.

Kadota, R., and Miura, M. (1982). Studies on lipids of citrus fruits. II. Effect of storage temperature on the lipid content and fatty acid composition of juice of satsuma mandarin (*Citrus unshiu* Marc.). *Bull. Faculty Agric. Miyazaki-University* 29(2), 275–283.

Kajiura, I. (1973). The effects of gas concentration on fruits VII. A comparison of the effects of CO_2 at different relative humidities, and of low O_2 with and without CO_2 in the CA storage of natsudaidai. *J. Japanese Soc. Hort. Sci.* 42, 49–55.

Karibasappa, G.S., and Gupta, P.N. (1988). Storage studies in Khasi mandarin.*Haryana J. Hort. Sci.* 17(3–4), 196–200.

Kaushal, B.B.L., and Thakur, K.S. (1996). Influence of ambient and evaporative cool chamber storage condition on the quality of polyethylene packed Kinnow fruit. *Adv. Hort. Sci.* 10(4), 179–184.

Kawada, K., and Kitagawa, H. (1986). Effects on juice composition of postharvest treatments for reducing acidity in citrus fruits. Kagawa Daigaku Nogakubu Gakuzyutu Hokoku Technical Bull. FacultyAgric., Kagawa-University 38(1), 1–4.

Ke, D., and Kader, A.A. (1990). Tolerance of 'Valencia' oranges to controlled atmospheres as determined by physiological responses and quality attributes. *J. Am. Soc. Hort. Sci.* 115, 779–783.

Khalifah, R.A., and Kuykendall, J.R. (1965). Effect of maturity, storage temperature and prestorage treatment on storage quality of Valencia oranges. *Proc. Am. Soc. Hort. Sci.* 96, 286–296.

Kim, W. (1988). Packaging, transportation and storage for selected fruit crops in Republic of Korea. In *Postharvest handling of tropical and sub-tropical fruit crops*, Food and Fertilizer Center for Asia and Pacific, Taiwan, pp. 56–66.

Kitagawa, H. (1981). Marketing of citrus fruits in Japan. *Proc. Int. Soc. Citric. Vol. 2,* pp. 848–852.

Kohli, R.R., and Bhambota, J.R. (1966). Storage of limes (*C. aurantifolia* Swingle). *Indian J. Hort.* 23, 140–146.

Kubo, N., and Heginuma, S. (1980). Effect of storage conditions on the quality and some of the constituents of satsuma mandarins. *J. Japanese Soc. Hort. Sci.* 49, 260–268.

Ladaniya, M.S. (2004a). Standardization of temperature for long term refrigerated storage of 'Mosambi' sweet orange (*Citrus sinensis* Osbeck). *J. Fd. Sci. Technol.* 41, 692–695.

Ladaniya, M.S. (2004b). Response of 'Kagzi' acid lime to low temperature regimes during storage. *J. Fd. Sci. Technol* 41, 284–288.

Ladaniya, M.S. (2004c). Reduction in postharvest losses of fruits and vegetables. NATP Project Final Report of NRCC, Nagpur, pp. 135.

Ladaniya, M.S. (1997b). Evaporative cool chamber for storage of 'Nagpur' mandarin. *Extension Bulle.* 13, NRCC, Nagpur, 10 pp.

Ladaniya, M.S. (2003). Citrus cold chain., In *Crop management and postharvest handling of horticultural products. Vol. 2. Fruits and vegetables.* Science Publisher, Enfield, NH, USA, pp. 239–276.

Ladaniya, M.S., and Sonkar, R.K. (1996). Influence of temperature and fruit maturity on 'Nagpur' mandarin (*C. reticulata* Blanco) in storage. *Indian J. Agric. Sci.* 66,109–13.

Ladaniya, M.S., Sonkar, R.K., and Dass, H.C. (1997). Evaluation of heat-shrinkable film wrapping of 'Nagpur' mandarin (*C. reticulata* Blanco) for storage. *J. Fd. Sci. Technol.* 34, 324–327.

Ladaniya, M.S., and Singh, S. (2000c). Response of 'Nagpur' mandarin to controlled atmosphere and refrigerated conditions. *Proc. Int. Symp. Citric.,* 23–27 November 1999, NRCC, Nagpur.

Lafuente, M.T., Martinez-Tellez, M.A., Gonzalez-Agullar, G., Mulas, M., and Zacarias, L. (1994). Physiological and biochemical responses associated with chilling sensitivity of 'Fortune' mandarin and temperature conditioning. Presented at workshop on *Postharvest Biology and Technology of Hort. Crops in the Mediterranean,* Chania, Greece, June, 16–21.

Li, J.Y., Shi, Y.P., Li, Z.Y., and Yuan, C.G. (1998). Effects of sugar ester and rhamnolipid on fresh-keeping of fruits. *Pl. Physiol. Commun.* 34, 115–117.

Liu, F.W., Pan, C.H., Hsueh, S.M., and Hung, T.H. (1998a). Influences of maturity at harvest and storage temperature on the storability of 'Ponkan' mandarin (*Citrus reticulata* Blanco). *J.Chinese Soc. Hort. Sci.* 44, 239–252.

Liu, F.W., Pan, C.H., and Hung, T.H. (1998b). Influences of harvesting date and storage temperature on the quality and storability of Tankan (*Citrus tankan* Hayata). *J. Chinese Soc. Hort. Sci.* 44, 253–263.

Lyons, J.M. (1973). Chilling injury in plants. *Ann. Rev. Pl. Physiol.* 24, 445–466.

Mahajan, B.V.C., and Sharma, R.C. (1997). Effect of different picking times on the storage behaviour of Kinnow. *Hort. J.* 10, 89–92.

Mann, S.S. (1978). Harvesting and storage of Kinnow fruits. *Punjab Hort. J.* 18, 154–155.

McDonald, R.E. (1986). Effect of vegetable oils, CO_2 and film wrapping on chilling injury and decay of lemons. *Hort Science* 2, 261.

McGlasson, W.G., and Eaks, I.L. (1972). A role of ethylene in the development of wastage and off flavors in stored Valencia orange. *HortScience* 7, 80–81.

Meir, S., Philosoph-Hadas, S., Lurie, S., Droby, S. Akerman, M., Zauberman, G., Shapiro, B., Cohen, E., and Fuchs, Y. (1996). Reduction of chilling injury in stored avocado, grapefruit, and bell pepper by methyl jasmonate. *Canadian J. Bot.* 74, 870–874.

Metlitskii, LV., Salkova, E.G., Volkind, N.L., Bondarov, V.I., and Yanyak, U.Y. (1985). *Controlled atmosphere storage of fruits.* Amerid Publishing Co. Ltd., New Delhi, 150 pp.

Miller, W.R., Chun, D., Risse, L.A., and Hatton, T.T. (1987). Influence of high-temperature conditioning on peel injury and decay of waxed or film wrapped Florida grapefruit after low-temperature storage. *Proc. Fla. Sta. Hort. Soc.* 100, 9–12.

Mukherjee, P.K., and Singh, S.N. (1983). Storage behaviour of citrus fruits grown in Northern India – A review. *Proc. Int. Citrus Symp.*, Bangalore, 17–22 December 1977. Horticultural Society of India, pp. 223–229.

Mulas, M., Gonzalez, A.G., Lafuente, M.T., Zacarias, L., Guardiola, J.L., Garcia-Martinez, J.l., and Quinlan, J.D. (1998). Polyamine biosynthesis in flavedo of 'Fortune' mandarins as influenced by temperature of post-harvest hot water dips. *Acta Horticul.* 463, 377–384.

Nagar, P.K. (1994). Effect of some ripening retardants on fruit softning enzymes of Kinnow mandarin fruits. *Indian J. Fruit Physiol.* 37, 122–124.

Nagar, P.K. (1995). Effect of different harvesting periods on fruit softening enzymes of Kinnow mandarin fruits. *Indian J. Pl. Physiol.* 38, 224–227.

Naik, N., Samiullah, R., Hittalmani, S.V., and Huddar, A.G. (1993). Studies on post-harvest treatments and physiological parameters of Kagzi lime fruit (*C. aurantifolia* Swingle). Paper presented in *Golden Jubilee Symposium on Horticulture Research – Changing Scenario,* Horticultural Society of India, Bangalore, pp. 358–359.

Oogaki, C., and Manago, M. (1977). Studies on the controlled atmosphere storage of Satsuma mandarin (*Citrus unshiu* Marc.). *Proc. Intl. Soc. Citric.* 3, 1127–1133.

Pal, R.K., Roy, S.K., and Srivastava, S. (1997). Storage performance of Kinnow mandarins in evaporative cool chamber and ambient condition. *J. Fd. Sci. Technol.* 34, 200–203.

Pantastico, Er. B. (1975). *Postharvest physiology, handling and utilization of tropical and subtropical fruits and vegetables*. AVI Publishing Co. Westport CT, USA, 560 pp.

Pantastico, Er. B., Soule, J., and Grierson, W. (1968). Chilling injury in tropical and subtropical fruits. II Limes and grapefruit. *Proc. Trop. Reg. Amr. Soc. Hort. Sci.* 12, 171–186.

Peretz, J., Rodov, V., and Ben-Yehoshua, S. (1998). High humidity packaging extends life of easily peeled citrus cultivars (*C. reticulata*). *Proc. 14th Int. Congress on Plastics in Agriculture*, Tel Aviv, Israel, March 1997, pp. 617–625.

Pesis, E., Marinansky, R., Zauberman, G., and Fuchs, Y. (1992). Reduction of chilling injury symptoms of stored avocado fruit by prestorage treatment with high nitrogen atmosphere. *Int. Symp. Postharvest Technology*, Davis, CA, August 1992.

Petracek, P.D., Dou, H., and Pao, S. (1998). The influence of applied waxes on post-harvest physiological behavior and pitting of grapefruit. *Postharvest Biol. Technol.* 14, 99–106.

Porat, R., Pavoncello, D., Peretz, J., Ben-Yehoshua, S., and Lurie, S. (2000). Effects of various heat treatments on the induction of cold tolerance and on the postharvest qualities of 'Star Ruby' grapefruit. *Postharvest Biol. Technol.* 18, 159–165.

Prange, R.K., Ramin, A.A., Daniels-Lake, B.J., DeLong, J.M., and Braun, P.G. (2006). Perspectives on postharvest bio-pesticides and storage technologies for organic produce. *HortScience* 41, 301–303.

Prasad, M. (1989). *Refrigeration and air-conditioning data book.* New Age International (P) Ltd. New Delhi, 304 pp.

Puppo, A.H., Zefferino, E., Bisio, L., Codina, J.C., and Supino, E. (1988). Study of ten different varieties of tangerines in Uruguay. *Int. Soc. Citric. Sixth Int. Citrus Congress*, Tel Aviv, Israel, 3, 499–1504.

Raison, S.K., and Orr, G.R. (1990). Proposal for a better understanding of the molecular basis of chilling injury. In *Chilling injury of horticultural crops.* (C.Y. Wang, ed.) CRC Press, Inc., Boca Raton, FL, pp. 145–164.

Rodov, V., Ben-Yehoshua, S., Albagli, R., and Fang, D.Q. (1995a). Reducing chilling injury and decay of stored citrus by hot water dips. *Postharvest Biol. Technol.* 5, 119–127.

Rodov, V., Ben-Yehoshua, S., Fang, D.Q., Kim, J.J., and Ashkenazi, R. (1995b). Preformed antifungal compounds of lemon fruit: citral and its relations to disease resistance. *J. Agric. Fd. Chem.* 43, 1057–1061.

Rosa, M.S. (1981). Physiological behavior of 'Marsh Seedless' grapefruit during short and long term cold storage. *Proc. Fla. Sta. Hort. Soc.* 731–734.

Roy, S.K., and Khurdiya, D.S. (1986). Studies on evaporatively cooled zero energy input chambers for the storage of horticulture produce. *Indian Hort.* 40, 26–31.

Ryall, A.I., and Pentzer, W.T. (1974). *Handling, transportation and storage of fruits and vegetables*, Vol. 2. AVI Publishing, CT, USA.

Rygg, G.L., and Harvey, E.M. (1959). Storage behaviour of lemons from the desert areas of Arizona and California. *USDA Mktg. Res. Rep.* 310.

Sale, J.M., and Lafuente, M.T. (2000). Catalase enzyme activity is related to tolerance of mandarin fruits to chilling. *Postharvest Biol. Technol.* 20, 81–89.

Sangay, D., Tanahashi, Y., Tsurusaki, T., and Vuthijumnok, K. (1999). Studies on storage houses for citrus fruits in south-western Japan. Memoirs College of Agriculture, Ehime University, 43, 105–111.

Schiffman-Nadel, M., Chalutz, E., Waks, Y. and Dagan, M. (1975). Reduction of chilling injury in grapefruit by thiabendazole and benomyl during long term storage. *J. Am. Soc. Hort. Sci.* 100, 270–272.

Schirra, M., (1993). Changes in fatty acid composition of flavedo tissue of 'Star Ruby' grapefruits in response to chilling and non-chilling storage temperatures. *Agricoltura Mediterranea* 123, 286–289.

Schirra, M., and Chessa, I. (1988). Physiological behaviour of Tarocco oranges during cold storage. *Proc. 6th Int. Citrus Congress*, Isreal, Vol. 3, pp. 1491–1498.

Schirra, M., and Cohen, E. (1999). Long-term storage of 'Olinda' oranges under chilling and intermittent warming temperatures. *Post-harvest Biol. Technol.* 16, 63–69.

Schirra, M., and Mulas, M. (1995). 'Fortune' mandarin quality following prestorage hot water dips and intermittent warming during cold storage. *HortScience* 30, 560–561.

Schirra, M.G., D'hallewin, P., Cabras, A., Angioni, S., Ben-Yehoshua, S., and Lurie, S. (2000). Chilling injury and residue uptake in cold-stored 'Star Ruby' grapefruit following thiabendazole and imazalil dip treatments at 20 and 50°C. *Postharvest Biol. Technol.* 20, 91–98.

Sealand (1991). *Shipping guide to perishables*. Sealand Services Inc., PO Box 800, Iselim, NJ, USA.

Sembdner, G., and Parthier, B. (1993). The biochemistry and physiological and molecular actions of jasmonates. *Ann. Rev. Plant Mol. Biol.* 44, 569–586.

Shaw, P.E., Moshonas, M.G., and Pesis, E. (1991). Changes during storage of oranges pretreated with nitrogen, carbon dioxide and acetaldehyde in air. *J. Fd. Sci.* 56, 469–474.

Sindhu, S.S., and Singhrot, R.S. (1994). Effect of different storage conditions and antifungal fumigants to enhance the shelflife of lemon (*C. limon* Burm.) cv. Baramasi. *Haryana J. Hort. Sci.* 23, 273–277.

Singh, A., and Singh, R., (1996). Quality of Kinnow mandarin as affected by modified atmosphere storage. *J. Fd. Sci. Technol.* 33, 483–487.

Singh, A., Singh, R., Dhiman, I.S., Singh, A., and Singh, R. (1992). Effect of modified atmosphere on rotting of stored Kinnow mandarins. *Folia Horticult.* 4, 43–57.

Singh Kirpal, K. (1971). Storage behaviour of sweet orange and mandarins. *Technical Bull.* (Agric Series), ICAR, No. 35, 106 pp.

Singh, S.B., Sinha, M.M., and Maurya, C.P. (1987). Effect of different pre and poatharvest treatments on storage life of Malta cv. Blood Red (*C. sinensis* Osbeck) fruits. *Progressive Hort.* 19, 10–16.

Singh, S.N. (1975). Storage of grapefruit under low temperature condition. *Pl. Sci.* 7, 76–79.

Smoot, J.J. (1969). Decay of Florida citrus fruits stored in controlled atmosphere and in air. *Proc. First Int. Citrus Symp.* Vol. 3, 1285–1294.

Spalding, D.H., and Reeder, W.F. (1976). Low pressure (hypobaric) storage of limes. *J. Am. Soc. Hort. Sci.* 101, 367–370.

Subedi, P.P. (1998). Effect of maturity stages and anti-fungal treatment on post-harvest life and quality of mandarin oranges in a cellar store. Working Paper. Lumle Agricultural Research Centre, Nepal, 1997–98, No. 98–27, 15 pp.

Sun, Y.T. (1998). Citrus fruit respiration and CA storage. *South China Fruits* 27, 16–17.

Testoni, A., Cazzola, R., and Ragozza, L. (1992). Storage behaviour of orange 'Valencia Late' in rooms with ethylene removal. *Proc. Int. Soc. Citric. VII Int. Citrus Congress,* (Italy), Vol. 3, pp. 1092–1094.

Thangaraj, T., Irulappan, I., and Rangaswamy, P. (1993). Pre-storage treatment of acid lime fruits under cool chamber storage. Paper presented in *National Seminar on optimization of production and productivity of acid lime.* The Horticultural Society of India and Horticultural Collage and Research Institute, Periyakulam, Tamilnadu, India.

Thompson, A.K. (1998). *Controlled atmosphere storage of fruits and vegetables*. CAB International, Wallingford, Oxon, UK, 278 pp.

Trivedy, G., and Mazumdar, B.C. (1982). Cold storage trial of Darjeeling orange. *Indian Agricult* 26, 51–56.

Tugwell, B.l., and Chvyl, W.L. (1995). The effect of storage temperature on development of rind blemishes on Washington naval orange. *Conference on Science and Technology for the fresh food revolution*, Melborne, Australia, 18–22.

Umbarkar, S.P., Bonde, R.S., and Kolase, M.N. (1991). Evaporatively cooled storage structure for oranges (*C. reticulata* Blanco). *Indian J. Agric. Engg.* 1, 26–32.

Underhill, S.J., Dahler, J.M., McLauchlan, R.L., and Barker, L.R. (1999). Susceptibility of 'Lisbon' and 'Eureka' lemons to chilling injury. *Aust. J. Exp. Agric.* 39, 757–760.

Vakis, N., Grierson, W., and Soule, J. (1970). Chilling injury in tropical and subtropical fruit III. The role of CO_2 in suppressing chilling injury symptoms of grapefruit and avocado. *Proc. Trop. Region Am. Soc. Hort. Sci.* 14, 89–100.

Valero, D., Martinez, R.D., Serrano, M., and Riquelme, F. (1998a). Influence of postharvest treatment with putrescine and calcium on endogenous polyamines,firmness,and abscisic acid in lemon (*Citrus limon* L. Burm Cv. Verna). *J. Agric. Fd. Chem.* 46, 2102–2109.

Wai-chin, S. (1990). Experimental orchard storage of unshu mandarin. *Proc. Int. Citrus Symp.*, China, pp. 743–744.

Waks, J., Chalutz, E., Schiffmann-Nadel, M., and Lomeniec, E. (1985). Relationship among ventilation of citrus storage room, internal fruit atmosphere, and fruit quality. *J. Am. Soc. Hort. Sci.* 110, 398–402.

Wasker, D.P., Damane, S.V., and Gaikwad, R.S. (1999). Influence of postharvest fungicidal application and packaging material on storability of sweet oranges at various environments. *Proc. Nat. Symp. Citric.* 17–19 November 1997, NRCC Nagpur, pp. 476–578.

Wild, B.L., and Rippon, J. (1973). Lemons held in storage. *Queensland Fruits Vegetable News* 44, 117–118.

Wild, B.L., McGlasson, W.B., and Lee, T.H. (1976). Effect of reduced ethylene levels in storage atmosphere on lemon keeping quality. *Hort Science* 11, 114–115.

Wild, B.L., McGlasson, W.B., and Lee, T.H. (1977). Long term storage of lemon fruit. *Fd. Technol.* Australia 29, 351–357.

Wills, R.B.H., Ku, V.V.V., Shohot, D., and Kim, G.H. (1999). Importance of low ethylene levels to delay senescence of non-climacteric fruit and vegetables. *Aust. J. Exp. Agric.* 39, 221–224.

Wojciechowski, J. (1989). Ethylene removal from gases by means of catalytic combustion. *Acta Hort.* 258, 131–139.

Ye, M.Z., Chen, Q.X., Xu, J.Z., Xu, X.Z., Zhao, M., and Xia, K.S. (2000). Some physiological changes and storability of citrus fruits during storage. *Pl. Physiol. Commun.* 36, 125–127.

Zhamba, A.I. (1987). Results of trials on storage of oranges in a gas medium. Snizhenie-Poter'-pri-Khranenii-Plodov, *Ovoshchei-i-Vinograda*, pp. 44–48.

Zhang, S.L., Chen, K.S., Liu, C.R., and Zhang, B.S. (1994). Studies on postharvest physiology and storage of natsudaidai fruits. *Acta Agriculturae Zhejiangensis* 6(2), 119–123.

Zheng, G.H., Pan, D.M., Qiu, Y.P., and Pan, W.G. (1999). Water content of tissues, protective enzyme activity and their relationship to juicy sac granulation in pummelo at postharvest stage. *J. Fujian Agric. Univ.* 28, 428–433.

13

TRANSPORTATION

Transportation of harvested citrus fruits takes place in two stages. The first stage is from orchard to packinghouse or local assembly market. The second stage comprises the transport of treated, packed, and cooled fruit to distant destinations within the country or to overseas markets, including retail distribution.

Several means are used to move fresh fruit from the point of production to packinghouses or nearby markets. The most common means are trucks of different capacities, as they can reach the interiors of production areas and very easily move 2–5 tons (in small trucks) or 10–15 tons (in big trucks) of fruit at a time. In remote areas of the hilly northeastern region of India, fruits are carried in conical bamboo baskets on the backs of men and women from orchards to nearby markets. Each basket contains 500–1000 fruits depending on size. Small consignments are also carried by bus. Fruits are transported even on animals. In Japan, harvested fruits from hill slopes are taken down to plains by ropeways and monorails.

In central part of India, mandarin fruits are transported to nearby wholesale markets in bullock carts using dry paddy straw as cushioning. Bullock carts and tractor-drawn trolleys can move on temporary field roads that are uneven and bumpy. Animal-driven carts with rubber tires as well as wooden wheels are also common in other parts of India. 'Nagpur' mandarins and sweet oranges are transported loose in trucks to nearby local assembly markets called '*Mandies*'. For distant transport, trucks of different sizes are used. Each large truck can carry 500–550 boxes. In 1994–1995, 14 100 truckloads of mandarins were sent from the Vidarbha region of Maharashtra state alone. About 3000 rail wagons have also been sent to distant markets (DMI, 1995). Each wagon carries 650–750 boxes. Vented plastic crates made of HDPE are also being used for transportation of mandarins to distant markets but on a limited scale. These crates, which hold 18–19 kg fruit, are quite sturdy and have been used in fields and packinghouses for last 10–15 years. Transporting fruit loose in trucks is cheaper than packing it in wooden boxes. Bulk or loose transport from farm to the terminal market is cheapest. Horizontal dividers are placed 90 cm high in the body of the truck to minimize transit damage.

In the well-established citrus industries of the world, transportation is done along more systematic lines. Standardization of transport vehicles with respect to size, capacity, and handling procedures in the advanced citrus-producing countries has already been done. Bulk bins/pallet boxes and hydraulic forklift equipment/trucks are used to handle fruits in the fields in the U.S. and many other countries. Bins/boxes are loaded onto trucks that carry the fruit to packinghouses. In Florida, filled pallet boxes in the orchard are loaded onto relatively smaller vehicles called grove trucks. These vehicles have hydraulic equipment (lightening loader) for transportation up to the main road/highway, where the boxes are then transferred to flatbed semi-trailers and carried to packinghouses (Grierson and Wardowski, 1986). From the packinghouse, fruit is taken to destination markets using various means of surface transport (road and sea) and air transport (Fig. 13.1).

I. ROAD AND VEHICLE CONDITION

The condition of the road and the way that the vehicle (truck or trailer) is driven determines the damage done to the fruit. The vehicle must have air-suspension for shock absorption and proper cushioning should be provided for the fruit. The shocks are transmitted in a standard steel-spring suspension system during transport from grove to packinghouse and can cause abrasion bruises to 20 percent of the fruit touching the sidewalls of the bins (Brown, 1995). In most developing and underdeveloped countries, roads to remote areas and even highways are not paved, and careless driving on such uneven roads often results in unsightly blemishes on fruit surfaces. Limes, lemons, and mandarins are very prone to bruises. Fruit damage occurs by the rubbing of the fruit against other surfaces such as the container or the truck; therefore fruit has to be immobilized. The bigger the container the lower the losses from bruises because the contact surface per unit load of fruit decreases. Containers with rough surfaces and sharp edges are bound to cause injuries.

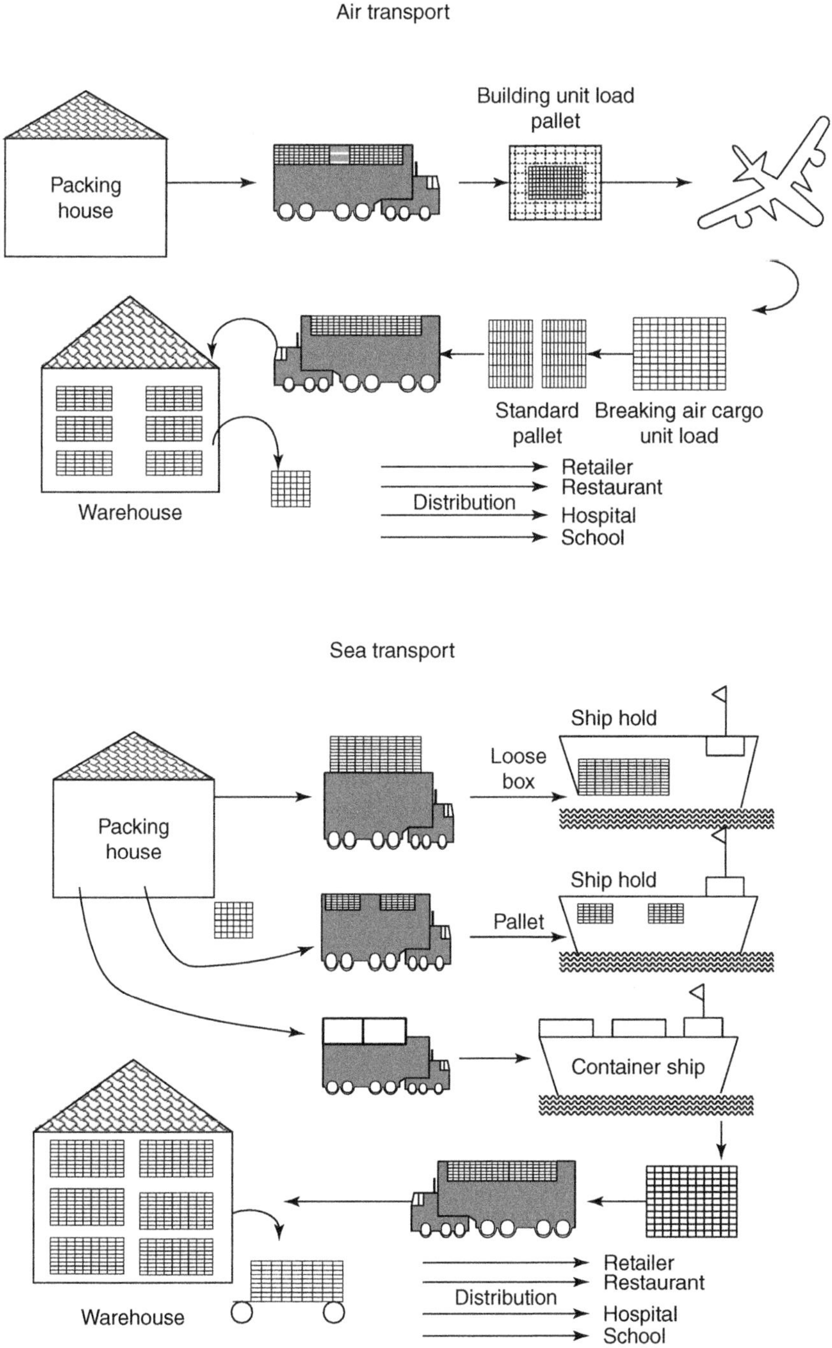

FIGURE 13.1 Air and Sea Transport Modes.

II. SURFACE TRANSPORT

Surface transport includes the movement of fruit through road-, rail-, and waterways (sea/ocean). All types of light and heavy vehicles, including trucks, five-wheelers and auto-rickshaws, are used for transporting fruit in developing countries. Trucks, highway trailers, and railcars are the main transport means for citrus fruits. For export to overseas countries, refrigerated containers and refrigerated ship holds are utilized. The trailer-on flat-car (TOFC) is utilized for export and domestic shipments. TOFCs can be used on railways and highways, and can be rolled on and rolled off for ocean transport on vessels. The regulation of temperature, relative humidity, and ventilation are important during shipment. The maintenance of the recommended temperature will reduce (1) senescence and related changes due to ripening, softening, and color changes, (2) respiration and metabolic activities, (3) decay, (4) water loss, (5) off-flavor. Obtaining and maintaining the desirable temperatures is affected by (1) ability of the refrigeration equipment to remove heat, (2) initial temperature of the fruit, (3) air distribution, (4) heat load. For best results, temperature has to be maintained as close as possible to recommended levels. The thermostat is set slightly above the optimum to avoid chilling the product.

Citrus fruits continue to lose water in vapor form in the process of transpiration. Excessive loss of water from fresh fruits results in wilting, shriveling, toughening of fruit rind, and lack of natural flavor. Moisture loss can be reduced by (1) lowering the temperature, (2) maintaining high relative humidity (80–90 percent), (3) waxing with a water-impermeable coating, (4) providing only enough air movement to remove the heat of the respiration, (5) maintaining the refrigerator coil at 1–2°C lower than the desired cargo temperature, (6) packaging in semi-permeable bags or films.

Refrigeration, air circulation, air exchange, and modified atmosphere systems are required to control storage and transport atmosphere around the fruit. Most systems can provide heating also when the vehicle is passing through areas with suboptimal temperatures. Humidifiers are also part of the cooling system. Fans of required capacity are necessary to operate under high static pressure for sufficient air circulation. Corrugated or T-beam floors and ribbed walls in the vehicles help improve air circulation. The precise temperature-control thermostats with sensors in the return air, in the discharge air or in both are available. Presently, controlled-atmosphere units are not used for citrus fruit shipments since the benefits are not significant enough to justify high freight costs. In reefers, equipment for air exchange to reduce undesirable concentrations of CO_2 and ethylene can be very desirable to maintain freshness of citrus fruits during long-distance voyages.

For domestic markets in many citrus-producing countries of the world, transit times are not more than 4–5 days. In hot tropical climates, maintenance of near-optimum temperatures becomes imperative to retain quality and reduce losses. The following factors need to be considered before loading fruit for the shipment:

(a) Fungicide and wax treatment to fruit to prevent decay and weight loss. Careful handling, good sanitation, and low temperature management.
(b) Packaging containers with required strength to prevent collapse.
(c) Sufficient ventilation in packaging containers, allowing air movement into the container to cool the fruit and to remove volatiles such as ethylene and carbon dioxide. This allows oxygen to reach the fruit and prevent anaerobiosis and the development of off-flavors. At the time of palletization and loading, the alignment of the vents need to be maintained.
(d) Selection of high-quality fruit to meet grade standards. Shipping inferior fruit can lead to serious losses if consignment is rejected since freights are high.

For overseas voyages, transport equipment has to be serviced prior to each loading to ensure that the desired temperature is achieved. Transport vehicles or reefers should be checked for the following points:

(a) The vehicle interiors should be clean and free of odors. Trash or debris in grooved floors can restrict airflow and adversely affect temperature maintenance.
(b) Transport vehicles should have special equipment or capabilities such as built-in load bracing, reverse airflow, ventilation, and heating. An air-exchange system is useful for ventilation of the load compartment with outside air to reduce off-flavor.
(c) The drains in the floor of the container or trailer should be clear to carry defrost water.
(d) The door and closing device should be checked for tight seals to prevent excessive heat transfer. The container body and insulation should also be checked for any damage.
(e) Refrigeration units and blowers should be working properly and thermostats set at the desired level for the intended cargo. Refrigeration units of truck trailers and TOFCs are powered by diesel motors. For marine containers, electricity is provided on docks and ships. These units are also powered by diesel motors or generator sets while on the road. Allow the unit to precool the vehicle before loading. Temperature monitoring and recording system/data loggers should be functional. Modern reefer containers have integral computer or electronic devices for automatic data recording.

To make economic use of space, transportation of mixed loads is common during local distribution from the wholesaler/supplier to retail stores and also during long-distance transit. The compatibility of the commodity is of prime importance in mixed-load transportation. The temperature requirements of the commodities, their response to relative humidity, and their sensitivity to ethylene should be considered. Citrus fruits are prone to chilling injury and hence should not be kept in contact with ice along with vegetables. Biphenyl-treated fruit should not be transported in a mixed load as other commodities are likely to develop off-flavors. Citrus fruits produce odors that are absorbed by animal and dairy products such as meat, milk, and milk products. Vegetables such as garlic, ginger, and onion produce odors that are picked up by citrus fruits.

Grapefruits from late season can be shipped at about 10–12°C with cucumbers, brinjals, limes, potatoes, pumpkin, winter squash, and watermelon. Early-season grapefruits are susceptible to chilling and hence should be shipped at 15–16°C with other commodities. Oranges and tangerines can be transported at 4.5–7°C with 90–95 percent RH with compatible commodities such as beans, lychees (litchi), *bhindi* (okra), green peppers, summer squash, pink tomatoes, and melons. Lemons can be transported with cantaloupes, cranberries, and lychees. Limes should not be shipped at temperatures below 7.5°C. For most perishables, RH in the range of 85–90 percent is desirable.

A. Trailers/Trucks (Refrigerated and Non-refrigerated)

In most citrus-growing countries, the bulk of fresh citrus is transported by refrigerated trucks and highway trailers. Truck sizes vary in different countries. A 12.2 m trailer usually carries about 1000 citrus cartons. Refrigerated trailers generally have a 22-ton fruit-holding capacity, $12.19 \times 2.26 \times 2.49\,m$ ($60.59\,m^3$) or $14.63 \times 2.43 \times 2.43\,m$ ($86.38\,m^3$).

Before loading into truck trailers, fruit has to be properly precooled because the capacities of refrigeration units in truck trailers are generally only enough to maintain the temperature. Trailers with 2–3 separate compartments are available to carry products with different temperature requirements. Refrigerated trailers are 12.19 and 14.63 m long, and refrigeration capacity is about 10080 Kcal/h.

Depending on the distance of production areas from major markets, transit times vary greatly. Trucks do not take more than 4–5 days between coasts in the U.S. Fruit from the southern states of Florida or Texas takes 2–3 days to reach New York and other markets in the northeast. Within the European Union, 2–4 days is needed for transit from production areas to consuming markets. Transportation time also depends on topography, terrain, and road conditions. In India, trucks take 3–6 days from production areas in Western or Central India to reach markets in distant corners of the northeast of the country.

Air distribution takes place from the refrigeration unit through a canvas or metal duct attached to the ceiling and extending one-half to two-thirds of the length of the trailer. The cold air moves over the top of the load to the rear until it reaches the doors. Air returns back through the horizontal channels of the load from the rear to the bulkhead, and then back to the evaporator coil. For securing unitized loads, dunnage is used against flat walls. Generally a center-line loading pattern is preferred.

TABLE 13.1 Capacity of Trucks Generally Used for Fruit Transportation in India to Carry Corrugated Cartons and Conventional Wooden Boxes

Truck dimensions (cm)	CFB box (50 × 30 × 30 cm)	Wooden box (47.5 × 32.5 × 32.5 cm)
200 L × 243 W × 243 H	256	175
300 L × 243 W × 243 H	336	280
510 L × 210 W × 210 H	490	336
524 L × 210 W × 210 H	497	360
540 L × 210 W × 210 H	518	390
560 L × 210 W × 210 H	546	408

Ladaniya et al. (1999).

There are different ways to arrange fruit cartons in the trailer. A pigeonhole and a modified, bonded-block load arrangement both provide good air distribution. In pigeonhole loads, boxes are arranged in straight lengthwise channels in alternate layers for longitudinal air circulation. In modified, bonded-block load arrangements, air channels are connected in alternating layers so that circulating air reaches each carton in the load. The return air inlet at the bulkhead must be open no matter which channel-load pattern is followed. In pigeonhole load arrangements, it is easy to check for open channels.

In non-refrigerated trucks, like those used in many Asian countries, wooden boxes are staked to use maximum space. The major problem in transportation of wooden boxes is that these containers are not of uniform size (as they are not machine-made and thus slight variations occur). During transportation, as bouncing and vibration occur, space is created and movement of boxes take place that damages both boxes and fruit.

In India, nearly 70–80 percent of produce is transported by road either loose or packed. The largest size of truck carries 400–500 boxes (Table 13.1). Road transport takes 50–60 h from Nagpur to Delhi; the same distance is covered by train in 120–130 h.

B. TOFC

TOFCs (Trailer-on-flat-car) are also called as piggyback trailers. TOFCs are 12–14 m trailers that remain on their chassis during loading on flat railcars and also during road transport. The advantage of using TOFCs is that they can start from the packinghouse where railway facilities are not available. Then they can be moved over highways to railroad ramps at railway stations to load onto flatcars for further transportation by rail. They can also be transported by road to warehouse or a final destination. An independent refrigeration unit with nearly 5000–9000 kcal capacity per hour is fitted to each TOFC. Air is distributed through ducts at the top for cooling from blowers in the front to the rear. The walls of the trailer are corrugated for air movement.

C. Rail Cars

The major advantage of transportation by rail mode is lower freight cost, although it is slow. Other advantages are the high cube capacity of each car, with high refrigeration capacity systems capable of cooling warm fruit. Mechanically refrigerated rail cars used in the U.S. for transportation have interior dimensions of 15.47 × 2.74 × 2.74 m. These cars can carry 2754 cartons, which weigh more than 45 tons. Railroad cars 18.28 m long are also used. Fruit is transported within a week from production areas in the west to markets on the east coast of the U.S. Railcars also carry citrus fruits from the U.S. to Canada.

Cartons are generally loaded onto railcars in an open-stack pattern called a chimney stack. Air moves from top to bottom through the load in this arrangement. Cold air is discharged from the evaporator coil (at one end of the car) into a ceiling plenum running along the length of the vehicle. The perforations/holes in the plenum allow the air to move down through the load, under the floor, and back to the evaporator coil of refrigeration unit. Fruit is loaded on pallets or stacked by hand directly onto the floor.

Unventilated rail wagons are also used in India, but such wagons are not suitable for transport of citrus fruit and using them can result in huge losses. Rail transport is also slow, taking 4–5 days as compared with trucks, which take 2 days to cover the same distance. A train wagon accommodates about 700–750 wooden boxes. Collaborative studies of CFTRI with Southern Railways in India showed that air-conditioned coaches, after appropriate structural changes, could be used for transportation of fresh produce including citrus (CFTRI, 1982). Transportation by rail is cheaper than road transport. For rail transport, fruit has to be taken to railway stations by trucks, which involves loading and unloading – unless inter-modal TOFC facilities are available.

D. Reefer Containers (Refrigerated Containers)

Van containers or marine containers or reefers carry maximum loads of citrus fruit in long-distance international trade through sea freight. The containers can be transported on road, rail, and ship, which makes them very versatile (Fig. 13.2, see also Plate 13.1). The containers can be transported on roll-on and roll-off ocean vessels and highways. While transporting on railway flatcars, these containers are stacked one-high or two-high. Containers are stacked on ocean container vessels also. For electrical supply, power plugs are provided aboard ship and also at depots. The mechanical refrigeration is a self-contained unit that makes these containers popular among exporters. Some newer ships have cooling towers to which containers are attached in the ship's holds. The major advantage of reefers is convenience in routing and inter-modal handling. There is no handling of cartons

FIGURE 13.2 Reefer Container with Cooling Unit.

en route when shipped by container. The loading takes place at the packinghouse; unloading takes place at the warehouse/port in the importing country. The handling costs are less, and there is labor savings with increased utilization of equipment. This reduces damage to fruit and cartons, and pilferage is also almost nil. With a continuous electricity supply, breaks in refrigeration are minimum, which maintains the fruit quality.

The time taken to reach the market is relatively short in reefers and consequently fruit quality is better. A voyage from Australia to the east coast of the U.S. takes 6 weeks, while to Singapore it takes just 1 week. For long voyages, a temperature of 10°C is recommended for early fruit of Washington Navel oranges. The normal recommended temperature for shipment is 5°C for Washington Navel oranges grown in Australia. At 5–10°C, weight loss is just 3.5–3.8 percent during a 5–6 week voyage, which is a manageable figure. Chilling injury is reported at 2°C (Tugwell and Chvyl, 1996). Exporting citrus fruits from Florida to Japan takes a total of 35–40 days. California to Western Europe requires about 28–32 days if routed via ship through the Panama canal. The same shipment, if routed by rail to Houston, Texas and from there by ship to Europe takes 25–26 days. If the shipment is routed by rail to Newark, New Jersey, and from there by ship to Europe, the time taken is 20–23 days. Only about 16 days are required for the same if routed by highway truck to New Jersey and loaded into van containers for further ocean shipment (Harvey and Houck, 1986).

Kinnows from Rajasthan and Punjab states packed in 10 kg–capacity boxes (three layers of fruit per box and 48–84 per carton count) are exported in refrigerated containers (4°C) to Sri Lanka and Malaysia. The containers have plug-in facilities at ports and also during the ship voyage. From the date the containers are loaded at the packinghouse, it takes 28–30 days to reach the Thames port at London, 10–12 days to reach the Colombo port in Sri Lanka, and 18 days to the ports of Saudi Arabia. Shipping by reefer container would take about 8–10 days to South East Asian markets.

Citrus fruit can be held under optimum conditions in van container during overseas voyages that may last up to 30 days or more. Van containers are reliable for delivery to distant markets anywhere in the world. In van containers, heating arrangements are also provided in case of very low temperatures en route. During sea transport of grapefruits between Cuba and the Baltic Sea lasting up to 19 days during January to April, external air temperatures decrease from approximately 22–24°C to 4°C (Cruz, 1995). The heating system must start automatically to maintain the required transit temperatures. Transit temperature varies depending on the type of citrus fruit and must be critically maintained (Table 13.2) if outside temperatures are below 0°C for considerable periods of time.

Grapefruits derived from cold storage are not recommended for shipping, especially with late-harvested fruit. Small fruits have been found to be more susceptible to decay and damage compared to large ones. For export shipments of Florida grapefruit, a transit temperature of 16°C is recommended from October to January and 10–11°C for the remaining seasons. Temperatures below 11°C leads to chilling injuries. Relative humidity should be maintained at 85–90 percent (Hale et al., 1981).

1. Air Distribution in Reefers

The cooling and air circulation capacity of the reefer is designed only to maintain fruit temperature, not for precooling. Precooling of fruit must be done before loading. In marine containers, the common air-distribution systems are under-the-load and over-the-load.

(a) **Over-the-load system:** This system is similar to systems used in truck trailers where air is delivered at the top through a duct. Air is distributed through open, horizontal channels at the rear. Then air passes longitudinally through the load from the rear of the van to the front.

(b) **Under-the-load-system:** Cold air from the evaporator coil passes under the load through the ribbed floor. Air is forced up through the bottom and top vents of the cartons to the top of the load and back to the evaporator coil. Cartons with ventilation holes at the top and bottom need to be arranged in register alignment to provide passage to the cool air through the vents. If such cartons are not used, vertical channels need to be provided by chimney-type arrangement.

TABLE 13.2 Relative Humidity and Temperature Requirements for Citrus Fruits During Transportation

Citrus fruit	Relative humidity (%)	Temperature (°C)
(1) Sweet oranges		
Valencia (California and Arizona)	85–90	4–9
Valencia (Florida and Texas)	85–90	2
Shamouti	85–90	5–6
Mosambi and Sathgudi	90	5–6
Navel (California)	85–90	5–6
(2) Mandarins and Tangerines		
Satsuma	80–85	2–3
Clementine	85–90	4
'Nagpur' mandarin	90	5–6
Dancy tangerine	90	4–5
Kinnow mandarin	85–90	4–5
Temple	85–90	4–5
Orlando tangelo	85–90	4–5
(3) Lemons		
Dark green	85–90	13–15
Light green	85–90	10–13
Yellow	85–90	9–10
(4) Limes		
Mexican, 'Kagzi' Dark green	85–90	10–11
Yellow	85–90	8–9
Tahiti	85–90	9–10
Bearss	85–90	7–13
(5) Grapefruits		
Florida (October–January)	85–90	15–16
Florida late	85–90	10
California – Arizona	85–90	15–16
(6) Pummelos	85–90	8–9

Hale et al. (1981), Hardenburg et al. (1986), Ladaniya and Sonkar (1996), Ladaniya (2004a,b).

E. Cargo Holds of Ships

Individual pieces of the cartons are stacked in the cargo hold of the ship, hence this is also called a break-bulk system. A large volume of citrus is conventionally shipped overseas in the break-bulk ships with refrigerated cargo holds. With advancements in technologies and facilities at ports, fruit is unitized on pallets and loaded into the cargo hold. The compatibility and temperature requirements of various commodities have to be considered in cargo holds when several types of fruits and vegetables are shipped together.

Because of loading and unloading of individual boxes, damage to cartons and fruit is relatively higher in break-bulk ships. The cooling is also not as efficient as in van containers. Ships stop en route for additional cargo. During the loading of break-bulk refrigerated ships, all boxes need to be stacked in register and in direct vertical alignment. Modern, refrigerated break-bulk ships have central refrigeration systems capable of quickly cooling large quantities of warm fruit. Temperature and relative humidity sensors and recorders are also fitted to monitor and record fruit temperature and surrounding RH throughout the load.

Chartered refrigerated ships for the exclusive use of one or more shippers can be used to avoid most of the problems in break-bulk shipping. Such an arrangement may also result in lower freight costs than other modes of shipping.

III. CONDITIONING AND QUARANTINE TREATMENTS DURING TRANSPORTATION

The Mediterranean fruit fly (*Ceratitis capitata* Wied.), which is a major post-harvest insect-pest, poses a great threat to exports from several citrus-exporting countries. The Mexican fruit fly and the Caribbean fruit fly are also major fruit flies that are destructive to fruit. Therefore, quarantine treatments are required for fruits shipped from infested areas to markets in non-infested areas. Low temperatures (chilling) have been tried to disinfest the fruit. Cold treatments of citrus fruit for the quarantine purposes can be performed commercially in the approved, instrumented ships or the van containers. Precise temperature control is crucial and must be monitored and recorded at several locations in the load at hourly intervals during the treating period. For cold treatment, citrus fruit are maintained at one of the following time–temperature combination: 0.6°C for 10–12 days, 1.6°C for 14 days, or 2.2°C for 17 days. Lower temperatures need shorter treatment time and higher temperatures require longer treatment time. The temperature has to be selected on the basis of fruit tolerance.

The beneficial effects of conditioning on lemons with respect to fruit quality are well known. Sale (1993) conducted a simulation study for 28 days (the transit period from a packinghouse in New Zealand to Japan by sea fright) with Genoa lemons packed in cartons with or without polyethylene liner. Fruits were held in a container at a constant 20°C temperature. (Lemons are normally shipped at around 8–11°C.) Smoothness of skin was noticeably improved, skin thickness was reduced by >10 percent, and juice content increased. No fungal rots were observed for 2 weeks, and the incidence was only 6 percent after 28 days, which falls in the normal range. After conditioning, very few fruits developed rots and their shelf life was well over 3 weeks under ambient conditions.

Grapefruit get chilling injuries when exposed to temperatures below 10°C. It is observed that fruit from the outer canopy of the tree is more susceptible to chilling injury than interior fruit. The injury leads to decay.

During shipment, quarantine treatment to Florida grapefruits has been tried in van containers with a fruit temperature at 2.0 or 2.2°C for 14 days or more (Houck and Hinsch, 1983). Solid stack load, under-the-load air distribution, and a sensor for air-temperature control located in the discharge air stream provided quicker cooling. The van with an air channel stacked load, over-the-load air distribution and the thermostat located in the return-air stream resulted in relatively slower cooling. Cooling citrus fruit from 17–22°C to 5°C took 3–4 days in both vans. An under-the-floor air-delivery system cooled warm fruit throughout the load more uniformly than an over-the-load air-delivery system. Ismail et al. (1986) reported that Florida grapefruits (*Citrus paradisi* Macf.) were cold-treated during shipment to Japan in 12 m refrigerated van containers for 14 days at 1.6°C. Cold-treated fruit appeared generally fresh with bright-yellow color. Firmness was retained and fruit was free of internal breakdown from cold injury. The incidence of decay, pitting and scalds were also low. In fruit that had been degreened and shipped with biphenyl pads, skin pitting and scald were noted. The combined incidence of decay, pitting, and scald was 6 percent 2 weeks after termination of cold treatment. Wax coating containing TBZ as fungicide and conditioning of fruit at 15°C for 7 days (at pulp temperature) has been recommended. Proper warming is necessary after cold treatment.

The Mediterranean fruit fly is also endemic in some areas of Argentina. Disinfestation with cold treatments is required for export of grapefruits from Argentina to the U.S. and Japan. In-transit cold treatment (16–20 days at 2.2°C) of Star Ruby grapefruit with preconditioning to reduce chilling has been recommended (Vazquez and Dalmazo, 2000). The quarantine treatment is also required for lemons exported to Japan. Torres (2000) suggested preconditioning for 7 days at 16°C, quarantine treatment for 18 days at 2°C, and a transit period of 17 days at 8°C for lemons.

IV. AIR SHIPMENTS

Shipments by air cargo are generally used for very perishable, high-value commodities. Although fruits can be taken to market in excellent condition because of the very short time in air transport, citrus fruits are rarely transported commercially by air as the freight costs are very high. Air transport may be useful for specialty citrus sent for gifts on special occasions. Generally, small consignments of produce are covered with nets and transported in the cargo compartments of passenger airplanes. The waiting periods before and/or after air transport are long and can be detrimental to quality if refrigeration facilities are not available at airports. Water loss takes place to some extent as a result of air exchange and variations in air pressure in the cargo area of the plane. The cargo and passenger compartments are at similar temperature and pressure in passenger airplanes. Fruit has to be sufficiently precooled before loading onto passenger airplanes.

Insulated air-transport containers of a particular size that fit in the cargo hold are also used. These containers are built on pallets. Most air-transport containers lack refrigeration systems and relative humidity is also not maintained. For transportation on cargo planes, fruit should be precooled and transported to the air terminal in a refrigerated vehicle. In case of delays, refrigerated storage facilities should be available at air terminals. Dry ice (solid CO_2) and liquid nitrogen are used for cooling, but these methods are not suitable for citrus fruits. Air cargo refrigerated containers generally have a weight of 1–1.5 tons and internal measurements of about $1.5 \times 1.45 \times 1.40$ m but bigger capacity containers are also being used.

REFERENCES

Brown, G.K. (1995). Harvesting fruit for fresh market. *Thirty-fourth annual citrus packinghouse day.* September 7, 1995. Cooperative Extension Service, Institute of Food and Agricultural Sciences, University of Florida, Gainesville.

CFTRI (1982). Research and Development at CFTRI – The first three decades (1951–80). Central Food Technological Research Institute, Mysore, India, 360 pp.

Chace, Jr. W.G., Harding, P.L., Smoot, J.J., and Cubbedge, R.H. (1966). Factors affecting the quality of the grapefruit exported from Florida. United States Department of Agriculture, Marketing Research Report 739.

Cruz, V.L. (1995). Evaluation of losses and damage in Valencia oranges, (*Citrus sinensis* L. Osbeck) during simulated transport under forced ventilation. *Alimentaria* 33(268), 69–75.

DMI (1965). Marketing of citrus fruits in India. Directorate of Marketing and Inspection, Government of India. Marketing Series No. 155.

DMI (1995). Existing status of postharvest management of fruits and vegetables. MPDC, Report No. 17, 18 pp.

Grierson, W., and Wardowski, W.F. (1986). Transportation to the packing house. *In* '*Fresh citrus fruits*' (W.F. Wardowski, S. Nagy, and W. Grierson, eds.), AVI Publishing, USA, pp. 227–242.

Hale, P.W., Smoot, J.J., and. Hatton, Jr. T.T. (1981). Factors to be considered for exporting grapefruit to distant markets. *Proc. Fla. State Hort. Soc.* 94, 256–258.

Hardenburg, R.E., Watada, A.C., and Wang, C.Y. (1986). The commercial storage of fruits, vegetables, and florist and nursery stocks. USDA, ARS. Agricultural Handbook No. 66, p. 128.

Harvey, J.M., and Houck, L.G. (1986). Transportation to markets In: *Fresh citrus fruits*. The AVI Publishing Co., Westport, Connecticut, USA. pp. 465–478.

Houck, L.G., and Hinsch, R.T. (1983). Evaluation of van containers for cold treatment of citrus fruits for quarantine purposes. *Proc. Fla. Sta. Hort. Soc.* 96, 340–344.

Ismail, M.A., Hatton, T.T., Dezman, D.J., and Miller, W.R. (1986). In transit cold treatment of Florida grapefruit shipped to Japan in refrigerated van containers: problems and recommendations. *Proc. Fla. Sta. Hort. Soc.* 99, 117–121.

Ladaniya, M.S. (2004a). Standardization of temperature for long term refrigerated storage of 'Mosambi' sweet orange (*Citrus sinensis* Osbeck). *J. Food Sci. Technol.* 41, 692–695.

Ladaniya, M.S. (2004b). Response of 'Kagzi' acid lime to low temperature regimes during storage. *J. Fd. Sci. Technol.* 41, 284–288.

Ladaniya, M.S., and Sonkar, R.K. (1996). Influence of temperature and fruit maturity on Nagpur mandarin (*Citrus reticulata* Blanco) in storage. *Indian J. Agric. Sci.* 66(2), 109–113.

Ladaniya, M.S., Sonkar, R.K., Shaikh, A.J., and Varadarajan, P.V. (1999). Corrugated fibre board containers for packaging, transport and storage of 'Nagpur' mandarin in domestic market. *Indian Fd. Packer* 53(4), 5–15.

Sale, P. (1993). Curing lemons in transit to Japan looks feasible. *Orchardist of New Zealand* 66(3), 16, 18–21.

Torres, L.G.J. (2000). Effect of cold quarantine treatments on quality of four commercial varieties of lemon (*Citrus limon* Burm). Abstracts, *Int. Soc. of Citric. Congress*, Orlando, Florida, 3–7 December, 2000.

Tugwell, B.L., and Chvyl, W.L. (1995). The effect of storage temperature on development of rind blemishes on Washington naval orange. *Science and Technology for the fresh food revolution*, Melborne, Australia, pp. 18–22.

Vazquez, D.E., and Dalmazo, J.J. (2000). Effects of cold disinfestation treatment on the quality of Star Ruby grapefruit. *Int. Soc. Citric. Congress*, Orlando, Florida, 3–7 December, 2000.

14

MARKETING AND DISTRIBUTION

After production, postharvest treatment, and packaging, fruit enters the marketing system, either locally or overseas. The condition of fresh citrus depends on the duration of the marketing period, and how and under what climatic conditions fruit is handled en route. Marketing can be cooperative-based, controlled by private functionaries, or state-organized. Although wholesaling is organized/regulated in developing countries, retailing is not so regulated or organized. The marketing of perishable commodities, including citrus, is not at all organized and regulated in underdeveloped small countries. Inefficiencies in distribution systems cause losses. There are leakages in tax collection; as a result of the informal nature of marketing and the lack of controls and regulations, producers and consumers incur losses. Facilities for storage and display are not available at the retail level in most parts of the developing and underdeveloped world. This situation is gradually changing and supermarkets are finding a high growth rate in developing countries, while growth is more sluggish in the developed world. The big chain stores in the developed world are entering developing countries because of their growing economies and liberalization.

I. MARKETING SYSTEMS AND DISTRIBUTION

Citrus fruit passes through various functionaries in different marketing channels of different countries. Considerable variety exists in product-handling practices at the distribution level. Well-known food stores in many developed countries have warehouses to service their own stores. The wholesalers, commission merchants, and various other functionaries also play important roles in distribution systems and selling the fruit further to retailers. Good management practices and facilities are essential to providing consumers with the fruit of the best possible quality.

Fruit received at a destination after refrigerated transit is stored at a low temperature in the wholesalers' warehouses and further distributed at the same temperature. The warehouses of chain stores are set at 8–12°C and 1–4°C. Fruit is displayed and sold at 18–20°C to minimize losses. Shelf life has to be at least two weeks in retail outlets, so the cool chain has to be critically maintained.

During transit and distribution, fruit warming and rough handling can increase losses. Fruit displayed for several hours at temperatures either too warm or too cold also deteriorates quality. Poor handling during marketing, particularly retailing, in developing countries is due to a lack of proper facilities and knowledge about commodity requirements. Sometimes retailers do have cold-storage facilities but they are inadequate. Moreover, mixed commodities – especially high ethylene–producing fruits and vegetables – can have a damaging effect on citrus fruits and their senescence begins early.

In unregulated and unorganized systems, the real challenge in maintaining fruit quality is during wholesale marketing and retailing. Sweet oranges, limes, and lemons have relatively longer shelf life, but tangerines, mandarins, and grapefruits cannot withstand handling abuse, as they soften and lose shape.

The imported fruit also has to pass through local channels of marketing before it reaches the consumer. Efforts are made in this chapter to discuss citrus-marketing channels and the role of functionaries. It is a brief overview of domestic marketing practices and requirements in major citrus-producing and citrus-consuming countries around the world.

A. Japan

In Japan, growers have formed large cooperatives through which they market their produce. The Nakagai (middlemen) and Seika Company (wholesale company) are involved in marketing fruit (Kitagawa, 1981). Domestic market shipments continue year-round as stored and imported fruits come into the market when fresh citrus harvesting season is over. The peak domestic marketing is from October to February in harvest season. The common channel of marketing is as follows:

Grower → Agric cooperative → Wholesale market
→ Nakagai → Retailer → Consumer

Wholesale markets are run by civic bodies in cities and towns. These markets have Seika companies and many Nakagai.

Imported citrus fruit is handled by fruit-trading companies and trading departments of Seika company. These companies import citrus and supply it to chain stores and supermarkets. Imported fruit is also traded in the wholesale markets of big cities such as Tokyo and other cities through the Seika companies, which sell the fruit to Nakagais who further sell it to greengrocers, fresh-fruit stores, and fresh-fruit vendors. Grapefruits are generally imported from April onward and oranges from June onward as the domestic citrus season is over by that time. Imported citrus available in Japan are mainly grapefruits, lemons, and sweet oranges. In Japan per capita per year citrus consumption was 10 kg for Satsumas alone among city residents during 1981 (Kitagawa, 1981). Citrus consumption increased to 23 kg per person per year in 1987–88 with lemons, grapefruits, and oranges contributing 1, 1.5, and 1.5 kg respectively, with the rest shared by Satsumas and other citrus produced in Japan (Tallent, 1988).

B. Taiwan

In Taiwan, cooperative setup is relatively strong and citrus growers are involved actively in marketing their produce. The farmers' associations, agricultural cooperatives, and the Taiwan Provincial Fruit Marketing Cooperative are the three important organizations that handle fruit in the cooperative marketing system. The Taiwan Provincial Fruit Marketing Cooperative handles citrus domestically and also exports the fruit (En-Usiung, 1991). It also provides input for cultivation to growers. The most common domestic channel is as follows:

Producer → Local shipper → Wholesaler → Retailer → Consumer

About 50 percent of the total produce passes through the wholesalers and about 30 percent is handled by farming cooperatives. The rest of the produce is consumed fresh locally. Cooperatives are growing stronger because of an increasing awareness among growers. The growers' cooperatives directly approach supermarket chains and thus reduce marketing costs.

C. China

Chinese people prefer fresh citrus fruits. The significant increase in citrus consumption is due to increases in income from a growing economy.

During last 50 years, citrus fruit production has expanded rapidly in this country. According to statistics, estimated growth was very impressive during last decade. In 1952, per capita production of citrus fruit was 0.36 kg, rising to 8.58 kg in 1999. Zhejiang, Fujian, Hunan, Sichuan, Jiangxi, Hubei, and Guangdong provinces produce more than 80 percent of the total citrus; surplus is sent to other parts of the country. Citrus is surplus in the northwest, the south, and the east; fruit distribution flow is toward the southwest, northeast, and north.

Citrus fruit is the most cultivated and highest-value crop in the above mentioned provinces. The major resources of farmers' income and revenues of some local governments are generated from citrus cultivation.

As per the strategy of China's Agricultural Development (by the State Development Planning Commission), in the years 2015 and 2030, the annual per capita consumption of citrus fruit is likely to be 11.7 and 16.0 kg and the total demand is likely to be more than 16.99 and 25.65 million tons respectively. In China, infrastructure facilities such as transportation, storage, and cool chain are lacking, but development is taking place gradually. The harvest time of October–January is very limited. There is a supply shortage; thus imported citrus has a good market in the country.

In the past, China had a centralized fruit-marketing system. Fruits were mainly bought and sold through state-owned channels. Now the marketing system is changing rapidly with economic reforms. Agencies and functionaries for marketing citrus are as follows:

1. Countryside Fairs or Bazaars: These are the oldest type of markets in rural and some urban areas. Hawkers and peddlers also do transactions with small volume of fruit at limited scale.
2. Primary Markets: Growers sell their produce directly in these markets, which are also called elementary markets. State-owned trading companies and rural marketing cooperatives are the main purchasers. The individual farmer distributors, organized farmer distributors, and processing enterprises procure fruit in these markets.
3. Wholesale Markets: The wholesale markets are of two kind – the producing center-based market and the consuming center-based market. Wholesalers in consuming center-based wholesale markets purchase fruit in producing center-based market. These wholesalers have sufficient funds and send their agents to the producing center-based markets for fruit purchasing. Fruit is transported to consuming center-based wholesale markets for direct or indirect sale. In Guangdong region two major wholesale markets, Huadu and Lishui, provide fruit to South China.
4. Retail Markets: Retail markets in cities contribute a great deal to fruit distribution for city dwellers. Retailers and vendors buy fruits in wholesale markets and transport it to retail markets for sale.
5. Large marketplaces, supermarkets, and chain stores: This is a new phenomenon in developing towns and cities of China for retail marketing. North Beijing International fresh fruit central wholesale market also deals in large quantities of citrus.

Traders/agents in Hong Kong deal in imports and sell the fruit in other markets of China. Supermarkets are increasing at a tremendous rate in this country and selling imported fruit. They purchase fruit from wholesalers and directly from importers. It is likely that supermarkets selling imported citrus will be common in all cities of China over the next 10–15 years. Supermarkets such as Park-N-Shop and Carrefour are marketing fresh fruit at present (Hanlon, 2001).

The agencies playing important role in citrus fruit marketing in China are:

1. Citrus growers: Growers on an individual or group basis doing business of their own.
2. Cooperatives and Associations: Growers' cooperatives and specialized associations tie-up with marketing or processing companies for the marketing. These companies provide pre- and postharvest services.
3. State-owned trading companies/enterprises: These companies procure the fruit in remote areas where private trade companies have not yet reached. Local government agencies are improving the quality of their services and increasing purchasing prices.
4. Privately owned companies: Fruit-marketing companies for distribution are coming up with chain stores. They are engaged in both wholesale and retail businesses. These companies buy produce in the producing center-based wholesale markets, pack it, and sell in their chain stores or resell in wholesale market.

In the entire fruit trade system, middlemen do exist in China. They have good contacts with buyers and producers.

Small private distributors are increasing and help the fruit growers to get a higher selling price. They buy fruit in producing center-based market during harvesting and sell in the consuming center-based market where there is a demand. A larger portion of the trade is now controlled by private producers and companies.

D. South Asia

The countries of South Asia – including India, Pakistan, Bangladesh, Sri Lanka, Bhutan, and Nepal – have some similarities in trading and marketing citrus fruit. The preharvest contractors, wholesalers, commission agents, retailers, hawkers, and small vendors play important roles in the distribution of citrus. The schematic diagrams (Fig. 14.1) as given below briefly describe the marketing system in this region.

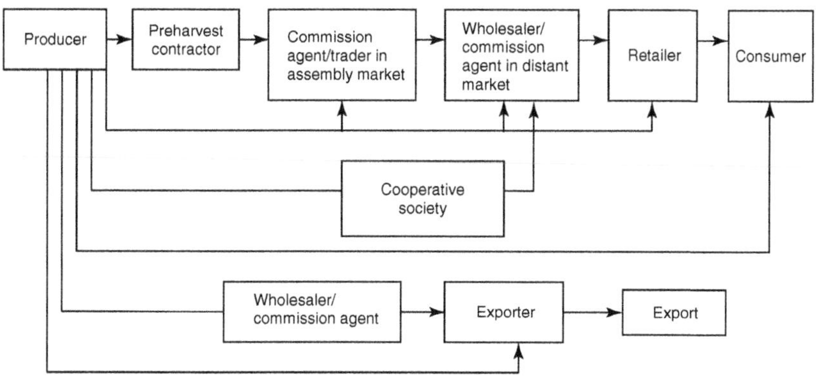

FIGURE 14.1 Channels of Marketing of Citrus Fruit in South Asia.

1. India

In India, marketing citrus fruits is organized mainly in three ways. (1) The preharvest contractors, or traders, purchase the orchard at flowering stage or just before harvest and pack the fruit in the orchard itself or at their packinghouses located in production areas, and then send it to distant markets. (2) Growers send the fruit directly to distant city markets as packed or loose by truck and train. (3) Growers sell the produce in local wholesale markets through an auction process where buyers from distant markets or local traders purchase it and send it to distant markets either loose or packed. This is the important channel of marketing for citrus fruit in the country. The commission agent is an important functionary in all marketing channels.

The commercially important channels of citrus fruit marketing in the country are:

1. Grower → Local trader/preharvest contractor/subcontractor → Wholesaler → retailer → consumer
2. Grower → Wholesaler → Retailer → Consumer
3. Grower → Commission agent → Wholesaler → Retailer → Consumer → Buyer (distant market) → Wholesaler → Retailer → Consumer

Growers, crop contractors, and wholesalers also sell produce to processing units (PKV, 1992; Satyanarayana and Ramasubbareddy, 1994). The abovementioned three marketing systems have existed for many decades. Unpredictable fluctuations in prices result in huge losses to growers in bumper-crop seasons. Traders hamper the correct assessment of fruit quality and price and hence the creation of an alternate buying agency and regulatory system is considered necessary (DMI, 1982). Lack of sufficient storage and processing facilities, poor transport, absence of a cooperative marketing structure, and organized retail marketing by growers have affected the acid lime, 'Nagpur' mandarin, Mosambi orange, and Sathgudi orange industries (Mallareddy and Kumar, 1990; PKV, 1992; Ladaniya and Wanjari, 2003; Wanjari et al., 2003; Ladaniya, 2004).

Mosambi and Sathgudi sweet orange growers do not directly market their produce. The major constraints are: (1) risk in transportation and marketing, (2) engagements in other field operations, (3) lack of market information/intelligence, (4) small quantity of produce, (5) lack of knowledge of marketing procedures, (6) language barriers, (7) lack of transportation facilities, (8) lack of managerial skills, (9) lack of finance facilities (Ladaniya and Wanjari, 2003). Market information systems need to be developed to serve the growers.

Postharvest treatment facilities also need to be created for the use of growers as well as traders at certain places in growing regions for reduction in losses and better returns. It has been observed that every rupee spent on chemical treatment to mandarins to minimize losses fetched Rs. 2.63–10.33 in return, depending on chemicals used. This has not only increased returns to growers/traders but also benefited consumers by minimizing transit loss and thus reducing transport cost and commodity price (Subramanyam, 1986).

Since early 1990s many state and central government agencies are participating actively in developing postharvest handling and marketing infrastructure and providing finances to growers for marketing their produce. Growers cooperative organizations are also engaged in marketing fruits and vegetables including citrus. State governments are promoting cooperative societies in their respective states for more remunerative prices to the growers and to improve infrastructure facilities such as packinghouses, pre-cooling units, cold storages, and refrigerated transport vehicles.

The fruit and vegetable division of National Dairy Development Board (NDDB) directly purchases fruit from citrus producers and sells it through retail outlets in cities such as Calcutta (Kolkata) and New Delhi. Producers' cooperatives – FRESH in Hyderabad and HOPCOMS in Bangalore, Karnataka – have started retailing, thus the growers' share in the price paid by consumers has increased more than 80 percent. Consumers get better-quality fruit at reasonable prices in this system. Punjab Agro-industries Development Corporation and Punjab State Export Development Corporation have provided facilities to Kinnow mandarin growers – thus losses are reduced and produce is being marketed in distant markets. Government organizations, specifically National Horticultural Board (NHB), National Cooperatives Development Corporation (NCDC), and National Agricultural Marketing Federation (NAFED), have played a vital role in improving marketing and postharvest handling infrastructure of horticultural produce including citrus in the country. But there are still miles to go to ensure orderly marketing based on scientific postharvest management of citrus fruits throughout the country. Standardization and implementation are lacking in many areas of marketing and fruit-quality assurance.

Oranges and mandarins grown in different parts of India are available from September to April. Acid limes are available year-round (Table 14.1). Coordinated efforts on the part of various state marketing corporations and growers' organizations are necessary to develop the program of marketing of produce for fair distribution with maximum returns to the growers. Careful planning is required to integrate the supply and demand of citrus fruit.

E. Europe

Europe is one of the biggest consuming markets for high-quality fresh citrus in the entire world. The European Union, comprising 27 countries, is a big block and member countries have certain privileges. Import duties are lower if fruit is imported from member countries. Supermarkets and chain stores control the most of the retailing.

1. Italy

The cooperative movement of growers is playing an important role in citrus marketing in Italy. There were 514 citrus producers' cooperatives by 1990. These cooperatives handled 2.1 million tons of citrus produced during 1988–91.

TABLE 14.1 Harvesting and Marketing Period of Important Citrus Fruits in India

Name	Variety	State/region	Period
Mandarin	'Nagpur'	Madhya Pradesh and Maharashtra	February–April, October–December
	'Coorg'	Karnataka	June–July, November
		Tamil Nadu	November
	'Khasi'	Assam, NEH region	November–December, January
	'Darjeeling' santra	West Bengal	November–December
	Kinnow	Punjab, Himachal Pradesh, Rajasthan, Haryana	November–March
Sweet Orange	Malta, Valencia, Jaffa	Punjab	January–February, March
	Mosambi	Maharashtra	October–December
			June–July
	Sathgudi	Andhra Pradesh	February–April, October–November
Acid Lime	Kagzi	Tamil Nadu	February–April, July–September
		Maharashtra, Gujarat Andhra Pradesh	March–August, October–November
Lemon	Galgal, Baramasi, Nepali; Assam lemon	Punjab	
		First crop	January–March
		Second crop	June–September
		Assam	October–December, March–May

Cooperatives deal in domestic and export marketing of fresh fruit and have their own processing units. Besides citrus, these cooperatives also deal in vegetables. Cooperatives in Sicily mainly deal in lemons, while in Calabria they mostly deal in oranges (Zarba, 1992). The Coop is the biggest cooperative food-supply chain store and *Esselunga* is the biggest supermarket. Wholesale markets are of significance in the Italian citrus trade. Multitudes of intermediaries operating in the various segments of commercial channels render them tortuous and complex (Sturiale, 1992).

Wholesale markets are crucial in fruit and vegetable marketing chains in Italy, offering services such as recording arrivals and sale prices and providing refrigerated storage. Packaging and wrapping facilities are also provided to growers. The support functions of these markets are surveillance of the supply and demand situation, weighbridge maintenance, banking, postal services, and providing hygienic and sanitary services. There is an increase in supermarkets and

Hypermarkets. In this transformative phase, Italian citrus growers – particularly in the north – have responded with improvements in their trade operations and better services (Sturiale, 1992).

2. France

France is not an important producer of citrus; however, there is some production in Corsica. Morocco, Spain, and other Mediterranean countries are the main suppliers of citrus.

The Paris–Rungis wholesale market is one of the most important markets for citrus trade in the region and it guarantees technical and organizational support to the people who are doing trade there. The producers and wholesalers are in the supply chain, while retailers, other intermediaries, and caterers are important in the demand chain. Every year thousands of tons of citrus fruit from many Mediterranean countries are traded here.

Marketing practices are similar to those in other European countries, and wholesalers play an important role in main terminal markets. Large supermarket chains procure fruits from exporters in other countries directly.

There is a considerable scope for new varieties and attractive packaging in the markets of France. The general trend of Europe supporting people's choice can be seen in this market as well. A rapid increase in citrus fruit consumption, particularly mandarins (seedless easy peelers), has been seen in this market during the 1980s (Rosa, 1988). Monreals, Topaz, Minneolas, and Ortanique, which have relatively few seeds, are acceptable when seedless easy peelers are not available. Citrus fruits are available in this country during September to April from the Northern hemisphere and from May to September from the Southern hemisphere. Clementines from Spain and Morocco are preferred and considered fruits of pleasure. (This can be eaten anywhere since no knife or plate is required and there are no pips or seeds.) Among oranges, Navelina and Navelate are preferred for their good taste and easy peeling quality. Maltese oranges also have a good market. Ruby Red and other intensely red-colored grapefruit are preferred. Similarly, green skinned Sweetie grapefruits from Israel are liked for their sweet taste.

3. Germany

In Germany, retailing chain stores such as Aldi, Rewe–Liebbrand, Co-op AG, and many others market fresh citrus fruit (Pape, 1988). New varieties, freshness of fruit, and attractive appearance and packaging are the needs of this market. In fact, juice consumption has increased because of its convenience when compared with fresh fruit.

The Frankfurt fruit and vegetable market in Germany for the northern areas of the country provides a dedicated infrastructure for trading and technical services such as storage facilities, packaging, and sorting. Wholesalers in this market cater to big institutions such as restaurants and government and private institutions/companies. Wholesale markets are becoming less important as big-market chain stores import citrus directly from producing countries.

4. United Kingdom

As in other European markets, wholesalers play an important role in the domestic and international trade of citrus fruits in this country. The United Kingdom does not produce any citrus and all fruit is imported from EC members and other sources in Africa, Israel, and South America (Argentina and Brazil). Spain is a major supplier of soft citrus (Clementines and mandarin hybrids). During harvesting season from October to March, fresh fruit is imported from Spain; and from April to September, South Africa and many Latin American countries are the suppliers. Lemons are mostly imported from Italy, Spain, and South Africa. Israel, Cyprus, and South Africa export grapefruit to the UK market.

Direct marketing by way of supermarkets accounted for about 44 percent of fresh citrus in UK with wholesalers accounting for about 56 percent during the 1986–88 season (Darby, 1988). Wholesalers are quite active in the overall trade of citrus and supply to greengrocers, market stalls, caterers, and distributive wholesalers. The distributive wholesalers further sell the fruit to chain markets, institutions, and fruit and vegetable shops. Wholesalers import directly from overseas sources or receive fruit from importers. Wholesalers have facilities such as cool storage, ripening chambers, and temperature-controlled transport and have to supply fruit year-round. The need for long credit facilities has been felt in trade (Darby, 1988). Some of the big wholesale markets such as Kent also deal in citrus.

Retailing is mainly through big supermarket chains, department stores, multiple grocers, greengrocers, and market stalls. In supermarkets and other retail outlets (multiple and greengrocers) both pre-packed and loose fruit is sold.
Channels of marketing are:

Importer → Wholesaler → Stores/green grocers → Consumers

Supermarkets → Consumers

5. Spain

The Spanish citrus industry is export-oriented. The Agricultural Confederation of Spanish Cooperatives (CCAE) in Spain became a registered organization in the year 2000; there are about 392 operating cooperatives with 93 000 farmer-members and a turnover of 10 000 million Euros, which is 42 percent of the final agricultural produce (Izquierdo et al., 2003). Cooperative activity is mainly in citrus production and few other crops. Agricultural associations of a socio-economic nature are a basic structural feature of agribusiness and rural development in Spain.

In Valencia, the Valencia Agricultural Cooperative Federation officially represents about 100 citrus cooperatives. The estimated production was roughly 1 102 628 tons in 2000–01. The Spanish citrus management committee (Valencia) is a professional association representing citrus-fruit exporters in Spain. Domestically, citrus fruits are marketed through outlets of associations and consumers co-operatives, supermarkets and weekly markets.

F. Israel

Israel is one of the leading fresh citrus producing and exporting countries of the world. The Citrus Marketing Board of Israel (CMBI) represents thousands of growers and the main brand is Jaffa (Israel). Fresh citrus fruit is mostly sold through supermarket chains. The CMBI sets quality standards and controls grade standards and fruit quality. The logistics, distribution, promotion, advertising, and market research is also done by this organization. Exotic fruits such as kumquats, Sweeties (grapefruit), red pummelos, and New Seedless easy peelers are exported to capture the market. The local consumption of citrus is 10–15 percent of the total production of about 1 million tons. Per capita consumption is 27–28 kg/year. The main distributors market the fruit in the country; individual growers also market directly from orchard to market.

G. South Africa

Citrus Growers Association of South Africa is an organization engaged in fruit production, handling, and marketing within and outside South Africa. Citrus South Africa is a newly established citrus growers' cooperative serving citrus producers, exporters, and international trade. South Africa has completed 100 years of citrus exports to U.K. In domestic market, produce is distributed through agents, whole salers, retailers, hawkers, and institutional buyers.

H. United States

The U.S. is a major producer of fresh citrus in the world and is itself a large consuming market for fresh and processed citrus. In the U.S., consumption of citrus fruit was 13.1 kg/year/person, which declined to 11.8 kg/person/year. If processed and fresh fruit are taken together, consumption varied as 36.2–43 kg/person/year (Tallent, 1988). Citrus has to compete with other fruits in domestic markets both cost- and quality-wise. There is a good amount of citrus fruit consumption in the U.S. and fruits are available year-round.

Service wholesalers, who receive fruit through shippers, supply fruit to various retail chains, grocers, institutions, and retail stores. The big chain stores purchase fruit directly from producers or their groups/organizations and retailing is done in their own stores across the country. Carlot receivers and commission merchants also service retail stores. Jobbers receive fruit from commission merchants and sell to small stores and independent countryside retailers. Mixers sell mixed loads of fruits and vegetables, including citrus, to wholesalers and retailers of distant markets.

Service wholesalers and modern chain distribution centers have cold stores set at various temperatures for different commodities. Because retail handling varies as much as wholesale handling, large retail stores and chains have their own cold rooms. Product temperature management is minimal at shops of small greengrocers and roadside markets. The supermarkets, chain stores, fast food chains, greengrocers, and roadside stands and small markets are mainly engaged in retail distribution.

I. Australia

Australian Citrus Growers Incorporated (Inc.) represents the industry and interacts with government on issues of concern such as prices, exports, and fruit standards, etc. Australian citrus growers have their own cooperatives and sell their produce to institutions, market chains, and also to wholesalers. Retailing of fruit is done mostly by chain stores and supermarkets in cities such as Sydney, Melbourne, Perth, and Adelaide.

REFERENCES

Darby, A.G. (1988). Trends in the UK markets for fresh citrus sector. *Proc. 6th Int. Citrus Congress*, Israel, pp. 1679–1690.

DMI (1982). 'Nagpur' mandarin orange grading and chemical treatment at assembling market. Directorate of marketing and inspection market planning and design centre. *Report No. 2*, Nagpur.

En-Usiung, H. (1991). Citrus co-op marketing in Taiwan. *Proc. Int. Citrus Symp.*, Guangzhou, China, pp. 857–864.

Hanlon, D. (2001). China's citrus and trade: observations and issues. *Proc. 2001 China/FAO Symposium*, Beijing, China, pp. 85–101.

Izquierdo, R.S., Martinez, G.G., and Esteve, E.S. (2003). The new technologies of information in the Spanish citrus co-operatives. Paper presented in *EFIT Conference*, 5–9 July 2003, Dabrecan, Hungary.

Kitagawa, H. (1981). Marketing of citrus fruits in Japan. *Proc. Int. Soc. Citric.*, Japan, Vol. 2, pp. 848–852.

Ladaniya, M.S. (2004). Reduction in post harvest loses of fruits and vegetables. Final Report, *NATP Project*, NRC for Citrus, Nagpur.

Ladaniya, M.S., and Wanjari, V. (2003). Marketing and post-harvest losses of 'Mosambi' sweet orange in some selected districts of Maharashtra. *Indian J. Agric. Market.* 17, 8–10.

Mallareddy, R., and Kumar, H.S.V. (1990). Problems and prospects in marketing of sweet oranges in Prakasam district of Andhra Pradesh. *Agric. Market.* 33(2), 44–45.

Pape, K. (1988). Changes and trends in the citrus market of Affluent countries. *Proc. 6th Int.Citrus-Congress*, Israel, pp. 1663–1677.

PKV (1992). *Twenty five years of research on citrus in Vidarbha (1967–1992).* Director of Extension, Punjabrao Krishi Vidyapeeth, Akola, 54 pp.

Rosa, A.L. (1988). Evaluation and trends of citrus consumption in France. *Proc. 6th Int. Citrus Congress*, Vol. 4, pp. 1651–1661.

Satyanarayana, G., and Ramasubbareddy, M. (1994). *Citrus cultivation and protection.* APAU and Department of Horticulture, Andhra Pradesh, Hyderabad. Wiley Eastern Ltd., New Delhi, 66 pp.

Sturiale, L. (1992). The role of wholesale markets in the trading of citrus fruit in Italy. *Proc. Int. Soc. Citric. Aciriole*, Italy, Vol. 3, pp. 1189–1194.

Tallent, D. (1988). Consumption of fresh fruit: Trends in the United States and a comparison with Japan. *Proc. 6th Int. Citric Congress* Israel, pp. 1693–96.

Wanjari, V., Ladaniya, M.S., and Singh, S. Use (2003). Marketing and assessment of post-harvest losses of acid lime in Andhra Pradesh. *Indian J. Agric. Market.* 16 (May–August), 32–39.

Zarba, A.S. (1992). Recent developments in the co-operative movement for marketing citrus produce in Italy. Proc. *Int. Soc. Citric. Acireale* Italy, Vol. 3, pp. 1185–1188.

15

IRRADIATION

The postharvest deterioration of citrus fruits occurs as a result of endogenous biochemical and physiological changes, microbial and insect attacks, desiccation, and mechanical injury. The problem of spoilage is often severe in the tropical regions where high ambient temperatures and relative humidity prevail during most of the year. Inadequate infrastructure for storage and improper handling of the produce during packing, transport, storage, and marketing also cause considerable deterioration in quality. Fruit shelf life can be prolonged by lowering the storage temperature to a level tolerated by the individual fruit species/variety without impairing quality. Some citrus fruits are, however, susceptible to chilling or cold injury and therefore can be stored for a short duration at low temperatures. Disinfection and disinfestations of fruit using ionizing radiation is one promising area. The response of citrus fruit to irradiation is discussed in this chapter.

I. IONIZING RADIATION

The technology for extending the shelf life of citrus fruits by treatment with ionizing radiation has been studied extensively since the 1950s. As the name indicates, ionizing radiation can be defined as radiation energy that removes electrons from atoms, thereby leading to the formation of ions:

$$H_2O \text{ in food} + \text{radiation} \rightarrow H_2O^+ \text{ and electron } (e^-)$$

In the cascading chemical reactions that follow, hydroxyl radicals and hydrogen peroxide are also formed, along with many other products. Gamma rays and electrons are able to break chemical bonds, which is called ionization. Ionization produces electrically charged particles called free radicals. These particles will react further, resulting in the changes associated with irradiation. Cobalt-60 is commonly used as a source of gamma rays in food irradiation and is produced as a byproduct of the nuclear power industry by adding more neutron to Cobalt-59. Cesium-137 can also be used.

Radioisotopes are atoms that contain an extra neutron in their atomic structure. Carbon, which is stable, contain six neutrons and six protons and is referred to as carbon 12. Another form of carbon has eight neutrons and six protons and is therefore called carbon 14. Both forms are isotopes of carbon, but carbon 14 is a radioisotope and is structurally unstable. The nucleus can disintegrate, releasing *alpha* and *beta* particles and the high-energy electromagnetic waves (gamma rays) known as radioactivity. These rays travel at the speed of light and are highly penetrative. Radioisotopes cannot be switched on and off, so they are immersed in a pool of water and covered with heavy concrete walls to allow operators to enter the processing area.

Irradiation is considered a cold treatment because it achieves its effect without significantly raising the temperature of the product. During irradiation, the energy waves affect unwanted organisms. Radiation as a postharvest treatment of fruit is used either to cause a desired physiological responses or to control insect-pests and disease. Radiation sterilizes or kills insect or microbes by damaging their genetic material and forming a substance toxic to the organism. Insect control by irradiation has been suggested by researchers for the elimination of fruit flies from oranges, papaya, and litchi and the elimination of seed weevils from mangos (Ross and Brewbaker, 1964). Probably the most interest and controversy has been generated by the use of irradiation as a fungicidal treatment for postharvest disease control (Sommer and Fortlage, 1966).

A. Irradiation Process

The irradiation process has two requirements: (1) a source of radiant energy, (2) a way to confine that energy while applying it. The effects of radiation depend on the amount of radiation absorbed. It is an energy absorbed in joules in the kg of material. The following four types of ionizing radiation are used: (1) Gamma (γ) rays produced by radio nuclides isotope Cobalt-60 (1.17–1.33 MeV), (2) Cesium-137 (0.662 MeV), (3) X-ray radiation generated by machine sources, and (4) electron-beam (β) particles. These are operated at or below an energy level of 5 MeV (million electron volts). X-rays produce 5 MeV energy.

1. Radioactive Source

It may be noted that these sources of ionizing radiation have energy thresholds that are unable to induce radioactivity in the treated fruit. In a food-irradiation facility using gamma rays, the radioisotope Cobalt-60 is stored underwater in a source storage pool 6 m deep built inside a chamber with thick (1.5–1.8 m) concrete walls. The concrete walls of the chamber prevent radiation from reaching the work area when the radiation source is raised above the water level for irradiation purposes. The material to be irradiated is placed on conveyors or carriers in packages, cartons, or crates and moved into the concrete-shielded irradiation chamber through a labyrinth. The radiation source is then raised from the storage pool by remote control. The radiation dose absorbed by the material is determined by the strength of the Cobalt-60 source and the length of exposure. The treated material is brought out by the conveyer system. The whole processing area is surrounded by thick concrete to contain the radiation field. Another way of applying gamma radiation is to load the fruit into the chamber and raise the radioactive source for a specific amount of time for the required dose to be absorbed. Electron doses of radiation that are absorbed by the commodity are measured in Grays (Gy). The rad was an earlier unit of gamma radiation absorbed. One rad = 0.01 Gy; or 100 rads = 1 Gy or 1000 rad (Kilo rad or 1 Krad) = 10 Gy or 100 Krad = 1 KGy. One rad is equal to energy absorption of 100 ergs per gram (0.01 joules per kg) of irradiated material.

2. Electron Gun

In this method, radiation is produced from a machine source that accelerates electrons produced in a heated cathode by a high-voltage electrostatic field. These electrons are guided to form a beam, which may be used directly on the commodity. Alternatively they may be used to strike a heavy metal such as tungsten to produce X rays, or gamma rays, which in turn can be used to irradiate the commodity. The main advantage of this form of irradiation is that it can be switched off when not in use. But it tends to be expensive and has limited penetration.

B. Possible Benefits of Irradiation in Citrus

There are advantages and limitations of the use of irradiation technology in fresh citrus fruit handling (Table 15.1). It has a potential for use as a fruit-fly disinfestation and disease-management tool, depending on the fruit species' tolerance of doses. Fruit conditioning and other treatments combined with irradiation may prove helpful in reducing deleterious effects of irradiation.

1. Fungicidal Action/Decay Control

Gamma rays, with their great penetrating power, seem to offer some promise as a physical fungicidal treatment for control of storage rots. The feasibility

TABLE 15.1 Applications and Limitations of Gamma Radiation for Fresh Citrus Fruits

Fruits	Objective	Minimum dose required (KGy)	Irradiation dose tolerated (KGy)	Detrimental effects	Alternatives, if available
Grapefruit, lemon, lime, orange, tangelo, tangerine	Insect disinfestations	0.15–0.30	0.25–1.0 depending on the commodity	Softening, discoloration Surface pitting, off-flavor	Cold and heat treatments
Grapefruit, lemon, lime, orange, tangelo, mandarin	Decay control	1.5–2.0	-do-	Softening, discoloration, surface pitting, off-flavor	Fungicides, Bioagents, hot water dip, GRAS chemicals and their combination

of this treatment depends upon the relative sensitivity of the hosts and insect/pathogen to radiation injury.

Considerable wastage from fungal diseases occurs in fruits during postharvest storage and marketing. Unlike chemical fungicides, gamma radiation can be used for treating deep-seated fungal pathogens within the fruit. Ionizing radiation permits therapeutic action. Chemicals do not usually achieve a therapeutic effect because there is little or no penetration of host tissues. Chemical fungicides and bactericides are usually entirely protective and act to prevent infection, not to stop disease after deep-seated infection has occurred. The effect of radiation on lesions is determined by the sensitivity of the fungus and the extent of its colonization of the host.

Radiation at the pasteurization dose level could be the means of extending the shelf life of fruits. The problems are related to physiological response of the citrus fruits being irradiated and the dose required to achieve their disinfection.

The dose level for disinfection can be intolerable to some citrus fruits, resulting in disorders. Temple mandarins have been found to be less susceptible to irradiation damage than Pineapple and Valencia oranges (Dennison et al., 1966). Low doses of gamma irradiation can reduce the population of microorganisms and the growth of incipient fungal infections sufficiently to reduce storage losses from decay. A radiation dose of 1.5–2 KGy prevents rotting in lemons and oranges inoculated with *P. digitatum* for about 12 days at 24°C and for 17 days at 12.8°C without causing any injury to the fruit. However, the development of other fungi, such as *Alternaria* rot at the stem end of irradiated fruits, indicates that irradiation predisposes the fruits to secondary decay (Beraha, 1964). Inactivation of pathogens is possible at higher irradiation doses only if the host is tolerant and fungi are sensitive to these doses (1.75–2 KGy).

Gamma radiation can often be advantageously combined with other physical (hot-water dips) or chemical (fungicides) agents to control postharvest diseases. Since such combinations may have cumulative effects, even lower irradiation doses could be effective. Heat and irradiation combinations are synergistic and are possibly the most promising means of obtaining a widely useful therapeutic treatment (Sommer, 1964).

Penicillium spp. (blue and green molds) are the major pathogen-causing rots of citrus fruits worldwide. The lethal effect of superficial electron radiation on *Penicillium digitatum* spores *in vitro* increased with the dose, causing about 90 percent inactivation at 100 Krad. At sub-lethal doses, delay in the initiation of spore germination was recorded. Combined treatments of 25–100 Krad and heating at 48–52°C react synergistically in reducing the viability of *Penicillium* spores. A combination of 25 Krad and 52°C resulted in 98 percent spore inactivation. Total inactivation can be achieved with a combined treatment of 50 Krad with 52°C, or of 100 Krad with 48°C (Barkai-Golan and Padova, 1981). Uncombined, a radiation dose up to 50 Krad and heating at 52°C were both ineffective in preventing disease development in superficially inoculated Valencia oranges held at 17°C and 88 percent RH. However, combined heat and radiation treatments markedly reduced fungal development *in vivo*. Increasing the radiation dose in the combined treatment up to 150 Krad did not increase the inhibitory effect. The effect is also reduced with prolongation of the time lag between inoculation and irradiation. Radiation-induced damage in irradiated Valencia fruit increased with radiation dose but not with storage duration. Mature fruit is less sensitive than immature fruit. The use of polyethylene wrappers, or waxing the fruit prior to irradiation, reduces peel damage.

Extensive studies have been conducted with respect to response of Shamouti oranges to gamma radiation. Doses up to 100 Krads appeared ineffective in controlling rots in these oranges inoculated with spores of *Penicillium digitatum* and *P. italicum*. Treatment with 150 and 200 Krad inhibited development of rot for 10–35 days as compared with 3 days for non-irradiated control fruits. Doses of 250 Krad prevented any development of rot during storage up to 80 days at 23°C. Stem-end rot caused by *Diplodia natalensis* and *Alternaria citri* occurred frequently in irradiated fruits, possibly because of an increase in susceptibility following irradiation treatments (Barkai-Golan and Kahan, 1966).

Similar results have been reported by Kahan and Monselise (1965) in the fresh Shamouti oranges inoculated with *Penicillium digitatum* 2–3 days after picking. Fruit remained in good condition for 80 days after doses of 100–250 Krad. Uninoculated fruit drawn from cold storage and irradiated with 100–360 Krad was not adversely affected by the treatments when examined after 6 weeks at 25°C. In the case of Valencia oranges, storage life increased linearly with the amount of irradiation (45–175 Krad of CO^{60} gamma rays) and was in the range of 7–20 days.

Irradiation (50–100 Krad) prolongs the incubation period of the *P. digitatum* and thus increases the storage (14–20°C) life of the Shamouti fruit. The 100 Krad dose increased the incubation period to 12–30 days in 95 percent of the fruits (compared with 7–9 days in biphenyl-treated fruits) but caused some peel damage. Irradiation with 150 and 200 Krad prolonged the incubation periods considerably but these dosages were associated with the development of secondary rot (Barkai-Golan and Kahan, 1967).

The combination of hot water (50°C for 5 min) and radiation result in significant reduction of green mold rot (*Penicillium digitatum* Sacc.) in Marsh seedless grapefruit over that of the control. Treatment effectiveness was less after a delay of 72 h. The percentage of green mold rot was lower with the combination treatment than with either treatment alone, but a synergistic effect of hot water and radiation was not observed. The reported synergism was minimal at 250 Gy and was probably based on research that showed an increase in the radio sensitivity of fungi as a result of the combination of heat and irradiation. Non-rotted fruit from the various treatments showed no visual signs of internal or external injury and no off-flavors (Spalding and Reedar, 1985). Earlier attempts to control post-harvest decay of oranges and grapefruit in Florida using gamma radiation were not successful. Severe peel injury resulted in both kinds of fruit (Grierson and Dennison, 1966).

In Satsuma mandarins, irradiation with 150 or 200 Krad controlled the blue and green molds *Penicillium italicum* and *P. digitatum* respectively during storage at 3°C for 3 months (Watanabe et al., 1976). Lower doses were not effective.

Sanguinello oranges inoculated with *Penicillium digitatum* and exposed to a 330 Krad dose of radiation for 3.7 h remained in good condition for 7 days under warm (21°C), humid conditions, compared with only 1 day in the non-irradiated controls. Inoculated Sanguinello oranges exposed to 150 Krad for the same period remained in good condition for 62 days at 3°C compared with 14 days in the control (Tamburino et al., 1959).

Washington Navel oranges proved particularly susceptible to irradiation injury if picked and irradiated at the beginning of the harvesting season. Doses below 100 Krad had only a slight effect on the taste, ascorbic acid and soluble-solids content, and the titratable acidity of the oranges, although the rinds were frequently damaged. Irradiation is not considered a satisfactory means of mold control in this cultivar because higher doses are required, which damage the fruit (MacFarlane and Roberts, 1968).

In 'Nagpur' mandarins (*C. reticulata* Blanco), radiation doses up to 1.5 KGy do not cause any rind disorder. Radiation treatments did not reduce decay percentage but *Penicillium* rot was delayed in fruit treated with 1.5 KGy while it appeared early in untreated ones. Irradiation doses were ineffective in controlling rots caused by *Botryodiplodia theobromae* and *Alternaria citi* (Ladaniya et al., 2003). In the case of Mosambi sweet oranges and Kagzi acid limes, peel damage occurred at the dose of 1.5 KGy and decay control was not satisfactory. Weight loss and decay were higher in treated yellow acid limes than in untreated fruit.

Gamma radiation at 120 Krad reduces the incidence of rot by 50 percent in peeled Shamouti oranges inoculated with *Penicillium digitatum* spores and stored at 17°C. The incubation period of the fungus in irradiated fruit is prolonged to

5 days as compared with 1–2 days for the control. Packaging the irradiated fruit in sealed, high-density polyethylene (HDPE) and polyvinyl chloride (PVC) wrappers resulted in 25 percent and 10 percent rot respectively, whereas all the non-irradiated wrapped fruit rotted after 4 days of storage. Radiation totally inhibited *Penicillium* decay in uninoculated peeled oranges, whether wrapped in plastic films or not. Irradiation and wrapping reacted additively in decreasing contamination caused by *Cladosporium herbarum* on the peeled fruit. Both HDPE and PVC were effective in delaying dryness of the peeled fruit at shelf-life conditions (Barkai-Golan and Karadavid, 1984).

2. Quarantine Treatment for Fruit Disinfestation

In international trade, fruit is subjected to quarantine regulations by the importing countries because of the presence of fruit flies in most growing areas. The current official quarantine procedures used for treating fruits and plant material for export markets are vapor heat and low temperature treatment. Some of the chemical treatments are effective but banned, while others cause damage to fruit. The usefulness of gamma irradiation as an alternative to chemical fumigation has been demonstrated against fruit flies. The use of ionizing radiation for disinfestations of citrus fruits is covered in detail in Chapter 18.

C. Physiological, Physico-chemical, and Organoleptic Properties of Fruit

1. Respiratory Activity, Ethylene Evolution, and Internal Gas Concentration

An immediate increase in the rate of carbon dioxide evolution can be seen in citrus fruits following irradiation. One day after treatment, the respiratory rate of green lemons increased directly with doses between 50 and 1000 Krad. This dose relationship persisted for about 4 days after treatment. Thereafter, the respiratory rate of fruit subjected to 1000 Krad declined rapidly. Fruits subjected to 400, 600, 800, and 1000 Krad developed severe bronze skin discoloration after 5, 4, 3, and 2 days respectively. The rapid degreening of green lemon fruits indicated that gamma irradiation may have stimulated C_2H_4 production to destroy the chlorophyll in the flavedo. Doses of 50, 75, and 100 Krad induced climacteric-like rates of C_2H_4 production in lemons at 20°C, with the peak rates occurring between the 5th and 6th days. The green color started to disappear and yellow color developed by that time. It appears that as with carbon dioxide evolution, the internal C_2H_4 concentration also increases, which is approximately directly related to the dose applied to lemon fruits. Doses higher than 400 Krad induce bronze discoloration of the fruit skins within a few days. The buttons (calices) of lemon fruits are extremely sensitive to gamma irradiation. Immediately after irradiation, the percent of CO_2 in the internal atmosphere has been found to be twice as much as that in non-irradiated fruits. This trend continued with prolonged storage but with smaller differences in CO_2 content between the irradiated and control fruits. Irradiated Naval and Temple oranges exhibited higher CO_2 and corresponding lower O_2 contents in their atmosphere as compared to controls. Changes in these gases occurred immediately after irradiation, reaching a maximum in the next two days then declining gradually as the storage period was increased (Ahmed and Dennison, 1966). Irradiation increased respiration rate in Nagpur mandarins, Mosambi oranges, and Kagzi acid limes considerably immediately after the treatment. Unlike mandarins and sweet oranges, acid limes continued to have an increased respiration rate with storage extension (Ladaniya et al., 2003).

2. Enzymatic Activity

Irradiation treatment induces several chemical changes and enzymatic activity is one of them. Increased peroxidase activity in orange and grapefruit peel tissue beginning 1–2 weeks after irradiation in the external layers has been reported (Monselise et al., 1968). There seems to be a relationship between increased peroxidase activity and visible damage to the peel. Enzyme activity depended on species, age of fruit, and environmental conditions. The effects of β and γ rays were similar. Catalase activity of the external peel layer was also affected, but while it clearly decreased in orange peel, it was enhanced initially in grapefruit peel.

The activity of phenylalanine ammonia lyase (PAL), a key enzyme in the synthesis of phenolic substances, increased several times over almost immediately after irradiation in grapefruits and Clementine mandarins (Monselise et al., 1968; Riov et al., 1968; Oufedjikh et al., 2000). Hydroxylated and methoxylated coumarin derivatives accumulate in gamma-irradiated citrus flavedo (Dubery, 1990). The phenolic accumulation in damaged peel cells and increased PAL activity as a direct effect of radiation seems to correlate. The PAL activity rapidly decreases to control values in fruit that do not show visible damage symptoms later upon storage but remains relatively high for several days in those fruits later showing peel damage.

Conditioning by heat treatment has been attempted to reduce the adverse effects of irradiation on fruit. Irradiation-induced PAL activity can be reduced by temperature conditioning for 2 h at 38°C or 42°C. Thus heat treatment before irradiation decreases the susceptibility of fruits to irradiation-induced pitting damage (McDonald et al., 2000). The heat treatment may increase phytoalexins and similar compounds that increase fruit resistance. Umbelliferone and scopoletin have been found to inhibit the phenylalanine ammonia-lyase activity in oranges (*C. sinensis*) by increasing the negative cooperativity between the substrate binding sites. Esculetin, scoparone, and fraxetin activated the enzyme, with fraxetin effecting a decrease in negative cooperativity (Dubery, 1990).

Irradiation doses and storage duration also affect pectinesterase enzyme (PE) activity. As the doses increased, the fruit exhibited higher activity. PE activity declined during storage (Dennison et al., 1967).

3. Compositional Changes

Changes in composition can be the result of the direct effect of radiation on the constituent or the result of an indirect effect of induced alterations in the physiology of the fruit. Changes in most chemical constituents as a result of irradiation are pronounced.

Vitamin 'C'

Vitamin C content declines with irradiation to citrus fruits. The drop seems to be related to the dose of irradiation, the species of citrus, and the temperature at which fruit is held after irradiation. In lemon fruits, any dose of 20 Krad and above markedly reduces the ascorbic acid content following 1 month storage at 15°C. Lemons subjected to 300 and 400 Krad are found to be almost devoid of the vitamin (Maxie et al., 1964).

The loss of ascorbic acid from a 1.5 KGy irradiation dose was observed to be 15.84 percent in Mosambi oranges, 26.80 percent in Nagpur mandarins, and 29.20 percent in Kagzi acid limes when compared with controls, indicating that the degree of diminution was species-dependent (Ladaniya et al., 2000). At lower doses (0.25–1 KGy) the loss is considerably less. The trend of changes in ascorbic acid content is more or less similar in treated and non-treated fruit during storage. The irradiation treatment can partially oxidize ascorbic acid to dehydroascorbic acid, which also has vitamin activity. Ascorbic acid is more stable than dehydroascorbic acid, but the two have nearly the same nutritional effect. It is also possible that dehydroascorbic acid may be reduced to form ascorbic acid in the intact biological system of the fruit; this matter needs further investigation. A change from ascorbic acid to dehydroascorbic acid is reversible, particularly in extracted juice (Ting, 1977). An increase in the vitamin C content of Clementine mandarins treated with 0.3–0.5 KGy has been reported during storage (Abdellaoui et al., 1995).

Carotenoids and Color

Irradiation doses of 400 Krad or more bleach out the orange color of the flesh of sweet orange fruits. At doses below 300 Krad, an intensification of the orange color has been recorded in the flesh of oranges after 2–3 weeks in storage at 2°C. Irradiated, colored Navel oranges and Temple fruits became lighter in color during storage, while non-irradiated fruit had a normal orange color. Radiation induced color loss was due to decreased concentration of a relatively few carotenoids. The chief losses of Temple carotenoids were in an esterified orange-colored xanthophyll and an unesterified yellow xanthophyll (Ahmed et al., 1966).

Color development, particularly carotenoid synthesis, was probably adversely affected by irradiation in Mosambi sweet oranges, while the same effect was not apparent in Nagpur mandarins and Kagzi acid limes (Ladaniya et al., 2003). Non-irradiated Mosambi fruit developed a yellow-orange color, while irradiated fruit remained lemon yellow. Irradiation had a pronounced adverse effect on peel and pulp carotenoids.

Total Soluble Solids, Titratable Acids, and Juice Content

Irradiation doses of 200 Krad increase juice pH and decrease the percentage of soluble solids and crushing resistance of the pulp of Washington Navel oranges stored at higher temperature (Guerrero et al., 1967).

Washington Navel oranges subjected to 200 Krad and stored for 3 weeks at −1°C showed a higher percentage of soluble solids, while significantly lower level of soluble solids was recorded at temperatures 1, 5, 15, and 25°C. Titratable acidity (expressed as percent citric acid) declined considerably in lemon fruits subjected to doses of 200 Krad or more following 49 days storage at 15°C. Oranges subjected to 200 Krad and stored for 1 month at 5°C and 20°C showed a significantly higher pH in their extracted juice than did non-irradiated fruit. Total soluble solids and acidity decreased in Marsh grapefruits with an increase in irradiation dose from 0.5 to 1 KGy (Miller and McDonald, 1998a).

Irradiating Baramasi lemon fruits (0–100 Krad) had no significant effect on specific gravity during storage, but juice content increased significantly as the radiation dose increased, reaching 39.56 percent at 100 Krad compared with 35.06 percent in the control (Borthakur and Ranjit Kumar, 1998).

Volatiles and Lipids

Irradiation (0.89–8.71 Gy) reduced the concentration of acyclic monoterpenes such as gerenial and neral during storage of juice but did not affect other monoterpenes. The reduction of acyclic monoterpenes increased linearly with the radiation dose. Most other volatile compounds were stable during 21 days of storage (Fan and Gates, 2001). Irradiation increased campesterol in the free sterol and steryl glycoside fractions and decreased isofucosterol in the free sterol fraction (McDonald et al., 2000).

Phenols

Irradiation stimulated the biosynthesis of hesperidin (a phenolic compound) corresponding with maximum PAL activity in Clementine mandarins after 14 days of storage. The p-coumaric acid content was particularly high in irradiated fruits after 49 days of storage (Oufedjikh et al., 2000). In grapefruits, irradiation (1 KGy) had no effect on total phenols or lignin deposition (McDonald et al., 2000).

Organoleptic Changes

Texture and flavor scores of Nagpur mandarins and Mosambi sweet oranges were not affected by irradiation as recorded after one week. In the case of Kagzi acid limes, however, texture and flavor scores were lower in treated fruit (0.5, 1, and 1.5 KGy). The pulp texture of treated acid limes appeared slightly melting and darker yellow-colored as compared with non-treated fruit, which had normal texture of juice vesicles and normal greenish-yellow color (Ladaniya et al., 2003).

Washington Navel oranges treated at 2 KGy of radiation dose were acceptable to a taste panel after storage for 30 days at 5°C (Maxie et al., 1969). Satsuma fruits irradiated with 0.05 or 0.2 KGy had off-flavor immediately after treatment, but the off-flavor decreased after 7 days of storage at 4°C. Peel browning occurred after storage in fruits irradiated with 0.2 KGy (Watanabe et al., 1976).

The organoleptic quality of Clementine mandarins treated with 0.3–0.5 KGy dose was maintained without adverse effects during storage for 3 weeks at 17°C and 46 percent RH (Abdellaoui et al., 1995). The pulp flavor of Hamlin and

Valencia oranges and Fallglo, Minneola, and Murcott mandarins was negatively affected at 0.45 KGy as recorded after 14 days under refrigerated conditions followed by 3 days at 20°C (Miller et al., 2000).

The difference in flavor among irradiated fruits could be due to changes in odor and taste, especially acidity content/sourness (O'Mahony et al., 1985). There was a progressive decline in taste-panel ratings of the aroma and flavor of orange juice, and of the flavor and texture of sections as irradiation dose increased from 0 to 1, 2, and 3 KGy. The juice and sections were acceptable at a 2 KGy irradiation dose.

D. Phytotoxic Effect of Irradiation on Peel and Flesh

The phytotoxic effect of irradiation in citrus fruit manifests in the form of peel pitting, which develops slowly during prolonged storage. The extent of damage varies with species, dose of irradiation applied, and even fruit position in the canopy (Miller and McDonald, 1998a). Orange fruits often develop rind pits resembling chilling injuries when subjected to two or more KGy. The development of rind pitting is accelerated when the fruits are stored at elevated temperatures.

It is necessary to discover the response of citrus fruits to various kinds of radiation. Effects of 25–240 Krad doses of four kinds of ionizing radiation – CO^{60} gamma rays, 0.5 MV (surface) electrons, 1.5 MV(deep) – electrons applied uniformly to the whole surface and 1.5 MV (localized) electrons applied to either stem or stylar end or to the equatorial regions of non-rotating fruits of Shamouti oranges have been studied by Kahan et al. (1968). Surface and deep electrons caused significantly more peel damage than gamma rays. The extent of peel damage was similar after 2 and 4 weeks of storage. Damage appeared only on that portion of the peel that had been irradiated and was much more extensive at the stem end than at the stylar end. The dose-absorption profile of the irradiation treatment within the peel layers differed markedly from each other. Fruit tolerance to irradiation is species-dependent. Mature orange-colored Nagpur mandarins appeared to have a tolerance to dosed up to 1.5 KGy. Mosambi oranges were not so tolerant to gamma radiation – pittings (in the form of brown, sunken areas) developed on the rind after 75 days under refrigerated storage conditions. Fruit firmness also declined in irradiated fruit. Mature yellow acid limes are also not tolerant to doses of 0.5–1.5 KGy, since the peel and pulp texture and other quality parameters were adversely affected (Fig. 15.1, see also Plate 15.1). Deteriorative effects of higher doses are apparent after prolonged storage (Ladaniya et al., 2003).

Pineapple and Valencia oranges are more susceptible to irradiation-induced peel injury, but Temple fruits were the least susceptible (Dennison et al., 1966). Navel orange fruits harvested in early January and irradiated 3 or 5 days later with doses of 0.62–0.85 KGy developed brown blemishes 4–6 weeks after storage at 5–8°C (O'Mahoney et al., 1984).

Kinnow mandarins irradiated with gamma rays at doses of 1, 2, or 3 KGy develop skin injury during subsequent storage at 20–25°C for 5 weeks. The

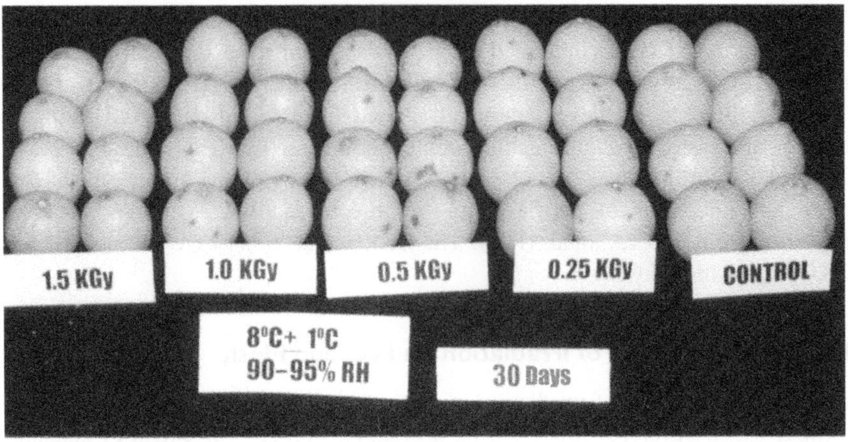

FIGURE 15.1 Peel Damage Due to Irradiation in Acid Lime.

degree of injury increased significantly with the radiation dose. Irradiation increased respiration and ethylene production during storage. After 5 weeks of storage, the irradiated fruits were unacceptable because of skin injury (Farooqi et al., 1987).

Grapefruit treated with 0.60 and 0.90 KGy showed rind breakdown and scald after storage for 28 days at optimum temperatures. Scald was the dominant injury in the fruit harvested in October and December, while rind breakdown was the dominant injury in fruit picked in February, April, and May. At the 0.15 and 0.30 KGy exposures, the injury was least and fruit were acceptable. Injured areas developed decay in storage and marketing conditions at 21°C (Hatton et al., 1982). Tissue necrosis, excessive softening, ripening abnormalities, and off-flavor are common injury responses, which drastically limit the dose.

Waxed fruits of Washington Navel oranges withstood irradiation damage better than unwaxed fruits (Guerrero et al., 1967). GA_3-treated Marsh grapefruits can tolerate an irradiation dose of 0.3 KGy without serious damage. At a dose of 0.6 KGy, serious peel damage detrimental to fruit quality is likely to develop during storage (Miller and McDonald, 1996). White Marsh grapefruits harvested from an exterior canopy position were less tolerant of 0.5 or 1.0 KGy gamma radiation than were interior fruits. Pitting was reduced by 30 percent with temperature conditioning (vapor heat at 38°C or 42°C for 2 h). Exterior canopy fruit had twice as much pitting, had greater weight loss, and were firmer than interior canopy fruits (Miller and McDonald, 1998b). McDonald et al. (2000) have also suggested heat treatment before irradiation in order to condition the fruit.

Irradiation has been suggested as a substitute for refrigeration in developing countries. Lemons, limes, and oranges offer no promise for commercial irradiation, since alternative, cheaper, non-injurious, and more effective procedures are available. Further, irradiation dosages required to retard decay result in the rind breakdown, impair sensory qualities, and effect many changes in normal metabolism. Irradiation cannot be a substitute for refrigeration and at best can only supplement it. In spite of the large amount of work done on the irradiation of citrus fruits, no commercial application of gamma irradiation for citrus fruit storage has been reported. Irradiation can be a very effective treatment for disinfesting citrus fruits affected by insects, especially fruit flies.

REFERENCES

Abdellaoui, S., Lacroix, M., Jobin, M., Boubekri, C., and Gagnon, M. (1995). Effect of gamma irradiation combined with hot water treatment on the physico-chemical properties, vitamin 'C' content and organoleptic quality of Clementines. *Sci. des Aliments*, 15, 217–235.

Ahmed, E.M., and Dennison, R.A. (1966). Irradiation effects on the respiratory activities of lemons and oranges. *Proc., 17th Int. Hort. Congress*, Madrid, Vol. 1, 518.

Ahmed, E.M., Knapp, F.W., and Dennison, R.A. (1966). Changes in peel color during storage of irradiated oranges. *Proc. Fla. State Hort. Soc.* 79, 296–301.

Barkai-Golan, R., and Kahan, R.S. (1966). Effect of gamma radiation on extending the storage life of oranges. *Plant. Dis. Rep.*, 50, 874–7.

Barkai-Golan, R., and Kahan, R.S. (1967). The effect of gamma irradiation on the development of *Pennicillium digitatum* rot on inoculated citrus fruits. Summaries lectures. *1st Israel Congress on Plant Pathology*, pp. 94–95.

Barkai-Golan, R., and Karadavid, R. (1984). Control of *Penicillium* rot peeled citrus fruit by gamma radiation and plastic wrapper. *Proc. Int. Soc. Citric.*, Vol. I, pp. 506–507.

Barkai-Golan, R., and Padova, R. (1981). Eradication of *Penicillium* on citrus fruits by electron radiation. *Proc. Int. Soc. Citric.*, pp. 799–801.

Beraha, L. (1964). Influence of gamma radiation dose rate on decay of citrus, pear, peach and on *Penicillium italicum* and *Botrytis cineria* in vitro. *Phytopathology* 54, 755.

Borthakur, P.K., and Kumar R. (1968). Effect of gamma radiation of specific gravity and juice percent of Baramasi lemon fruit during storage. *J. Agric. Sci. Soc. North-East India* 11, 253–254.

Dennison, R.A., Grierson, W., and Ahmed, E.M. (1966). Irradiation of Duncan grapefruit, Pineapple and Valencia oranges and Temples. *Proc. Fla. Sta. Hort. Soc.* 79, 285–92.

Dennison, R.A., Ahmed, E.M., and Martin, F.G. (1967). Pectinesterase activity in irradiated 'Valencia' oranges. *Proc. Am. Soc. Hort. Sci.* 91, 163–8.

Dubery, I.A. (1990). Effect of hydroxylated and methoxylated coumarins on the regulatory properties of phenylalanine ammonia-lyase from *Citrus sinensis*. *Phytochem.* 29, 2107–2108.

Fan, X., and Gates, R.A. (2001). Degradation of monoterpenes in orange juice by gamma irradiation. *J. Agric. Fd. Chem.* 49, 2422–2426.

Farroqi, W.A., Ahmed, M.S., and Zain-ul-Abadin (1987). Physiological and biochemical studies on irradiated citrus fruits. *Nucleus Pakistan* 24, 31–35.

Grierson, W., and Dennison, R.A. (1966). Irradiation treatment of 'Valencia' orange and 'Marsh' grapefruit. *Proc. Fla. Sta. Hort. Soc.* 78, 233–7.

Guerrero, F.P., Maxie, E.C., Johnson, C.F., Eaks, I.L., and Sommer, N.F. (1967). Effects of postharvest gamma irradiation on orange fruits. *Proc. Am. Soc. Hort. Sci.*, 90, 515–528.

Hatton, T.T., Cubbedge, R.H., Risse, L.A., and Hale, P.W. (1982). Phytotoxicity of gamma irradiatiion on Florida grapefruit. *Proc. Fla. Sta. Hort. Soc.* 95, 232–234.

Kader, A.A. (1986). A summary of potential applications and limitations of ionizing radiation for fresh fruits and vegetables, fruits and flower paper presented at a seminar on *Food irradiation for the produce industry*, Newport Beach, California 21–22, May.

Kahan, R.S., and Monselise, S.P. (1965). Extension of storage life of citrus by irradiation. *Food Technol.* 19, 122–124.

Kahan, R.S. et al. (1968). Comparison of the effect of radiation of various penetrating powers on damage to citrus fruit peel. *Radiat. Bot.* 8, 415–423.

Ladaniya, M.S. Singh S., and Wadhawan, A.K. (2003). Response of 'Nagpur' mandarin, 'Mosambi' sweet orange and 'Kagzi' acid lime to gamma radiation. *Radiat. phys Chem.* 67, 665–675.

Macfarlane, J.J., and Roberts, E.A. (1968). Some effects of gamma radiation on Washington Navel and Valencia oranges. *Aust., J. Exp. Agric. Anim. Husb.* 8, 625–629.

Maxie, E.C., and Kader, A.A. (1966). Food irradiation – Physiology of fruits as related to feasibility of the technology. *Adv. Food Res.* 15: 105–108.

Maxie, E.C., Caroll, F.J., and Boyed, C. (1965). U.S.A.E.C. Rpt. UCD-34P80, p. 69.

Maxie, E.E., Eaks, I.L., and Sommer, M.F. (1964). Some physiological effects of gamma irradiation on lemon fruits. *Radiat. Bot.* 4, 405.

Maxie, E.E., Sommer, M.F., and Eaks, I.L. (1969). Effect of gamma radiation on citrus fruits. *Proc. First Int. Citrus Symp.* Vol. 3, 16–26 March, 1968. Riverside, California, UC, pp. 1375–1387.

McDonald, R.E., Miller, W.R., and McCollum, T.G. (2000). Canopy position and heat treatments influence gamma-irradiation-induced changes in phenylpropanoid metabolism in grapefruit. *J. Am. Soc. Hort. Sci.* 125, 364–369; 38.

Miller, W.R., and McDonald, R.E. (1998a). Short-term heat conditioning of grapefruit to alleviate irradiation injury. *HortScience* 33, 1224–1229.

Miller, W.R., and McDonald, R.E. (1998b). Amelioration of irradiation injury to Florida grapefruit by pre treatment with vapor heat or fungicide. *HortScience*. 33, 100–102.

Miller, W.R., and McDonald, R.E. (1996). Postharvest quality of GA treated Florida grapefruit after gamma irradiation with TBZ and storage. *Postharvest Biol. Tech.* 7, 253–260.

Miller, W.R., McDonald R.E., and Chaparro, J. (2000). Tolerance of selected orange and mandarin hybrid fruits to low dose irradiation for quarantine purposes. *HortScience* 35, 1288–1291.

Monselise, S.P., Riov, J., and Kahan, R.S. (1968). Some changes in the enzymatic activities of citrus peel tissue after fruit radiation. *Proc. FAO/IAEA Panel on Enzymological Aspects*. Food Irrad., Vienna., pp. 71–81.

O. Mahoney., M, Wong, S.Y., and Odbert, N. (1984). Sensory evaluation of Navel oranges treated with low doses of gamma radiation. *J. Fd. Sci.* 50, 639–46.

Oufedjikh, H., Mahrouz, M., Amiot, M.J., and Lacroix, M. (2000). Effect of gamma-irradiation on phenolic compounds and phenylalanine ammonia-lyase activity during storage in relation to peel injury from peel of *Citrus clementina* Hort. ex. Tanaka. *J. Agric. Fd. Chem.* 48, 559–565.

Riov, J., Monselise, S.P., and Kahan, R.S. (1968). Effect of gamma radiation on phenylamine ammonia-lyase activity and accumulation of phenolic compounds in citrus fruits peel. *Radiation Bot.* 8, 463–6.

Ross, E., and Brewbaker, J.L. (1964). Ionizing radiations. *Fourth Ann. Contr. Meet Conf.* 641002 U.S.A.E.C. TID-4500, p. 49.

Ross, R.T., and Engeljohn, D. (2000). Food irradiation in the United States: irradiation as a phytosanitary treatment for fresh fruits and vegetables and for the control of microorganisms in meat and poultry. *Radiat. Phy. Chem.* 57, 211–214.

Sommer, N.F., and Fortlage, R.J. (1966). Ionizing radiation for the control of post harvest diseases of fruits and vegetables. *Adv. Fd. Res.* 15:147.

Spalding, D.H., and Reeder, W.F. (1985). Effect of hot water and gamma radiation on postharvest decay of grapefruit. *Proc. Fla. Sta. Hort. Soc.* 98, 207–208.

Tamburino, S.M. et al. (1959). Orange storage by means of CO^{60} gamma Irradiation. *Tech. Agric.* **11**, 631–635.

Ting, S.V. (1977). Nutritional labeling of citrus products. In *Citrus science and technology*. (S. Nagy, P.E. Shaw, and M.K. Veldhuis, eds.). AVI Publishing Co, Inc., Westport, CT, USA, pp. 401–444.

Watanabe, H., Akoi, S., and Sato, T. (1976). Radiation preservation of *Citrus unshiu*. Part I. Effect of gamma radiation on organoleptic properties and reduction of spoilage during storage. *J. Fd. Sci.* 23, 300–305.

16

POSTHARVEST DISEASES AND THEIR MANAGEMENT

I. POSTHARVEST DISEASES

Losses from postharvest diseases caused by various pathogens account for nearly 50 percent of the total wastage in citrus fruits. Infection and contamination occur at different stages in the field and after harvest during marketing. Apart from actual losses from wastage, further economic loss occurs if the market requirement necessitates sorting and repacking partially contaminated consignments. All of these operations require labor, packing facilities, and packaging containers. The flavor of intact healthy fruits of oranges, especially Valencias, is also adversely affected by the volatiles from decaying fruit in the box.

Most postharvest pathogens are weak (*Penicillium, Alternaria, Diplodia, Phomopsis*) and invade through wounds and when the host defense is weak. *Colletotrichum* is able to penetrate through the skin of healthy produce.

Micro-organisms are quite selective in their host tissues and pH. Citrus fruits have a pH lower than 4, so most of the fungi attack these fruits. No bacterial postharvest disease of commercial importance is reported on citrus fruit. On the contrary, bacterial diseases are more common in vegetables, which have a pH of more than 4–4.5.

A. Decay from Field Infection

1. *Alternaria* Rot

It is better to control this pathogen in the field rather than at the packing-house. *Alternaria* is found on citrus fruits as SER (stem-end rot) and ICR (internal core rot) or black rot (Fig. 16.1, see also Plate 16.1). *Alternaria citri* Ellis & Pierce or *A. alternata* Fr. (Keissler) cause these rots. It also causes foliage disease known as *Alternaria* leaf spot. This is a serious problem of Minneola tangelos in South Africa. Fungus survives on infected twigs, stems, and leaves from one year to another. Therefore, pruning dead twigs and wood, prompt disposal from field, and timely sprays of chemicals all give effective control of the pathogen. *Alternaria* spp. colonize flower parts during bloom and after petal fall. *Alternaria* can be recovered from sound fruits (stylar and stem-end) but because of the long incubation period, disease does not develop until fruit becomes susceptible during long storage. Stem-end or stylar-end rots develop during storage, or black-colored fungus develop in the core of fruit (without external symptoms) – hence it is also called core rot. The core rot disease is very common in 'Nagpur' mandarins stored for long periods under refrigerated conditions.

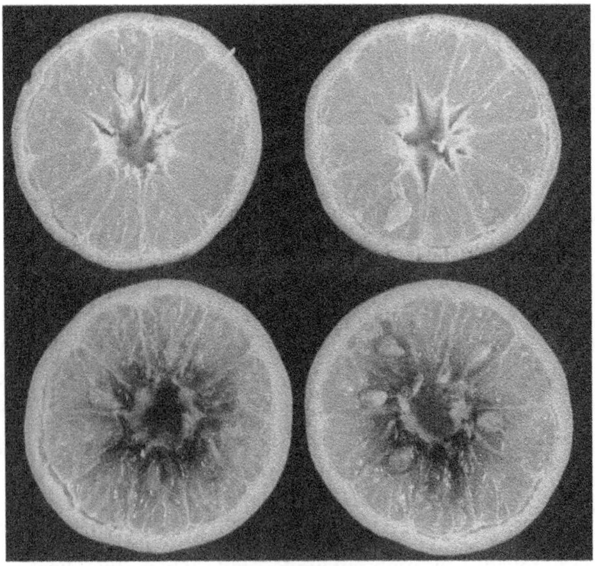

FIGURE 16.1 *Alternaria* Rot of 'Nagpur' Mandarin.

Decay can be effectively reduced by using 2.4-D (500 ppm) together with Imazalil (2000 ppm) in wax coating (water-emulsion wax). Using 2.4-D not only effectively controlled SER and ICR in fruit with buttons, but also reduced the incidence in debuttoned fruit (Schiffman-Nadel et al., 1981). Preharvest sprays with prochloraz have been found to reduce field infection of *A. alternata.*

2. *Phytophthora* Rot

This is caused by *P. citrophthora*, which survives in soil. It belongs to the same group of pathogens that cause the dreaded *Phytophthora* root and collar rots leading to decline of citrus trees worldwide. It is not common in high hanging fruit or fruit that is handled carefully without touching the ground. Fruit that hangs low and touches the ground gets infected from the splash of water from rains or irrigation systems. Damp grove conditions also cause this disease. Aliette (1000 ppm) as a preharvest spray 30 days before harvest controls this pathogen.

3. *Diplodia* and *Phomopsis* Rots

The commonly known *Diplodia* stem-end rot is caused by *Botryodiplodia theobromae (Physalospora rhodina Berk & Curt.) Cooke. Phomopsis* stem-end rot is caused by *Phomopsis citri* Fawcett. These fungi are waterborne and spread through rain splash from dead twigs. The *Botryodiplodia* is very active in warmer seasons and its symptoms are characteristic (Fig. 16.2, see also Plate 16.2), while *Phomopsis* is active in cooler seasons. The spores are lodged at the stem end because of wind or rains when fruit is very small and growing in the field. After picking, the fungus enters the fruit through scars at the stem-end. These stem-end rots are a serious problem in humid areas. In India, particularly the central and southern parts, postharvest losses of citrus are mainly due to these pathogens. These diseases are not common in arid areas. That is why fruits grown in California are less affected by these rots than fruit from Florida. Similarly in India, these diseases are much more common in 'Nagpur' mandarins grown in Central India than on Kinnows grown in arid parts of Punjab (Abohar).

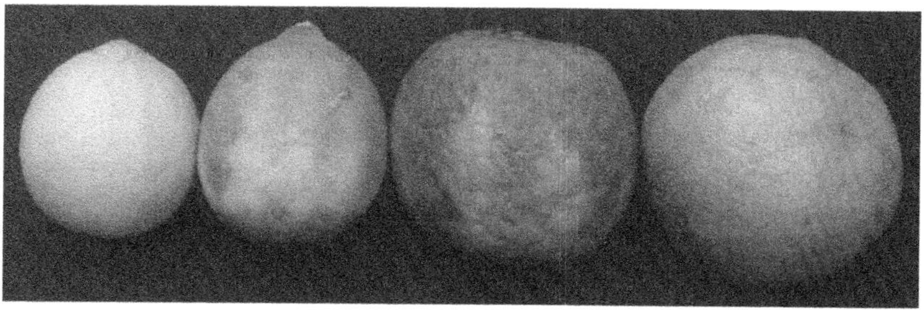

FIGURE 16.2 *Botryodiplodia* (*Physalospora rhodina* Berk & Curt) Stem-End Rot.

4. Anthracnose Rot

This rot is caused by *Colletotrichum gloeosporioides* (Penz) Sacc. This fungus exists in a quiescent state on the surface of immature fruit as ungerminated appresoria. During degreening, ethylene causes appresoria to germinate; that is how infection hyphae penetrate intact, healthy fruit tissue. Some citrus fruits, such as tangerines (Robinson and Dancy), are more susceptible. The penetration and disease manifestation are greater in green-colored fruit, while orange-colored fruit rind resists hyphal penetration by forming a hypersensitive response. In studies conducted by Brown (1992), ethylene has been shown to enhance spore germination on rind and glass. Water waxes and resin solutions increased anthracnose in degreened fruit. Thus high rates of ethylene and higher concentrations of water waxes contribute to a greater incidence of anthracnose. Anthracnose is also observed in fruit injured by oleocellosis. Disease also appears at the button of tangerine fruit as tear stains.

B. Decay from Postharvest Infection

1. *Penicillium* Rots

Penicillium digitatum Sacc. and *Penicillium italicum* Wehmer are the two most significant and widely reported postharvest pathogens in citrus. The major menace of these pathogens is due to their spores, which appear as fine powder and are airborne. The stem end is the most common entry site for the *Penicillium* species (Kaul and Lall, 1975). In infected fruit, very profuse sporulation can be seen – fruit is completely covered by white mycelium followed by green and bluish spores of *Penicillium digitatum* and *Penicillium italicum* respectively. The typical terpenous odor spreads in the surrounding area where these fungi infect the fruit. It is quite possible that these fungi produce ethylene in sufficient quantities, resulting in the rapid senescence of adjacent fruits. Citrus volatiles and even the synthetic mixtures of ethanol, limonene, acetaldehyde, and CO_2 at certain concentrations stimulate the growth of *P. digitatum* (Eckert et al., 1992).

Blue mold is more harmful because it spreads in the box and healthy fruits are directly attacked, regardless of injury (Fig. 16.3, see also Plate 16.3). Blue mold is a nesting-type pathogen, meaning that it produces enzymes that soften the adjacent fruit and thus allow fungus to enter. Green mold does not spread by nesting; thus, if a single fruit is affected it remains as such without contaminating adjacent fruit. However, spores lead to soiling of fruits and thus require repacking with a box change.

Green mold (*P. digitatum*) is quite common in India and grows rather slowly at lower temperatures. At higher temperatures (25–30°C) it grows very rapidly. Green mold infects fruit through wounds (Fig. 16.4, see also Plate 16.4). Orchard and packinghouse sanitation is required to restrict sporulation of *Penicillium* on fruits in orchards and packinghouses to minimize decay losses. Benomyl is used as preharvest spray in South Africa and many other citrus-growing countries to prevent *Penicillium* rots.

FIGURE 16.3 Blue Mold Rot.

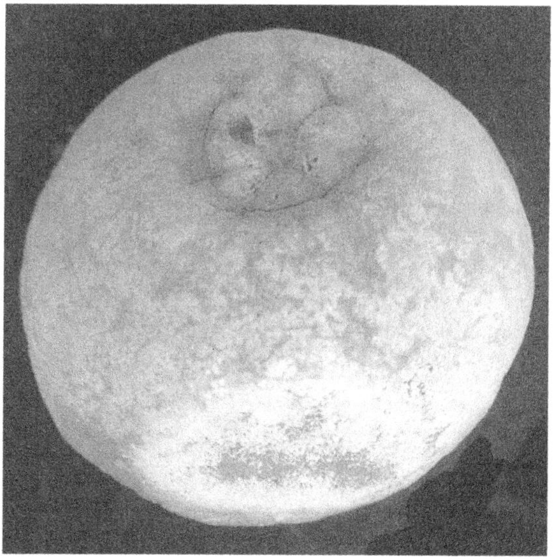

FIGURE 16.4 Green Mold Rot.

2. *Aspergillus* and *Rhizopus* Rot

Aspergillus niger V. Tieghem and *Rhizopus oryzae* are parasitic fungi that penetrate citrus fruit tissue through micro-wounds and bruises. *Aspergillus* rot covers the fruit with black mold and even adjacent fruits are infected, as the spores contaminate the whole lot. High temperatures and high relative humidity favor the growth of these organisms. The most favorable temperature is 25–40°C. *A. niger* causes rapid decay and spreads very fast at 30–35°C. It is interesting to note that at this temperature *Penicillium* growth slows down. *Aspergillus* does not grow below 15°C and at refrigerated conditions there is no growth at all. The fungus is very temperature-specific. It is interesting to see that acid lime fruit stored at 30–35°C develop very high decay levels and sporulation when infected, while fruit from the same lot stored under refrigerated conditions (8–10°C) are free from decay. *Rhizopus* is troublesome in that it spreads and infects neighboring fruit with myselium as it comes into contact with it ('nesting' type). *Rhizopus* grows rapidly at 15°C. Out of the two pathogens, *A. niger* is more common on citrus fruits in India.

3. Sour Rot

This rot is caused by the pathogen *Galactomyces citri-aurantii* (formerly *Geotrichum candidum* Link.), which resides in soil. In warm or hot areas of India where fruit is handled on the ground or where packaging is done on paddy straw, this rot is a major problem. The pathogen enters through wounds on the fruit surface. Fungus is also reported to enter through the stylar-end in limes, lemons, and Kinnows (Kaul and Sharma, 1978). Most of the fruit of the Nagpur mandarin is spoiled by this rot in the March–April harvest season. The disease is not as common on crops harvested in November–December when temperatures are low. Temperatures lower than 10°C are known to suppress fungal development. This disease develops very rapidly at ambient temperatures of 28–30°C. The fruit becomes a soft, stinking, semi-solid mass as the fungi secrete very active extracellular enzymes that rapidly degrade the tissue. This disease spreads rapidly inside the package, and corrugated cartons may also lose strength as they get wet.

4. *Fusarium* Rot

The causal organisms are *Fusarium moniliforme, F. oxysporum*, and *F. solani.* The rots are reported in India, South Africa, and Israel.

C. Physiological Effects of Postharvest Infection

Fungal attacks have a significant effect on postharvest fruit maturation and other physiological functions. Fungal infections of lemons with *Penicillium italicum, Diplodia natalensis*, and *Alternaria citri* accelerate degreening of the fruit at 17°C. *Penicillium*-inoculated fruit develop a yellow color after 4 days and *Diplodia*-inoculated green lemons after 9 days, while the non-infected fruits hardly change color. Degreening of the infected fruit is accompanied by an increase in ethylene evolution and respiration rates.

In yellow lemons the incubation periods of the fungi were shorter and rot development was quicker than in green lemons; ethylene evolution and respiration rate were markedly lower than in green fruit. The enhanced rates of ethylene evolution and respiration occurring during the incubation period of the rot continued during its development and began to decrease when the rot covered about half of the fruit. The shorter incubation period of the fungus was related to greater ethylene evolution and higher respiration rates in the infected fruit. Fungal infection has a similar effect to that of exogenous ethylene treatment of sound fruit, when the respiration rate of the fruit is correlated to the ethylene application (Schiffmann-Nadel, 1975).

In kumquat (*Fortunella margarita*) fruit stored at lower storage temperatures, chilling injuries result in increased decay. With the progression of the picking season and the corresponding change of the peel color from yellow-green to yellow and orange, the susceptibility of kumquat fruit to postharvest rots decreased (Chalutz et al., 1989).

Cell wall degradation and pectin dissolution occur in the infected fruit. The amounts of cell walls recovered from the *Penicillium spp.*–infected fruit after 20 days of incubation has been found to decrease by 9.3 percent of fresh weight compared to 12.4 percent in control fruit. Enzymes extracted from the fruit inoculated with *P. digitatum* exhibited relatively high activities in polygalacturonase (PG), α- and β-galactosidase, α-glucosidase, and β-arabinosidase, while *P. italicum* increased PG and β-glucosidase activities. When cell walls and imidazole-soluble pectins derived from citrus peels were incubated with the crude enzymes extracted from the infected tissues, the composition of polyuronides decreased. The β-galactosidase and α-arabinosidase are also active in hydrolyzing pectin side chains during the *Penicillium* infection (Hwang et al., 2002).

It is well-known that fruits have natural defense mechanisms that result in the formation of lignin, enzymes, phytoalexins, and so forth at the wound site. A significant increase in the inhibition percentage has been observed when the time between fruit wounding and inoculation increased as compared to the control. Varietal difference also occur in natural resistance to postharvest diseases. Fairchild, Carvalhal, and Biondo fruits – harvested completely ripe – showed a greater defense capacity compared to Satsuma fruits (Arras and Sanna, 1999).

II. PESTICIDES, RESIDUES, AND TOLERANCES

Pesticide sprays (mostly insecticides and fungicides) are applied to citrus trees right from fruit set until a few weeks or months before harvest. These sprays leave their residues on and in the fruit. Pesticides are inherently toxic to human beings, and because the etiology of some human disorders is associated with the residues of these pesticides, it is naturally a matter of concern to keep residues well below tolerance limits. Preharvest sprays of various insecticides are applied to control different insect-pests in the field (see Chapter 4). The tolerance limits of these residues are given below. Fungicides are also applied pre- and postharvest. The treatments and residues of fungicides are as follows.

A. Disinfectants and Fungicides

Chlorinated water is generally used to disinfect fruit after arrival in packinghouse. Chlorine has fungicidal and bactericidal action. Chlorine activity depends on free/available chlorine. (Hypochlorus acid, or HOCl, is formed when chlorine gas reacts with water.) Sodium or calcium hypochlorite is used to chlorinate the water. The HOCl dissociation depends on the pH and temperature of water. Usually 50–200 ppm active chlorine concentration with 2 min of contact time is sufficient for disinfection of fruit. In dump tanks it is used at 200 ppm concentration as hypochlorite; pH is set at 7–8. Brushing with chlorinated water (1000 ppm) followed by rinsing with plain water on nylon brushes provided satisfactory disinfection of mandarin, sweet orange, and acid lime fruits (Ladaniya and Sonkar, 1997; Ladaniya, 2001). Minimum decay in Nagpur mandarins was recorded in the fruit dip treated with chlorine solution (1000 ppm dipped for 2 min) and stored under ambient condition up to 30 days. Hydrogen peroxide dip (500 ppm) was not as effective. The decay was mainly due to *Botryodiplodia theobromae* and *Alternaria* sp., which infect the fruit and establish deep inside the fruit in the field. This indicates that infection inside the fruit cannot be controlled by surface disinfection. The *Penicillium* spp. that infect the fruit through surface injuries were not observed in treated fruit (Ladaniya and Shyam Singh, 2001). Generally, non-recovery sprays of disinfectants are preferred over dip treatment in commercial applications. Chlorine-based halogenated disinfectants are also available for treatment. Commercial products such as Brilloclor (8 percent sodium hypochlorite) from Fomesa in Spain are also available. Ozonized water can be used in high-pressure sprays to dislodge spores or infection from fruit surfaces.

1. SOPP (Sodium Ortho (*O*)-Phenyl Phenate)

This chemical is used in soak tanks, as a drench, and also applied in wax. SOPP is also used in foam machines together with detergent and is applied in concentrations of 1.5–2 percent (and in 1 percent in wax) on brush beds. The contact time is 30 s to 1 min when applied on brushes. Dip treatment requires a large quantity of water and chemical; hence non-recovery sprays are generally preferred. Its undissociated form – ortho phenyl phenol (OPP) – is phytotoxic, while SOPP is not. SOPP and OPP can control *Penicillium* rots, *Geotrichum* rot, and stem-end infection to a certain extent (Eckert et al., 1981; Johnson, 1991). Usually a concentration of 2 percent is sufficient for SOPP (SOPP tetrahydrate form) as a dip treatment with 1–2 min contact time at 30–35°C with pH of the solution at 11.7–12. If the pH is lowered, epidermal burn injuries can occur. The pH is controlled at 11.5–11.8 in some countries; in this range residues are not found to be more than 2–3 ppm (Hayward and Grierson, 1960). SOPP is active at wounds and injuries in the form of free phenol. The OPP (HOPP) diffuses to injury sites and thus prevents infection. Fruit needs to be rinsed with plain water after SOPP treatment. Its use is restricted in South Africa (Pelser, 1977) to tolerant citrus fruits only because it increases the incidence of oleocellosis and darkens injuries on the rind. In Israel, SOPP at 0.25–0.5 percent is a widely used fungicide at 32°C. Dipping in 2000 ppm OPP at ambient temperature or 600 ppm at 46°C is as effective as SOPP (Cohen, 1991). OPP has several advantages over

SOPP. OPP can be applied as a spray, replacing the SOPP soak tank. There is no need for pH adjustment at 11.8 or rinsing fruit after treatment (Cohen, 1991). OPP is phytotoxic, so buffer hexamine is used along with it. Currently the disinfection of fruit is achieved with SOPP in packinghouses, followed by fungicide treatment with TBZ, Guazatine, or Imazalil. SOPP is not used as stand-alone treatment. Steribox-p (10 percent SOPP) from Fomesa in Spain is available as a commercial product.

2. Biphenyl or Diphenyl

These have a fumigant action. Phenol vapors are released when impregnated paper is placed among fruit layers. Most of the oranges in world trade were wrapped with biphenyl paper before the introduction of Imazalil. Satisfactory control of decay and sporulation of *Penicillium* rots in transit and in storage were obtained by placing one biphenyl pad over the bottom layer of fruit and one pad between the upper layers (Hale et al., 1981). Biphenyl provided additional decay control in transit after SOPP and TBZ treatment in the packinghouse.

Biphenyl-impregnated paper has 40 mg biphenyl per 600 cm^2, or it is used at the 3.5 g/16 kg carton. Wrapping individual fruit gives better control than placing impregnated paper between the layers of fruit. The tolerance was fixed at 70 ppm in Europe for biphenyl. The disadvantage is that fruit absorbs the phenol odor. It is considered oil-soluble and dissolves in the oil in the peel. This odor dissipates slowly at ambient conditions within few days as the fruit is removed from the wrap. It is not suitable for mandarin-type fruit, which has a very delicate flavor. Biphenyl is not effective against sour rot, stem-end rots, and *Phytophthora* (Eckert et al., 1981).

3. Benzimidazoles (MBC Generating)

This group includes thiabendazole, benomyl, and carbendazim. These fungicides have been used extensively since the 1970s, and some of them, particularly carbendazim and benomyl, have been discontinued for postharvest use in many countries, including the U.S. and Japan. They are not phytotoxic and are systemic in action with long residual activity. They are also used as preharvest sprays to reduce infection of *Botryodiplodia* and *Phomopsis* rots. These fungicides are very effective against stem-end rots and *Penicillium* rots but ineffective against *Alternaria, Rhizopus*, and *Geotrichum* (Eckert et al., 1981). These compounds interfere with mycelial growth and affect conidia formation and also inhibit spore germination. TBZ in benomyl suspension has to be agitated continuously during application because it settles. It is more effective in water and emulsion wax than in solvent waxes. About 2000–3000 ppm TBZ is brushed on fruit in water wax or 1000 ppm as water suspension on washed fruit. To control preharvest infection, sprays of 500 ppm ai. of benomyl or 500 ppm of carbendazim are applied in the field 2–3 weeks before harvest. The USEPA has proposed cancellation by January1, 2008 in the U.S. because of the tolerance of benlate for citrus.

4. *Sec*-butylamine (2 AB)

It is used as water dip (1 percent) or for drenching and can also be added to wax. This compound is used as a phosphate salt and effective against *Penicillium* and stem-end rots. It does not accumulate in oil cells. The tolerance limit in fresh citrus is set at 30 ppm.

5. Imidazoles

This group includes Imazalil and prochloraz. Imazalil (1-β–allyloxy)–2.4-dichlorophenethyl imidazole (trade name Fungaflor and Fungazil) is the best antisporulent against *Penicillium* rots and has replaced biphenyl paper wraps. Imazalil is effective against *Diplodia* stem-end rot but less effective against *Alternaria* rot. It is not effective against *Phytophthora* rot and sour rot (Eckert et al., 1981). The imidazoles give complete control of *A. niger* and are not effective against *Rhizopus oryzae* (Tuset et al., 1992). Imazalil is intensively used in California packinghouses as aqueous spray (1 g/lit) or in water wax (2 g/lit). The residues with this treatment are 1–2 ppm (mg/kg whole fruit). Prochloraz is a imidazole derivative like Imazalil; it interferes with ergosterol biosynthesis in fungal cell walls, thereby preventing normal development of the pathogen. Prochloraz (available as Omega 450 EC) is incompatible with alkaline water–emulsion waxes. The concentration of 1000 ppm is recommended for controlling *P. digitatum* in South Africa. The application has to be at earliest after fruit harvest (Lesar and Du Toit Pelser, 1997). It can penetrate a short distance into the peel and the fruit button and can prevent decay even some time after infection.

6. Triazoles (Analog of Imidazole)

There are more than 30 different compounds in this group. Etaconazole, penconazole, propiconazole, myclobutanil, and tebuconazole are some of the important ones. Tebuconazole (0.021 percent) as a preharvest spray can control *Alternaria* leaf spot in the field and postharvest *Alternaria* decay in tangelos (Schutte et al., 1992). The postharvest use of these fungicides is subject to registration and permission for use in various countries.

7. Morpholines

Fenpropimorph comes under this group. Fenpropimorph (Corbel–Mistral, marketed by Maag Agro-chemicals) is an ergosterol biosynthesis inhibitor and is widely used for control of powdery mildew. It has been found effective against *Geotrichum candidum* within 24 h of infection and *Penicillium* molds within 48 h of infection when used at 1000–2000 ppm concentration (Cohen, 1991). The disease lesions and sporulation are arrested; thus it is both preventive and curative.

8. Guazatine

Guazatine (9–aza-1,17-diguanidinoheptadecane acetate) is effective against *Penicillium* and *Geotrichum* spp. Strains of *Penicillium* doubly resistant to guazatine have been found (Cohen, 1991).

9. Metalaxyl

This fungicide at 1000–2000 ppm ai. in wax or aqueous drench reduces *Phytophthora citrophthora* (Cohen, 1991). Metalaxyl-M applied at 2.5 g per ton of fruit in wax gives residues of approximately 1.2–1.6 mg/kg. The rate of wax application has to be controlled in terms of fungicide per unit weight of fruit. Metalaxyl is marketed as Ridomil by Ciba and is quite widely used for preharvest spray or tree root zone drenching. Postharvest use is restricted.

10. Fludioxonil

This is a contact fungicide and is effective against *Penicillium, Diplodia, Botrytis,* and *Phomopsis* when applied 4–12 h after harvest. It is applied at 500–600 ppm in aqueous solution as drench or at 1000–6000 ppm in wax. Residues of 1–2 ppm provide decay control; 2–4 ppm can check sporulation (Syngenta, 2006). USEPA has designated it as a reduced-risk pesticide.

11. Pyrimenthanil

This new postharvest fungicide (aniline-pyrimidine) is effective against *Penicillium* and *Botrytis* rots. It inhibits cell-wall degradation enzymes of fungus such as protienases, cellulases, pectinases, and laccase. It also inhibits the biosynthesis of Methionine (Bylemers and Goodwin, 2004).

12. Strobilurin (Azoxystrobin)

This belongs to the methoxyacrylate class of chemicals and is effective against *Penicillium* and stem-end rots (*Diplodia*). It is marketed by Syngenta, Inc. as Abound.

B. Postharvest Fungicide Treatments

After surface disinfection, fruit has to be treated with fungicides for long-lasting effect, preventing rotting until fruit is marketed and consumed. Several compounds have been used for chemical control of citrus postharvest rots in India (Tables 16.1 and 16.2). The use of chemicals in citrus postharvest disease management has a history going back 100 years or so in the U.S. and Europe.

Presently, the chemical control of various rots is the integral part of the postharvest handling system in citrus. Several chemicals – sodium carbonate, sodium o-phenyl phenate (SOPP), imidazole, and benzimidazole group of fungicides – are being used commercially and many new fungicides are being evaluated and registered (Table 16.3). Prior to the mid-1960s, borax, sodium carbonate, SOPP, and biphenyl were used extensively and prevented harvest-related infection by *Penicillium*. Some of these chemicals are still being used. Thiabendazole (TBZ), benomyl, thiophanate methyl, and carbendazim are the important benzimidazoles used for postharvest treatment of citrus. Some of these chemicals are fungistatic in action (not killing the pathogen but inhibiting spore germination or arresting its growth by some other mechanism). The fungistat should be in contact with

TABLE 16.1 Storage Life of Mandarin Fruits as Influenced by Fungistatic Compounds

Mandarin cultivar	Treatment	Storage condition	Storage (days)
'Nagpur'	Sorbic acid (Saturated Solution)	4.5°C	20
'Nagpur'	0.1% Benlate + Calixin + 6% wax	21–32°C; 35–65% RH	20
'Nagpur'	Wax 2.5% (Highshine) + carbendazim (2000 ppm) + vented PE	20–35°C; 50–60% RH	21
'Coorg'	4% wax + 1% SOPP two coatings	26–30°C	24
'Coorg'	Vented PE liner + diphenyl wrapper	4–5°C; 85–90% RH	135
Coorg	SOPP 1% + hexamine	23–29°C; 45–65%	20
Coorg	0.1% TBZ	5–7°C	60
Kinnow	Wax 6% + Benlate 0.1% + PE	0–3.5°C	60
Kinnow	Waxol + Rovral	24°C	42
Kinnow	Diphenyl paper (40 mg/paper) wrapper	20 + 4°C	28
Kinnow	Diphenyl paper wrapper	20°C; 85–90% RH	120
Kinnow	Wax 6% + diphenyl paper wrapping	19 + 8°C	90
'Darjeeling'	2% phenol	4.4°C	20
'Khasi'	Wax 3% + Benlate (350 ppm) + PE	11–190°C	24

Dutt et al. (1960), Laul et al. (1972, 1976), Ladaniya and Sonkar (1997a), Dalal et al. (1962b), Subba rao et al. (1967), Dalal et al. (1961), Subramanian et al. (1973), Dhillon et al. (1976), Singh and Gupta (1979), Mishra (1989), Gupta (1986), Singhrot et al. (1987c), Dutt et al. (1960), Karibasappa and Gupta (1988).

the organism to be effective. These chemicals are used in various forms: in soak tanks or as water sprays, as fumigants or in wax, or impregnated in paper for keeping in fruit layers or as wraps. The effectiveness of fungicides is known to be greater in water than in wax; therefore, to achieve the same results, the concentration of fungicides applied in wax is generally increased (Brown, 1983). The results of decay-control treatment further depend on the kind of wax and the kind of fungicide used (Radnia and Eckert, 1988). Generally, two or three types of fungicides are applied with the wax to combat a variety of fungi.

In early days, citrus packinghouses in the U.S., South Africa, Australia, and many other citrus-growing countries used alkaline inorganic salts such as sodium tertaborate (borax), sodium carbonate, and sodium bicarbonate (baking soda) to control *Penicillium* rots. Borax was later discontinued because of boron residues. Lately there has been an interest in the use of sodium bicarbonate because it is nontoxic compared to chemical pesticides.

Satisfactory decay control has been achieved in Florida when SOPP treatment is followed by thiabendazole and Imazalil treatment. Different countries follow different fungicide treatment schedules depending on the major problem

TABLE 16.2 Storage Life of Sweet Orange Fruits as Affected by Fungicide Treatment

Sweet orange cultivar	Treatment	Storage condition	Storage (days)
'Mosambi'	SOPP (2%) + hexamine dip followed by wax (9%)	13–20°C, 55–95% RH	24
'Mosambi'	Wax (9%) + 2.4-D (500 ppm) + Bavistin (0.1%) + non-perforated PE	Ambient condition	63
Sathgudi	6% wax + SOPP (0.5%)	26–29°C	30–34
Sathgudi	SOPP (2%) + hexamine (2:1) +wax(6%)	5.5–7°C	127
Blood red	Benomyl (1000 ppm) and packing in wooden box	Zero energy cool chamber	80
		Room temperature	60

Garg and Ram (1972), Tarkase and Desai (1989), Dalal et al. (1962a), Dalal and Subramanyam (1970), Rana and Kartar Singh (1992).

pathogens in those areas. Australian protocol is based on the Australian Horticulture corporation's quality-management guide for citrus (Tugwell and Chvyl1, 1997). Fruits are carefully harvested, handled, and delivered to packinghouses within 24 h for dipping in Guazatine (500 ppm) followed by Imazalil (500 ppm). Alternatively these fungicides are applied on the packing line followed by wax coating with 2.4-D (500 ppm). This treatment is common for lemons. The problem of sour rot occurs in lemons, and Guazatine effectively controls this rot. Lemons are stored up to 10 weeks at about 20°C in Australia (Tugwell and Chvyll, 1997).

The control of wound pathogens depends on the prompt application of fungicides at required concentrations. If the application of fungicide is delayed, the pathogen gets established. In such cases, the pathogen can escape the protective action of fungicide. Fruit temperature and surrounding conditions that affect pathogen growth rates are important. The warm mix of TBZ, sodium carbonate, and chlorine as a drench to fruit prior to degreening effectively controlled 82 percent of decay (Smilanick, 2004).

The intensive and indiscriminate commercial use of fungicides alternatively and in mixture has been followed by the industry since the mid-1980s. All fungicides are not effective against all important pathogens and there are problems related to use of these fungicides. The treatments have to be used in mixtures of fungicides to achieve effective postharvest decay control (Tables 16.1 and 16.2). The resistance of fungal pathogens to antifungal compounds necessitates the screening of new antifungal compounds. Simultaneously, new chemicals,

TABLE 16.3 Some Disinfectants and Fungicides, Their Trade Names and Suppliers

Common and chemical name of the fungicide	Trade name	Supplier	Remarks
2-aminobutane (2AB), *sec*-butylamine (phosphate salt) or (hydrochloride salt)	Butafume, Tutane, Frucote	BASF (FRG)	Postharvest use, mostly discontinued
Benomyl, methy1 1-(buty1carbamoy1) benzimidazole-2y1-carbamate	Benlate, Benosan, Benex	DuPont agricultural Products, Washington DE (USA)	Postharvest use discontinued and allowed for preharvest use
Carbendazim,	Bavistin	BASF (FRG)	-do-
SOPP, sodium *0*-pheny1phenate tetrahydrate	Dowicide A, Steribox-p, 10%SOPP (FOMESA)	Dow Chemical Co., (Dow Agroscience) Midland, MI, (USA)	Currently being used as postharvest disinfectant/fungicide
Dipheny1 or biphheny1, 1,1′-bipheny1	United Diphenyl pads, Flavouseal	–	Discontinued
Borax, sodium tetraborate decahydrate, disodium octaborate	Polybor	Rio Tinto Borax, Boron, California (USA)	Being used to some extent as postharvest, GRAS
Sodium carbonate		FMC Corp. (USA), Pharmacia Corp. (Monsanto)	Being used as postharvest, GRAS
Sodium bicarbonate	MSH-50%-WSC	Hodogaya Chemical Co., Ltd. (Japan), Pharmacia Corp.	Being used as postharvest, GRAS
Thiabendazole, 2-(thiozal-4-y1) benzimidazole	Mertect, Tecto, Freshgard-598	Merck and Co. Inc. (USA), FMC, USA	Widely used as postharvest fungicide, no resistance report from Florida
Imazalil, 1-[2-(2,4-dichloropheny1)-2-(2-propenyloxy)ethy1]-1H-imidazole	Fungaflor	Janssen Pharmaceutica (Belgium)	Widely used as postharvest fungicide, no resistance report from Florida
Guazatine, di-(8-guanidino-octy1) amine acetate salt	Panoctine, Kenopel-20 (20% EC)	Makhteshim-Agan, Israel, KenoGard AB (Sweden),	Widely used as postharvest fungicide in Australia and Europe

Etaconazole, 1-[2-(2,4-dichloropheny1)-4-ethy1-1, 3-dioxolan 2y1 methy1] –1H-1,2,4-trizole	CGA-64251, Vanguard, Sonax	Ciba–Geigy (Switz)	Very limited use as postharvest fungicide especially on citrus
Propiconazole, 1[2-(2,4-dichloropheny1)-4-propy1-1,3-dioxolan-2y1 methy1)]	CGA-64250, Tilt, 'Break EC' Banner, Orbit	Ciba-Geigy Corp. (Switzerland), Novartis	–
Fosty1-Al, Aluminum tris-o-ethysphonate	Aliette	Rhone-Poulenc Agrochemie (France), Aventis	Preharvest use for *Phytophthora* brown rot of fruit
Fenpropimorph, 4-(3-(4-(1,1, –dimethylethy1) phene1) –2methy1) propy2,6 (cis) – dimethy lmorpholine	Corbel	Dr. R. Maag Ltd. (Switz), BASF (FRG)	–
Tridemorph, 2,6-dimethy 1-4-tridecy1 morpholine	Calixin	BASF (FRG)	–
Prochloraz, 1-[N-propyt1-N-2-(2,4,6-trichlophenoxy) ethylcarbamoly] imidazole	BTS-40542, Sportac	Boots Co. Ltd. (UK)	Not registered in USA but registered in S.Africa and Europe
Metalaxy1, *N*-(2,6-dimethy1pheny)-*N*-methoxycety1)alanine, methy1 ester	CGA 48988, Ridomil	Ciba-Geigy Corp. (Switz)	Preharvest use for *Phytophthora* brown rot of fruit
Potassium sorbate, 2-4hexadienoic acid, potassium salt	Monitor	–	Being used as postharvest, GRAS
Fenapronil, α–buty-α-pheny1-1 H-imidazazole –1-propane nitrile	RH 2161, Sisthane	Aventis, Syngenta	Fostharvest fungicide recently registered in US
Pyrimenthanil, (4,6-dimethyl–N-phenyl–pyrimidineamine)	Penbotec-400 SC	Janssen Pharmaceutica (Belgium)	Postharvest fungicide recently registered in US
Fludioxonil, (4-(2,2-difluoro -1,3– benzo-dioxol-4yl)-1H-pyrrole-3-carbonitrile)	Graduate, Scholar, Medallion	Syngenta Inc., Novartis Inc.	Non-systemic fungicides, protective and contact activity
Azoxystrobin	Abound, Heritage	Syngenta Inc.	Pre- and postharvest fungicide

application methods, and combined treatments to inhibit sporulation need to be investigated (Table 16.3). Because sporulation from decaying fruit causes spoilage of healthy fruit in the carton and serves as a source of infection in addition to reducing the commercial value of healthy fruit, the search for antisporulent fungicides continues. Imazalil has already been commercialized, and some other fungicides – Fenapronil, Prochloraz, and Fenpropemorph – have been established to be highly effective antisporulents. They are effective against green mold and to some extent sour rot. To be effective, these fungicides have to be applied within 24 h of inoculation; therefore the earliest treatment after harvest is effective (Gutter 1981). Continuous and intensive use of biphenyl, SOPP, TBZ benomyl, and *sec*-butyl amine has led to problems of fungicidal resistance in citrus postharvest pathogens, mainly *Penicillium* spp. Triple-resistant biotypes of *Penicillium digitatum* (resistant to Imazalil, TBZ, and OPP) have been isolated in California packinghouses. Such resistant strains were rare in *P. italicum* (Holmes and Eckert, 1999). Zhang (2003) reported that three new fungicides – PH066 (pyrimenthanil), Scholar (fludioxonil), and Abound (azoxystrobin) – are likely to be registered in U.S. for postharvest use to combat decay. PH066 is a product from Janssen Pharmaceutica, Inc., while Scholar and Abound are the products of Syngenta, Inc.

C. Strategies against Decay and Resistance Development

1. Rotating fungicides (TBZ, Imazalil, guazatine, SOPP).
2. Fungicide aqueous spray or drench and then separately wax coating (without fungicide).
3. Effective application with residues that are lethal to pathogens.
4. Proper hygiene in packinghouses and cold storage.
5. Avoiding prolonged storage and reprocessing of fruit on the packing line.
6. Proper harvesting without injury to fruit.

Fruits must be free of skin breaks, mechanical injuries, and bruises if they are destined for distant markets and storage. Eckert (1991) has suggested the sequential use of fungicides with unrelated fungicides and suppression of spore production and dispersal as a strategy to reduce development of resistant strains.

D. Residues and Tolerances

As a result of the increasing awareness among public and findings on toxic effects of pesticides on human health, countries are imposing stricter regulations regarding the registration of fungicides and residue monitoring so as to restrict residues in fruits. Many fungicides are not re-registered because: (1) the registration process is cumbersome and expensive, (2) fungicide residue levels are beyond tolerance limits, and (3) Pathogens may have developed resistance.

Many fungicides used in the past have been discontinued and many more are on the list of discontinuance. Benomyl was withdrawn for postharvest use in the U.S. during late 1980. However, preharvest spray of benomyl and postharvest use of TBZ are still allowed (Brown and Craig, 1989).

In Japan postharvest fungicides are considered food additives and are regulated by food laws. Hence residue limits are set in a lower range. Many fungicides allowed in other countries are already banned in Japan and thus not permitted in fruit imported to that country. The EU sets its own standards for maximum residue limits (MRLs).

During the export of fruit, pesticide residue analysis is mandatory. Developing plant-protection spray schedules to minimize blemishes and at the same time restrict pesticide residues within limits (specified by various countries) is imperative.

Because fungicides are applied in the field or after harvest, residue persists in or on the fruit. These fungicides are harmful to human health if intake is greater than a specified quantity. Hence, permissible residue limits have been set. Imazalil and thiabendazole (TBZ) are important postharvest fungicides used in most countries. Their maximum permissible residue levels in citrus fruits as set by Codex (FAO/WHO) and EEC are 5 and 10 ppm respectively. The residue limit for Pyrimethanil is 3 ppm per Codex MRL.

Residues of carbendazim in Nagpur mandarins were within tolerance limits (5 ppm) when carbendazim was applied at 500 ppm as a preharvest spray followed by postharvest treatment at 2000 ppm (Naqvi, 1993). The initial residues of carbendazim were 2.25–2.40 ppm in the peel and 1.55–1.70 ppm in the pulp of Nagpur mandarins after two preharvest sprays (500 ppm) at an interval of 15 days before harvest (Ladaniya, 2005). Residues declined in treated and packed fruit during storage under ambient conditions up to 28 days (Table 16.4). Thiabendazole residues were also within tolerance limits when applied at 1000 ppm with wax (6 percent) as a postharvest treatment (Ramana et al., 1983).

TABLE 16.4 Residues of carbendazim in 'Nagpur' mandarin fruit stored under ambient condition

Treatments (wrapping film)	Days in storage: Initial		15		28		% reduction after 28 days	
	Peel	Pulp	Peel	Pulp	Peel	Pulp	Peel	Pulp
D955(15 μm)	2.30	1.70	2.5	1.7	2.1	1.5	8.69	11.76
D955(25 μm)	2.40	1.50	2.5	1.7	1.7	1.4	29.16	6.66
Control	2.25	1.55	2.1	1.8	1.7	1.4	24.44	9.67

Source: Ladaniya (2005); Two preharvest sprays of carbendazim (500 ppm) at 15 days interval (last spray 15 days before harvest).

The use of SOPP in foam washers at 1.86 percent concentration resulted in residue of 0.5–2 ppm as against the maximum standard of 10 ppm in the U.S. The use of Imazalil and TBZ as a drench, a spray, or in wax at 1000 ppm resulted in residue of 0.2–1.0 ppm against the maximum standard of 5 ppm (Johnson, 1991).

Benomyl is usually applied as a preharvest spray. In South Africa, Washington Navel oranges and grapefruit are harvested 90–120 days after the last benomyl spray. Valencia oranges and lemons are harvested 150 and 60–70 days after the last spray respectively. Residues have been found to be 0.2 ppm at 50 days after orchard spraying (Pelser, 1977).

The benzimidazoles are systemic in nature and persist in tissues. When benomyl is applied in an oil spray emulsion, methylbenzimidazole carbamate (MBC) in the pulp of Valencia fruit averaged only 0.14 ppm after 21 days, during which migration from peel to pulp continued (Brown, 1974). For application of 2.4-D after harvest, sodium salt is used (500 ppm) as an aqueous solution. The tolerance limit for this chemical is 2 ppm in the fruit as set by Codex (FAO/WHO) and EEC.

Residues are strictly dependent upon the concentration of fungicide used and the treatment temperature and storage temperature maintained. TBZ residue uptake by lemon fruits is the same following dipping in 1200 ppm TBZ at 20°C and dipping in 150 ppm TBZ at 50°C. Thiabendazole residues show great persistence during storage. The residue contents of fruits averaged 70 percent of their initial values after 13 weeks at 8°C and after 1 week at 20°C (Schirra et al., 1998).

The length of fungicide treatment also affects uptake and persistence. Imazalil (IMZ) uptake in citrus fruit is related to treatment duration, whereas thiabendazole (TBZ) residue is not. Residues of IMZ and TBZ fungicides were significantly correlated with dip temperature. Treatment at 50°C produced a deposition 8 and 2.5 times higher than when treatments were carried out at 20°C in IMZ and TBZ respectively. No significant differences in terms of IMZ deposition were detected after treatments carried out alone or in combination. Uptake of the two fungicides was associated with their physicochemical characteristics as well as different formulation types. IMZ residues showed a great persistence during storage when applied separately, and >83 percent of the active ingredient was present after 9 weeks of storage. IMZ residues increased with dip length, doubling when dip time increased from 0.5 to 3 min. In contrast, TBZ residues did not change with different dip times. Following postharvest dip treatments of citrus fruit at 20 or 50°C, home washing removed 50 percent of the IMZ and 90 percent of the TBZ (Cabras et al., 1999).

The reduction in doses of fungicides is possible so as to reduce residues by heating treatment solutions (Cohen et al., 1992). Doses of TBZ from 4000 ppm to 250 ppm and Imazalil from 1000 ppm to 250 ppm are possible when the solution is heated to 48°C. SOPP 5000 ppm can be reduced to OPP 400 ppm at 45°C. It can be applied as a hot water dip or a hot spray.

Insecticides are used prior to harvest (See Chapter 4). Their application also leaves hazardous residue. This aspect is beyond the purview of present chapter.

However, since it is a matter of concern for consumers of fresh citrus it is covered here very briefly. The application of insecticides has to be scheduled in such a way that keeps residues within specified limits. If insecticides are applied and a safe waiting period for their metabolism and disintegration in fruit tissue is observed, residue hazard can be greatly minimized. In Mosambi sweet oranges, waiting periods of 32 and 34 days were required for monocrotophos and quinalphos residues to be within 0.2 and 0.25 ppm tolerance limits respectively when trees were sprayed at fruit maturity (Awasthi and Anand 1986). In acid limes, waiting times for monocrotophos, quinalphos, and phosphamidon were 14–15, 13–19, and 7–10 days respectively (Awasthi and Ahuja, 1989), whereas synthetic pyrethroides dissipated much faster (1–4 days) excluding permethrin (8–13 days). Pyrethroides do not move from pericarp to pulp (Awasthi, 1989).

Every country enacts laws to prevent food contamination (including fruits) with pesticides. The MRL values of some insecticides and fungicides as approved in some countries/group of countries are given in Table 16.5.

In the U.S., the U.S. Food Quality Protection Act (FQPA) of 1996 established new human health standards, stating that no harm (with reasonable certainty) could result from aggregate exposure to pesticide chemical residues. The new FQPA amended the earlier FIFRA (Federal Insecticides, Fungicides, and Rodenticide Act) and FFDCA (Federal Food, Drug, and Cosmetics Act). The new law emphasized concerns about the health of infants and children with regard to pesticides.

In the U.S., three federal government agencies are responsible for regulation of pesticides in food (including citrus fruits). The Environmental protection Agency (EPA) approves and registers pesticides under FIFRA and FFDCA and establishes maximum permissible residues in food. The Food and Drug Administration (FDA) enforces tolerances in domestic and imported food (including fruit) in cooperation with states. It also conducts surveys of residues in fruit. The USDA's Agricultural Marketing Service (AMS) conducts residue-testing programs and reports data. In imported lots, the FDA samples fruit consignments and, based on EPA tolerance limits, either permits or detains the consignments for the entry.

III. ORGANIC POSTHARVEST MANAGEMENT

It is very important to maintain the organic integrity of organic citrus fruit once the fruit enters the packinghouse. In organic produce, the chain of its organic integrity from farm to consumer has to be maintained. During preharvesting, the organic grower (who is certified by a government-accredited agency) is responsible for the organic nature of the produce. The treatments should be organic in nature and there can be no commingling of organic and inorganic inputs/ingredients. Proper records in this regard are to be maintained to show to the certifying agency if required (Plotto and Narciso, 2006). It is important that food

TABLE 16.5 Pesticide Residue Tolerance Limits in Fresh Citrus as Set by Codex and Some Countries

	Tolerance (parts per million)					
Pesticide	**Codex**	**EU, EEC**	**USA**	**Canada**	**Japan**	**India**
Chlorpyriphos	1	2	1	–	1	0.5
Malathion	4	2	–	–	–	4
Imazalil	5	5	5	5	5	–
Imidacloprid	1	1	–	–	–	–
Methidathion	2 (orange)	2			–	–
TBZ	10	6	6	10	10	–
Dicofol	5	2	–	–	–	
SOPP	10	12	10	10	10	–
Carbendazim	5	5				10
Pyrimethanil	–	–	10	–	–	–
Fludioxonil	–	–	10	–	–	–
Pyraclostrobin	–	–	2	–		
2.4-D	1	–	2	–	–	2
Guazatin	5	5	5	–	5	–
Methylparathion	–	1				
Dimethoate	2	1				5
Captafol	0.05	0.02				–
Carbofuron	–	0.50				–
Metalaxyl	5	5				–
Monocrotophos	0.2	–	–	–	–	0.2
Prochloraz	10	–	–	–	–	–
Carbaryl	7					5
Dichlorvos	0.1		–			0.1
Methylthiophanate	–	5	7	–	5	–

EEC (2000), FAO/Codex standards for pesticide tolerances (2005), USEPA (2004), Anonymous (1994).

safety should not be compromised in the name of 'organic produce'. In the U.S., the National Organic Program (NOP) has been fully implemented since October 2002.The rules and regulations of NOP contain guidelines for processing, handling, and packaging organic produce. The Organic Materials Review Institute (OMRI), a separate organization, provides lists of organic materials/ingredients with brand names and manufacturers. The mechanical and biological processes/methods are emphasized in processing/handling to minimize the use of inorganic materials. Biopesticides can be used on organic produce. These are derived from natural sources such as microbes, minerals, plants, and animals. Relatively safe chemicals such as GRAS chemicals/materials can be used. They are not organic but are still less polluting and less hazardous to humans.

Several other countries and the European Union have their own set of rules and regulations for purely organic produce. The EUREPGAP (European Retail Parties Good Agricultural Practices), an organization of growers, exporters, and importers of Europe, has developed its own set of standards based on GAP. It is a partnership and growers are certified by accredited agencies. The production and postharvest practices included in EUREPGAP protocol are organic to some extent, although they are not 100 percent organic. This program has both environmental and social dimensions.

A. GRAS Chemicals

Chemicals that are 'generally regarded as safe' (GRAS) by the FDA are recognized so when used in accordance with FDA's 'good manufacturing practices' and contain no residue of heavy metals or other contaminants in excess of tolerances set by the FDA. For example, sodium carbonate and bicarbonate, borax (E285, used as a food additive), sorbic acid, and biphenyl (E 230), are GRAS chemicals and used for postharvest treatment. These chemicals pose relatively little health hazard. Chlorine (sodium hypochlorite), sodium carbonate, borax, and peroxyacetic acid are used as disinfectants/sanitizers for washing fruit. The disadvantage of the continuous use of chlorine is that it is corrosive to metals and weakens pallet box wood by delignification. Strictly speaking, GRAS chemicals do not fall into the category of organic ingredients/materials, but they are much less harmful than many other inorganic compounds.

1. Sodium Carbonate (Na_2CO_3 $10H_2O$) and Sodium Bicarbonate (NaH CO_3)

Sodium carbonate, a common washing soda, is soluble in water and has some bleaching properties. At 43°C it is effective against *Penicillium* and *Geotrichum* infection, while at 48°C it can control *Phytophthora* brown rot (Eckert et al., 1981). Sodium carbonate may be phytotoxic to some citrus and as a hot solution has to be used carefully. It forms scales on equipments Treatment with sodium bicarbonate (2000 ppm) with thiophenate methyl (700 ppm) is less toxic. Sodium carbonate is compatible with chlorine and TBZ.

2. Borax

Borax has detergent and bleaching action. It has biostatic activity and is used in wash water. Sodium tetraborate decahydrate ($N_2B_4O_7$ $10H_2O$) is soluble in water.

3. Peroxyacetic Acid (PAA)

PAA is used as a sanitizer and readily decomposes into acetic acid and hydrogen peroxide. It has been tried in some U.S. packinghouses and is equally effective as chlorine. The generally used concentration is 100 ppm. Plotto and Narciso (2006) reported that PAA is consistent and more efficient than chlorine as a sanitizer.

4. Hydrogen Peroxide

Hydrogen peroxide is a strong oxidizing agent and commonly used as disinfectant/sanitizer. It is very corrosive and has to be used in low concentrations with a care. Its advantage is that it decomposes into water and O_2. It is commonly used in minimally processed food items.

5. Sorbic Acid

Sorbic acid is commonly used as a food additive and has antimicrobial activity.

6. Chlorine

Calcium hypochlorite, sodium hypochlorite, chlorine dioxide, and chlorine gas can be used for the chlorination of water. The residue left should not be more than 4 ppm free chlorine.

B. Bioagents (Microbial Antagonists)

The microbial control of postharvest disease is an effective and viable alternative to chemical pesticides. The research and development of biocontrol agents has recently gained momentum and commercial products are already available for citrus postharvest disease management. These biocontrol agents can be used along with chemical fungicides for more than 90 percent control. Another alternative is to use physical means in combination with bioagents for effective, eco-friendly, and non-hazardous control. Yeasts and bacteria are being evaluated as bioagents against postharvest fungal pathogens of citrus.

A strain (43E) of *Candida famata* (*Torulopsis candida*) when used together with 0.1 gTBZ/liter at a concentration of 106 cells/ml gives significantly better disease control than either TBZ or the yeast alone (Arras et al., 1997). The mechanism of action of the yeast is primarily by competition for space and nutrients through colonization of pathogen hyphae, as seen in scanning electron micrographs (SEM). The isolate 43E of *C. famata* was found to be compatible with high TBZ concentrations (5 g/liter). Detection of antibiotic or toxic substance production by the yeast has not been possible. The yeast *Debaryomyces hansenii* has been reported to reduce incidence of *Penicillium* rots (Mehrotra et al., 1996) and sour rot of orange fruits (Mehrotra et al., 1998). A water suspension of yeast cells applied to wounds on the fruit surface prior to inoculation with spore suspensions of pathogens reduced disease by 80–90 percent. The efficacy of the antagonist in reducing sour rot was affected by the concentration of both the yeast cells and pathogen spore suspension. The strain of commonly used yeast in wineries, *Sacchromyces cerevisiae,* was found to inhibit the growth of *Aspergillus niger* in Nagpur mandarins and acid limes (Naqvi, 1998).

The combination of yeast *Candida saitoana* with 0.2 percent glycolchitosan (bioactive coating) as a biocontrol treatment of postharvest diseases of citrus fruit has also been tried. The bioactive coating alone was also superior to *C. saitoana* in controlling the decay of oranges (Washington Navel, Valencia, Pineapple, and Hamlin cultivars) and Eureka lemons, and the control level was equivalent to that with Imazalil. The bioactive coating and Imazalil treatments offered consistent control of decay on Washington Navel oranges and Eureka lemons in early and late seasons. *C. saitoana* or 0.2 percent glycolchitosan were most effective on early-season fruit (El-Ghaouth et al., 2000). The combination of this yeast with 0.2 percent glycolchitosan also reduced the incidence of stem-end rot on Valencia oranges.

Aspire, a product of yeast *Candida oleophila* (which competes with the pathogen for nutrients released by injuries) is a promising biological control agent that is commercially registered for control of postharvest rots of citrus. The combination of Aspire with thiabendazole (200 ppm) reduces the incidence of decay caused by *Penicillium digitatum* and *P. italicum* as effectively as a conventional fungicide treatment. Furthermore, Aspire is highly efficacious against sour rot caused by *Galactomyces citri-aurantii*, a decay not controlled by conventional treatment (Droby et al., 1998).

A major factor affecting the efficacy of Aspire is how quickly and how well the yeast colonizes injuries on the fruit surface, including minor injuries involving only oil vesicles. Peel oil was toxic to cells of *C. oleophila* but not to *P. digitatum* spores. *C. oleophila* colonized punctures more uniformly than individually damaged oil glands and provided more effective control of *P. digitatum* originating at punctures than at oil-gland injuries. Ruptured oil glands were colonized more effectively if treated 7 h after injury rather than immediately (Brown et al., 2000).

The bacterial strains ESC-10 and ESC-11 of *Pseudomonas syringae* produce syringomycin, which controls green and blue molds. Syringomycin is shown to inhibit the fungi *in vitro* (Bull et al., 1998). The yeast *Candida oleophila* (Aspire) and bacteria *Pseudomonas syringae* (Bio-save 10) have shown activity against green mold but not against disease developing from quiescent infection. These pioneering biocontrol products were registered in the U.S. by the EPA in 1995 and are commercially available.

The fungus *Muscodor albus*, a biofumigant that produces certain low molecular weight volatiles, has been used to fumigate whole rooms of lemons to control pathogens during storage. It is reported to be effective on green mold and sour rot (Mercier and Smilanick, 2005).

C. Botanicals and Other Organic Compounds

Fungicide treatments result in hazardous residue; hence non-hazardous eco-friendly treatments are being developed to control the decay of citrus fruits. Recently, plant-leaf extracts and essential oils have been shown to reduce rotting. Several alternative chemicals – Methyl jasmonate (MJ) and jasmonic acid, vapors of acetic, formic and propionic, acids, and acetaldehyde and alcohol – have also been found to be effective.

The most effective concentration of jasmonates for reducing decay in cold-stored fruits or after artificial inoculation of wounded fruit at 24°C is 10 μMol/l. It is suggested that jasmonates probably reduce green mold decay in grapefruits indirectly by enhancing the natural resistance of the fruits to *P. digitatum* at high and low temperatures (Droby et al., 1999). Decay by *P. digitatum* can be reduced from 86 to 11 percent by vapors of acetic acid (1.9 or 2.5 μl/l), formic acid (1.2 μl/l) and propionic acid (2.5 μl/l). Formic acid causes browning of the fruit peel in grapefruits and oranges when fumigated (Shelberg, 1998). Acetaldehyde and octanol are naturally occurring citrus volatiles and can be effective fungicides. Exposure of oranges, lemons, and limes to 0.5–1 percent acetaldehyde vapor for 24 h did not cause rind injury or off-flavor, but exposure at 4 percent for 1 h or 5 percent for 30 min induced severe rind injury (Prasad, 1975).

Geraniol and Mentha oil were effective in checking growth of *P. italicum* (Arora and Pandey, 1977). Leaf extract of *Ageratun conzyoides* exhibited broad fungistatic spectrum and completely halted *P. italicum* (Dixit et al., 1995), while extracts of *Eucalyptus globulus, Punica granatum, Lawsonia innermis*, and *Datura species* were effective against *C. gloeosporioides* and *Botryodiplodia theobromae* (Babu and Reddy, 1986). Tulsi (*Ocimum* sp.) leaf extract is also effective against sweet orange rots (Godara and Pathak, 1995). An antimicrobial plant extract treatment based on essential oils and phenolics has been found to reduce losses and extend the shelf life of Mosambi sweet oranges up to 25 days as compared to 10 days in control (CFTRI, 1984). *Chenopodium* oil has the most fungitoxic properties, followed by *Tengetes* oil. Palmarosa and citrus oils are also fungistatic. *Aspergillus nidulans* and *A. niger* show extreme resistance to *Mentha piparata* oil (Srivastava and Walia, 1996). *Thymus capitata* essential oil, citral, and acetaldehyde have been found to be effective against *P. digitatum* (Arras and Piga, 1997). Steam-distilled oils of *Lavandula angustifolia, Mentha piperita* at concentrations of 500, 1000, 5000, and 10 000 ppm completely inhibited *Alternaria citri* (*A. alternata*) under *in-vitro* studies after 2–6 days of post-inoculation evaluation (Poswal, 1997).

The GAs, cytokinins, kinetins, and N6-benzyladenine are USEPA-approved biopesticides in the category of growth regulators and can be used in organic postharvest management of diseases and disorders.

D. Physical Methods

The presence of antifungal substances in plant tissues has been shown to play a role in resistance to plant diseases. The level of the antifungal activity in fruit flavedo gradually declines during storage, correlating with a decrease in postharvest disease resistance. Various treatments, such as curing, UV treatment, and heat treatment are reported to reduce decline in antifungal activity (Ben-Yehoshua et al., 1988; Kim et al., 1991). Antifungal materials can be pre-formed in the peel at harvest or can be induced by certain treatments. In lemon fruit flavedo, citral and limettin are pre-formed and act as a first line of defense. UV illumination induces the production of the phytoalexins scoparone and scopoletin in citrus-fruit flavedo (Ben-Yehoshua et al., 1992).

1. Ultraviolet Rays

Fruit treatment with UV radiation has also been shown to control some fungal pathogens. Being eco-friendly and non-hazardous, this method of controlling rot is considered very promising. As early as 1978, treatment of citrus fruits with UV rays for 10 min was shown to control *Penicillium* rots (Yadav et al., 1978). The combination of ultraviolet light (254 nm, UV-C) and yeast (*Debaryomyces hansenii*) is suggested as an alternative to chemical control of green mold (*Penicillium digitatum*) in tangerines (Stevens et al., 1997). UV irradiation enhances the resistance of fruits against the development of decay pathogens at the fruit wound sites by promoting production of certain phytoalexins and proteins.

Chitinase and β-1.3-endoglucanase play a role in the UV-induced resistance of grapefruit against *P. digitatum*. Immunoblotting analysis using specific citrus chitinase and β-1.3-endoglucanase antibodies showed that UV irradiation, wounding of the fruit, or their combination induced the accumulation of a 25 kD chitinase protein in the fruit's peel tissue. In contrast, UV irradiation or wounding of the fruit alone was unable to induce the accumulation of 39 and 43 kD β-1.3-endoglucanase proteins (Porat et al., 1999).

The sensitivity of various sweet orange cultivars to UV-C radiation varied with dosage (0.5,1.5, 3 kJ m^2) and season of harvest. The phytoalexin concentration also varied with the sensitivity of the cultivar to UV rays. Concentrations of phytoalexins rose as the irradiation dose increased. No scoparone or scopoletin could be detected in untreated fruits. The highest concentration of phytoalexins was recorded in Valencia Late and the lowest in Tarocco, with intermediate accumulation in Washington Navel and Biondo Comune. UV-C dose at 0.5 kJ m^2 effectively reduced decay development compared with untreated fruits (D'hallewin et al., 1999).

2. Corona Discharge and Ozone Treatment

Ozone (O_3) gas contains three atoms in the molecule (an allotropic form of O_2, or diatomic molecule). Ozone is chemically very active as a strong oxidizing agent. It is produced by subjecting air to electric discharge and is commonly used in disinfecting and purifying water and air and in bleaching processes. The ozone concentration in air that effectively kills the pathogen spores exceeds 0.1 ppm; this is also the limit of exposure to workers as per U.S. OSHA regulations. Ozonized water can be used as a sanitizer because it readily decomposes into oxygen leaving no residue. Maximum solubility of ozone in water is 10 μg per ml but most systems produce up to 5 μg per ml. At 1.5 μg/ml and a contact time of 2 min, 95–100 percent of fungi, bacteria, and viruses are killed (Smilanick et al., 1999). Ozone has been placed in the category of generally regarded as safe (GRAS) for use in food in the U.S. (Palou et al., 2001). The incidence of blue and green mold was delayed by 1 week under 0.3 ppm ozone at 5°C in Valencia oranges that were artificially inoculated. Sporulation was prevented or reduced by gaseous ozone without noticeable ozone phytotoxicity to the fruit. A synergistic effect between ozone exposure and low temperature was observed for the prevention of sporulation. The proliferation of spores of fungicide-resistant strains of *Penicillium digitatum* and *P. italicum*, which often develop during storage, can probably be delayed, thus prolonging the useful life of postharvest fungicides.

Exposure of Valencia oranges and Eureka lemons continuously to 1.0 ± 0.05 ppm ozone (vol/vol) at 10°C in an export container for 2 weeks resulted in the delay of incidences of blue and green mold, and infections developed more slowly under ozone (Palou et al., 2001).

Ozone reduces the ethylene level in storage rooms, and ozonizing the air (0.4 ppm) significantly reduced decay (Skog and Chu, 2001).

The negatively charged air ions (negative air ions, or NAI), also known as corona discharge, have been reported to reduce airborne mold and bacteria. The combination of NAI and O_3 is more effective (Farney et al., 2001). An ozone concentration of 200 ppm is required to kill the spores of *Penicillium.* The lemons and oranges were not injured at 300 ppb level of ozone treatment during nighttime for up to 6–8 weeks. The treatment reduced sporulation but did not stop infection of *Penicillium* spp. Fruit color, acidity, and TSS content were not affected (Smilanick, 2004).

3. Hot-Water Treatment

Physical treatment with hot water can eradicate quiescent infection of *Diplodia*. Studies by Smoot and Melvin (1963 and 1965) revealed that hot water (53°C for 5 min) controlled green mold and SER by *Phomopsis* and *Diplodia natalensis* in oranges. The treatment has the greatest success if the fruit is not degreened. The least success was reported with grapefruit, especially if degreened. Hot-water dip (52 ± 1°C for 5 min) with 10 percent ethyl alcohol before degreening minimized *Diplodia natalensis* in Hamlin sweet oranges (Brown and Baraka, 1997).

Hot-water treatment (40°C water for a 5-min dip) has been reported to increase levels of phytoalexin (scoparone) in the peel, thus controlling *P. digitatum* (de villiers et al., 1997). The treatment is eco-friendly and cheaper than chemical treatments. The addition of bioagents such as *Bacillus subtilis* to warm water is as effective as chemical treatment. A 2-min dip in water at 53°C or 56°C markedly reduced decay caused by green and blue molds in kumquats (Nagami cultivar). Higher temperatures (59°C or 61°C) increased decay. Drenching the fruits for 30 sec with water at 53°C or 56°C gave similar results to dipping for 2 min and is recommended as the preferred commercial treatment because of the enhanced quality of uniform, continuous heat application. Brushing was less effective at reducing decay but improved fruit gloss. The decay reductions allowed the kumquats to be exported by sea rather than air (Ben-Yehoshua et al., 2000). In Marsh grapefruits, a combination of hot water (52°C) and growth regulators (50 ppm Progibb + 200 ppm 2.4-D) reduced decay compared with hot-water treatment alone. In Valencia oranges, Imazalil (400 ppm) at 55°C achieved similar decay control to 1000 ppm non-heated Imazalil.

Hot-water treatment (50–55°C for 2–3 min) also results in the redistribution of the epicuticular wax layer and a reduction of cuticular microlesions (cracks), thus improving physical barriers to pathogen penetration. Scanning electron microscopy studies indicated that exposure to 50–52°C delayed spore germination of *P. italicum* but did not affect spore vitality. The mycelium that develops from conidia subjected to 50°C hot-water treatment is thinner and unable to colonize fruit (Dettori et al., 1997).

Studies carried out by rinsing and brushing citrus fruits 24 h after artificial inoculation with a *P. digitatum* spore suspension indicated that hot-water brushing (HWB) at 56°C, 59°C, and 62°C for 20 s reduced decay development to only 20 percent 5 percent, and less than 1 percent respectively. HWB reduced epiphytic microflora on the fruit surface. HWB at 56°C for 20 s smoothed epicuticular waxes and thus covered and sealed stomata and cracks on the fruit. Storage experiments with organically grown Minneola tangerines, Shamouti oranges, and Star Ruby grapefruits showed that HWB at 56°C for 20 s reduced decay by 45–55 percent without causing surface damage or affecting weight loss and internal composition (Porat et al., 2000).

4. Curing (High-Temperature Treatment Under Dry or Moist Conditions)

At higher temperatures and relative humidity, curing citrus fruits facilitates wound healing through the formation of lignin at the wound site. This helps in reducing decay. In tuber and bulb crops such as potato, sweet potato, and many tropical vegetables, a surface layer of protective, suberized, wound periderm tissue is formed over the product at the wound site. The curing process allows the bruised and injured surface to heal. The wound periderm thickens from lignin formation, resulting in a reduction of water loss and decay during postharvest storage. This process generally involves storing the commodity at 15–40°C and 70–90 percent RH. According to Ben-Yehoshua (1991), curing reduces fruit decay by (1) thermic inhibition of pathogens, (2) enhancing activity of PAL (phenylalanine ammonia lyase enzyme), which catalyzes lignification and wound healing and thus causing a better barrier against pathogens, (3) maintaining a higher concentration of antifungal chemicals in flavedo.

The changes in the fruit surface caused by high temperatures include the repair of cuticular wax, thereby reducing weight loss and softening and alleviating chilling injuries. Heat treatment also results in formation of heat-shock proteins, lignins, and wound gums to prevent the growth of pathogens (Ben–Yehoshua et al., 1988, 1992).

In Marsh grapefruits and Oroblanco, either curing or hot-water dip (52°C for 2 min) followed by individual seal packing or waxing (or packaging in PE-lined cartons) controlled decay without fungicide use. Addition of GA (50 ppm) and 2.4-D (500 ppm) in hot dip improved the effectiveness of the dip in grapefruit. In curing, fruit is sealed before introduction into a high-temperature chamber or high RH is maintained (Rodov et al., 1997).

Curing sealed pummelo fruits at 32–36°C enhanced lignification, antifungal metabolite production, and resistance of fruit to decay. Curing prevented the development of several pathogens, including *Penicillium digitatum*. Curing sealed pummelos enabled 6 months of storage free of chemical treatment or toxic residues without any reduction in quality (Ben-Yehoshua et al., 1988).

The effect of preheating (30°C, 35°C, or 40°C for 24 h before storage at 0°C and 10°C) on the storage of *Citrus junos* fruits was investigated by Park and Jung (1996). Preheated fruits exhibited less decay caused by *P. digitatum, P. italicum,* and *G. candidum* than unheated fruits. Respiration rate and ethylene production were not affected.

Shellie and Skaria (1998) demonstrated that the development of green mold on grapefruit caused by *P. digitatum* is inhibited by the same time and temperature regimes of moist forced air (300 min at 46° C) that are known to provide quarantine security against Mexican fruit flies.

In organic management of postharvest diseases, the approach should be integrated – including cultural practices in the field to minimize infection – followed by the use of above mentioned eco-friendly, less risky, and less polluting organic practices. Careful handling, correct temperature management, and sanitation can reduce the use of inorganic chemicals and their residues. With integrated pest management (as practiced in California and some other parts of the world) the use of pesticides and residues can be reduced.

REFERENCES

Anonymous (1994). *Pesticide residues in foods and their safety limits*. AFSTI education Trust, CFTRI, Mysore, India.

Arora, R., and Pandey, G.N. (1977). The application of essential oils and their isolates for blue mold decay control in *C. reticulata* Blanco. *J. Fd. Sci. Technol.* 14, 14–16.

Arras, G., and Piga, A. (1997). Effect of TBZ, acetaldehyde, citral and *Thymus capitata* essential oil on Minneola tangelo fruit decay. *Proc. Int. Soc. Citric*, S. Africa, pp. 406–409.

Arras, G.G., and Sanna, P. (1999). Resistance of citrus fruits to *Penicillium italicum. Proc. 51st In national symp. crop protection*, Gent, Belgium, 4 May 1999, Part II, 64: 3b, 527–530.

Arras, G., Dessi, R., Sanna, P., Arru, S., and Michalczuk, L. (1997). Inhibitory activity of yeasts isolated from fig fruits against *Penicillium digitatum. Acta Horticul.* 485, 37–46.

Awasthi, M.D., (1989). Dissipation of synthetic pyrethroids and monocrotophos residue on acid limes *(C. aurantifolia)* fruits. *Indian J. Agric. Sci.* 59, 660–664.

Awasthi, M.D., and Ahuja, A.K. (1989). Dissipation pattern of monocrotophos, quinalphos and phosphamidon residue on acid lime fruits and their movement behaviour to fruit juice. *Indian J. Hort.* 46, 480–484.

Awasthi, M.D., and Anand, L. (1986). Dessipation pattern of monocrotophos and quinolphos residue from sweet orange fruits. *J. Fd. Sci. Technol.* 23, 97–99.

Babu, K.J., and Reddy, S.M. (1986). Efficacy of some indigenous plant extracts in the control of lemon rot by two pathogenic fungi. *Nat. Acad. Sci. Lett* 9(**5**), 133–134.

Ben-Yehoshua, S. (1991). New developments in seal-packaging of fruits and vegetables. *Proc. Int. Citrus Symp. China*, pp. 757–771.

Ben-Yehoshua, S., Rodov,V., Kim, J.J., and Carmeli, S. (1992). Preformed and induced antifungal material of citrus fruit in relation to the enhancement of decay resistance by heqat and ultraviolet treatments. *J Agric. Fd. Chem.* 40, 1417–1421.

Ben-Yehoshua, S., Kim, J.J. Rodov, V., Shapiro, B., and Carmeli, S, (1992). Reducing decay of citrus fruits by induction of endogenous resistance against pathogen. *Proc. Intn. Soc. Citric.*, Italy, pp. 1053–1056.

Ben-Yehoshua, S., Shapiro, B., Kim, J.J., Sharoni, J., Carmeli, S., and Kashman, Y. (1988). Resistance of citrus fruit to pathogens and its enhancement by curing. *Proc. Sixth Int. Citrus Congress*, Tel aviv, Israel, Vol. 3, pp. 1371–1379.

Ben-Yehoshua, S., Peretz, J., Rodov, V., Nafussi, B., Yekutieli, O., Wiseblum, A., and Regev, R. (2000). Postharvest application of hot water treatment in citrus fruits: the road from the laboratory to the packing-house. *Acta Horticul.* 518, 19–28.

Brown, G.E. (1974). Postharvest citrus decay as affected by benlate application in the grove. *Proc. Fla. Sta. Hort. Soc.* 87, 237–240.

Brown, G.E. (1983). Control of Florida citrus decay with Guazatine. *Proc. Fla. Sta. Hort. Soc.* 96, 335–337.

Brown, G.E. (1992). Factors affecting the occurrence of Anthracnose on Florida citrus fruits. *Proc. Int. Soc. Citric., Italy*, pp. 1044–1048.

Brown, G.E., and Baraka, M.A. (1997). Effect of washing sequence and heated solution for degreened Hamlin oranges on *Diplodia* SER, fruit colour and phytotoxicity. *Proc. Int. Soc. Citric.* S. Africa, pp. 1164–1170.

Brown, G.E., and Craig, J.O. (1989). Effectiveness of aerosol fungicide application in the degreening room for control of citrus fruit decay. *Proc. Fla. State Hort. Soc.* 102, 181–185.

Brown, G.E., Davis, C., and Chambers, M. (2000). Control of citrus green mold with Aspire is impacted by the type of injury. *Postharvest Biol. Technol.* 18, 57–65.

Bull, C.T., Wadsworth, M.L., Pogge, T.D., Le, T.T., Wallace, S.K., Smilanick, J.L., Duffy, B., Rosenberger, U., and Defago, G. (1998). Molecular investigations into mechanisms in the biological control of postharvest diseases of citrus. In *Molecular approaches in biological control.* (B. Duffy, and U. Rosenberger, eds.) Delemont, Switzerland, 15–18 September, 1997. Bulletin-OILB-SROP, 1998, 21: 9, 1–6.

Bylemers, D., and Goodwin, B. (2004). Penbotec 400SC: A new postharvest fungicide. Washington tree fruit postharvest conference, December 2004, Yakima, WA. http://postharvest.tfrec.wsu.edu/pc2004/pdf.

Cabras, P. Schirra, M., Pirisi, F.M. Garau, V.L., and Angioni, A. (1999). Factors affecting imazalil and thiabendazole uptake and persistence in citrus fruits following dip treatments. *J. Agric. Fd. Chem.* 47, 3352–3354.

CFTRI (1984). Prevention of micro-organisms by use of antimicrobial agents of plant origin. Central Food Technological Research Institute, *Annual Report for 1982–83*, pp. 68–69.

Chalutz, E., Lomaniec, E., and Waks, J. (1989). Physiological and pathological observations on the post-harvest behaviour of kumquat fruit. *Trop. Sci.* 29, 199–206.

Cohen, E., (1991). Investigations on post harvest treatments of citrus fruits in Israel. *Proc. Int. Citrus symp*, Guangzou, China, pp. 32–35.

Cohen, E., Chalutz, E., and Shalom Y. (1992). Reduced chemical treatment for post harvest control of citrus fruit decay. *Proc. Int. Soc. Citric*, pp. 1064–1065.

Dalal, V.B. and Subramanyam, H. (1970). Refrigerated storage of fresh fruits and vegetables. *Clim. Cont.* 3(3), 37.

Dalal, V.B., Subramanyam, H. and D'Souza, S. (1961). Postharvest fungicidal treatment to control storage diseases in mandarins. *Fd. Sci. Mysore* 10, 283–286.

Dalal, V.B., D'Souza, S., Subramanyam, H., and Srivastava, H.C. (1962a). Wax emulsion for extending storage life of Sathgudi oranges. *Fd. Sci. Mysore* 11, 232–235.

Dalal, V.B., Subramanyam, H., and Srivastava, H.C. (1962b). Studies on effect of repeated wax coating on storage behaviour of Coorg oranges (*Citrus reticulata* Blanco). *Fd. Sci.* Mysore 11, 240–244.

Dettori, A., D'hallowin, G., Agabbio, M. Marceddu, S., and Schirra, M. (1997). SEM studies on *Penicillium italicum*- 'Star- Ruby' grapefruit interaction as affected by hot water dipping *Proc. Int. Soc. Citric*., Sun City, South Africa, pp. 1158–1163.

De-Villiers, E.E., Van Dyle, K., Korsten, L., Swart, S.H., and Smit, J.H. (1997). Potential alternative decay control strategiesfor South African citrus pack-houses. *Proc. Int. Soc. Citric*, South Africa, pp. 410–412.

D'-hallewin, G., Schirra, M., Manueddu, E., Piga, A., and Ben-Yehoshua, S. (1999). Scoparone and scopoletin accumulation and UV-C induced resistance to postharvest decay in oranges as influenced by harvest date. *J. Am. Soc. Hort. Sci.* 124, 702–707.

Dhillon, B.S., Bains, P.S., and Randhawa, J.S. (1976). Studies on the storage of Kinnow mandarins. *J. Res.*, PAU 14(4), 434–438.

Dixit, S.N., Chandra, H, Ramesh, T., and Dixit, V. (1995). Development of botanical fungicide against blue mould of mandarins. *J. Stored Products Res* 31(2), 165–172.

Droby, S., Cohen, L., Daus, A., Weiss, B., Horev, B., Chalutz, E., Katz, H., Keren, T.M., and Shachnai, A. (1998). Commercial testing of Aspire: a yeast preparation for the biological control of postharvest decay of citrus. *Biol. Contr.* 12(2) 97–101.

Droby, S., Porat, R., Cohen, L.,Weiss, B., Shapiro, B., Philosof-Hadas, S., and Moir, S. (1999). Suppressing green mold decay in grapefruit with postharvest jasmonate application. *J. Am. Soc. Hort. Sci* 124, 184–188.

Dutt, S.C., Sarkar, K.P., and Bose, A.N. (1960). Storage of mandarin orange. *Indian J. Hort.* 17, 60–68.

Eckert, J.W. (1991). Resistance of citrus fruit pathogen to postharvest fungicides. *Proc. Int. Citrus Symp*., China, 698–701.

Eckert, J.W., Bretschneider, B.F., Ratnayake, M. (1981). Investigations on new postharvest fungicides for citrus fruits in California. *Proc. Int. Soc. Citric. Japan*, Vol. 2, 804–810.

Eckert, J.W., Ratnayake, M., and Wolfner, A.L. (1992). Effect of volatile compounds from citrus fruits and other plant materials upon fungus spore germination. *Proc. Int. Soc. Citric*., Italy, pp. 1049–1052.

EEC (2000). *Pesticide residues in food*. European Economic Council, Council Directives.

El-Ghaouth, A., Smilanick, J.L., Brown, G.E., Ippolito, A., Wisniewski, M., and Wilson, C.L. (2000). Application of *Candida saitoana* and glycolchitosan for the control of postharvest diseases of apple and citrus fruit under semi-commercial conditions. *Pl. Disease* 84, 243–248.

FAO (2006). Codex Alimentarius official standards for pesticide residues in fresh citrus fruit. FAO/WHO, JMPR Report 2006.

Farney, C.F., Fan, L., Hidebrand, P.D., and Song, J. (2001). Do negative air ions reduce decay of fresh fruits and vegetables? *Act. Hort* 553, 421–423.

Garg, R.C. and Ram, H.B. (1972). Effect of waxing on the storage behaviour of Kagzi lime (*Citrus aurantifolia* Swingle), oranges (*C. reticulata* Blanco) and Mosambi (*Citrus sinensis* Osbeck). *Progr. Hort.* 4(3–4), 35–44.

Godara, S.L. and Pathak, V.N. (1995). Effect of plant extracts on Post-harvest rotting of sweet orange fruit. Global conference on advances in research on plant diseases and their management. Rajasthan College of Agriculture, Udaipur, 12–17 February, 172 pp.

Gupta, O.P. (1986). Effect of diphenyl and packaging on the storage behaviour of Kinnow fruit at low temperature followed by room temperature. *Progr. Hort.* 18(3–4), 181–188.

Gutter, Y. (1981). Investigation on new postharvest fungicide in Israel. *Proc. Int. Soc. Citric*, Japan, Vol. 2, pp. 810–811.

Hale, P.W., Smoot, J.J., and Hatton, Jr. T.T. (1981). Factors to be considered for exporting grapefruit to distant markets. *Proc. Fla. State Hort. Soc.* 94, 256–258.

Hayward, F.W., and Grierson, W. (1960). Effects of treatment conditions on o-phenyl phenol residues in oranges. *J. Agric. Fd. Chem.* 8, 308–310.

Holmes, G.J., and Eckert, J.W. (1999). Sensitivity of *P. digitatum* and *P. italicum* to post harvest citrus fungicide in California. *Phytopathology* 89, 716–721.

Hwang, B.H., Kim, J., Yang, Y.J., and Kim, K.J. (2002). Cell wall degradation from mandarin fruit by Penicillium spp. during postharvest storage. *J. Korean Soc. Hort. Sci.* 43, 195–200.

Johnson, T.M. (1991). Citrus postharvest technology to control losses. *Proc. Int. Citrus Symp.*, China, 1991, pp. 704–708.

Karibasappa, G.S., and Gupta, P.N. (1988). Storage studies in Khasi mandarin. *Haryana J. Hort. Sci.* 17(3–4), 196–200.

Kaul, J.L., and Lall, B.S. (1975). Mode of entry of *Penicillium italicum* and *P. digitatum* in different citrus fruits. *Sci. Cult.* 41, 29–30.

Kaul, J.L., and Sharma, R.S. (1978). Mode of entry of *Geotrichum candidum* causing sour rot of citrus fruits. *Indian Phytopath.* 31, 77–79.

Kim, J.J., Ben-Yehoshua, S., Shapiro, B., Hanis. Y., and Carmeli, S. (1991). Accumulation of scoparone in heat treated lemon fruit inoculated with Penicillium digitatum Sacc. *Pl. Physiol.* 97, 880–885.

Ladaniya, M.S. (2001). Response of 'Mosambi' sweet orange (*Citrus sinensis*) to degreening, mechanical waxing, packaging and ambient storage conditions. *Indian J. Agric. Sci* 71, 234–239.

Ladaniya, M.S. (2005). Packaging of 'Nagpur' mandarin in permeable films: modified atmosphere, fruit quality and carbendazim residues. Paper presented in *12th Vasantrao Naik memorial National Seminar on Value addition of Agro-Horti-Medicinal Produce and its marketing*, 17–19, October 2005, College of Agriculture, Nagpur.

Ladaniya, M.S., and Sonkar, R.K. (1997). Effect of curing, wax application and packaging on collar breakdown and quality in stored Nagpur mandarin (*Citrus reticulata* Blanco). *Indian J. Agric. Sci.* 67 (11), 500–503.

Ladaniya, M.S., and Singh, S. (2000). Effect of dip treatment with disinfectants on shelf-life of Nagpur mandarin fruit. Ann Rep, Nat. Res., Centre for Citrus, Nagpur, India.

Laul, M.S., Bhalerao, S.D., Dalal, V.B., and Amla, M.L. (1972). Effect of fungicides and other treatments on the post-harvest storage of Nagpur mandarin oranges (*C. reticulata*). *Indian Fd. Packer* 26(3), 42–49.

Laul, M.S., Bhalerao, S.D., and Amla, M.L. (1976). Effect of some treatments with fungicides and wax emulsion on the storage behaviour of Nagpur mandarin oranges (Ambia) packed in ventilated wooden cases. *Indian Fd. Packer* 30(3), 31–36.

Lesar, K.H., and Du Toit Pelser, P. (1997). The efficacy of prochloraz against postharvest diseases of citrus fruits. *Proc. Int. Soc. Citric.* South Africa, pp. 365–368.

Mehrotra, N.K., Sharma, N., Ghosh, R., and Nigam, N. (1996). Biological control of green and blue mold disease of citrus fruits by yeast. *Indian Phytopath.* 49, 350–354.

Mehrotra, N.K., Sharma, N., Nigam, M., and Ghosh, R. (1998). Biological control of sour-rot of citrus fruits by yeast. *Proc. Nat. Acad. Sci.* India. Section B, *Biol. Sci. Vol.* 68(2), 133–139.

Mercier, J., and Smilanick, J.L. (2005). Control of green mold and sour rot of stored lemon by biofumigation with *Muscodor albus. Biol. Control.* 32, 401–407.

Mishra, B.P. (1989). A note on control of blue mold of Kinnow fruits caused by *Penicillium italicum* through diphenyl wrappers. *Haryana J. Hort. Sci.* 18, 65–66.

Naqvi, S.A.M.H. (1993). Pre-harvest application of fungicides in Nagpur mandarin orchards to control postharvest storage decay. *Indian Phytopath.* 46, 190–193.

Palou, L., Crisosto, C.H., Smilanick, J.L., and Zoffoli, J.P. (2001). Evaluation of the effect of ozone exposure on decay development and plant physiological behaviour. *Acta Hort* No. 553, 429–430.

Palou, L., Smilanick, L., Crisosto, C.H., and Mansour, M. (2001). Effect of gaseous ozone exposure on the development of green and blue molds on cold stored citrus fruit. *Pl. Disease* 85, 632–638.

Park, Y.S., and Jung, S.T. (1996). Effects of storage temperature and preheating on the shelf life of yuzu during storage. *J. Korean Soc. Hort. Sci.* 37(2), 285–291.

Pelser, P.du T., (1977). Post harvest handling of South African citrus fruit. *Proc. Int. Soc. Citric.*, Florida, Vol. 1, pp. 244–249.

Plotto, A., and Narciso, J.A. (2006). Guidelines and acceptable postharvest practices for organically grown produce. *HortScience*. 41, 287–291.

Porat, R., Lers, A., Dori, S., Cohen, L., Weiss, B., Daus, A., Wilson, C.L., and Droby, S. (1999). Induction of chitinase and beta-1,3-endoglucanase proteins by UV irradiation and wounding in grapefruit peel tissue. *Phytoparasitica* 27(3), 233–238.

Porat, R., Daus, A., Weiss, B., Cohen, L Fallik, E., and Droby, S. (2000). Reduction of postharvest decay in organic citrus fruit by a short hot water brushing treatment. *Postharvest Biol. Technol.* 18, 151–157.

Poswal, M.A.T. (1997). Antifungal activity of natural plant oils against A. citri – a navel-end rot pathogen. *Proc. Int. Soc. Citric.* South Africa, pp. 387–381.

Prasad, K. (1975). Phytotoxicity of citrus and subtropical fruit to acetal dehyde vapor. Abstract, *Proc. Am. Phytopath. Soc.* 2, 37.

Radnia, P.M., and Eckert, J.W. (1988). Evaluation of Imazalil efficacy in relation to fungicide formulation and wax formulation. *Proc. 6th Int. Citrus congress*, Israel, Vol. 3, 1427–1434.

Ramana, K.V.R., Moorthy, N.V.N., Radhakrishnaiah-Shetty, G., Saroja, S., and Nanjundaswamy, A.M. (1983). Post-harvest fungicidal treatments to control spoilage in monsoon crop of Coorg mandarins (*C. reticulata* Blanco). *Proc. Int. Citrus Symp*, Bangalore, 17–22 December (1977). Horticultural Society of India, pp. 258–265.

Rana, G.S., and Singh K. (1992). Storage life of sweet orange fruits as influenced by fungicides, oil emulsion and package practices. *Crop Res.* 5, 150–153.

Rodov, V., Peretz, J., Ben–Yehoshua, S., Agar, T., and D'hallowin, G. (1997). Heat application as complete or partial substitute to postharvest fungicide treatments of grapefruit and Oroblanco fruts. *Proc. Int. Soc. Citric.*, Sun City, pp. 1153–1157.

Schiffmann-Nadel, M. (1975). Relation between fungal attack and postharvest fruit maturation. Colloques Internationaux du Centre National de la Recherche Scientifique, No. 238, 139–145. Editions du CNRS, Paris, France.

Schiffmann-Nadel, M., Waks, J. Gutter, Y., and Chalutz, E. (1981). Alternaria rot of citrus fruits. *Proc Int. Soc.Citric. Tokyo*, Japan, Vol. 2, pp. 791–793.

Schirra, M., Angioni, A., Ruggiu, R., Minelli, E.V., Cabras, P. (1998). Thiabendazole uptake and persistence in lemons following postharvest dips at 50ºC. Italian *J. Fd. Sci.* 10, 165–170.

Schutte, G.C., Lesar, K.H., Pelser, P. du T., and Swart, S.H. (1992). The use of Tebuconazole for the control of *A. alternata* an Minneola tangelos and its potential to control postharvest decay when applied as preharvest spray. *Proc Int. Soc. Citric.*, Italy, pp. 1070–1074.

Shelberg, P.L. (1998). Fumigation of fruit with short-chain organic acids to reduce the potential of post-harvest decay. *Pl. Disease* 82, 689–693.

Shellie, K.C., and Skaria, M. (1998). Reduction of green mold on grapefruit after hot forced-air quarantine treatment. *Pl. Disease* 82, 380–382.

Singh, J.P., and Gupta, O.P. (1979). Control of post-harvest decay in mandarin hybrids and sweet orange during storage. *Indian J. Agric. Sci.* 49(11), 862–866.

Singhrot, R.S., Singh, J.P., Sharma, R.K., and Sandooja, J.K. (1987). Use of diphenyl fumigant in wax coating with different cushionings to increase the storage life of Kinnow fruits. *Haryana J. Hort. Sci.* 16, 31–39.

Skog, L.J., and Chu, L.L. (2001). Ozone technology for shelf life extension of fruit and vegetables. *Acta Hort.* No. 553, 431–432.

Smilanick, J.L. (2004). Commercial and experimental developments in California for control of postharvest citrus. *Disease Rep*, 7 pp.

Smilanick, J.L., Crisoto, C., and Mlikova, F. (1999). Post harvest use of ozone on fresh fruit. *Perishables Hand. Quart.* 99, 10–14.

Smoot, J.J., and Melvin, C.F. (1963). Hot water as a control for decay of oranges. *Proc. Fla. State. Hort. Soc.* 76, 322–327.

Smoot, J.J., and Melvin, C.F. (1965). Reduction of citrus decay by hot water treatment. *Pl. Dis., Rep.* 49, 463–467.

Srivastava, S., and Walia, D.S. (1996). Fungitoxicity test of certain essential oils against storage fungi. *Int. J. Trop. Pl. Dis.* 14, 227–228.

Stevens, C., Khan, V.A., Lu, J.Y., Wilson, C.L., Pusey, P.L., Igwegbe, E.C.K., Kabwe, K., Mafolo, Y., Liu, J., Chalutz, E., and Droby, S. (1997). Integration of ultraviolet (UV-C) light with yeast treatment for control of post-harvest storage rots of fruits and vegetables. *Biol. Contr.* 10(2) 98–103.

Subbarao, K.R., Narasimham, P., Anandswamy, B., and Iyengar, N.V.R. (1967). Studies on the storage of mandarin oranges treated with wax or wrapped in diphenyl treated papers. *J. Fd. Sci. Technol.* 4: 165–69.

Subramanian, T.M., Sadasivam, R., and Raman, N.V. (1973). Screening of some fungicides for the control of storage decay in mandarin oranges. *Indian J. Agric. Sci.* 43, 284–287.

Syngenta (2006). Graduate-contact fungicide for citrus. Syngenta crop protection. http://www.syngentacropprotection.us.com.

Tarkase, B.G., and Desai, U.T. (1989). Effects of packaging and chemicals on storage of orange cv. Mosambi. *J. Maharashtra Agric. Univ.* 14, 10–13.

Tugwell, B.L., and Chvyl, W.L. (1997). Modified atmosphere packaging for citrus. *Int. Citrus Congress,* Sun City, South Africa, pp. 1150–1152.

Tuset, J.J., Portilla, M.T., Hinarejos, C., and Buj, A. (1992). *Aspergilus niger* and *Rhizopus oryzae* causing post harvest decay of citrus. *Proc. Int. Soc. Citric.*, Italy, pp. 1042–1043.

USEPA (2004). The United States Environment Protection Agency, Federal Register, Pesticide Tolerances. http://www.epa.gov/fedregister/EPA-PEST/2004.

Yadav, G.R., Nirwan, R.S., and Sharma, B.P. (1978). Control of citrus molds during cold storage. *Progressive Hort.* 10, 9–12.

Zhang, J. (2003). New fungicide registration to combat postharvest decay. *Citrus and Vegetable Magazine*, December, p. 13.

17

PHYSIOLOGICAL DISORDERS AND THEIR MANAGEMENT

Physiological disorders are the result of dysfunction or malfunction of the physiological processes of the fruit tissues due to abiotic stresses (temperature, RH, moisture/water stress, chemicals, and nutrient excesses and deficiencies) and are therefore distinct from disorders caused by biotic factors such as disease-causing pathogens and insect-pests. Physiological disorders are also different from mechanical bruises/injuries inflicted during handling. Most physiological disorders are incurable once they develop: prevention is the best solution.

These disorders reduce the market value of citrus fruits. The postharvest development of disorders depends on management of temperature, humidity, and handling practices. Injuries and disorders on the surface and inside the fruit are manifested slowly, as in the case of chilling injury, oleocellosis, and rind breakdown.

Some disorders develop as a direct consequence of postharvest handling practices, while some develop after harvest as a consequence of preharvest causes such as a disruption in normal water relations in fruit tissues, deficiency of nutrient elements, or the effect of the chemicals applied in the field. Citrus fruits are much less perishable than many other fruits and therefore can tolerate some adverse handling and poor environmental conditions. However, their apparent sturdiness should not excuse carelessness in applying beneficial postharvest procedures that are economically feasible (Eckert and Eaks, 1989).

Most disorders appearing on the surface of various citrus fruits are related to the rupture of oil glands, phytotoxic injury to tissues, and subsequent water loss. Sunburn, granulation, and fruit cracking are preharvest disorders related to climate and management practices during fruit growth in orchard. If fruits with sunburn injuries or granulation are stored they deteriorate rapidly. Fruits with even very slight splits/cracks will rot eventually. Granulation and regreening generally take place because of a delay in harvesting (Bakhshi et al., 1967, Nath and Roy, 1972). Other physiological disorders caused by pre-harvest factors include boron and copper deficiency, zebra skin, and water spotting.

The symptoms and control measures of some important disorders such as chilling injury, oleocellosis, rind staining, granulation, puffiness, fruit cracking, kohansho, peteca, stem-end rind breakdown, stylar-end breakdown, and creasing are as follows.

I. DISORDERS CAUSED BY POSTHARVEST FACTORS

A. Chilling Injury

Chilling injury develops mainly in tropical and subtropical fruits when held below 10–15°C for certain period of time. In citrus fruits, which are also tropical, the critical temperature at which injury develops varies with species and varieties and even within same variety grown under different climatic conditions. In general limes, lemons, and grapefruits are more susceptible than mandarins and oranges. Chilling is a result of the time–temperature relationship. Chilling is different than freezing: there is no hardening or ice crystals in the tissues in former, though there can be some resemblance in symptom development. Frozen tissues collapse completely after thawing, and freezing takes place at a much lower temperature than chilling. At chilling temperature, tissues actually take a long time to develop injury symptoms, and symptoms usually develop very quickly when fruit is taken to warmer temperatures, while freezing takes place at freezing temperatures. One common chilling-injury symptom in citrus fruit is a pitting of the peel. Cell collapse creates irregular areas that are sunken and brown. The blemishes, although superficial, render the fruit esthetically undesirable, resulting in loss of revenue. Chilling-injury development can be caused by (1) changes in physical properties of cell membrane as a result of the physical state of cell-membrane lipids and (2) changes in structural proteins such as tubulin and enzymes, resulting in changes in enzyme kinetics. The chilling injury could be due to either or both of these factors. Membrane lipid characteristics were found to change at a characteristic temperature range of 7–15°C, which is also the critical temperature range below which most tropical and subtropical plant tissues show chilling symptoms. About 10 percent or less of the membrane lipids undergo a physical change, which is probably a phase separation (Wills et al., 1998). With changes in the physical properties of the membrane, the movement of lipids, ions, enzymes, and metabolites is likely to be affected in and out of the cell. These changes in turn cause metabolic imbalances and the eventual disruption of cell membranes, breakdown of cellular compartments, cell death, and the appearance of chilling injury. The increase in respiration of chilling-injured tissues can be caused by imbalances in metabolism that lead to increased energy requirements to cope with disruptive processes in the cell. The imbalances of metabolism from changes in enzyme kinetics may lead to the production of toxic substances such as acetaldeyde (Eaks, 1980). Respiratory enzymes are affected, and some multimeric enzymes split into their component sub-units with a consequent loss of enzymic activity. Tubulin, which is structural protein of cell cytoskeleton, is cold-labile and likely to undergo dissociation at low temperatures, leading to an adverse effect on protoplasmic streaming. Research is in progress in these areas and various treatments are being developed, including conditioning of the cells and tissues to acclimatize for exposure to chilling temperatures.

It is suggested that the greater the chlorophyll content, the more intact the photosynthetic system and the greater the potential for free-radical formation and damage during chilling. Carotenoids may act as free-radical scavengers; thus, the greater the contents in relations to chlorophyll, the greater the resistance to chilling. This hypothesis appears to correlate with results obtained and with field experience in that better-colored fruits are usually less subject to chilling injury (Bower et al., 1997). Contradictory findings to this hypothesis have also been reported. Green fruits of mandarin cultivars Fortune and Nova show no pitting on the tree. In cold storage (4°C) only non-green fruits developed pitting (up to 90 percent were pitted after 4 weeks). It has been suggested that pitting susceptibility develops with fruit pigmentation (Duarte et al., 1995).

In grapefruits, susceptibility to chilling injury varies significantly with cultivar and picking date. The grapefruit hybrid Oroblanco is more resistant, while Red Blush and Star Ruby both show an intermediate susceptibility when stored at 4°C for 5 weeks and at 20°C for an additional week (Schirra et al., 1998a). Marsh Seedless grapefruits show greater susceptibility to chilling injury over the harvesting season in Florida. Mid-season (February–March) grapefruits are more resistant to chilling than early- or late-season fruit. Grapefruits harvested from the interior canopy of the tree have a higher proline content in the peel and are more resistant to chilling injury than those harvested from the exterior canopy that contain lower proline (Purvis, 1981).

The suboptimal low temperature (4°C or below) caused chilling injury in the form of irregular, brown, pitted lesions and dull yellow, watery breakdown in

Nagpur mandarins (Ladaniya and Sonkar, 1996) and grapefruit (Singh, 1975). The watery breakdown symptoms were similar to injury caused by freezing and thawing: the fruit became dull yellow in color and soft. Different chilling symptoms, such as complete dull yellow–brown watery breakdown in some fruits and peel pitting in others indicated the low temperature that fruit attained, but they could also be due to different physiological conditions of the fruit. Sometimes albedo and carpellary membrane turn brown in lemons at chilling temperatures. The development of these chilling symptoms seems to be related to a particular low temperature in a particular citrus fruit (Smoot et al., 1971).

Every citrus cultivar has a lowest safe temperature for maximum storage life and the fruit must be stored at that temperature to avoid chilling. If fruits are to be stored at suboptimal temperatures (chilling temperatures), techniques discussed earlier such as conditioning or intermittent warming may be used to avoid chilling (see Chapter 12, I. B). The treatment of fruit with thiabendazole fungicides, PGRs, and wax coating – particularly in grapefruits and mandarins – is also reported to reduce chilling injury to some extent (Grierson, 1971; Schiffman-Nadel et al., 1972; Grierson et al., 1982; Ladaniya et al., 2005). Application of 1-MCP at 50–500 ppm is reported to reduce chilling injury and peel pitting in Fallglo tangerines and grapefruits (Dou et al., 2005).

B. Oleocellosis (Rind-Oil Spot)

Oleocellosis is found in all citrus fruits; and early-harvested, green-colored fruits are more prone to this disorder. Early-morning harvesting of limes, lemons, mandarins, and Navel oranges followed by immediate transport can cause oil spotting. When released, phytotoxic oils (terpenes) cause injury to surrounding living cells, resulting in oleocellosis. This is mainly due to bruises at harvest and rough handling. Rupture of the oil glands results in necrosis of the adjacent epidermis, inducing the formation of irregularly shaped yellow, green, or brown spots in which the oil glands of the skin stand out prominently because of a slight sinking of the tissues between them. Oleocellosis appears while the temperature is high. Biochemical and physiological characteristics have been studied as they relate to the occurrence of oleocellosis in citrus. The level of antioxidant compounds seems to be related to the development of oleocellosis. In *Citrus junos*, flavedo with natural rind spot had less antioxidative activity than sound flavedo. Total tocopherol contents comprising about 8.2 mg/100 g α-tocopherol and 1.0 mg/100 g γ-tocopherol were lower in flavedo with rind spot than in sound flavedo, which had 7.2 and 4.5 mg/100 g, respectively (Sawamura et al., 1988). After storage for 90 days at 10°C under low RH, increased rates of CO_2 and ethylene production are associated with increased severity of rind-oil spot. Total non-structural carbohydrate content was also related to severity of rind-oil spot (Kanlayanarat et al., 1988).

In Washington Navel oranges, the development of oleocellosis symptoms was affected by temperature, with temperatures below 10°C reducing the rate of development of the disorder. Other environmental factors such as oxygen and carbon dioxide concentrations were also shown to affect the rate and extent of color development of the rind blemish. The degree of surface exposure to sunlight on the tree also increased the rind's susceptibility to the oil. Artificial wax coatings were shown to reduce oil damage by up to 35 percent. The tape method of sensitivity testing demonstrates that variations in oleocellosis development are not only due to changes in the pressure required to break the oil glands, but also to the ability of the peel to tolerate the released oil (Wild, 1998).

Harvesting fruit in the afternoon and retaining the produce in sheds for 1–2 days before transport reduces their turgidity and susceptibility. Significant variation in the incidence of oleocellosis on fruit harvested from different orchards has also been observed, indicating that harvesting procedures determine the severity of rind blemish.

C. Rind Staining

As a result of slight abrasion during harvesting, packing, and transportation, citrus fruits develop a brown or reddish-brown discoloration on the damaged areas. To control rind staining, careful handling of the soft mature fruit is necessary. Rind staining increases in matured peel, and GA_3 applications may prove helpful in reducing this disorder.

D. Kohansho

Symptoms of Kohansho include pitting on the stem-end, stylar-end, or periphery of the fruit. Spots are sunken and discolored. This disorder is observed mainly in Hassaku (*C. hassaku*) grown in Japan. In Kohansho-affected fruit, sugar content decreases and respiration and ethylene evolution increase in the peel (Hasegawa and Iba, 1981). Low temperatures are responsible for Kohansho in Navel orange fruits. Reducing light intensity by shading decreases the incidence in stored Navel orange fruits. An annual fluctuation in the incidence of the disorder during storage in fruits taken from the same groves has also been observed. Lighter crop loads tended to increase the incidence on the tree. Spots were mostly found on the stylar-ends of fruit. Even fruits harvested in October and November that had not been exposed to low temperatures developed the injury symptoms during storage at 10°C (Chikaizumi et al., 1999). Individual seal-packaging with LDPE film and treatment with thiabendazole (TBZ) are effective methods of reducing the occurrence of this disorder.

E. Peteca

The symptoms of Peteca are rind pitting, sinking rind, and darkening oil glands. Peteca is common in lemons. Increased brushing time induces the incidence of this disorder. Polyethylene-based waxes induce more Peteca than carnauba wax emulsion. Peteca develops at very high humidity. Nutritional imbalance has also been shown to cause this disorder. Control of as many factors as possible can reduce the incidence.

The Peteca disorder of lemons in Lebanon is reportedly caused by high oxalate levels in the leaves. Increasing levels of N and, to a lesser degree, moisture stress contributed to high oxalic acid levels in the leaves. High Ca and low available P levels in the soil further contribute to this disorder (Khalidy and Nayyal, 1974).

F. Red Blotch or Red-Colored Lesions

Superficial wounds on flavedo with a reddish color in the lesion and the surrounding peel have been reported in degreened Florida citrus (Grierson, 1986). With wound healing, lignin formation takes place and colored polyphenolic compounds are considered responsible. A similar physiological disorder in degreened lemons has been reported from Israel (Cohen, 1991). An oxidative enzyme system is considered the causative factor. Temperatures of 30°C during degreening and dip treatment with antioxidants (2500 ppm) resulted in reducing the red blotch.

G. Stylar-End Breakdown

Stylar-end breakdown occurs as dull-colored, water-soaked areas at the nipples of limes. Rough handling of Tahiti or Persian limes leads to stylar-end breakdown. Fruit size, turgor pressure, temperature, and humidity during storage affect the susceptibility of fruits. Large fruits are more prone to this disorder. Secondary infection by *Penicillium, Aspergillus*, and *Colletotrichum* follow after stylar-end breakdown.

H. Stem-End Rind Breakdown (SERB)

SERB is also referred to as aging and occurs in most citrus fruits. The rind around the stem wilts and dries down in sunken, brown, irregular areas during storage. In overripe fruit it may appear in the orchard itself. Nutritional imbalance of N and K in the orchard may also render fruit susceptible to this disorder (Grierson, 1986). Fruits with severe symptoms have an off-flavor. Affected fruits are prone to stem-end rot or other fungi. The incidence varies from season to season, but it is more prevalent in fully mature and overripe fruit; hence prompt and careful harvesting and handling prevent the disorder.

Curing (holding fruit for 1–2 days after harvest), brushing, and packing without polyethylene liners had a cumulative effect on the development of brown-colored, sunken, breakdown of collars in Nagpur mandarins (Ladaniya and Sonkar, 1997). The raised collar of the Nagpur mandarin is more susceptible to handling abuse than the rest of the fruit surface. Careful harvesting and handling followed by quick transportation of fruit to the packinghouse and holding it at 80–90 percent RH with minimum brushing on the packing line can minimize the disorder. The exceptions are limes and lemons, which need holding for 2–3 days in the orchard in the shade for curing.

II. DISORDERS CAUSED BY PREHARVEST FACTORS

A. Freeze Injury

This is a common problem in areas where citrus is grown above 30ºN and 30º South latitudes or higher. Freezes have even damaged entire citrus plants. In India, in areas of Punjab and Rajasthan where temperatures fall below 0ºC during winter for considerable periods of time, Kinnow fruit damages have occurred. In the Mediterranean region, the U.S., Latin America, China, and Japan, citrus-growing areas suffer damages from freezes. The frozen fruits contain distorted intercarpellary membranes and white hesperidin crystals between segments can be seen. Desiccated areas appear at the stem end, where moisture loss has taken place through damaged membranes (Grierson and Hayward, 1959). The acceptance of frozen fruit varies in different countries depending on the extent of the damage. In Florida, oranges of Grade No. 1 should not have freeze damage of more than a 1/4th-in. slice at the stem end, while in California, fruit damage up to 20 percent in the center cut is allowed (Grierson, 1986). The internal membranes and juice vesicles are more susceptible to freeze damage than the peel is.

B. Granulation

As per the available literature, granulation was first reported by Bartholomew et al. (1935) in Valencia oranges. In granulated fruits of oranges, tangerines, and tangerine hybrids, extractable juice is severely decreased because of gel formation within the juice vesicles (Fig. 17.1, see also Plate 17.1). Fruit weight, percentage of pulp, percentage of extractable juice and TSS content are lower and the peel percentage higher in granulated fruits compared with normal fruits. In Ponkan fruits, granulation is characterized by enlarged, hardened, and nearly colorless juice vesicles that first appear at the stem end. In badly granulated fruit sugars and acids decrease greatly and the fruit becomes tasteless (Xingjie, 1990). Granulation is apt to develop in fruits with advanced maturity, harvested late in the season or from young trees. The freezing of fruits on the tree also causes dryness and symptoms similar to those of granulation. Ranjit Singh and Room Singh (1980) tried to determine the relationship between nutrient status and the development of granulation in fruit tissue. In areas with a high incidence of granulation, plant tissues contained large amounts of Ca and Mn and low amounts of P and B. Granulation seemed to have no definite relationship with N, K, Mg, Zn, Cu, and Fe contents of the plant. Studies of the physiological characteristic of granulated fruit indicated that ethanol content increased before granulation, followed by a decrease as granulation proceeded in Ponkan mandarins and Satsumas

FIGURE 17.1 Granulation of Nagpur Mandarin Fruit.

(Hu et al., 1997). Electrical conductivity in the peel increased steadily during postharvest storage. The more that granulation developed, the more the electrical conductivity increased. Respiration rate and weight loss of the fruit also increased with the development of granulation. Pectinmethylesterase (PME) catalyses the hydrolysis of the ester bonds in citrus fruit pectins. Polygalacturonase (PG) acts on the de-esterified region of pectins by hydrolytic cleavage of the 1,4-1,-glycosidic bonds. PME peaks appeared before PG peak in granulated fruit. As soon as the fruit PG peak disappeared, the fruits began developing granulation. As the second PG peak disappeared, severe granulation of fruit developed. This indicates certain correlation between PME and PG activities and granulation in Ponkan mandarin fruits (Xingjie, 1990). Levels of alcohol-insoluble solids (AIS) largely composed of pectin and other cell wall materials increased significantly in the granulated juice vesicles. It was suggested that storage itself is not responsible for the marked accumulation of AIS in granulated juice vesicles. Some interactions of fruit size with maturation as well as other factors such as tree age and rootstock were suggested as contributing to the development of granulation (Burns and Albrigo, 1998). In South Africa granulation is reported mainly in Valencias and Navels (Gilfillan and Stevenson, 1977).

Granulation in the pummelo cultivar Jinyou increased in intensity when harvesting was delayed. As harvest was delayed, peroxidase activity in the peel increased and the role of this enzyme in pulp granulation has been suggested (Chen et al., 1994). Recently Room Singh (2001) reviewed the literature on granulation and opined that the specific cause of this disorder is still obscure and several factors are responsible, including climate, cultivation practices, harvesting time, and storage period. He also suggested the term *citrus sclerosis* as a more relevant term for this disorder because hardening of the juice cells takes place as a result of lignification.

Harvesting at the proper maturity level can minimize losses from this disorder. In Mosambi oranges the incidence of granulation was reduced with GA_3 (15 ppm) treatment. 33.3 percent granulation was observed in treated fruit compared with 72.5 percent in the control (Harminder Kaur et al., 1991).

C. Fruit Splitting or Cracking

Citrus fruits, especially acid limes and mandarins, develop cracks in the rind or even in the pulp and become unmarketable. Wide fluctuations in water content of soil and consequently plant tissues could be one possible reasons. In India, this problem occurs more in developing fruits of acid limes and oranges during May, June, and July when atmospheric humidity and soil moisture status changes abruptly with the onset of monsoon rains. Abrupt heavy irrigation or a sudden heavy downpour of rain can cause fruit splitting. Mature fruits are more susceptible and hence such fruits should be harvested before the commencement of monsoon rains. In the Valencia orange-growing areas of South Africa, as much as 20 percent of the crop is lost before picking, and losses further increase as a result of postharvest bursting of fruit and decay (Bower et al., 1992). The development of this disorder is related to the inability of the rind to stretch sufficiently during fruit expansion. Early harvesting and crop thinning to improve rind thickness have been suggested as remedies. In Navelina oranges, fruit with greater rind thickness and a higher ratio of longitudinal/transverse diameter had less splitting (De Cicco et al., 1988). Foliar application of $CaCl_2$ (0.5 percent) at the half-grown stage of fruit development has been highly useful for reducing fruit cracking in Kagzi Kalan lemons without any adverse effect on fruit size, yield, and quality (Sharma et al., 2002).

D. Puffiness

Ponkan and Satsuma mandarins (*Citrus reticulata* Blanco and *C. unshiu*) are very prone to puffing, and fruits of advanced maturity develop puffiness in the orchard itself. The rind becomes thick and separates from the pulp (segments) creating an air gap between peel and segments. In stored fruit, puffiness develops because of high humidity as the storage period is extended. Puffing is rare in sweet oranges, limes, and lemons. Prestorage curing of the fruit can reduce the incidence of puffiness (Murata, 1981). Kinnow mandarin hybrid fruits do not get puffy, while most *C. reticulata* fruits, including Nagpur mandarins become puffy (Ladaniya et al., 1990). Wei et al. (2000) studied the mechanism of puffing and its physiological characteristics and found that with an increased puffiness percentage, the content of soluble sugar decreased and cell membrane permeability increased. The application of GA_3, spermidine, and Ca $(CH_3COO)_2$ regulated the sugar content and membrane permeability and controlled puffiness. Harvesting at the proper stage of maturity and avoiding long storage at high RH can minimize puffing.

E. Superficial Rind Pitting (SRP)

Superficial rind pitting (SRP) is a serious disorder of Shamouti oranges, and damage to fruits causes tremendous losses to growers. SRP develops on the tree also. SRP symptoms develop mainly 3–5 weeks after harvest, during shipment and marketing. Ethylene increases the disorder. Fruits with SRP had lower rind K content than healthy fruits. Spraying trees with potassium fertilizer increases leaf K concentration and reduces the incidence of SRP. Storage at 5°C also restricts the development of SRP. A combination of preharvest K spray with postharvest low-temperature storage provides good control.

Soil and climatic conditions seem to affect the incidence of SRP in Shamouti orchards. Tamim et al. (2000) hypothesized that K deficiency leads to a malfunction of biomembranes, causing water loss followed by cell collapse and necrosis and resulting in SRP. The postharvest appearance of SRP can be reduced if potassium is applied to fruit in the packinghouse together with wax.

F. Creasing

Creasing is also called albedo rind breakdown. A definite cause is not yet known but in general factors such as heavy cropping, old trees, low N and K with high P, or water stress that result in reduction of peel thickness may have potential in increasing susceptibility to creasing (Sneath, 1987). Calcium deficiency may also cause creasing. This may be caused by moisture fluctuation and shading of fruit, which can have an adverse effect on transpiration and calcium uptake (Storey, 1989). Delayed harvesting also leads to creasing (Grierson, 1986). The difference in growth of albedo and flavedo tissues may also result in creasing. The cell division of albedo tissue ceases after 8–9 weeks and subsequently grows with expansion only, while flavedo cell division continues until fruit maturity.

In Riverland, Australia, creasing or albedo breakdown is a major defect of Navel and Valencia oranges, reducing pack-out by up to 20 percent. Navels are usually harvested in July–September and Valencias are harvested during October–December, but with delays in harvest, creasing increases. A significant reduction in creasing of Valencia and Navel oranges has been achieved with GA 20 ppm spray (pH 4.0) when fruit is 30–50 mm in diameter (early January) using high-volume applications (Tugwell et al., 1997).

G. Sunburn/Sunscald

This disorder is common in tropical arid or semi-arid areas where light intensity is very high and peel develops burn injury at particular spots facing high

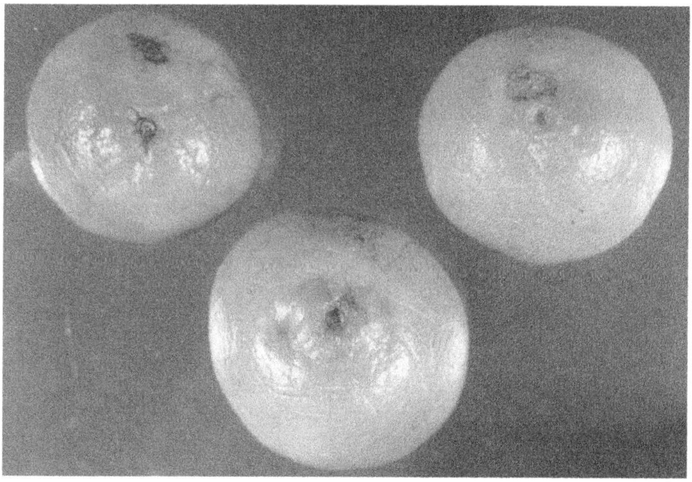

FIGURE 17.2 Sunburn or Sunscald on Peel of Nagpur Mandarin Fruit.

solar radiation. Cultivars that bear fruit inside the canopy (such as Kinnows) are observed to be less prone to this disorder, while the Nagpur mandarin is quite susceptible (Fig. 17.2, see also Plate 17.2).

REFERENCES

Bakshi, J.C., Singh, G., and Singh, K.K. (1967). Effect of time of picking on fruit quality and subsequent cropping of Valencia Late variety of sweet orange (*C. sinensis*). *Indian J. Hort.* 24, 67–70.

Bartholomew, E.T., Sinclair, W.B., and Raby, E.C. (1935). Granulation of Valencia orange. *Calif. Citrog.* 21, 5, 30.

Bower, J.P., Leser, K., Farrent, J., and Sherwin, H. (1997). Parameters relating to citrus chilling sensitivity. *Citrus J.* 7, 22–24.

Burns, J.K., and Albrigo, L.G. (1998). Time of harvest and method of storage affect granulation in grapefruit. *HortScience* 33, 728–730.

Chen, K.S., Zhang, S.L., Chen, Q.J., Bei, Z.M., and Ye, Y.Y. (1994). Effects of harvest date on fruit granulation during storage of *Citrus grandis*. *Pl. Physiol. Commun.* 30(3), 196–198.

Chikaizumi, S., Hino, A., and Mizutani, F. (1999). Influences of environmental factors affecting fruit growth on incidence of 'Kohansho' disorder in Navel orange (*Citrus sinensis* Osbeck var. brasiliensis Tanaka) fruit. *Bull. Exp. Farm* College Agriculture, Ehime University, No. 20, 7–14.

Cohen, E. (1991). Investigations on postharvest treatments of citrus fruits in Israel. *Proc. Int. Citrus Symp.*, Guangzou, China, pp. 32–35.

De-Cicco, V., and Other, S. (1988). Factors in Navelina orange splitting. *Proc. 6th Int. Citrus Congress*, Israel, Vol. 1, pp. 535–538.

Dou, H., Jones, S., and Ritenour, M. (2005). Influence of 1-MCP application and concentration on postharvest peel disorders and decay in citrus. *J. Hort. Biotechnol.* 80, 786–792.

Eaks, I.L. (1980). Effect of chilling on respiration and volatiles of California lemon fruit. *J. Am. Soc. Hort. Sci.* 105, 865–869.

Eckert, J.W. and Eaks, I.L. (1989). Post-harvest disorder and diseases of citrus fruits. In The citrus Industry (W. Reuther, E.C. Calavan, and G.E. Carman, eds.) Vol. 5. Division of Agricultural, and Natural Resources UC, California, pp. 179–260.

Grierson, W. (1971). Chilling injury in tropical and subtropical fruit IV. Role of packaging and waxing in minimizing chilling injury in grapefruit. *Proc. Trop. Reg. Am. Soc. Hort. Sci.* 15, 76–87.

Grierson, W. (1986). Physiological disorders. *In* Fresh citrus fruits (W.F. Wardowski, S. Nagy, and W. Grierson, eds.) AVI Publishing, USA, pp. 362.

Grierson, W., and Hayward, F.W. (1959). Evaluation of mechanical separators for cold damaged oranges. *Proc. Am. Soc. Hort. Sci.* 73, 278–288.

Grierson, W., Soule, J., and Kawada, K. (1982). Beneficial aspects of physiological stress. *Hort. Rev.* 4, 247–271.

Gilfillan, I.M., and Stevenson, I.A. (1977). Postharvest development of granulation in export oranges. *Proc. Int. Soc. Citric*, Vol. 1, pp. 299–303.

Harminder Kaur, Chanana, Y.R., and Kapur, S. (1991). Effect of growth regulators on granulation and fruit quality of sweet orange cv. 'Mosambi'. *Indian J. Hort.* 48, 224–227.

Hu, X.Q., Shao., P.F., Wang, R.K., Zhu, R.G., and Qin, S.X. (1997). Mandarin granulation related to some physiological characteristics during fruit storage. *Acta Hort. Sinica.* 24, 133–136.

Kanlayanarat, S., Oogaki, C., Gemma, H., Ogaki, C. (1988). Biochemical and physiological characteristics as related to the occurrence of rind-oil spot in *Citrus hassaku*. *J. Japanese Soc. Hort. Sci.*, 57, 521–528.

Kahalidy, R., and Nayyal, A.W. (1974). Effect of nitrogen and irrigation regime on leaf oxalic acid formation in Eureka lemon. *J. Res.* 18, 26–30.

Ladaniya, M.S., Huchche, A.D., and Dass, H.C. (1990). The influence of pre-harvest sprays of gibberellic acid on post-harvest storage life of 'Nagpur' mandarin (*Citrus reticulata* Blanco). *Proc. Int. Citrus Symp.*, Guangzou, China, pp. 718–722.

Ladaniya, M.S., and Sonkar, R.K. (1996). Influence of temperature and fruit maturity on Nagpur mandarin (*Citrus reticulata* Blanco) in storage. *Indian J. Agric. Sci.* 66, 109–113.

Ladaniya, M.S., and Sonkar, R.K. (1997). Effect of curing, wax application and packaging on collar breakdown and quality in stored 'Nagpur' mandarin (*Citrus reticulata* Blanco). *Indian J. Agric. Sci.* 67, 500–503.

Ladaniya, M.S., Singh, S., and Mahalle, B. (2005). Sub-optimum low temperature storage of 'Nagpur' mandarin as influenced by wax coating and intermittent warming. *Indian J. Hort.* 62, 1–7.

Murata, T. (1981). Physiological disorders of citrus fruits in Japan. *Proc. Intn. Soc. Citric.*, Vol. 2, pp. 776–778.

Nath, N., and Roy, S.K. (1972). Granulation studies in mandarin (*C. reticulata* Blanco). In Abstracts, *Third international symposium on tropical and subtropical horticulture*, Bangalore, Horticultural Society of India, pp. 76–77.

Sawamura, M., Kuriyama, T., and Li, Z. (1988). Rind spot, antioxidative activity and tocopherols in the flavedo of citrus fruits. *J. Hort. Sci.* 63, 717–721.

Sharma, R.R., Saxena, S.K., Goswami, A.M., and Shukla, A.K. (2002). Effect of foliar application of calcium chloride on fruit cracking, yield and quality of Kagzi Kalan lemon. *Indian J. Hort.* 59, 145–149.

Singh, R. (2001). 65-year research on citrus granulation. *Proc. Seminar on New Horizons in production and postharvest management of tropical and sub-tropical fruits*. Indian Soc. Horticulture Special Issue. *Ind. J. Hort.* 58, 112–144.

Singh, R., and Singh, R. (1980). Relationship between granulation and nutrient status of Kinnow mandarin at different localities. *Punjab Hort. J.* 20(3–4), 134–139.

Schiffman–Nadel, M., Chalutz, E., Waks, J., and Lattar, F.I. (1977). Reduction of pitting in grapefruit by TBZ during long term storage. *HortScience* 7, 394–395.

Smoot, J.L., Houck, L.G., and Johnson, H.B. (1971). Market diseases of citrus and other subtropical fruit. *USDA Agric Handbook*, 398 pp.

Sneath, G. (1987). Albedo breakdown in citrus. *Australian Citrus News* December 1987, 4–5.

Storey, R. (1989). Albedo breakdown research. *Australian Citrus News*. November 1989, pp. 15–16.

Tamim, M., Goldschmidt, E.E., Goren, R., and Shachnai, A. (2000). Potassium reduces the incidence of superficial rind pitting (nuxan) on 'Shamouti' orange. *Alon Hanotea*. 54, 152–157.

Tugwell, B.L., and Chvyll, W.L. Moulds, G., and Hill, J. (1997). Control of albedo rind breakdown with gibberellic acid. *Proc. Int. Soc. Citric.*, Sun City, South Africa, pp. 1147–1149.

Wei, Y.R., Zhang, Q.M., and Zheng, Y.S. (2000). Mechanism of puffiness and the methods of control in citrus fruit: I. The rule of early-maturing variety citrus and its physiological characters. *J. Hunan Agric. Univ.* 26, 267–270.

Wild, B.L. (1998). New method for quantitatively assessing susceptibility of citrus fruit to oleocellosis development and some factors that affect its expression. *Australian J. Exp. Agric.* 1998, 38: 3, 279–285.

Wills, R., McGlasson, B., Graham, D., and Joyce, D. (1998). *Postharvest: an introduction to the physiology and handling of fruits and vegetables and ornamentals*. CAB International, Willingford, Oxon, UK, 262 pp.

Xingjie, T. (1990). Relationship between pectinmethylesterase and polygalacturonase activities and granulation in citrus fruit. *Proc. Int. Citrus Symp.*, Guangzhou, China, pp. 796–798.

18

POSTHARVEST TREATMENTS FOR INSECT CONTROL

Several insects infest the fruits in the orchard, leading to deterioration of fruit quality. This is also a matter of concern in international trade from a phytosanitary standpoint. Therefore, during export, insect infestation of citrus fruit is viewed quite seriously by importing countries as well as exporting countries. With the World Trade Organization (WTO) in place and most citrus-growing countries signatory to it, sanitary and phytosanitary issues are raised time and again as trade barriers. Fruit flies of the family *Tephritidae* are considered the most important insect-pest risk carried by exported fruits worldwide, and fruits suspected of harboring fruit fly eggs and larvae must be treated to control virtually 100 percent of any tephritids present (Figure 18.1, see also Plate 18.1). False codling moths can also be a problem in the export of grapefruit from some areas.

Fruits have a natural resistance to insect infestation to a certain extent. The resistance of various citrus fruits to attack is attributed to allelopathic essential oils in the flavedo region of the peel. Marsh and Florida Ruby Red grapefruits are more susceptible to larval development than the mandarin cultivar Temple. Eureka and Lisbon lemons are virtually immune. Fruit-fly problems are also not generally observed in acid lime fruit, which are very acidic. Resistance also depends on the stage of fruit maturity. Fruits allowed to remain on the trees until they are overripe are somewhat more susceptible than early-season fruits. Resistance is correlated with (1) flavedo thickness, (2) a high concentration of linalool in relation to limonene in the peel oil, and (3) the absolute amount of oil per unit area of peel. Volatile components of the peel oil, rather than high boiling fractions, appear to account for oil toxicity (Greany et al., 1983).

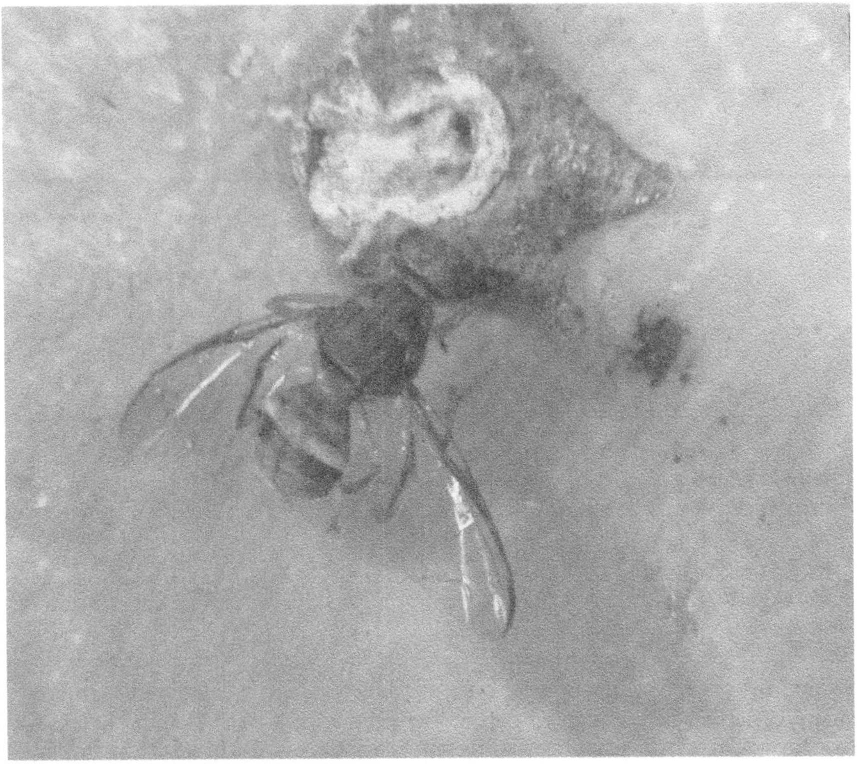

FIGURE 18.1 *Bactrocera dorsalis* (Oriental Fruit Fly) Adult Male (Without Ovipositor) on Citrus Fruit.

Essential oil in citrus fruit glands is phytotoxic, and it is toxic to many insects as well. Gland-oil compounds that are toxic to the Caribbean fruit fly (*Anastrepha suspensa*) – such as pinine, limonene, terpeneol, and some aldehydes – decrease with time in grapefruit, suggesting that fruit becomes increasingly susceptible to the fly (Qing and Petracek, 1999). Wax application was also found to decrease normal levels of β-pinine and γ-phellandrene in oil of grapefruit peel.

During postharvest handling and international trade of oranges, mandarins, and grapefruits in major citrus-growing regions (including North and Central America, Latin America, Australia, South Africa, and the Mediterranean countries), fruit flies cause serious problems – mainly the Mediterranean fruit fly (*Ceratitis capitata* Wied.), the Caribbean fruit fly (*Anastrepha suspensa* Loew.), the Mexican fruit fly (*Anastrepha ludens* Loew.), and the West Indian fruit fly (*Anastrepha obliqua* Macquart). To meet the challenge posed by fruit flies in international citrus trade, quarantine measures must be strictly followed during export.

I. FLY-FREE ZONE PROTOCOL

This program is accepted by many countries as a way to keep the fruit-fly level monitored and under control (below economic threshold levels). In Florida, some regions of 22 citrus counties are fly-free; all fruit exported to Japan must be from these counties (Kender, 2004). Growers must apply to Florida Department of Agriculture and Consumer Services (FDACS) to participate in the program. Growers must allow inspections, must remove host plants of the Caribbean fruit fly (Loquat, Rose-apple, Guava, and Surinam cherry), and must undertake trap surveys. Even packinghouses are monitored. The fly-free-zone protocol is administered by FDACS and the USDA Animal and Plant Health Inspection Service (APHIS). Agents of Florida citrus growers, University of Florida, USDA, and FDACS comprise the technical committee. With participation of stakeholders – especially growers – and the proper support of local agriculture departments, state policies, and legislation, these fly-free zones will be very effective in meeting the challenge of fruit flies worldwide.

II. FRUIT-FLY DISINFESTATION

A. Chemical Control

Citrus fruits from infested areas need to be disinfested before they are shipped. Fumigation with ethylene dibromide (EDB), ethylene oxide, and methyl bromide (MB) had long been the practice for disinfestations, but these treatments are now banned for reasons of food safety and environmental protection. EDB was last used for shipment of Florida grapefruit to Japan in 1987–1988. In South Africa, EDB was withdrawn for use as a fumigant in 1984.

B. Physical Treatments

Alternative physical procedures that are effective, non-damaging, and non-polluting are being developed to combat fruit flies.

1. Low Temperature

Regular atmosphere cold storage at 0°C for 28 days is known to be effective against insects, particularly fruit flies infesting grapefruit (Paull, 1990). For treatment to be practical, fruit has to tolerate the effective low temperature without an adverse effect on external and internal fruit quality. Different citrus fruits have different tolerance limits to low temperatures.

Cold treatment is designed to disinfest fruit from possible fruit-fly infestation. Increased restrictions on the use of ethylene dibromide in disinfestation of citrus, has intensified the use of cold treatment. Since Japan is not infested with fruit flies (Mediterranean or Caribbean), citrus fruit exported to Japan must be quarantined (Kitagawa et al., 1988). The normal schedule for cold treatment is recommended as follows: 0.6°C for 14 days, 0.8°C for 16 days, 1.1°C for 17 days, 1.4°C for 19 days, and 1.7°C for 20 days after the fruit pulp reaches this temperature. These different timings are for different citrus fruits that can tolerate the temperature without chilling injury. For areas with a low density of Caribbean fruit fly, the time–temperature schedule is 0.6°C for 10 days, 1.1°C for 12 days, and 1.7°C for 14 days.

Grapefruit is susceptible to chilling injury when exposed to temperatures below 10°C. Fruit on the outer canopy is more susceptible than those on the interior. Chilling injury leads to decay. Large losses were reported (some 150000 cartons) in grapefruit shipped in large break-bulk cargo when subjected to cold-temperature treatment (Ismail et al., 1986). To save time, fruit was treated during shipment in refrigerated van containers (40ft) at low temperatures during the voyage. Fumigation with hydrogen cyanide (HCN) before and after the voyage has also been reported. Cold-treated fruit (14 days at 1.6°C) appeared fresh, bright yellow in color, firm and free of internal breakdown from cold injury with low decay incidence. Two weeks after the treatment, total losses were 6 percent (Ismail et al., 1986). The recommendations that emerged from these studies were:

1. Selection of good-quality fruit.
2. Avoiding prolonged degreening, which weakens the fruit and enhances stem-end decay.
3. Proper application of fungicides (TBZ) before low-temperature treatment.
4. Use of water wax.
5. Avoiding the use of biphenyl pads in grapefruit shipments subjected to cold temperature.
6. Uniform cooling within the load (proper stacking and positioning of pallet).
7. Proper and uniform conditioning of fruit (7 days at fruit pulp temperature of 15°C as the minimum temperature).
8. Proper warming at 12°C after completion of cold treatment.
9. Avoiding fumigation with hydrogen cyanide until fruit pulp temperature reaches 10°C. (In 1999, USEPA revoked use of HCN in citrus).
10. Avoiding unnecessary delays in marketing and storage of cold-treated grapefruit at 15.6°C in November–December and at 10–12°C in January–June.

The U.S., Japan, Israel, South Africa, Spain, Korea, and Australia have approved cold quarantine treatment (1.5–1.7°C for 19–20 days) for *C. capitata* for exported and imported citrus. In Concordia, in northeastern Argentina, *Anastrepha fraterculus* Weid infests oranges, grapefruits, and mandarins. Cold treatment of Star Ruby grapefruit at 2 ± 0.5°C for 18 days resulted in complete (100 percent) mortality of third instar larvae by the tenth day. This treatment is effective for *C. capitata* as well as *A. fraterculus*, and Argentine grapefruit can be exported to Europe and North America (Putruele and Marty, 2000).

Because the low temperatures required to kill larvae or eggs of fruit fly in the fruit are suboptimal temperatures (0–1°C) and cause chilling injuries, pre-storage treatments are being developed to increase the tolerance of fruits to the chilling temperature necessary for insect disinfestation. Techniques that reduce chilling injury include temperature conditioning, step-down temperature regimes, controlled-atmosphere storage, chemical treatments, and the application of coatings and plant-growth regulators. Temperature conditioning (high, low, or stepwise) and controlled-atmosphere storage manipulate and modify the storage environment, while the other treatments are applied directly to the commodity (see Chapter 12 I. D). These treatments either increase the tolerance of fruit to chilling temperatures or simply retard the development of chilling injury symptoms (Wang, 1994).

2. Vapor Heat

Application of heat treatment in the form of vapor to disinfest the fruit and the study fruit tolerance to this treatment has long been an area of research. The standard condition for heat treatment (moist or dry) is 43.5°C for 8 h. Moist forced air at 46°C for 4 h is promising for Valencia oranges. Texas N33 Navel oranges tolerate exposure to moist forced air at 46°C for 4.5 h as a quarantine treatment against *Anastrepha ludens* without deleterious effects on fruit marketing quality (Shellie and Mangan, 1998). The treatment of high-temperature forced air (HTFA) at 45°C for 3.5–4 h to kill *Anastrepha ludens* larvae is a viable alternative to methyl bromide for the disinfestation of Dancy tangerines (Shellie and Mangan, 1995). Vapor heat (VH) or hot water (HW) treatment is not as successful for disinfestation of grapefruits. VH and HW treatments at the temperatures and durations necessary to control the Caribbean fruit fly (*Anastrepha suspensa*) cause peel injury to Marsh grapefruits, regardless of treatment with gibberellic acid (Miller and McDonald, 1997).

Vapor heat can be used for conditioning grapefruit to withstand low temperatures and/or low doses of irradiation as a disinfestation treatment. Pre-storage vapor heat treatment at 48°C (>90 percent RH) for 120 min is most effective in reducing chilling injury during disinfestation of Star Ruby grapefruits at low temperatures (Wright et al., 1997). Fruits can be stored at 1°C for 28 days + 20°C for 14 days with minimal chilling injury. Vapor heat (2 h at 38°C with thiabendazole (TBZ) at 0.4 percent or TBZ 0.1 percent + imazalil 0.1 percent) reduces the severity and incidence of peel injury caused by irradiation at 0.5 or 1.0 KGy for Caribbean fruit fly sterilization in Marsh grapefruit by 50 percent without adversely affecting other quality attributes (Miller and McDonald, 1998b).

3. Gamma Radiation

Gamma irradiation of fruit is also effective for disinfestation. For more than 50 years irradiation has been studied as a potential means of disinfestation for agricultural commodities, including citrus, but public reluctance to accept irradiated material for consumption made its use on a commercial scale unfeasible. The efficacy of irradiation has been defined as the prevention of adult fly emergence

(Ohta et al., 1985), the prevention of the emergence of flies capable of flight, and the prevention of flies capable of reproducing (APHIS, 1989).

Von Windeguth (1982) studied the mortality of Caribbean fruit fly after irradiation. Pupae were recovered from grapefruits following irradiation at 15 and 30 kilorads (0.15–0.30 KGy). No adult insects emerged from fruit irradiated at 60 and 90 kilorads. Adults emerged from the recovered pupae, but they died before reproducing. In another study of grapefruit infested with immature stages (eggs and larvae) of the Caribbean fruit fly, *Anastrepha suspensa* (Loew), viable adult flies have not been recovered at any of the irradiation dose (43–415 Gy) and very few larvae survived to pupate at the 150 Gy range (Von Windeguth and Ismail, 1987).

The third instar (last stage of larva) and pupa of most fruit flies are the most tolerant to irradiation doses that can be tolerated by fruit. Quarantine treatments against tephritids generally targeted only eggs and larvae because the larvae leave the fruit to pupate in the soil. Pupae are generally more tolerant of irradiation than are larvae and eggs. However, irradiation would usually be applied after fruits are packed, and some packed fruits remain at ambient conditions for a few days before being irradiated, and by that time third instars emerge and pupate in the package. Although tolerance to irradiation generally increases with increasing stages of development, the insect immediately preceding pupation and larval to pupal molt was usually more susceptible than 24 h earlier. It is recommended that fruits to be irradiated should not remain at ambient temperatures for enough time to enable third instars to develop to the pupal stage, which is 3 days at 25°C. Mexican fruit fly third instars inside grapefruit were more tolerant of irradiation than third instars in ambient air (Hallman and Worley, 1999). Irradiation does not cause acute mortality of insects, but does prevent them from maturing or sterilizes them. Tephritids are the most studied group of quarantined pests as far as irradiation is concerned. Minimum absorbed doses confirmed with large-scale testing to provide 99.99 percent control have ranged from 150 to 250 Gy in different species of fruit fly (Table 18.1). Insects remain alive for some time after irradiation, which is one of the major obstacles to its use. Irradiation may be the most widely applicable quarantine treatment from the standpoint of fruit quality, but when doses as low as 150 Gy are applied on a commercial scale, much of the fruit load may receive >300 Gy. The application is not uniform in irradiation and some of the significant tephritids attacking citrus fruits, such as *Anastrepha* spp., can be controlled with lower doses. Hallman (1999) reported that the current recommendation of 150 Gy for several species of fruit flies infesting grapefruits may result in significant damage to fruit because the uniformity ratio is usually 2:1 in commercial applications.

According to Kader et al. (1984), if the objective of irradiation is to controlling insect-pests by stopping development or reproduction, a dose of 5–75 Krad (0.5–0.75 KGy) would be sufficient. On the lower end of this dose, some life stages of insects may survive and damage the product, and even develop into adults. However, those adults will be sterile or, in the case of some moths, their offspring will be sterile.

TABLE 18.1 Dose of Ionizing Radiation Absorbed (as Minimum Dose) for Quarantine of Fruit Flies

Species of fruit fly	Dose (Gy)	Distribution in the world
Anastrepha ludens (Loew), Mexican fruit fly	150	Southern Texas, Mexico, Guatemala
A. oblique (Macquart), West Indian fly	150	Caribbean islands, Mexico, Brazil
A. serpentina (Wiedemann), Zapote	150	Mexico, central America, Argentina
A. suspense (Loew), Caribbean fruit fly	150	Florida, Caribbean
Bactrocera cucurbitae (Coquillett), Melon fly	210	Asia, Eastern Africa, Hawaii
Daucas dorsalis or B. dorsalis (Hendel), Oriental fruit fly	250	India, China, Hawaii
B. jarvisi (Tyron)	150	Australia
B. latifrons (Hendel), Malaysian fruit fly	150	India, China, Southeast Asia, Hawaii
B. tryoni (Froggatt), Queensland fruit fly	150	Australia and nearby islands
Ceratitis capitata (Wiedemann), Mediterranean fruit fly	225	Mediterranean region, Middle-East, Africa, Australia, Central and South America, Hawaii islands

APHIS, USDA (1996).

Citrus fruits are less tolerant of radiation than many other fruits. The percentage of rind injuries to grapefruit irradiated with 600, 300, 150, and 0 (control) Gy was 20, 9, 4, and 1 percent respectively (Hatton et al., 1982). The dose of 300 Gy was considered the maximum that grapefruit could tolerate without peel injury (Miller and Mc Donald, 1996).

The 300 Gy dose is lethal to all stages of *Ceratitis capitata* without altering the chemical composition of Tarocco and Sanguinello sweet orange fruits. Advantages include deep and uniform penetration of energy and guaranteed disinfestation, even in the center of the fruit, with no undesirable effects. The treatment is effective in wrapped fruit also, thereby preventing recontamination (Adamo et al., 1996). In Florida, an irradiation dose of 1.0 KGy has been approved by the FDA and found effective for control of the Caribbean fruit fly in grapefruit shipped to Japan and Europe. Mainland United States has begun to use irradiation as a quarantine treatment for some fruits imported from Hawaii. The U.S. is the first country to use irradiation as a quarantine treatment, although on a very limited basis. Irradiation offers some additional risk-abatement advantages over other quarantine treatments (Hallman, 1999). Short-term vapor-heat treatment reduces the severity and incidence of peel injury in fruit irradiated for disinfestation (Miller and McDonald, 1998a). Irradiation of Clementines (Marisol) from Spain at doses of 0.3 and 0.5 KGy has been suggested as an alternative to quarantine treatment that is presently followed (Abdellaoui et al., 1995).

Gamma irradiation is also reported to be effective for control of Fuller rose beetle eggs on citrus fruits from Florida (Coats et al., 1990).

REFERENCES

Abdellaoui, S., Lacroix, M., Boubekri, C., and Gagnon, M. (1995). Effect of gamma irradiation combined with hot water treatment on the physico-chemical properties, vitamin 'C' contents and organoleptic quality of elementines. *Sci. des Alim.* 15(3), 217–235.

Adamo, M., D'lio, V., and Gionfriddo, F. (1996). The technique of ionization of orange fruits infested by Ceratitis capitata. *Informat. Agrario* 52(49), 73–75.

APHIS (1989). Animal and plant health inspection service, USDA, Use of irradiation as a quarantine treatment for fresh fruit of papaya from Hawaii (Final rule). *Federal Reg.* 54, 387–393.

APHIS (1996). Animal and plant health inspection service, USDA, The application of irradiation to phytosanitary problems. *Federal Reg.* 24, 433–439.

Coats, S.A., and Ismail, M.A. (1990). Ovicidal effects of gamma radiation on eggs of the Fuller rose beetle. *Florida Entomol.* 73, 237–242.

Greany, P.D., Styer, S.C., Davis, P.L., Shaw, P.E., and Chambers, D.L. (1983). Biochemical resistance of citrus to fruit flies. Demonstration and elucidation of resistance to the Caribbean fruit fly, *Anastrepha suspensa. Entomologia Experimentali et Applicata.* 34, 40–50.

Hallman, G.J. (1999). Ionizing radiation quarantine treatments against tephritid fruit flies. *Postharvest Biol. Technol.* 16, 93–106.

Hallman, G.J., and Worley, J.W. (1999). Gamma radiation doses to prevent adult emergence from immatures of Mexican and West Indian fruit flies (*Diptera:* Tephritidae). *J. Econ. Entom.* 92, 967–973.

Hatton, T.T., Cubbedge, R.H., Risse, L.A., and Hale, P.W. (1982). Phytotoxicity of gamma irradiation on Florida grapefruit. *Proc. Fla. Sta. Hort. Soc.* 95, 232–234.

Ismail, M.A., Hatton, T.T., Dezman, D.J., and Miller, W.R. (1986). In-transit cold treatment of Florida grapefruit shipped to Japan in refrigerated van containers: Problems and recommendations. *Proc. Fla. Sta. Hort. Soc.* 99, 117–121.

Kader, A.A. (1984). Irradiation of plant products. Comments from CAST. Council for Agricultural Science and Technology.

Kender, W. (2004). Fruit fly facts. *Citrus Ind. (Florida)* 85(5), 22–24.

Kitagawa, H., Matsui, T., and Kawada, K. (1988). Some problems in the marketing of citrus fruits in Japan. *Proc. 6th Int. Citrus Congress*, Israel, Vol. 4. pp. 1581–1587.

Miller, W.R. and McDonald, R.E. (1996). Postharvest quality of GA treated Florida grapefruit after gamma irradiation with TBZ and storage. *Postharvest Biol. Technol.* 7, 253–260.

Miller, W.R., and McDonald, R.E. (1997). Comparative response of pre-harvest GA treated grapefruit to vapor heat and hot water treatment. *HortScience* 32, 275–277.

Miller, W.R., and McDonald, R.E. (1998a). Short-term heat conditioning of grapefruit to alleviate irradiation injury. *HortScience* 33, 1224–1229.

Miller, W.R., and McDonald, R.E. (1998b). Amelioration of irradiation injury to Florida grapefruit by pre-treatment with vapor heat or fungicide. *HortScience* 33, 100–102.

Ohta, A.T., Kaneshiro, K.Y., Kurihara, J.S., Kanegawa, K.M., and Nagamine, L.R. (1985). Is the probit 9 security appropriate for disinfestations using gamma radiation? In *Radiation disinfestations of food and Agricultural products.* University of Hawaii (J.H. Moy, ed). Manoa, Honolulu, pp. 111–115.

Paull, R.E. (1990). Post-harvest heat treatments and fruit ripening. *Postharvest News Informat.* 1(5), 355–363.

Putruele, G., and Marty, N.P. (2000). Cold treatment quarantine of south American fruit fly (*A. fraterculus* Weid). *ISC congress presentation.* 3–7 December 2000, Florida, Abstract, p 69, p 103.

Qing, S., and Petracek, P.D. (1999). Grapefruit gland oil composition is affected by wax application, storage temperature and storage life. *J. Agric. Fd. Chem.* 47, 2067–2069.

Shellie, K.C., and Mangan, R.L. (1995). Heating rate and tolerance of naturally degreened Dancy tangerine to high temperature forced-air for fruit fly disinfestations. *Hort. Technol.* 5, 40–43.

Shellie, K.C., and Mangan, R.L. (1998). Navel orange tolerance to heat treatment for disinfesting fruit fly. *J. Am. Soc. Hort. Sci.* 128, 288–293.

Von-Windeguth, D.L. (1982). Effect of gamma irradiation on the mortality of the Caribbean fruit fly in grapefruit. *Proc. Fla. Sta. Hort. Soc.* 95, 235–237.

Von-Windeguth, D.L., and Ismail, M.A. (1987). Gamma irradiation as a quarantine treatment for Florida grapefruit infested with Caribbean fruit fly. *Proc. Fla. Sta. Hort. Soc.* 100, 5–7.

Wang, C.Y. (1994). Chilling injury of tropical horticultural commodities. *HortScience* 29, 986–996.

Wright, M.J., Kaiser, C., Bard, Z.J., and Wolstenholme, B.N. (1997). Prestorage vapour heat effects on grapefruit 'Star Ruby', *Proc. Int. Soc. Citric.*, Sun City, South Africa, pp. 1136–1138.

19

FRUIT QUALITY CONTROL, EVALUATION, AND ANALYSIS

I. FRUIT QUALITY

Fresh citrus fruit must have required external and internal quality when it is harvested and until it reaches the consumer. The meaning and perception of quality is different for different people. The fruit quality could be defined as the combination of fruit attributes or characteristics that have significance in determining the degree of consumer acceptance. This means that better the quality, the higher the rate of acceptance and vice-versa. While *quality* refers to the quality of all food items, it is particularly relevant to fresh citrus fruits and other perishables. For the sake of quality control, certain standards are fixed. Based on these standards, quality inspectors judge the quality and decide its grade, utility, and marketability. Quality evaluation and control is also essential for deciding fruit price.

Quality control can take place at various stages of fruit handling and marketing. In fact, it starts in the field. Quality attributes can be evaluated by both subjective and objective methods. The objective methods are precise and involve the use of instruments, while subjective methods make use of human senses. Quality evaluation methods could also be grouped by destructive and non-destructive methods depending on whether fruit is destroyed (cut or punctured) during analysis or remains intact. The advantage of non-destructive methods is that they can be used when fruit is still attached to the tree and quality can be monitored. Moreover, non-destructive methods can be used after harvest to make sure that every fruit sent to the market meets the quality norms, in contrast to the destructive method, in which a representative sample is taken and it is lost during analysis. (Thus it cannot be marketed.) Most of the conventional methods of quality evaluation are destructive. The quality attributes and evaluation methods could be grouped into three categories: physical, chemical, and physiological, on the basis of analytical process and principles involved. Microbial quality is evaluated in processed citrus fruit, but physiological attribute evaluation is not done in processed fruit. These are mostly objective methods, while the sensory evaluation is subjective and based on the response of human senses to external and internal fruit quality. For example, the external quality parameters are mostly related to the senses of vision and touch as influenced by the fruit appearance. As many as 12 quality factors affect appearance. These parameters are: (1) rind color, (2) rind texture, (3) discoloration, (4) blemishes (scars resulting from wind or twig rubbing, scab, scales, melanose, stem-end injury, or any other insect damage), (5) bruises resulting from transport or other factors, (6) oleocellosis, (7) oil-spray injury, (8) hailstorm injury, (9) sunburn, (10) creasing, (11) rind breakdown, and (12) puffing.

The internal quality of fruit is governed by taste, aroma (the combination of taste and aroma produces flavor), color of juice/flesh, appearance, and mouthfeel. All these quality attributes are influenced by the chemical composition. The acceptable taste of citrus fruit results mainly from the proper blending of sugars and acids. Mouth-feel, which is also known as body, is a result of viscosity and the presence of water-insoluble solids. The firmness of citrus sections is also affected by pectin and water-insoluble solids and fibers.

A. Physical Parameters

These attributes of quality are measured by applying principles of physics and measuring the response of fruit to light, weight, force, time, space distance, and so forth. The measurements that come under this category are: (1) firmness, (2) rind color, (3) fruit size/shape, (4) fruit weight, (5) fruit volume, (6) rind thickness, (7) rag percentage, (8) juice percentage, (9) total soluble solids, and (10) specific gravity.

B. Chemical Parameters

These attributes of quality are measured by applying certain principles of chemistry and based on response of fruit internal parts/composition to chemical reactions. These quality attributes for citrus include total titratable acidity, ascorbic acid, total sugars, reducing and non-reducing sugars, and pH. These are a few chemical attributes which are routinely analyzed and provide fairly good information about fresh fruit quality. The most commonly and easily determined attributes are titratable acidity and ascorbic acid contents, which provide fairly good information about taste and nutritive value. Many other quality attributes of biochemical nature, such as total carbohydrate, amino acid, pigment content, and so on may also fall in this category but they are beyond the scope of this book. Most flavor and aroma compounds are also determined by chemical analysis and also by applying principles of physics and chemistry, as in the case of gas chromatography and portable sensors developed for gases, aroma compounds, and detecting specific rots/pathogens on fruit after harvest.

C. Physiological Parameters

The following parameters indicate physiological functions/processes going on inside the fruit and as a consequence determine shelf life of the fruit: CO_2 evolution or O_2 consumed (rate of respiration), ethylene evolution, rate of transpiration or water loss from fruit, rate of enzymatic reaction, and growth regulator/hormonal content and gaseous content of fruit (CO_2, O_2, and C_2H_4). Analysis of these parameters provides fairly good information about fruit quality, particularly stage of maturity and aging/senescence.

II. ANALYSIS OF QUALITY ATTRIBUTES AND INSTRUMENTS

A. Physical Attributes

1. Fruit Weight

Fruit weight is a basic parameter of quality and has to be measured precisely on digital balance/weighing machine. This data is cross-checked and calibrated with known weights for its accuracy. To calculate loss in weight of fruit, the weight of same fruit has to be taken during storage over a specific time interval. It should be in replicates of 5–10 fruits.

2. Fruit Size

Fruit size – mainly length (stem end to stylar end) and breadth (equatorial diameter) – is measured with Vernier calipers in millimeters. Diameter is measured in two places on the equatorial plane so that an average diameter can be taken.

3. Peel Thickness

Peel thickness is measured with the help of Vernier calipers. The peel is removed, a piece is placed between the two measuring jaws, and a reading is taken in millimeters.

4. Fruit Volume

Volume is an important parameter that indicates the growth of fruit and also space required for the handling and packing of fruit. The ratio of weight and volume indicates internal quality such as juiciness or dryness. The water-displacement method of measuring volume is common and accurate but also time-consuming; therefore, it can be used only for a small number of samples. Moreover, fruit has to be picked for measurement; therefore this method cannot be used for growing fruit. Fruit has to be thoroughly wet before it is dipped into a cylindrical, beaker-shaped container with a spout. The displaced water is collected and measured in the measuring cylinder and a reading is taken in milliliters or cubic centimeters.

The volume of citrus fruits can also be determined by measuring diameter or circumference and then calculating the volume using mathematical formulae (Zhang, 1992). Diameter (equatorial) can be measured in two to three directions by using Vernier calipers and taking an average measurement, since fruit diameter is not uniform from all directions. (A cross section of fruit is not usually circular.) The longitudinal diameter (stem end to stylar end) can also be measured with Vernier calipers. For a circumference, a measuring tape can be used. The formula is volume $V = 4/3\pi r^3$. This gives a value close to actual in the case of globose to spherical fruit. In case of elliptical or oval-shaped fruit, volume calculated by this formula may not be accurate. When these methods are used, measurement has to be very accurate to avoid multiplying errors during calculation.

5. Fruit Firmness

The firmness of the fruit is also related to mechanical properties. It can be measured in terms of puncture resistance, compression, creep, impact, and sonic properties. Citrus fruits are soft compared to apples and pears and have viscous components in addition to elastic properties. This viscous component contributes to hand-feel. Fruit firmness testers and universal texture-testing instruments can give some useful predictions about firmness. Researchers have used penetrometers to measure firmness of citrus fruits as an indication of maturity and general condition of the fruit.

Firmness of citrus fruit is an important quality characteristic; penetrometer readings can provide some information about firmness. As the fruit becomes soft, firmness decreases. Firmness can be expressed as resistance to puncturing the rind. Citrus fruits are like berries (hesperidium): albedo and flavedo constitute the peel or rind. Peel resists the force to puncture. Inside there are segments with juice sacs that exert very little resistance. When using penetrometer testers to measure firmness, the following information is needed: the type of instrument used, dimensions and geometry of probe, penetration distance, and rate of loading (speed).

With progressive, rapid, moisture loss in excessively hot and dry conditions, peel becomes tough and leathery in a short time. Thus the pressure required to puncture or penetrate the peel increases, possibly giving erroneous information about firmness measurement with a penetrometer. (Firmness should generally decrease with increase in softness during extended storage.) Therefore Abbott

(1999) suggested that for soft, juicy fruit such as citrus – which has significant viscous component in the texture – measuring the creep (deformation under constant load for certain time) can be useful. The loads and time need to be standardized for various kinds of citrus fruits. The relaxation measurement – the decrease in force with time at a constant deformation – can also provide information on firmness.

Different mechanized gauges are used to measure the force to puncture. One such gauge is the Hunter Spring mechanical force gauge (Model L-30-M) with 10 mm circular cylindrical probe head. It measures the force required to penetrate the flavedo and albedo.

The Effegi penetrometer FT0-11 is a handheld pressure tester with 5/16-inch plunger. The pressure can be measured in pounds or kilograms. The kilogram values are multiplied by 9.80 to convert into Newtons (N). Penetrometers are not used to measure maturity or ripeness of lemons and limes as a force. Rind-oil rupture pressure is measured with firmness testers in limes and lemons.

Instron texture-analyzing system or Stable Micro-system are very advanced, software-controlled systems that are operated through computers. The system can display the speed of probe/plunger, force measured, and distance. These texture-analyzer systems cover a wide range of food products. They have different load cells of various weights (5, 10, and 25 kg) and during a puncture test a 5- or 6-mm, flat-headed, stainless-steel cylindrical probe travels down the fruit tissue at certain speed (1 mm, for example) and at a certain distance. Firmness is used as a first peak force value to puncture as maximum force required to puncture the rind since inside part is very soft made of juice sacs.

To measure equatorial and residual deformation force, pressure is exerted perpendicular to the longitudinal axis. A 5 kg weight is placed on lemons, and after 30 s deformation is measured in millimeters (Ben-Yehoshua et al., 1981). The weight is removed and a second reading is taken when fruit shape is partially restored. This residual deformation is scored 30 s after removing the weight (when the shape becomes constant). The firmer the fruit, the lower the first reading of full deformation and second reading of residual deformation. In this way, softness – which is the shrinkage in fruit diameter (mm) – can be measured after compression (30 s, for example) with different loads (3 kg (29 N) or 5 kg (49 N)). The load placed on the fruit depends on the kind of citrus fruit being evaluated.

Compression force is also measured as an indicator of firmness. A 65 mm diameter flat plate is brought in contact with an orange at 1 mm/s speed to compress fruit for 10 mm from contact point. A 5000 N load cell is used. The firmness is expressed as the force required to compress the fruit to 10 mm distance. Higher force values indicate higher firmness (Singh and Reddy, 2006).

6. Fruit Detachment Force

Fruit detachment force indicates how strongly fruit is attached to the peduncle. This force can be measured with Chatillon dynamometer and expressed in Newtons. This measurement indicates whether fruit adheres to its peduncle or is going to abscise. The detachment force is measured to see whether fruit of a particular variety can be retained on the tree or needs to be harvested.

7. Juice Percentage

Juice percentage of citrus fruit is generally expressed on the basis of weight. Fresh citrus fruits are valued for their juice content. Juice content decreases with advancement of maturity and extension of storage period. Lime, lemon, and grapefruit maturity and quality is also based on the amount of juice in a fruit besides TSS content. Fruit of different sizes have different juice requirements at different times of the season.

Juice percentage is determined by taking the weight of the whole fruit or fruits. Juice is extracted with an electrically operated rimmer after halving. Juice is filtered through a muslin cloth so that rag and seeds are removed. Weight of juice is taken. Percentage of juice is calculated as the weight of juice divided by the weight of fruit multiplied by 100.

In the case of lemons and limes, juice percentage is determined on volume basis. The volume of the fruit can be determined by the water-displacement method. Juice is squeezed and measured in milliliters or cubic centimeters and percentage is calculated as the volume of juice divided by the volume of the fruit multiplied by 100.

8. Specific Gravity

The palatability of the citrus fruit is associated with its internal quality. Soluble solids, juice content, and weight of the fruit are affected by freezes in very cold weather. These quality factors are associated with the specific gravity of the whole fruit (Ting and Blair, 1965). Freezes also affect juice content, texture, and the appearance of juice vesicles. Measuring specific gravity can provide clues to internal quality. The ratio of the weight of the fruit to its submerged weight indicates specific gravity (the weight of fruit in air divided by weight in water). The specific gravity also indicates the relationship between fruit weight and volume. The specific gravity changes with freezing of fruit because the juice-vesicle tissues are damaged and become dry. To determine specific gravity, fruit is placed in a 1-lit beaker containing 600–700 ml of water on a top-loading balance. The total weight is noted and the weight of the submerged fruit is calculated. Fruit may float, and hence a thin, stainless-steel rod with a loop at one end is used to submerge it so as to more accurately record the weight. If the fruit sinks, the loop or thread tied to the fruit is used to hold the fruit up in a submerged condition. In this case the specific gravity will be greater than one when calculated as before. Specific gravity of a multiple-fruit sample can also be measured by water displacement (Ting and Rouseff, 1986). The separation of citrus fruits damaged by freeze injury is possible by measuring specific gravity. The freeze-damaged fruit has to be separated from the good fruit at the packinghouse. Frozen-fruit separators have been developed; they work on the principle of difference in specific gravity of sound and desiccated frozen fruit (Grierson and Hayward, 1959).

9. Color

The quality of fresh citrus fruit is determined by appearance – mainly peel color. Although peel color bears no relation to palatability, maturity, or flavor of citrus fruit, consumers expect characteristic color for specific fruit. Fruit color is also one of the criteria for sorting into commercial grades. Consumers are reluctant to purchase green fruit except for limes and lemons.

a. Subjective Color Measurement: Colour measured by visible means (human eye) can be subject to error; color perception can differ among individuals. In spite of this, subjective color analysis is helpful when instruments are not available. In this method, seven standard colors of citrus fruit rind – deep green, light green, yellowish-green, greenish-yellow, yellowish-orange, orange, and deep orange – are given numerical values 1–7 respectively. (Deep green is 1 and deep orange is 7.) The index is calculated by segregating fruits of the sample in the different color categories and multiplying by respective numerical value. These values are summed up and divided by total number of fruits in the sample.

b. Objective Color Measurement: To express color in numerical dimensions and values, complete quantitative measurement is necessary, including attributes of color such as hue (red, blue, green); saturation or chroma (intensity or strength of hue); and lightness (brightness or darkness) of the color. Color is associated with luminous flux transmitted/reflected from fruit surface. Therefore, to measure fruit color as the human eye perceives it, a controlled source of light is thrown onto the object (the fruit surface).

In colorimetry, color and its concentration are measured by determining relative absorption of light with respect to known concentration of light. Visual colorimetry uses white light with filters. A photoelectric colorimeter is one example. When the light falls on the object that is shining, that light is reflected as both a specular reflection (glare) and a diffused reflection (scattered light). The diffused scattered light gives us an idea of an object's color. The light intensity is reflected to determine the color intensity. If all light is diffused through an object's rough surface, the color is perceived as diluted and appears faint, as in a matte finish. If the object is glossy, the color is perceived as dark.

The appearance of an object is influenced by basic color and surface characteristics such as gloss and texture. Therefore, while perceiving the color of an object, a human eye tries to avoid glare and perceives the color as a real color composition or pigments. Therefore, glare has to be excluded to a certain extent while color is measured. If total reflectance (specular + diffused) is measured, samples with different gloss can be measured. Selecting the right type of equipment is therefore necessary depending on the purpose of the work.

Color-difference meters measure light in terms of a tristimulus color space that relates to human vision. The light penetrates a very short distance in the tissues and much of it is reflected.

Several color-coordinate systems describe color of the fruit viz. Hunter and Harold (1987) and Minolta (1994). The common and popular systems are RGB (red, green, and blue), which is used in color video monitors. RGB data provide some clue about color status but the distribution is not unique to give very clear picture. The RGB color mode is not an efficient model with which to accurately explain human perception. The HSI (hue, saturation, and intensity) model is more accurate for describing color. Therefore, this model forms the basis of most color-processing algorithms used in modern vision systems. Generally, vision systems developed in the world detect fruit color in RGB space. The transformation of RGB data to HSI space is needed.

The color space symmetry and the coordinate system used to define points within that space differ in the Hunter '*L*', '*a*', '*b*' and CIE (Commission International de l'Eclairage), *L** *a** *b**, XYZ, *L** *u** *v**, *Yxy*, and LCH systems. Most instruments measure color using the tristimulus methods of the CIE and the Hunter system. The CIE system is based on the concept that human eye recognizes three primary colors through receptors that are the red, green, and blue of the light spectrum; all other colors are combinations of these three. Commonly used notations are CIE Yxy color space (devised in 1931), the Hunter '*L*' '*a*' '*b*' (developed in 1948) for photoelectric measurement, and the CIE L* *a** *b** color space (developed in 1976) to provide more uniform color difference in relation to human perception (Abbott, 1999). To measure given color, a relative amount of red, blue, and green are required to match the color. The standard color determination by CIE called *L**, *a** and *b** – where *L** is a light factor, *a** and *b** are the chromaticity coordinates – more closely represents human eye sensitivity to color. Therefore, it is best adopted to fruit-color imaging or color measurement of citrus.

'*L*', '*a*', '*b*' values are measured with the help of the Hunter lab color meter; color index is calculated as ratio of a/b. The –ve value indicates green color while +ve value indicates yellow color. With advancing maturity of fruit, green color decreases and therefore –ve value (score/reading) also decreases. As the +ve value of this ratio increases, the rind color turns toward yellow orange and red:

(i) Portable Color Meter: A handheld, battery-operated digital colorimeter with liquid crystal display and optical unit is very handy for measuring fruit color in citrus. This instrument can be calibrated with white tile/board with known '*L*', '*a*', '*b*' values. The user of this equipment has choice of mode – *L**, *a**, and *b**; Hunter '*L*', '*a*', '*b*'; *x*, *y*, *z*, and so forth. Fruit is held against the opening of the instrument; the reading is taken with press of the button.

(ii) Hunter Colorimeter (Color Difference Meter): To measure the color with the Hunter color difference meter, fruit is placed on the instrument's opening. An area of the fruit's surface is marked with a marking pen so that the reading can be taken from same spot periodically. From '*L*', '*a*', and '*b*' values, a ratio of *a*/*b* is calculated that gives the extent of orange color. The Hunter values are in close approximation to *x*, *y*, and *z* functions of CIE system, in which *a* value is function of '*x*' and '*y*'. The '*b*' values are functions

of z and y (Nickerson, 1964). The values of 'L', 'a', and 'b' are related to the tristimulus 'x', 'y', and 'z' by the following equations:

$$L = 100y;\ a = 175 \times 1.20x - y/y;\ b = (y - 0.847z/y).$$

The Hunter values can be converted to CIE system.

For particular color (hue), a/b ratio is calculated or Hue angle ('ϑ' or theta $=$ $\tan^{-} b/a$) is calculated. The saturation, or chroma, is calculated as root of $a^2 + b^2$. The Hunter L values are directly comparable with Y of the CIE system.

To determine the interior, or flesh, color of the fruit, the cut surface is placed on a clean glass plate to protect the instrument from juice. The fruit sample is rotated on the glass plate so that several readings can be taken. Care should be taken so as not to include the central core in the measurement. Aperture size can be changed to accommodate small fruit. Averages from two or three readings are used in the calculations.

10. Fruit Gloss

Fruit gets natural gloss if natural wax is rubbed gently on its surface. To provide an artificial sheen, a coating is applied commercially in packing houses. Gloss can be measured with a gloss meter, which operates on the principle of light reflectance. This *reflectometer* measures the gloss units when the fruit is placed close to the circular opening of the instrument. Gloss meter models such as the Micro TRI gloss (BYK Gardener, Silver Springs, MD, U.S.A) measure gloss and are calibrated with a standard surface.

11. Total Soluble Solids

Total soluble solids (TSS) of sweet orange, mandarin, grapefruit, and pummelo juices constitute mainly sugars (80–85 percent). Citric and other acids and their salts, nitrogenous compounds, and other minor soluble substances such as water-soluble vitamins constitute the remaining composition of TSS. °Brix of citrus juices indicates all the soluble solids. It is not a measure of sugars only. Soluble solids in juice can be measured from the refractive index, and refractometers are calibrated to give °Brix or percent total soluble solids values directly. Hydrometers can also be used to measure solids in the juice.

Handheld, portable refractometers are very easy to use. A drop of juice is placed on the prism. After putting on the cover and holding the instrument against light one can see a reading as a percent of solids through an eyepiece. Refractometer reading changes with the temperature as the refractive index of sugars changes with the temperature. Automatic compensating type handheld refractometers have a built-in mechanism that automatically corrects the scale. So as the ambient temperature changes from standard 20°C, the borderline and scale deviate to compensate this temperature change. In non-compensating-type refractometers, the addition or deletion from the reading is necessary depending on the temperature. A correction factor is added to the reading above 20°C, while below 20°C it is deleted. A table is provided with the instrument by the manufacturers.

A refractometer calibrated with a °Brix scale gives values in °Brix. If a °Brix scale is not built-in, the refractive index values of the solution can be obtained and the relative °Brix found in a table. Some of the more sophisticated °Brix refractometers have built-in temperature-correction circuits as well as digital readings of °Brix instead of an analog scale. The °Brix value can be found by placing a drop of the liquid on the clean prism and taking a reading.

Hydrometers for measurement of °Brix of citrus juice are usually calibrated from 5 to 15°Brix at 17°C or 20°C. Air has to be removed by placing juice in a large filtering flask and applying vacuum. The hydrometer is placed in the cylinder containing juice. This cylinder is larger in diameter than the hydrometer and almost identical in height. By carefully lowering the hydrometer into the juice, one can record a reading on the scale at juice level.

B. Chemical Attributes

1. Total (Titratable) Acidity

Citric acid is the most predominant acid in citrus, accounting for 80–95 percent of the total acids in various citrus fruits. In limes, lemons, and other acid citrus fruits, citric acid constitutes most of the total soluble solids, as the sugar content is very low. Organic acids are weak acids and when titrated with a strong base, the equivalence point is not neutral (pH 7) but slightly basic, because of the salt.

Phenolphthalein is generally used as a visual endpoint indicator; it gives a pink-colored endpoint. Total acidity can be determined by using 0.1 NaOH (or 0.3125 N NaOH) for a large number of samples (Ting and Rouseff, 1986). Total acidity of citrus juice and concentrates is determined as anhydrous citric acid and expressed as percent by weight. The endpoint can also be ascertained with a pH meter. The titration is complete when the pH reaches 8.2. The amount of sample used in the titration depends on the acid concentration in the sample. A much smaller sample is used with lemons or limes than with oranges.

To take this reading, add a 5 ml pipette of filtered, single-strength orange or grapefruit juice into a 150 ml Ehrlenmeyer flask and carefully add 20 ml of distilled water. Take 5 ml from this in another flask and add 2 drops of a 1 percent phenolphthalein solution (prepared in 50 percent isopropyl alcohol). Titrate with 0.1 N NaOH to the endpoint. Calculate the total titratable acidity by putting the values in the following formula:

$$\text{Acidity (\%)} = \frac{\text{Titre} \times \text{Normality of alkali} \times \text{Volume made up} \times \text{Equivalent weight of citric acid (64)} \times 100}{\begin{array}{c}\text{Volume of sample} \\ \text{taken for estimation}\end{array} \times \begin{array}{c}\text{Weight or Volume of} \\ \text{juice/pulp taken}\end{array} \times 1000}$$

To convert the percent citric acid by volume to percent citric acid by weight, divide the results by the specific gravity of the juice. For most single-strength juice, a specific gravity of 1.04 is used, assuming the average soluble solids to be 10°Brix.

2. TSS-to-Acid Ratio

The relative sweetness or sourness of citrus fruit is determined by its ratio of sugars to acids. Since most soluble solids in oranges, mandarins, and grapefruits are constituted by sugars, the ratio of the soluble solids to acid is used for convenience. This is a maturity index and used to determine the legal maturity of oranges, mandarins, grapefruit, pummelos, and their hybrids for fresh-fruit (des sert) purposes. In case of limes and lemons, the TSS:acid ratio is not a criterion of maturity. Grapefruit with a ratio of 8–10 is considered relatively sweet, whereas oranges with same ratio would be sour for many consumers. The ratio gives relative measure of the fruit maturity and this standard varies with preference of people in different countries. It is calculated by dividing TSS (Brix) with titratable acidity.

Example 1: If 12% TSS ÷ 1% acidity the ratio is 12.

Example 2: If 12% TSS ÷ 0.8% acidity the ratio is 15.

As in example 1, for every 12 part of TSS, 1 part acid exists. In other words, for every 1 part acid there are 12 parts TSS. Similarly, as in example 2, for every 1 part acid there are 15 parts TSS. Thus the fruit in example 2 is sweeter because more sugars are present in relation to acid. A higher ratio indicates decreasing acid content. There is a limit of this higher ratio. Usually fruit with ratios higher than 19–20 are not liked by the people because the taste is quite sweet or flat with so much less acid. TSS:acidity ratio provides a clear picture only when presented along with TSS percent. Fruit with higher TSS:acidity ratio need not be very tasty. For example, a ratio of 14 can be achieved with 14 percent TSS and 1 percent acidity and also with 8 percent TSS and 0.57 percent acidity, but fruit in the former case would be tastier than the latter. With just 8 percent TSS, oranges do not have required sugars for the good sugar-to-acid blend and taste.

3. Juice pH

Juice pH is a hydrogen-ion concentration and provides an estimate of the extent of acidity in the juice. Juice pH can be measured with a pH meter – either with a desktop model or handy portable model.

4. Ascorbic Acid

Ascorbic acid can be estimated using 2.6-dichlorophenol indophenol dye that is reduced by ascorbic acid. Most citrus fruits have a pH range of 3–3.5 and this method gives fairly accurate results. The known weight of pulp or juice (5 ml or 5 g) is taken in a small quantity of a metaphosphoric acid (HPO_3) solution (strength 3 percent). The volume is increased to 25 ml with metaphosphoric acid and filtered through muslin cloth. In case of pulp, blending is done in homogenizer using metaphosphoric acid; volume can then be made. Five ml of aliquote is taken from 25 ml of volume and titrated with standard dye until a pink endpoint appears. The endpoint should be persistent for at least 15–20 s. Ascorbic acid is calculated as:

$$\text{Ascorbic acid in mg/100 ml juice or 100 g pulp} = \frac{\text{Titre value} \times \text{Dye factor} \times \text{volume made} \times 100}{\text{Aliquote taken for estimation} \times \text{juice/pulp taken}}$$

The dye solution is prepared by dissolving 50 mg of sodium salt of 2.6-dichlorophenol indophenol dye in 150 ml of hot distilled water to which sodium bicarbonate (40 mg) is added. Volume is increased to 200 ml in a glass-stoppered volumetric flask after cooling. Dye is stored in a refrigerator.

To standardize dye, an ascorbic acid solution of known concentration is necessary. L-ascorbic acid (100 mg) is dissolved in 100 ml metaphosphoric acid and kept in a colored, stoppered, volumetric flask in a cool, dry place. For standardization, it is diluted by taking 10 ml stock and adding 90 ml of HPO_3. (This gives 0.1 mg of l-ascorbic acid for 1 ml of solution.) Five ml of diluted standard ascorbic acid solution is taken and titrated with dye until a pink endpoint appears. The dye factor can be calculated by dividing mg ascorbic acid (0.5 mg in this case) with ml of dye (titre). Use glass distilled water for preparing reagents and making up volume.

5. Added Color on Fruit Surface

To induce chlorophyll loss, harvested green fruit is treated with low concentrations of ethylene at elevated temperature and humidity. Degreened grapefruits become yellow, but degreened oranges have little of the orange color that consumers prefer. For late-season oranges (maturing in March through June, particularly in Florida), the warm weather causes a reappearance of chlorophyll in the peel. Degreening of these fruit is not as successful, although it is commercially practiced. Citrus Red 2, an FDA-approved dye for citrus fruit, is applied to washed fruit as an emulsion. Concentrations on fruit up to 0.5 ppm are allowed. The amount of artificial-coloring material can be determined spectro-photometrically (Ting and Deszyck, 1960; Ting and Rouseff, 1986).

C. Physiological Attributes

1. Respiration

Respiration rate provides clue about the fruit's rate of metabolic activity and the substrate utilized in the respiration process. When sugars are consumed, the oxygen utilized is equal to carbon dioxide produced, indicating respiratory quotient (RQ) as 1. Respiration rate is expressed as mg carbon dioxide per kg fruit per hour. Respiration rate of fruit tissues as well as whole fruit can be measured. Usually infrared CO_2 analyzers are used for measuring CO_2 produced by fruits in flow through system. The instrument has three parts: the readout/control unit, the sensor, and the flow-control pump (Fig. 19.1). The flow-control

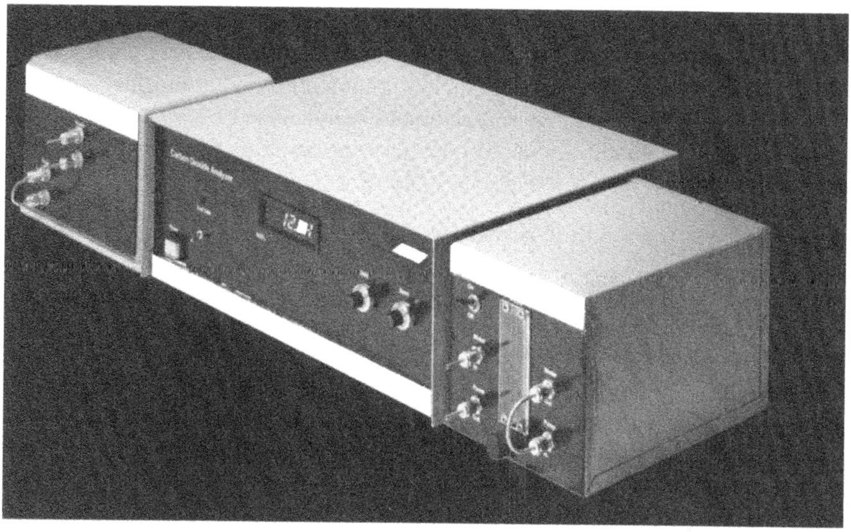

FIGURE 19.1 Infrared Carbon Dioxide Analyzer for Measuring Respiration Rate of Fruits.

unit (pump) can be used to draw or force the sample air through the sensor. The control unit displays the reading as $\%CO_2$. The flow rate of air is generally set at 100 ml per minute. To measure the respiration rate, the fruits are kept in an airtight jar and air is passed through the fruit. To calculate respiration rate the following formula can be used:

$$CO_2\,(\text{mg/kg/h}) = \frac{\%CO_2\,(\text{reading}) \times \text{flow in ml/min} \times 1.104\ (\text{factor at } 20°\text{C})}{\text{Fruit weight in kg}}$$

The factor is calculated for different temperatures as the gas volume would change accordingly. At NTP (normal temperature and pressure), such as 273° K (0°C temperature) and 760 mm Hg pressure, the 44 000 mg CO_2 is equal to 22 400 ml, or 1.976 mg CO_2 per ml volume (density of CO_2 at 0°C per ml of air). The pressure is generally taken as a constant since there are no changes on plains; otherwise, it can be taken as actual pressure divided by 760. The density has to be calculated for prevailing ambient temperature at the time of measuring respiration. It is calculated as:

$$\text{The density is calculated as } \frac{\text{K (i.e. 273)} \times 1.9769}{273 + °\text{C (prevailing temperature in °C)}}$$

For 30°C, the density would be 1.78 and for 35°C it is 1.75. For 5°C, 10°C, 15°C, 20°C, and 25°C values are 1.94, 1.90, 1.87, 1.84, and 1.81, respectively.

The percent of CO_2 values of the instrument are ml CO_2 per 100 ml of air. As we need mg values of CO_2, per ml of air, the density values are required. As the flow is in 100 ml per minute and the respiration rate is in terms of CO_2 per hr, 60 min are converted to 1 h. Hence factor can be calculated for various temperatures (at which commodity is placed for measuring respiration) using density values to put into a formula as above so as to get direct values of mg CO_2/kg/h.

$$\text{Factor } \frac{1.84\ (\text{density}) \times 60}{100} = 1.104\ (\text{factor at } 20^{\circ}\text{C})$$

2. Ethylene and Other Volatiles

Because citrus fruits are non-climacteric, they generally produce very small amounts of ethylene, but the quantity varies in many fruits of different citrus species. Accumulated ethylene in storage areas can accelerate deterioration. High ethylene concentrations in storage areas indicate rotting and senescing fruit. Ethanol and acetaldehyde are the major volatiles emanated by fruit after harvest. These volatiles are analyzed in citrus postharvest research since they impart an off flavor to fruit after harvest depending on handling, treatments applied, and storage practices.

A gas chromatograph is used to measure ethylene and other volatiles, but it is a cumbersome, expensive machine. A syringe with a detector tube is handier, less expensive, and fairly accurate. Detector tubes for different volatiles are available for detection in packages of fruit.

a. Gas Chromatograph: Ethylene and N_2, O_2, CO_2, gas analysis is routinely done on gas chromatograph. There are several models/makes of GC. The 80/100 mesh-activated alumina, Porapak-Q column, and flame-ionization detector are the essential requirements for ethylene analysis. Nitrogen is used as a carrier gas. The instrument has to be calibrated with a standard gas sample of a precise concentration. To measure ethylene production (evolution) by the commodity, 1 ml of a gas sample drawn from the sample jar with syringe is sufficient. The aroma volatile compounds are analyzed and identified using GC coupled with a mass selective detector. The high-resolution capillary column is used to separate the compounds and it is connected to mass spectrometer (MS). The mass spectra of the aroma compounds are compared with the known standard compounds. GC is in fact a versatile instrument and can be used to analyze various other gases in the atmosphere. An advantage of using this technique is that it can analyze small samples (0.2–5 ml). A thermal conductivity detector (TCD) with a molecular sieve column is required for N_2 and O_2 analysis. The carrier gas is helium. CO_2 can be measured with Porapak column with TCD detector.

b. Ethylene Detector Tubes: The handy and rapid technique of ethylene analysis involves a syringe (100 ml capacity) and a detector tube of glass (Fig. 19.2). It is very useful for outdoor analysis. The detector tube is packed with a proprietary

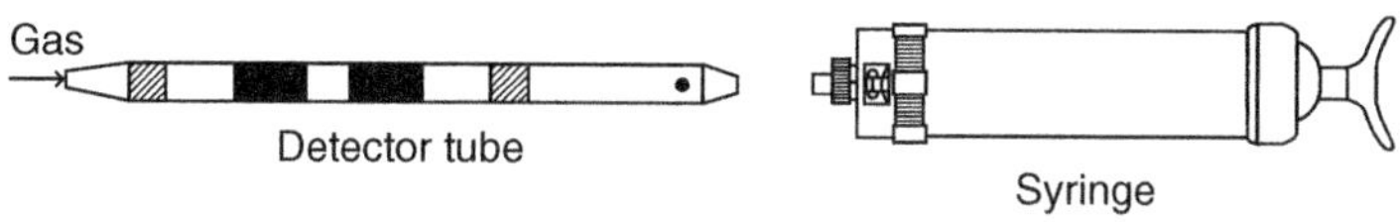

FIGURE 19.2 An Ethylene Detector Tube with Syringe Is Handy and Fairly Accurate for Determining Ethylene Level in Storage Areas. Detector Tubes for Acetaldehyde, Ethyl Alcohol, and Many Other Compounds and Gases Are Also Available.

compound that changes color on graduated scale on the tube that indicates concentration. The ends of the tubes are broken before use so that it can be fitted on the syringe to pull the sample through it. The tubes are disposable and the per-sample analysis cost is high.

c. Ethylene Analyzer: The battery-operated, portable instrument is very handy. Using an electro-chemical detector, it can measure ethylene up to maximum 100 ppm concentration with lower limit of 1 ppm. The resolution of 0.1 ppm is displayed on an LCD readout. The built-in pump can draw the sample air from few meters. The limitation of such instruments is that hydrocarbons similar to ethylene and other carbon compounds may interfere in the reading.

d. Sensors for Volatiles: The biosensors are also available to detect some specific volatiles. In these biosensors, enzymes react with substrates and create an electrochemical signal or a color change. One such biosensor, ALCO-SCREEN, is available from Chematics in North Webster, IN, USA. This biosensor measures ethanol in solutions. Biosensors can be used to measure ethanol in extracted citrus juices. They are simple to use. Ethanol biosensor has immobilized alcohol oxidase, peroxidase, and chromagen. Alcohol oxidase catalyzes oxidation of ethanol. The extent of color change depends on ethanol concentration, exposure time, and temperature. The color charts can be used to compare color strength and concentration of ethanol. This can give sufficient gross determination between high and low levels of ethanol. Biosensor strips are available in aluminum foil to protect them from atmospheric oxygen, as alcohol oxidase is sensitive to O_2.

A biosensor to measure ethanol in vapor phase is also available. This consists of immobilized alcohol oxidase and peroxidase with a 2.6-dichloroindophenol dye to give it color (Barzana et al., 1989).

Neotronics Sci. Inc., GA, U.S.A, is marketing an 'electronic nose' (e-nose) that uses polymer sensors. At room temperature, polymer sensors are exposed to the volatile compounds in the sample head-space (the space left above the juice sample kept in a vial or small container whose volume is known). Gas sensors are useful for aroma discrimination since their electrical resistance properties are altered by the adsorption of volatiles produced by the sample.

e. Sensors to Detect Fruit Decay in Pallets and Boxes: The sensors are developed to monitor the condition/quality of citrus fruits during handling and storage because the emission of certain volatile substances indicates the loss of quality (Conesa et al., 1993). Two of the most frequent instances in which quality deteriorates during transportation are (1) decay caused by *Penicillium digitatum, Botrytis cinerea, and Galactomyces citri-aurantii* (*Geotrichum candidum*) and (2) damage caused by bad handling. Ethanol and ethyl acetate emissions are related to decay caused by *Penicillium digitatum.* A strong emission of limonene is noticed in case of *Geotrichum candidum.* The concentration of these volatiles indicates the extent of the damage and the type of pathogen responsible for spoilage.

D. Sensory Attributes

It is said that fruit is purchased with the eye and the nose. Out of the five senses – vision (eye), touch (feeling with fingers of hand), olfaction (nose), hearing (ear), and taste (tongue) – humans mostly use the visual, olfactory, and touch senses for judging maturity of fruit. Whatever instrument we may develop to measure the quality, consumers will continue to judge fruit quality using their own senses at the time of purchase. Thus sensory evaluation is equally important for deciding the marketability of fruit. Sensory evaluation is generally done to determine consumer response to a newly developed variety or a treated and stored fruit in postharvest handling. A score sheet can be prepared, taking into consideration important characteristics that may change depending on the objectives of the study or evaluation. For a breeder of a new variety, fruit bitterness, sourness, and sweetness may be important to evaluate separately. Postharvest quality personnel may include this in flavor evaluation. These attributes are evaluated on a certain scale (10 to 100 and 1 to 9 or 1 to 10, with each numerical score having a certain phrase such as low, medium, high) judges or panelists are asked to give the score or numerical value on the score sheet. These scores are then statistically analyzed using analysis of variance or other statistical tools.

Generally studied and evaluated attributes of quality in sensory evaluation include:

1. External appearance: This includes skin defects/blemishes, rind color, freshness of fruit (whether it is wilted or shriveled, for example). In the evaluation of a new variety, size and shape can be given as separate attributes. Consumers demand that fruits have a characteristic shape. When it comes to grapefruits and mandarins, consumers do not like oblong or 'sheep-nosed' fruit. Most commercial citrus fruits should be globose to spherical, which also helps in machinability. Appearance is the most important and only criterion available to the buyer for judging the quality of fruit, since taste testing is not possible at the market.

2. Rind texture: This can be separately evaluated in case of preharvest PGR applications. Generally, a smooth peel texture is preferred. Therefore texture evaluation on score sheet could be for smooth, rough, and wrinkled surface.

3. Flavor: Sweet-sour taste and characteristic aroma of citrus fruit make its flavor. There should not be any bitterness. Some people like sweeter fruit while others like acidic. The sweetness, sourness, bitterness can also be evaluated separately. Similarly, the presence or absence of an off flavor can be evaluated on certain numerical scale on score sheet (0 for none, 1 for very slight, 2 for slight, 3 for medium, 4 for high, and 5 for very high).

4. Mouth-feel, juiciness, texture of pulp: Whether fruit sections (orange, grapefruit, pummelo) or segments (mandarins) are juicy, melting, or crisp. The crisp juice sacs are considered to contain less juice, as in case of some pummelos. Tongue, teeth, and internal mouth walls are sensitive to fruit we eat. Consumers do not like a coarse texture.

5. Color of rind and pulp: Consumers prefer bright colors such as red or blood-red or orange. The intensity and preference for color can be evaluated in a sensory test along with objective color measurement using a color-difference meter. Some experimental postharvest treatments may be likely to cause some discoloration; rind color can be evaluated by sensory test.

Scoring during a sensory test has to be done in separate cells/compartments to avoid bias. A trained panel is asked to evaluate in a sensory laboratory under controlled temperature and light. In sensory consumer tests a representative sample is taken from group of people belonging to a variety of categories of age and sex. The sample should be representative and sufficient replications should be taken to analyze the data statistically.

E. Fruit Sampling

This is the most important aspect in analytical work since wrong sampling can lead to erroneous results. Usually random samples are taken. About 1 percent of the boxes are sampled randomly from the given lot/consignment. Fruit is taken (10–20 fruit randomly from each box) and required analysis is carried out. Boxes have to be segregated as per size of the fruit in them and thereafter sampling is usually done so that analysis for different sizes is done separately. In quality analysis, usually 10 fruits are randomly taken per replicate. The storage experiments have to be laid out in proper statistical design.

F. Instruments Useful in Storage Atmosphere Management

Measurement of atmospheric variables such as temperature, relative humidity, air velocity, and gases that control postharvest fruit quality is very important to manage environment around the fruit.

1. Thermometers

High sensitivity is required to measure the temperature difference of ±0.5°C. For a high degree of accuracy, sensors with platinum resistance elements are recommended. At high RH, a small fluctuation in temperature (±0.5°C) can result in condensation on cool surfaces.

2. Thermographs

These instruments are meant for recording temperatures on charts (in °C or °F as the case may be) for days or months continuously in the fruit-storage environment, reefer container, or trailer/truck vans. For on-board temperature recording, thermograph units that are sealed and shockproof (such as those supplied by Ryan Instruments Inc.) are used. The sensors are fairly sensitive, battery-operated bimetallic strips/thermocouples. At the destination, the instrument can be hooked up to a computer and provide data and graphs of the temperature variations, if any, during the voyage.

3. Thermohygrographs

These instruments record temperature and relative humidity during certain set periods; this data can be retrieved and printed out when the instrument is hooked up on to a personal computer. Some instruments also continuously record readings on charts.

4. Wet- and Dry-Bulb Hygrometers

Dry-bulb thermometers measure ambient air temperature. Wet-bulb thermometers, which have a wet wick around the bulb, measure wet-bulb temperature. Evaporation of water from the wick into the atmosphere results in cooling. The drier the air, the greater the rate of evaporation and hence the greater the depression of temperature. The temperature depression can be translated to a percentage of relative humidity, water-vapor pressure, and dew point from a psychrometric chart.

5. Anemometer

The air velocity inside the cool storages and in the air plenum of the stack during forced-air cooling is usually measured with digital anemometer. Different models are available; Lutron model AM 4201 with a vane probe is also useful. The velocity is measured in m/s or ft/min and so on – the instrument can be placed in different modes (Fig. 19.3).

6. Infrared Analyzer

An infrared CO_2 analyzer, as mentioned above (see *fruit respiration*), can be used to measure percentage of the CO_2 in the atmosphere. The advantage of this instrument is that with long tubing, gas samples can be drawn from storage chambers and recorded at certain intervals as desired.

7. Firite Gas Analyzers

Handheld Firite gas analyzers use the 'Orsat' method of volumetric analysis of gases (Fyrite gas analyzer, Bacharach, Inc., Pittsburgh, PA, U.S.A.). Although it is

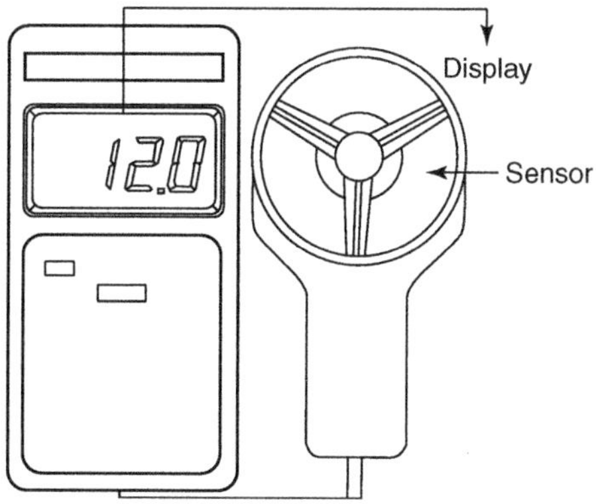

FIGURE 19.3 An Anemometer, a Handy and Portable Instrument to Measure Air Velocity in Storage Chambers and Precooling Units.

an old method, it can be useful when advanced instruments are not available. The chemical absorption of sample gases (CO_2 and O_2) is done using reagents such as potassium hydroxide (red) for CO_2 and chromus chloride (blue) for O_2. (These gases are selective in absorption.) In the same vessel absorption and indication is possible. The body of instrument is made of a highly transparent, high-strength plastic with top and bottom reservoirs. The bottom reservoir is filled with the indicator fluid. The aspirator rubber bulb and tube are used to draw the sample into the instrument through the plunger valve on the top. By turning the instrument upside down, gas is absorbed into the fluid and the level of fluid in the tube indicates CO_2 (0–7 percent or 20 percent or 60 percent). The same range of measurable concentration is available for O_2. There are limitations: For example, fumes of acetone, acetylene, and other unsaturated hydrocarbons can interfere in CO_2, analysis. In O_2, analysis H_2S and SO_2 may interfere.

8. Gas Analyzer

Custom-made gas-analysis equipment is now available per the requirements for individual gas concentration or all in one (ethylene, CO_2, and O_2). The data can be recorded at a certain period of time with sensors fitted into the storage area or the van containers.

9. CO_2 and O_2 Analyzer

CO_2 and O_2 analyzers are very useful instruments in studies concerning modified-atmosphere packaging. Both tabletop and portable models are available to measure gas concentrations. With a small needle probe, which is pierced through a window and into the package, an instant reading of percent CO_2 and O_2 inside the package can be obtained. The infrared CO_2 sensor and ceramic solid-state sensor (zirconium detector) can detect gas concentration from almost 0 to 100 percent. The instrument is fairly accurate with $\pm$1–2 percent deviation.

III. RAPID NON-DESTRUCTIVE QUALITY EVALUATION AND APPLICATION

For non-destructive quality evaluation of citrus fruits, several technologies are being used (Bellon, 1993). Cameras are used routinely to detect fruit color. Internal aspects (absence of internal defects) are detected by electromagnetic waves (visible spectrum and X-ray radiation and tomography), and nuclear magnetic imaging. The detection of degree of maturity on the basis of sugar content, firmness, and/or aroma is possible with rapid non-destructive methods that can be used online during sorting and grading in the packinghouse.

A. External Quality Evaluation

External quality refers to external physical characteristics of the fruit in terms of size, shape, color, and defects. Shape sorting is used to discard malformed fruits. Color creates an important psychological effect on consumers, who usually think something along the lines of 'the nicely colored one is the better one.' For homogeneous fruits, color determination is easy. Color cameras overcome the problem of color sorting when a large volume of fruit is to be handled. Automatic conversion among color-coordinate systems is possible and automated sorting for color, size, and shape is used on commercial packing lines.

B. Internal Quality Evaluation

Internal defects are by far the most damaging to a fruit/brand reputation. Internal defects include abnormalities such as internal breakdown, internal rots, dehydration, granulation, presence of seeds, and so on. The different techniques available deal with the change of transmitted electromagnetic waves (visible, X-rays, sounds) according to the effect. X-ray transmission detects variations in water density and is therefore useful for measuring dehydration. Dehydration or granulation in citrus has been identified using this technique (Sunkist Growers, California, U.S.A.).

Images from nuclear magnetic resonance imaging (MRI) can be processed in the same way as visible ones. MRI detects differences in viscosity of liquids such as oil and water. Worm holes, dry regions resulting from frost, or granulation in citrus are also visible by MRI (Chen et al., 1989). This is the most successful technique for internal-defect detection (Bellon, 1993). X-ray techniques are successful only in the case of large voids in the fruit or in case of freeze injury and drying of juice sacs. NMR imaging is the most suitable technique to view the depth of bruising or the location and presence of small seeds deep inside the fruit. However, its use in industry is currently not possible because of its high cost and the high level of technical expertise needed. Future innovations are likely to come from visible light spectrum coupled with either intensified cameras or focused lasers.

Light reflectance and light transmittance are technologies of electromagnetic wave spectrum that have potential and can be used for online quality evaluation in packinghouses and sorting. They can be used for color sorting and defect recognition. Light transmittance can measure internal fruit composition, while fluorescence can be used for damage identification (Kawano et al., 1992; Miyamoto et al., 1997).

To assure internal quality in terms of desired total soluble solids (sugars) in Satsuma mandarins, a non-destructive online measurement technique using Near Infra Red (NIR) spectroscopy in the transmittance mode has been developed (Kawano et al., 1992; Miyamoto et al., 1997) for use in packinghouses. NIR transmittance spectra is taken through the equator of the fruit and maximum absorption is between 770 and 900 nm (Miyamoto et al., 1997). These researchers have developed a high-speed system using a halogen lamp for a single beam of light. The transmitted light passes through the lens and falls on the mirror, which reflects it to the grating. The grating is used according to the detector. An accumulated light charge at the photo detector is measured and a signal is sent to the computer. The accuracy of the system is reported to be quite high. It is also possible to measure citric acid by same system, although with less accuracy.

Non-destructive methods can measure firmness of every individual fruit and can be implemented in automatic sorting machines in packinghouses. An impact sensor that can evaluate fruit firmness in real time has been developed (Molto et al., 1998). The sensor can discriminate between 'puffed' and 'non-puffed' mandarins (puffed fruits have a layer of air between the skin and the flesh) for commercial purposes; its values are comparable with those of the Universal Testing Machine.

Citrus fruits destined for the fresh fruit market must be seedless. Consumer response to seeded fruit is generally negative. It is desirable to remove such fruit with some non-destructive method before it reaches the market. Efforts are underway to develop real-time, online, non-destructive methods to integrate into current packing-line setup. Real-time radiography using X-ray systems can be used for this purpose (Sarig et al., 1992). The seeds are detected as bright gray areas (low absorption in relation to environment).

Miller et al. (1986) used an automatic weight and dimensional sizer to differentiate freeze damaged fruit from the healthy fruit based on specific gravity. The volume was computed from a Digitized Video Camera image. The weight of the fruit was obtained from a load cell that sensed the weight. The specific gravity measured online can provide information about freeze damage and enable online removal of such fruit.

IV. QUALITY CONTROL AND ASSURANCE SYSTEMS

In the WTO era, the world market has opened up for fresh citrus growers since many citrus growing countries are signatory to it. This has increased the competition that is compelling countries to tap their potential for higher exports. The scenario is leading to quality awareness and increased demand for quality and novelty fruit on the market shelf. Therefore it is imperative for all citrus-producing countries to have quality-assurance programs in place at the national level to compete worldwide.

The quality is always based on certain standards to be practically evaluated by both quality-control personnel and consumers. The setting of standards and quality norms varies from people to people and region to region. Every industry has to set certain standards for quality of raw materials and finished products in the same way the world citrus-fruit industry has certain quality standards, which are outlined in Chapter 7. The grade and fruit-maturity standards should have legal standing for enforcement and need to be monitored by inspectors so that they are meaningful, thereby helping to avoid and settle disputes during trade and consequently avoiding the exploitation of growers and consumers.

Quality assurance is a planned activity for systematic action at all the above-mentioned steps. It involves quality auditing and recordkeeping during cultivation and all handling steps. Total Quality Management (TQM) is a concept that has emerged from this philosophy. TQM has to have the participatory approach of growers, handlers, packers, transporters, exporters, and all those concerned to make it successful.

Quality assurance is done through program such as Hazard Analysis Critical Control Point (HACCP) in ISO 9000 System-certified industry. HACCP is a preventive system in which the safety of product (fruit) is ensured. The HACCP protocol consists of:

1. Hazard analysis and risk assessment: Hazards are identified during process of cultivation and postharvest handling and prevention measures are taken up.
2. Determination of critical control points (CCPs): The points where hazards can occur or fruit quality is likely to be affected, such as pesticide application in the field or in the packinghouse, harvesting operation, and brushing operation.
3. Monitoring of CCPs: The activities at these CCPs are critically observed and controlled to avoid hazards. Methods of application and the quality of inputs are well defined.

The first step for quality assurance is to develop procedures to control the product quality at different stages of production, handling, and packaging. The principles of fruit quality management include quality control in field, during delivery at the packinghouse, and at the time of packing. Plant hygiene and fruit quality are important in field. Sanitary conditions of machinery, containers of handling, material of packaging, and hygienic conditions of the people in the packinghouse are important.

The HACCP quality-management system is implemented to encompass all activities to meet objectives of TQM. It is a comprehensive quality-management program with series of standards to be followed in the field, the packinghouse, and also during distribution.

In India, keeping the concept of TQM in view, the Agricultural and Processed Food Products Development Authority (APEDA) brought out 'Quality assurance manual for export of Kinnows' to establish guidelines for production, handling, and packaging of Kinnows (APEDA, 1997). The 'Certificate of Registration of Premium Quality Exporter' is issued to the packinghouses adhering to the guidelines. The exporter has to apply for the registration, and after a quality audit by the team from APEDA, a certificate is issued to packinghouses that comply with the requirements. It is voluntary for the exporter to adopt these procedures and practices in their organization for quality to be assured. The requirements of ISO 9002 quality system model for quality assurance in production, installation, and servicing have been considered in developing these guidelines.

The quality assurance program should emphasize the following points:

1. Company and its management: It should be a registered business firm for fruit handling and export. The management structure of the company must have staff for quality management and company must have responsibilities and commitment to quality. The firm must follow a policy about fruit quality.

2. Procurement of raw material: Quality, hygiene, and safety of fruit primarily depend on source of raw material (fruit, pesticides, water, packaging material, and other chemicals). The personal hygiene and sanitary conditions inside the packinghouse are very important. Fruit that is procured must be from farms following Good Agricultural Practices (GAP), which comprise procedures, practices, and methods to minimize risk or hazard to public health from agronomic and harvesting practices. If the fruit is organically grown and to be labeled as organic, guidelines of the importing country regarding organic fruit have to be complied with. European Union Council regulations in this regard can be referred. The grower has to comply with procedures and practices, and an inspection procedure has to be in place to ensure that fruit is organically grown. Organic postharvest treatments must be applied to such fruit.

3. Process control and quality management: The monitoring of operations in packinghouses as per the guidelines for sanitary condition of material, machines, personnel, and process-control worksheets has to be critically followed. Personnel must wear clean and protective covers/clothing (such as gloves and proper covering of mouth and hairs). The final quality with respect to residues, microbial contamination, and wholesomeness must be checked. Fruit and packaging material must meet quality criteria/demands of the importing country. The technical guidelines of the entire process of handling and treatment must be followed for good results.

4. Labeling: The labels, glues, inks, and packing material used must meet international safety guidelines and requirements of the importer.

5. Testing facilities: The field and laboratory testing facilities must be in place and regular testing of materials and fruit must be performed. The personnel should be trained and equipment and instruments must be calibrated for accuracy in measurement.

6. Documentation and data control: The procedures followed have to be documented, process data has to be recorded in pro forma, and worksheets have to be filled out. Records of raw material quality, pesticide application records, and quality control reports have to be maintained.

7. Quality audit: This has to be carried out by the staff other than those associated with the operation. Auditing should be done with respect to quality of material, building, and machinery.

The quality system emphasizes food (fruit) safety because fruit that is not safe to consume cannot be called good quality fruit, no matter how good it may be to look at or how well it is presented. International standards (Codex Alimentarius Commission of the FAO), as listed in Chapter 7, and the standards of the importing country have to be followed. The sanitary and phytosanitary (SPS) measures from a quarantine point of view must also be taken care of (see Chapters 18 and 22).

Food safety programs are being observed in the Florida citrus industry; GAP is the cornerstone of the same since 1998 (Ritenour et al., 2003). Many packinghouses and fruit growers are following these practices by adopting FDA guidelines. The important practices and procedures followed include: testing soil and water quality, sanitizing orchards, avoiding contamination, monitoring workers' health and hygiene, using covers (gloves, etc.), maintaining sanitary conditions in packinghouse areas, cleaning and maintaining equipment, and managing temperature. Another dimension of food safety per the United States Bioterrorism Act of 2002 is preventing any terrorist attack (bioterrorism) through food (fresh citrus fruit falls under the category of food and hence packinghouse facilities must register with FDI) (Martin, 2003).

The Florida Department of Agriculture and Consumer Services (division of plant industry) requires that packinghouses sign the compliance agreement under the citrus health response program and follow the stipulated rules in the agreement. The sanitation aspects, prevention of spread of pests and diseases, and maintaining healthy citrus fruit are all emphasized in the program. Growers and packers have to conduct citrus business only with those entities that hold valid compliance agreements for farming, propagating, and handling citrus or providing related services. Only approved disposal sites for waste are to be used under the program.

REFERENCES

Abbott, J.A. (1999). Quality measurements of fruits and vegetables. *Postharvest Biol. Technol.* 15, 207–225.

APEDA (1997). Quality assurance manual for export of Kinnow. Agricultural and Processed Food Products Development Authority, New Delhi, 88 pp.

Barzana, E., Klibanov, A.M., and Karle, M. (1989). A colorimetric method for enzymatic analysis of gases: the determination of ethanol and formaldehyde vapors using solid alcohol oxidase. *Anal. Biochem.* 182, 109–115.

Bellon, V. (1993). Tools for fruit and vegetable quality control: a review of current trends and perspectives. In *Postharvesting operations and quality sensing*, Proc. IV Int. Symp. Fruit and Vegetable Product Engineering. Valencia, Spain (F. Juste, ed.), Vol. 2, pp. 1–12.

Ben-Yehoshua, S., Shapiro, B., and Even-Chen, Z. (1981). Mode of action of individual seal packaging in HDPE film in delaying deterioration of lemons and bell pepper fruit. *Proc. Int. Soc. Citric.* Japan, Vol. 2, pp. 718–721.

Chen, P., McCarthy, M.J., and Kauten, R. (1989). NMR for internal evaluation of fruits and vegetables. *Trans. ASAE* 32, 1747–1753.

Conesa, E., Puig, J., Hoddell, S., and Emmanoulopoulus, G. (1993). Citrus fruit life tracking system. In *Postharvesting operations and quality sensing* (F. Juste, ed.), Vol. 2, pp. 165–172.

Grierson, W., and Hayward, F.W. (1959). Evaluation of mechanical separators for cold damaged oranges. *Proc. Am. Soc. Hort. Sci.* 73, 278–288.

Hunter, R.S., and Harold, R.W. (1987). *The measurement of appearance.* Wiley Interscience, New York.

Kawano, S., Sato, T., and Iwamoto, M. (1992). Determination of sugars in Satsuma orange using NIR transmittance. *Proc. 4th Int. NIR Conf.*, pp. 387–393.

Martin, J. (2003). Citrus packing houses must register facility. *Citrus Veg. Mag,* September, p. 33.

Miller, W.M., Peleg, K., and Briggs, P.L. (1986). Computer based inspection of freeze-damaged citrus. American Society of Agricultural Engineers, Publication No. 86, 6554, 12 pp.

Minolta (1994). *Precise color communication.* Minolta Co., Ramsey, NJ.

Miyamoto, K., Kitano., Y., Yamashita, S., Honda, H., and Nakanishi, Y. (1997). Near infra-red spectroscopy for on-line sugar evaluation in Satsuma mandarin. *Proc. Int. Soc. Citric*, Sun City, South Africa, pp. 1126–1128.

Molto, E., Selfa, E., Pons, R., Fornes, I., and Juste, F. (1998). A firmness sensor for quality estimation of individual fruits. *Acta Hort.* 421, 65–72.

Nickerson, D. (1964). *Color measurement and its application to agricultural products.* USDA Misc. Publication No. 580, Washington, DC.

Ritenour, M., Miller, B., and Goodrich, R. (2003). Forty-second Florida packing house day. http//postharvest.ifas.ufl.edu/events/packinghouse/day/2003.

Sarig, Y., Gayer, A., Briteman, B., Israeli, E., and Bendel, P. (1992). Non-destructive seed detection in citrus fruits. *Proc. Int. Soc. Citric.*, Italy, pp. 1036–1039.

Singh, K.K., and Reddy, B.S. (2006). Measurement of mechanical properties of sweet orange. *J. Fd. Sci. Technol.* 42, 442–445.

Ting, S.V., and Blair, J.G. (1965). The relation of specific gravity of whole fruit to the internal quality of the oranges. *Proc. Fla. Sta. Hort. Soc.* 78, 251–260.

Ting, S.V., and Deszyck, E.J. (1960). Determination of mixture of Red No. 32 and citrus Red No. 2 dyes on oranges. *Citrus Mag.* 22, 18, 30, 31.

Ting, S.V., and Rouseff, R.L. (1986). *Citrus fruits and their products: analysis and technology.* Mercel Dekker, Inc., New York, 293 pp.

Zhang, L.R. (1992). A rapid and exact method for measuring orange fruit volume. *Proc. Int. Citric. Soc.*, Vol. 1, pp. 346–350.

20

NUTRITIVE AND MEDICINAL VALUE OF CITRUS FRUITS

The importance of citron (*Citrus medica*) as a fruit with certain medicinal values has been known to man since ancient times. The consumption of *Citrus medica, Citrus jambhiri*, and *Citrus limon* has been emphasized in general for nutritive value for ages in India. Besides fruits of the genus *Citrus*, the fruit of the genus *Aegle* of Rutaceae family Bael (*Aegle marmelos*) is also known to possess medicinal values, and people consume it in India for therapeutic purposes. It is covered in this chapter because it is a close relative of citrus, and its tree, fruit, and leaf characteristics are similar to those of citrus. Citrus fruits provide adequate nutrition with respect to vitamin C; however their role in providing other nutrients and factors of medicinal value cannot be underestimated. Nutrients from a fresh source are immediately available to the body, although in a small amount. Even a small amount of vitamins can prevent the appearance of sub-clinical signs of deficiencies.

I. NUTRITIVE VALUE OF CITRUS FRUIT IN THE HUMAN DIET

Citrus fruit or juice can be an excellent source of health-promoting substances at breakfast. A 150–200 ml glass of orange juice provides many nutrients required for good human health every day. In citrus fruits, carotenoids are mainly associated with pulp and its particles extracted in the juice; hence too much filtration of juice is likely to remove provitamin A activity from the juice. It is advisable to eat fruit rather than to drink its juice since many antioxidants present in the pulp or fibrous part are removed when juice is sieved. Vitamin C content decreases with fruit maturity in oranges, tangerines, and grapefruit. Fruits of early harvest or normal harvest are therefore more healthful than fruits of late season as far as this vitamin is concerned. Conversely, thiamin (vitamin B1) increases from about 0.5–0.6 μg to about 0.75–0.8 μg per gram of juice with maturity in oranges. Similarly, niacin content also increases (Ting, 1977). Folic acid, folate, and folacin are interchangeable forms of the chemical compound pteryl glutamic acid; its deficiency is known to cause a type of anemia. Folic acid is prone to oxidation and is generally protected in fresh fruit because of the antioxidant property of ascorbic acid (vitamin C) (Streiff, 1971). Vitamin E, or α–d–Tocopherol has been reported to be present in oranges (Newhall and Ting 1965; Ting and Newhall, 1965). Citrus juices also provide minerals that are part of the vital enzyme system of the human body. In addition, several compounds – flavonoids, limonoids, and other health-promoting substances such as dietary fibers and pectin – are present in citrus fruit. The recommended dietary allowance for average adults in the United States and India in terms of nutrients available in citrus juices are given in Table 20.1.

The life-extending qualities of citrus fruits (oranges) were endorsed by Hendrikje van Andel-Schipper, a Dutch woman who celebrated being the oldest living person in the world on her 114th birthday in June, 2004. Ms. van Andel-Schipper reported that she drank a glass of orange juice daily (Annon, 2004a). When fresh grapefruit is consumed daily with meals without changing diet results in reduction of weight, the average weight loss is 3.6 lbs in 3 months (Anon, 2004b).

A. Calorific Value

Citrus fruits contain carbohydrates in the form of sugars: sucrose, glucose, and fructose. In mandarins and tangerines, the ratio of sucrose, glucose, and fructose is generally 2:1:1. Total soluble solids in juice consist mainly of sugars. Fibrous rag and many other polysaccharides, which may also provide calories, are also eaten when fresh citrus fruit is consumed. A 150-gram edible portion of orange provides 17 g of carbohydrate that can supply up to 73 kilocalories (Church and Church, 1970). The unit of energy used is physiological calories, or kilocalories. (One calorie is the amount of heat required to raise the temperature of 1 kg of water by 1°C.) Whenever a calorie as a unit is used in nutrition, it is a physiological calorie or a Kcal, which is 1000 times larger than a calorie. One gram of carbohydrate or protein yields 4 Kcal (16.8 KJ). A factor of 4.2 is generally used for conversion of Kcal to KJoule, or KJ. Currently the Kcal is being replaced by KJ. Commonly grown citrus fruits in India provide calories as follows: Bael

TABLE 20.1 Recommended Dietary Allowance and Nutrient Content of Citrus Juices (One Glass)

Nutrient	Recommended dietary allowance essential in human nutrition (USRDA)	Recommended dietary allowance for average adult Indian (60 kg weight)	Orange juice	Tangerine juice	Grapefruit juice
Vit C	90 mg	40 mg	45–50 mg	30–31 mg	30–35 mg
Vit A	5000 IU (900 RAE)	600 μg retinal, β- carotene 2400	190– 400 IU	350–420 IU	Trace–21 IU
Vit D	200 IU	–	–	–	–
Vit E	15 mg	25 μg	100 μg	–	–
Thiamin (B1)	1.5 mg	1.4 mg	50–80 μg	50–80 μg	30–40 μg
Riboflavin	1.7 mg	1.6 mg	20–40 μg	20–40 μg	–
Niacin	20 mg	–	300–600 μg	200–250 μg	200 μg
Calcium	1 g	400–1000 mg	10–11 mg		9–10 mg
Iron	18 mg	28 mg	0.1–0.2 mg	–	–
Vit B6 (Pyridoxine)	2 mg	2.0 mg	47–66 μg	40–50 μg	18–20 μg
Folic acid	0.4 mg	100 μg	34 μg	21 μg	8 μg
Vit B12	5–6 μg	1 μg	–	–	–
Phosphorus	1 g		14–20 mg	16–18 mg	15–18 mg
Iodine	150 μg		0.25 μg	–	–
Magnesium	400 mg		8–12 mg	10–15 mg	8–10 mg
Zinc	15 mg		25–30 μg	–	–
Copper	2 mg		50–160 μg		
Biotin	0.3 mg		–	–	
Pantothenic acid	10 mg		130–150 μg	–	280–300 μg

Food and Nutrition Board (2004); Anon. (1962a); Streiff (1971); Rakieten et al. (1951); Braddock (1972); Watt and Merrill (1963); Birdsall et al. (1961); Ting (1977); Gopalan et al. (1999).

fruit, 137 Kcal; grapefruit (Marsh) 32–45 Kcal; lemon, 57 Kcal; Malta orange, 36 Kcal; Mosambi orange, 43 Kcal; and Nagpur mandarin, 48 Kcal (Gopalan et al., 1999).

Organic acids present in citrus fruits, such as citric acid, malic acid, oxalic acid, succinic acid, and malonic acid also provide calories, and are easily metabolized as they are the part of metabolic pathways in the human body. Citrus fruits do not increase the body's acid content. These acids are very mild compared to the hydrochloric acid present in stomach. Most of the acids are present in the form of salts of potassium (with K or Na cation). Sodium and potassium are alkaline metals; their salts are excreted by the body in the form of sweat or urine.

Citrus fruits have very low fat content and can substitute the snack eaten between meals. Thus a person can avoid eating saturated fats and cholesterol, which increase the risk of heart disease. However, citrus fruits are not a good source of proteins like many other foods, and thus from a nutrition point of view citrus fruits do not form a protein source in the diet.

B. Minerals

Citrus fruits have very high K content (300 mg in 178 ml of orange juice and 200 mg in grapefruit juice), while sodium content is relatively low (3–4 mg/178 ml orange juice and 4.5 mg/178 ml tangerine juice). The ratio of K and Na in orange juice plays an important role in maintaining electrolyte balance. Sodium is understood to play a role in water retention and edema. Dietary levels are 50–150 meq (milliequivalent) per day for potassium (Araujo, 1977). Sodium is lost in urine and particularly in sweat as sodium chloride. In hot, tropical climates, requirements of K and Na may be higher. K and Na are important constituents of fluids present within and outside the cells. A proper balance of electrolytes also maintains osmotic balance of cells and keeps them in shape. A glass of chilled orange juice or few fresh oranges/mandarins are very refreshing in summer and also provide required electrolytes. The RDA for iron in the Indian diet is 28–30 mg per day and should be higher for growing boys and girls. An orange (200 g) provides about 2 mg of iron. Two oranges a days can give 4 mg, which would be more than 10 percent of the RDA in the U.S. The RDA is set assuming a 10 percent rate of intestinal absorption. If the recommendation is 20 mg, the body is expected to get only 2 mg. In this context, iron from an orange is more available than from other sources as the ascorbic acid present in the digestive system increases iron absorption (Layrisse, 1975).

The RDA for calcium in the diet of adult Indian men is 400 mg per day; an orange provides about 2 percent of this amount (Table 20.1). The citric acid in orange juice may act as chelating agent and thus increase calcium absorption by preventing the formation of insoluble salts. In addition to K and Na, calcium, phosphorus, and magnesium are required in higher amounts. Like calcium and phosphorus, magnesium is sequestered in the bone. Magnesium is present in mitochondria and other enzymes important in energy transfer. Zinc, manganese, and copper are also important for the body and supplied by citrus fruits. In an average diet, phytates are sometimes included through some cereals and vegetables. These phytates inhibit zinc and calcium absorption. Ascorbic acid and citric acid increase the absorption of calcium and other minerals. Zinc and manganese are required as a part of various enzymes (prosthetic group). The effects of the chemical composition of fruit juices on the absorption of iron from a rice (*Oryza sativa*) meal were studied by Ballot et al. (1987). The study indicated that ascorbic acid was not the only organic acid responsible for promoting the effects of citrus fruit juices on iron absorption. Iron absorption from laboratory 'orange juice' containing 100 ml water, 33 mg ascorbic acid, and 750 mg citric acid was significantly better than that from 100 ml water and 33 mg ascorbic acid alone (0.097 and 0.059 g, respectively). There was a close correlation between iron absorption and ascorbic acid content. A weaker but still significant correlation with the citric acid content also exists. Although this may reflecte a direct effect of citric acid on iron absorption, it should be kept in mind that fruits containing citric acid also contain ascorbic acid. The negative correlation between iron absorption and the malic acid content of fruits may be due to the fact that fruits with a high malic acid content tend to have low values of ascorbic acid. The presence of citrus fruit is expected to increase iron absorption markedly in diets low in iron.

C. Dietary Fiber and Pectin

An average-size orange (7–8 cm diameter) can provide 0.8 g of fiber in the diet. Drinking of a cup of fresh orange juice provides 0.3 g of fiber. A half grapefruit eaten at breakfast adds 0.2 g of fiber to the meal, thus replacing fiber that has been removed from the breakfast cereal by the milling process (Church and Church, 1970).

Fiber has its own importance for the people of industrialized nations who eat high-fat, low-fiber diets full of highly refined and processed carbohydrates that move slowly through the intestines. In fresh citrus fruit, fiber contains cellulose, hemicellulose, lignin, and pectin – all found in citrus segments, membranes, and other parts of the albedo. Most of these are carbohydrates, except lignin. Lignin is a complex polymer of aromatic compounds and is linked by propyl units. Dietary fibers are not digested because humans cannot produce the required enzymes to break down the polymers. Starch and cellulose are formed from D-glucose, but their bonding differs: the starch comprises α-1.4 linkage, which is hydrolyzed by amylase in the human saliva. Cellulose is formed with β-1.4 linkage, but cellulase enzyme is not produced by humans. Similarly, pectins and hemicellulose cannot be broken down by humans. This is helpful – intact fiber prevents gastrointestinal disorders by easing the motion and rapid passage through the intestine that also minimizes the absorption of harmful compounds in the diet.

Dietary fiber also adsorbs calcium and may increase its availability. The total dietary fiber per 100 g of edible fruit is 1.9 g for Konatsu (an orange grown in Kochi Prefecture, Japan), 1.0 g for mandarin, and 0.9 g for grapefruit. The ratio of insoluble to soluble dietary fiber is about 2 for Konatsu, 1 for mandarin, and 0.3 for grapefruit. Calcium adsorbed in dietary fiber of Konatsu and mandarin oranges at pH 8.0 is 45 and 16 mg, respectively (Nishimura et al., 1992).

II. THERAPEUTIC/MEDICINAL VALUE

Several research findings have indicated that in addition to their thirst-quenching ability and refreshing taste, citrus fruits have therapeutic value. Most citrus fruits have one or more therapeutic values. The acid lime has antiseptic, astringent, and restorative properties. It is a digestive stimulant and encourages the appetite when eaten with meals. It is useful in treating anorexia. Acid lime is a tonic to skin. The fresh essence or aroma of lime is very refreshing and uplifts a tired mind, thereby energizing and revitalizing a depressed person. Mandarin fruit is antispasmodic, sedative, cytophylactic, and digestive. Fresh mandarin calms the intestines and aids in digestion. It is tonic to the liver and its gentle action is suitable for treating hiccups. Mandarin fruit promotes cell generation and its aroma is inspiring and strengthening (Watson, 1994). Fresh, cool orange acts as an antidepressant. It is antispasmodic, stomachic, and sedative as well. Eating orange fruit is good for digestion, eases constipation, and promotes peristalsis. The orange is also good for the circulatory system. Consumption of citrus fruits can lower the risk of cancer and heart diseases.

The methoxylated flavones that occur in low concentrations in citrus have been reported to decrease vascular adhesion of erythrocytes, or the sludging of blood, which is observed in different pathological conditions (Robbins, 1973). Methoxylated flavonoids are much more effective anti-adhesives for the clumping of red blood cells and blood platelets than those with hydroxyl groups. Therefore, tangeritin and nobiletin are more effective than hesperidin. Naringin lowers the increased hematocrits in humans; this can be achieved by simply including grapefruit in the daily diet (Robbins, 1988).

Besides reducing or preventing cell clumping, citrus phenols, flavonoids, and limonoids are also anti-carcinogenic, anti-inflammatory, and anti-allergen (Manthey et al., 2000).

Phenols, flavonoids, and limonoids act as a prototype of a variety of substances present in food. Their mechanism of action in preventing certain human disorders and diseases is being studied worldwide. The effects of flavonoids from citrus juices – particularly those found in oranges and grapefruit – on blood circulation, as well as their anti-allergenic, anti-carcinogenic, and antiviral properties are discussed by Filatova and Kolesnov (1999). Rice Evans et al. (1997) reviewed the antioxidant properties of phenolic compounds. Antioxidative properties have been observed in the polyphenols, phenolic acids, flavonoids, and ascorbic acid of Jaffa Sweeties and grapefruits. The correlation coefficient between the polyphenols and antioxidative activity varies from 0.73 to 0.99. Sweeties have higher antioxidative activity than grapefruits. Diets supplemented with Sweeties, and to a lesser extent with grapefruit, increased the plasma antioxidative potential and improved lipid metabolism, especially in rats fed with added cholesterol (Gorinstein et al., 2003).

Three coumarins from lemon fruit peel – 8-geranyloxypsolaren, 5-geranyloxypsolaren (bergamotin), and 5-geranyloxy-7-methoxycoumarin – have been found to be promising chemopreventive agents by inhibiting radical generation (Miyake et al., 1999). All these isolates markedly suppressed superoxide (O_2-) generation in differentiated human promyelocytic HL-60 cells, while 8-geranyloxypsolaren and 5-geranyloxy-7-methoxycoumarin reduced lipopolysaccharide and interferon-gamma-induced nitric oxide generation in mice.

Lemons, grapefruits, tangerines, and oranges are rich in pectin content. Anhydrogalacturonic acid content is highest in pectin from the segment membranes of tangerine and the flavedo/albedo of grapefruit. Lemon pectin contains the highest methoxyl content (MC). Pectin has been found to significantly inhibit the binding of fibroblast growth factor (FGF-1) to fibroblast growth factor receptor (FGFR1) in the presence of 0.1 μg/ml heparin (Liu et al., 2000). The pectin from the segment membranes of lemons was the most potent inhibitor. Kinetic studies have revealed a competitive nature of pectin inhibition with heparin, which is a crucial component of the FGF signal transduction process. Thus pectins can be effectively utilized as anti-growth factor agents in fibroblasts.

A. Vitamins

During long sea voyages in the 15th and 16th centuries, sailors who carried citrus fruits knowingly or unknowingly for consumption did not develop the disease later called as scurvy, while those who did not consume vitamin C either through citrus fruits or other sources did suffer from the disease. By the 18th century, sailors realized the importance of limes in reducing the risk of scurvy on voyages, and carried limes, oranges, and lemons. The extreme symptoms of scurvy are bleeding gums, cutaneous hemorrhages, and keratinized hair follicles. The milder symptoms are cracking of lip corners, weakness, and dry skin.

Significantly higher lycopene and total carotenoids are reported in red-fleshed grapefruit cultivars such as Star Ruby and Rio Red, while white-fleshed cultivars Duncan and Marsh had a negligible amount of carotenoids. Rio Red grapefruits had higher levels of lycopene, limonin 17-β-D glucopyranoside, total flavonone, and vitamin C levels compared to other cultivars. Fruits at optimum maturity are more nutritious because lycopene content decreases as the harvest season advances. The p-carotene content increases during the harvest season (Patil, 2000). Orange juice contains 0.5 mg/kg β–carotene, which is equivalent to 0.8 IU/g vitamin A (Bauernfeind, 1972). In tangerines, provitamin A is high. Cryptoxanthin, which also has provitamin A activity, is high in mandarins and tangerines. Recent studies at the University of Manchester, U.K. (by Dr. A.J. Silman and his colleagues) indicated that people consuming fruits with more carotenoids (β–cryptoxanthin and zeaxanthin) had 20–40 percent lower risk of arthritis (multiple joint inflammation). The anthocyanins, which are the pigments of blood oranges, also have therapeutic value. It has been observed that the consumption of the juice of blood oranges (cultivar Moro) can modulate the permeability of the blood vessels and induce a protective effect on the gastric mucosa (Saija et al., 1992). On the basis of studies conducted on rats, this juice is reported to elicit an immuno-stimulatory effect. The juice is desirable because it can act as co-adjuvant in the therapy of some circulatory system pathologies. It also increases the capability to react to unfavorable conditions (possibly infections) promptly.

It is estimated that a glass of orange juice (177.4 ml) provides about 100 percent of the recommended daily allowance of vitamin C to the average American diet. The RDA set in 2000 for an average adult of good health is 75–90 mg (Food and Nutrition Board, U.S.A., 2004). A WHO/FAO expert group has recommended a daily allowance of 30 mg vitamin C (FAO/ WHO, 1974). There is a catabolism of stored vitamin C in the body, and the store of vitamin C declines if there is no fresh intake. Baker et al. (1977) reported that scurvy was produced when reserves of ascorbic acid in the body decreased to 300 mg and catabolism was 9 mg per day. As per FAO/WHO standards, the body stores about 1000 mg, or more than 3 times that required to prevent clinical signs of scurvy. A standard serving of orange juice provides 81 mg of vitamin C and results in body reserves of 2700 mg (Araujo, 1977). A higher intake of citrus fruits or any other fruit having very high vitamin C should be avoided, as it may result in disruption of the normal metabolism of vitamin C itself along with other vitamins, especially B-12 and thus side effects may appear (Araujo, 1977).

Consumption of citrus fruits is lower in elderly and lonely (isolated) people who cannot manage a balanced diet, and also in infants. The distribution of citrus fruits to such people can minimize their health problems. Elderly people who drink alcohol and smoke risk damaging their overall metabolism and thus the absorption of vitamins and minerals. Many other sources of vitamins lose vitamins when cooked. A fresh orange or a glass of freshly extracted orange juice provides many healthful compounds in abundance without any loss of vitamins. The acidic pH of citrus fruits increases the absorption of calcium from the intestinal lumen. In elderly people, demineralization of bones is thought to be a result of low levels of calcium in the diet. The low absorption results in osteoporosis. Ascorbic acid is not present in cow's milk and its deficiency is likely in infants who take only milk. Cases of infantile scurvy have been reported by Krehl (1976) in his studies in the U.S. as the Vit 'C' content in breast milk is low.

Among other vitamins present in fresh citrus fruits are compounds of vitamin B complex. Folic acid, which is heat-sensitive and lost in food processing, can be obtained from fresh oranges, mandarins, or grapefruits. Fresh vegetables, pulses, and liver are also good sources, but they are processed and folic acid is lost. Folic acid or folate is required for the multiplication and maturation of red blood corpuscles; its deficiency can result in a type of anemia. Folic acid plays a role in the metabolic pathways in the formation of purine and pyrimidine nucleotides as well as certain amino acid conversions. Since growth requires proteins and nucleic acid synthesis, growing children and pregnant women are more sensitive to its deficiency. The FAO/WHO (1974) recommendation is 200 μg daily. Folacin contains at least one molecule of glutamic acid. As far as absorption is concerned, monoglutamate forms are more absorbable. Since citrus fruits contain monoglutamate forms, they are likely to provide a more absorbable vitamin species than other sources. The 5-methyltetra hydrofolic acid is the natural form of folate in citrus juices. Approximately 50 percent of the folate is in the monoglutamate, which can be directly absorbed by humans. The remaining 50 percent is present as several polyglutamate forms ranging from 3 to 6 glutamic acid residues attached to the pteridine ring. Intestinal enzymes cleave the polyglutamate form to monoglutamate prior to absorption and utilization. Folate reduces neural tube birth defects by up to 75 percent when taken by pregnant women. Folate has also been associated with a reduced risk of heart disease by lowering blood serum homosystine levels. Women of childbearing age and those at the greatest risk of coronary heart diseases need to take enough folate in their diet (Widmer and Stinson, 2000). In general, depending on variety and fruit maturity stage, folic acid in orange juice ranges from 30.90 μg per 177.4 ml serving, which is 7.5 to 22 percent of the USRDA of 400 μg. Thiamin (Vitamin B1) content ranges from 90 to 280 μg in a serving of 177 ml of orange juice, which is 6–18 percent of the USRDA.

Citrus fruits are also a source of the B6 vitamins known as pyridoxal, pyridoxamine, and pyridoxine. These are interchangeable in the body. The coenzyme form of the vitamin i.e. pyridoxal phosphate is required for the metabolism of amino acids, proteins, and fats in the body. Some types of stomatitis and a type of anemia have been shown to be cured by the administration of pyridoxine. The average intake of 0.6–2.5 mg of B6 is considered sufficient for all age groups of Indians (Gopalan et al., 1999). Orange juice (100 ml) contains 25–80 μg of vitamin B6 (Krehl, 1976). The B6 requirement varies depending on the amount of protein eaten, since this vitamin has a role in protein metabolism. The U.S. RDA has been set at 2 mg per day (Food and Nutrition Board, 1974).

B. Role of Citrus Fruit in Reducing Risk of Human Diseases

1. Heart Diseases

Dietary fiber is reported to lower the incidence of ischemic heart diseases (Trowell, 1972). Synthetic fibrous substances have been utilized to promote the lowering of blood cholesterol from the intestinal lumen (Garvin et al., 1965). Resnicov et al. (1991) reported that a type of strict vegetarian diet that is typically very low in saturated fat and dietary cholesterol and high in fiber can help children and adults maintain or achieve desirable blood lipid levels. Grapefruit pectin has been reported to lower the plasma cholesterol nearly 30 percent and improve the ratio of LDL/HDL by 31 percent in swine (Backey et al., 1988). This pectin also reduced plaque formation on the surface of aortas and decreased the narrowing of coronary arteries. There is a 7.6 percent decrease in plasma cholesterol, a 10.8 percent decrease in low-density lipoprotein (cholesterol-carrying lipoprotein, or LDL), and a 9.8 percent decrease in ratio of low-density lipoprotein to HDL in human volunteers given a pectin diet (Cerda et al., 1988). Each 1 percent reduction in blood cholesterol yields an approximately 2 percent reduction in risk of coronary heart disease; thus 7.6 percent decrease translates to a 15 percent decrease in the risk of coronary heart disease. Thus without drastically changing lifestyle and/or diet, cholesterol levels can be decreased and LDL/HDL ratio can be improved by adding fresh grapefruit or its juice to the diet (Attaway and Moore, 1992). Established high cholesterol levels have been lowered and established atherosclerosis has also been retarded in pigs fed on a grapefruit pectin atherogenic diet (Attaway and Moore, 1992). These findings are important for people who have established atherosclerosis and whose only remedy at present is bypass surgery.

Hypocholesterolaemic effects of Kabosu (*Citrus sphaerocarpa*) juice residue (an ethanol-precipitated portion of Kabosu juice) have been reported in lipid metabolism in the serum and liver of stroke-prone, spontaneously hypertensive male rats. The increase in serum total cholesterol was suppressed by Kabosu juice in comparison with the control as a result of the suppression of increases in the VLDL and LDL fractions. Increases in the contents of cholesterol and triglyceride were suppressed in the liver. The activities of enzymes such as cholesterol 7-alpha-hydroxylase [EC 1.14.13. 17] and acyl-CoA: cholesterol acyltransferase [EC 2.3.1.26] were also significantly lower (Ogawa et al., 1998). The principal flavanones of orange and grapefruit juices – hesperetin and naringenin – have been found to lower the overall production of apoB, the structural protein of LDL, in human liver cell line HepG2. This indicates a direct cholesterol-lowering action of minor citrus components in the liver. In humans with moderate hypercholesterolemia, a 4-week treatment with 750 ml of orange juice increased beneficial HDL cholesterol. In HepG2 cells, tangeretin and limonin are effective in lowering apo-B. Tangeretin and limonin were more effective than hesperetin and naringenin (Kurowska et al., 2000). These studies all indicated beneficial aspects of eating oranges and tangerines, showing that certain compounds, mostly flavonones, are promising for lowering LDL cholesterol (so-called 'bad cholesterol') without causing any harmful side effects. Studies conducted by USDA with KGK Synergise Inc, Canada, have also shown that antioxidant compounds (called polymethoxylated flavans, or PMF) are the most potent in reducing cholesterol (Anon, 2004a).

Citrus flavonoids affect capillary fragility, act as anti-platelet agents, and may be important in blood cell clumping, which can lead to coronary thrombosis (Robbins, 1980).

2. Cancer

Limonoids (a group of triterpenoids) in citrus fruits possess important biological activity. The similar chemical structure of citrus limonoids to those of recognized anti-tumor agents led to their evaluation as potential anti-tumor agents in mammalian systems. Studies in mice have shown that citrus limonoids induced a significant amount of the chemical carcinogen–detoxifying enzyme system glutathione–s-transferase in the liver and intestinal mucosa. In-vitro tests with human breast cancer cells have shown that limonoids have significant anti-tumor activity (Manners and Hasegawa, 2000). Flavonoids – nobiletin, tangeretin, hesperetin, and naringenin – are potent inhibitors of cellular proliferation in both estrogen receptor-positive (ER+) and negative (ER−) human breast-cancer cell lines. Animals fed on citrus juices develop fewer and smaller tumors than those given diets containing pure bioflavonoids or control diets (compared to controls, tumor incidence was less than 50 percent). The greater inhibitory actions of the juices is attributed to flavonoids, limonoids, and other components such as vitamin C and hydroxy cinnamic acids. Citrus limonoids such as limonin, nomilin, and limonoids-glycoside mixtures are more potent inhibitors of ER cancer cell proliferation (Gunthrie et al., 2000). Professor B.R. Das and his colleagues at Molecular Biotechnology Division, Institute of Life Sciences, Bhubaneshwar, India, have shown that d-limonene – a chemical found in citrus peel oils – could combat the carcinogenic effects of a chemical such as NDEA (N-nitrosodiethylamine). The injection of d-limonene effectively reversed the carcinogenic transformation of cells in mice.

In general, cancerous growth of tissues takes place with tumor formation; however, the invasiveness of a tumor and the metastatic capability of a tumor's cell population can have greater effect on growth of the disorder. Malignant tumors not only grow but also invade surrounding tissues, and during invasion these tumors cross membranes and enter the circulatory system. Through blood vessels they gain access to various other organs (Poste and Fidler, 1980). Among citrus flavonoids, tangeretin has potential to inhibit the invasion of tumor cells into normal cells. This could hinder or prevent the metastatic cascade effect (Bracke et al., 1989). Detectable levels of tangeretin have been found in liver and kidneys after oral administration via drinking water. In tumor-bearing mice, tangeretin has been found to affect liver invasion. In-vitro study by Middleton (1991) showed that tangeretin and nobiletin markedly inhibited growth of a human squamous-cell carcinoma line (Attaway and Moore, 1992).

Consumption of peel products is associated with reduced risk of squamous-cell carcinoma of the skin as researched by I.A. Hakin and R.B. Harris of the Arizona Cancer Center. Modified citrus pectin was also observed to interfere with the metastasis of other cancer cells. The perillyl alcohol of citrus peel is active in inducing apoptosis in tumor cells without affecting normal cells and can revert tumor cells back to a differentiated state. Applying peel extracts can inhibit the growth of melanoma and other skin cancer cells (Anon, 2004b).

3. Urinary Disorders

Orange juice is also reported to prevent the formation of kidney stones because of the presence of citrates (Anon, 2006). Kidney stones are formed when urine is too concentrated, causing minerals and other chemicals in the urine to bind together. The studies of Dr. Clarita Odvima, UT, South-Western Medical Centre, indicated that orange juice increased the level of citrates in urine and reduced the crystallization of uric acid and calcium oxalate. The study indicated that potassium citrate, a generally prescribed medicine, had gastrointestinal side effects in some patients and therefore orange juice was the better option.

4. Other Ailments

In hay fever and other allergic reactions histamines are released in the blood and antihistamines are given to treat these disorders. Hesperidin, tangeritin, and nobiletin have slight to moderate antihistamine activity (Middleton and Drzewiecki, 1982).

Research reported in the Archives of Ophthalmology reflects that eating oranges and bananas can ward off a form of disease caused by macular degeneration. In this disorder, the blood vessels behind the eye bleed, causing fluid buildup and scarring. The report also stated that about 10 million Americans are suffering from some form of this disease (Anon, 2004c).

III. NUTRITIVE AND MEDICINAL VALUE OF BAEL *(AEGLE MARMELOS)*

The varietal information, availability and fruit characteristics of bael are covered in Chapter 2. Fruits of bael have a hard cover that must be broken to extract a thick pulp with seeds. The fruit contains 56–74 percent pulp of a yellow-orange color, 2–7 percent fiber, 24–40 percent TSS and 0.35–0.75 percent titratable acidity. Vitamin C content varies from 18–21 mg/100 g pulp (Singh, 2001). Bael contains 1110 μg niacin, 55 mg carotene, 30 μg riboflavin, 130 μg thiamin and 1.2 g minerals per 100 g pulp. The pulp is scooped out with a knife and blended with water, diluted, and strained for seeds. The fruit is attributed with therapeutic and nutritional properties (Aiyer, 1956; Anon, 2001). The medicinal properties come from the marmelosin content. Mature fruit is astringent, digestive, and stomachic. The ripe fruit is considered to be healthy for both heart and brain. Bael fruit works like a tonic and is restorative. Consumption is good when a person is suffering from gastrointestinal disease and chronic diarrhea. Traditionally it is used in every Indian household for the treatment of digestion-related problems.

REFERENCES

Aiyer, A.K. (1956). The antiquity of some field and forest flora of India. Bangalore Printing and Publishing Co. Pvt. Ltd., Bangalore.

Anonymous (1962a). Chemistry and technology of citrus. Citrus products and by-products. *USDA, Agric. Handbook,* 68.

Anonymous (1962b). *USDA, Agric. Handbook*, Washington, DC.

Anonymous (2001). Herbs related patents. Bael. Bilva (*Aegle marmelos*) *Bull. Intellectual Property Rights*, 7(3–4), 1–2.

Anonymous (2004a) Healthy citrus. *Citrus Industry* (Florida) 85(8), 40.

Anonymous (2004b) Citrus buzz. *Citrus Industry* (Florida) 85(3), 5–6.

Anonymous (2004c) Citrus and health update. *Citrus Industry* (Florida) 85(9), 10.

Anonymous (2006). Orange juice prevents recurrence of kidney stones. *The Hitavad.*, Daily from Central India. September 2, 2006.

Araujo, P.E. (1977). Role of citrus fruit in Human Nutrition. In '*Citrus science and technology*'. (S. Nagy, P.E. Shaw, M.K., and Veldhuis, (eds.) AVI Publishing Co., WestPort, CT, pp. 1–32.

Attaway, J.A., and Moore, E.L. (1992). Newly discovered health benefits of citrus fruits and juices. *Proc. Int. Citrus. Congress*, Italy, Vol. 3, pp. 1136–1139.

Backey, P.A., Cerda, J.J., Burgin, F.L., Robbins, R.W., and Baumgarmer, T.G. (1988). Grapefruit pectin inhibits hypercholesterolemia and atherosclerosis in miniature swine. *Clin. Cardiol.* 11, 595–600.

Baker, E.M. et al. (1971). Metabolism of 14C and 3H-labeled L-ascorbic acid in human scurvy. *Am. J. Clin. Nutr*. 24, 444–454.

Ballot, D., Baynes, R.D., Bothwell, T.H., Gillooly, M., Macfarlane, B.J., Macphail, A.P., Lyons, G., Derman, D.P., Bezwoda, W.R., Torrance, J.D., Bothwell, J.E., and Mayet, F. (1987). The effects of fruit juices and fruits on the absorption of iron from a rice meal. *Bri J. Nutr.* 57, 331–343.

Bauernfeind, J.C. (1972). Carotenoid vitamin 'A' precursors and analogs in foods and feeds. *J. Agric. Fd. Chem*. 20, 456–473.

Birdsall, J.J., Derse, P.H., and Teply, L.J. (1961). Nutrients in California lemons and oranges. II. Vitamins, minerals and proximate composition. *J. Am. Dietet. Assoc.* 38, 555–559.

Braddock, R.J. (1972). Vitamin 'E' content of commercial citrus juices. *Twenty-third Citrus Processors Meeting*, AREC, Florida.

Church, C.J. and Church, W.A. (1970). *Food values of commonly used portions of fruits and vegetables*. Lippincot Publishing Company, Philadelphia, PA.

FAO/WHO (1974). *Handbook on human nutrition requirements*. FAO Nutritional Study. WHO Monograph Serial No. 61, Rome.

Filatova, I.A., and Kolesnova, Y. (1999). The significance of flavonoids from citrus juices in disease prevention. *Pishchevaya-Promyshlennost*. 8, 62–63.

Food and Nutrition Board (2004). *Recommended dietary allowances*, 9th Revised Edition. National Academy of Sciences, Washington, DC.

Garvin, J.E., Forman, D.T., Eiseman, W.R., and Phillips, C.R. (1965). Lowering of human serum cholesterol by an oral hydrophilic colloid. *Proc. Soc. Exp. Biol. Med*., Vol. 120, pp. 744–746.

Gopalan, C., Sastri, B.V.R., and Balasubramaniam, S.C. (1999). Nutritive value of Indian foods. NIN, ICMR, Hyderabad, p. 156.

Gorinstein, S., Yamamoto, K., Katrich, E., Leontowicz, H., Lojek, A., Leontowicz, M., Ciz, M., Goshev, I., Shalev, U., and Trakhtenberg, S. (2003). Antioxidative properties of Jaffa sweeties and grapefruit and their influence on lipid metabolism and plasma antioxidative potential in rats. *Biosci. Biotechnol. Biochem*. 67, 907–910.

Guthrie, N., Vandenberg, T., Manthey, J.A., Hasegawa, S., and Manners, G. (2000). The effectiveness of the citrus juice components on the inhibition on human breast cancer proliferation. *Int. Citrus Congress*, Florida, December 3–7, 2000, Abstract 72, p. 61.

Krehl, W.A. (1976). *The role of citrus in health and disease*. University of Florida Press, Gainesville.

Kurowska, E.M., Vandenberg, T., Manthey, J.A., Hasegawa, S., and Manners, G. (2000). Regulation of cholesterol metabolism by citrus juices, flavonoids and limonoids. *Int. Citrus Congress*, Florida, December 3–7, 2000, Abstract 69, p. 61.

Larrauri, J.A., Ruperez, P., Borroto, B., and Saura-Calixto, F. (1997). Seasonal changes in the composition and properties of a high dietary fiber powder from grapefruit peel. *J. Sci. Fd. Agric*. 74, 308–312.

Liu Y., Hassan, A., Luo, Y.D., Gardiner, D.T., Gunasekera, R.S., McKeehan, W.L., and Patil, B.S. (2001). Citrus pectin: characterization and inhibitory effect on fibroblast growth factor-receptor interaction. *J. Agric. Fd. Chem*. 49, 3051–3057.

Manners, G.D., and Hasegawa, S. (2000). Citrus limonoids: potential chemoprevetative agents. *Int. Citrus Congress*, Florida, December, 3–7, 2000, Abstract 70, p. 61.

Manthey, J.A., Grohmann, K., and Manthey, C.L. (2000). Anti-inflammatory properties of citrus flavonoides. *Int. Citrus Congress*, Florida, December 3–7, 2000. Abstract 71, p. 61.

Middleton Jr., E., and Drzeweicki, G. (1982). Effects of flavonoids and transitional metal cations on antigen induced histamine release from human basophils. *Biochem. Pharmacol*. 31, 1449–1453.

Miyake, Y., Murakami, A., Sugiyama, Y., Isobe, M., Koshimizu, K., and Ohigashi, H. (1999). Identification of coumarins from lemon fruit (*Citrus limon*) as inhibitors of *in-vitro* tumor promotion and superoxide and nitric oxide generation. *J. Agric. Fd. Chem*. 47, 3151–3157.

Newhall, W.F., and Ting, S.V. (1965). Isolation and identification of α-tocopherol, a vitamin E factor from orange flavedo. *J. Agric. Fd. Chem*. 13, 281–282.

Nishimura, K., Yoshida, N., and Kosaka, K. (1992). Adsorption of calcium on dietary fiber from 'Konatsu', a local orange grown in Kochi Prefecture. *J. Japanese Soc. Nutr. Fd. Sci.* 45, 545–550.

Ogawa, H., Mochizuki, S., and Meguro, T. (1998). Effect of kabosu juice precipitate on lipid metabolism in stroke-prone spontaneously hypertensive rats (SHRSP) fed a cholesterol diet. *J. Japanese Soc. Nutr. Fd. Sci.* 51, 273–278.

Patil, B.S. (2000). Enhancing citrus phytochemicals. *Int. Citrus Congress*, Florida, December, 3–7, 2000, Abstract 67, p. 60.

Poste, G., and Fidler, I.J. (1980). The pathogenesis of cancer metastasis. *Nature* 283, 139–146.

Rekieten, M.L., Newman, B., Falk, K.B., and Miller, J. (1951). Comparison of some constituents in fresh, frozen and freshly squeezed orange juice. *J. Am. Diet. Assoc.* 27, 864–868.

Resnicow, K., Barone, J., Engle, A., Miller, S., Haley, N.J., Fleming, D., and Wynder, E. (1991). Diet and serum lipids in vegan vegetarians: a model for risk reduction. *J. Am. Dietetic Assoc.* (USA). 91, 447–453.

Rice-Evans, C.A., Miller, N.J., and Paganga, G. (1997). Antioxidant properties of phenolic compounds. *Trends Pl. Sci.* 2, 152–159.

Robbins, R.C. (1973). Specificities between blood cell adhesion in human disease and anti adhesive action *in vitro* of methoxylated flavones. *J. Clin. Pharm* 13, 401–407.

Robbins, R.C. (1980). Medical and nutritional aspects of citrus bioflavonoides. In *Citrus nutrition and quality*. (S. Nagy and J.A. Attaway, eds.) ACS Symposium Series 143. Washington, DC, pp. 43–62.

Robbins, R.C. (1988). Ingestion of grapefruit lowers elevated hematocrits in human subjects. *Int. J. Vit. Nutr. Res.* 58, 414–417.

Saija, A., Scalese, M., Lanza, M., Imbesi, A., Princi, P., and Di-Giacomo, M. (1992). Anthocyanins of 'Moro' orange fruit juice: pharmocological aspects. *Proc. Int. Soc.Citric.*, Italy, pp. 1127–1129.

Singh, I.S. (2001). Minor fruit and their uses. *Indian J. Hort.* 58, 178–182.

Streiff, R.R. (1971). Folate levels in citrus and other juices. *Am. J. Clinic. Nutr.* 24, 1390–1392.

Ting, S.V. (1977). Nutrient labeling of citrus products. In Citrus Science and Technology, (S. Nagy, S. Shaw and Veldhuise, Eds.) AVI Publishers, Connecticut, Vol. 2, pp. 401–444.

Ting, S.V., et al., and Newhall, W.F. (1965). Occurrence of natural anti–oxidants in citrus fruits. *J. Fd. Sci.* 30, 57–62.

Ting, S.V. et al. (1974). Nutrient assay of Florida frozen concentrated orange juice for nutrition labeling. *Proc. Fla. Sta. Hort. Soc.* 87, 206–209.

Trowell, H. (1972). Ischemic heart disease and dietary fiber. *Am. J. Clin. Nutr.* 25, 926–932.

Watson, F. (1994). Aromatherapy blends and remedies. Thorsons, An imprint of Harper Collins Publishers, San Francisco, CA, pp. 270.

Watt, A.K., and Merill, B.L. (1963). Composition of foods: fresh and processed. *USDA Handbook* 8, 190 pp.

Widmer, W. W. and Stinson, W.S. (2000). Health benefits of folate in orange juice. *Int. Citrus Congress*, Florida, December 3–7, Abstract 73, p. 62.

21

BIOTECHNOLOGICAL APPLICATIONS IN FRESH CITRUS FRUIT

Recent developments in the fields of biotechnology, biochemistry, and molecular genetics have opened up avenues for genetically modified citrus cultivars with improved fruit size, better organoleptic qualities (appearance, flavor, and firmness), higher nutritive value (vitamin content) and physiological benefits (reduced respiration rate or increased wax deposition for reduced water loss). Metabolic pathways can be modified with certain enzymes, forming desired metabolites to get rid of certain problems related to fruit flavor and eating quality. Markets are demanding new fruit varieties with improved functional attributes in the form of nutritional, organoleptic, chemical, and physical properties. The potential is great but its realization depends on our understanding of these traits at the biochemical and genetic level and our ability to modify them. This chapter briefly discusses these possibilities.

To pave the way for transgenic citrus varieties and genetically modified (GM) citrus, genome mapping can reveal information about genes related to fruit quality and the development of damage and disorders (Roose et al., 2000). This will allow for the manipulation of genes and produce varieties having desired color, flavor, handling ability. and shelf life. In the GM plant, the specific sequence of DNA from one plant or organism is modified in the laboratory and transformed into a new plant with a specific DNA sequence that does not occur naturally. Therefore it is a plant with new genetic material (DNA) in it. Expression of interest gene is also over or down regulated. 'Genetically engineered' (GE) or 'genetically modified' (GM) are the preferred terminologies; they are used interchangeably and basically have the same meaning. Transgenic plants contain some genetic material from other plant species introduced through process of transformation.

Molecular approaches to crop improvement using t-DNA of *Agrobacterium tumefacience* to introduce foreign DNA into the genome of higher plants in a permanent and heritable fashion has opened up new vistas for the stable integration of foreign genes into other plants (transformation). Desired DNA sequences can be isolated from libraries. After the polypeptide that conditions the desired trait has been isolated and sequenced, it becomes possible to deduce the nucleotide sequence of an oligomer, which could encode a protein. The oligomer is then used to locate the genomic sequence based on binding affinity (Romig and Orton, 1989). There could be several other approaches for GM or GE citrus plants.

This new tool of biotechnology can be very helpful in improving existing fresh and processed citrus varieties in number of ways. Limonoids are one of the bitter principals in citrus fruit. Excessive bitterness lowers the quality and value of citrus. The citrus species *Citrus ichangensis* contains very high concentrations of ichangensin and deacetylenomilin (both are non-bitter) and very low levels of limonin (bitter). The ratio of non-bitter to bitter compounds is 50:1 (Hasegawa, 1989). Ichangensin and deacetylenomilin are formed from nomilin; this reaction is catalyzed by the enzyme nomilin acetyle esterase, which is unique to this species. In most other citrus species, nomilin is converted to obacunone by the enzyme nomilin acetyle-lyase, and then the obacunone is converted to limonin. A pathway of ichangensin formation from nomilin through acetyle esterase transcription is possible.

$$\text{Acetate} \rightarrow \text{Nomilin} \rightarrow \text{deacetylenomilin} \rightarrow \text{Ichangensin}$$

If the genes for the pathway from nomilin to ichangensin or from nomilin to deacetylenomilin (nomilin acetyle esterase protein) were transferred to commercial citrus fruits such as Navel or Hamlin oranges or some varieties of mandarins with relatively high bitterness, the non-bitter ichangensin or deacetylenomilin might accumulate, resulting in non-bitter fruits. Research on the creation of transgenic citrus trees that produce fruits free of the limonoid bitterness problem is in progress (Hasegawa and Miyake, 1996).

The quality of grapefruits can be improved through the enhancement of levels of β-carotene, lycopene, key flavor components, and through control of the ripening and senescence processes. The isolation of genes responsible for pigment formation/accumulation in plant materials and subsequent transformation with active genes for up-modulation could enhance specific colors (Romig and Orton, 1989). Recently, Ikoma et al. (2001) isolated a cDNA clone encoding a protein homologous to phytoene (carotenoid) synthase (PSY, CitPSY1) from a Satsuma mandarin. Expression analysis revealed that such a transcript is present in the fruits and its accumulation coincides with increased carotenoid synthesis. This kind of work may help lead to development of better-colored citrus fruits.

The enzymes involved in depolymerization of cell wall components could be modulated. Polygalacturonase (PG) is the major enzyme responsible for the depolymerization of cell walls and the softening of fruit tissues. Inhibition of the expression of the endogenous developmentally regulated gene for PG in transgenic citrus expressing antisense mRNA might be useful. Similarly, studies to understand molecular mechanisms that underlie chilling injury of some cultivars of citrus fruits have been conducted (Sanchez-Bellesta et al., 2001). The full-length cDNA, 3c1, has been obtained. This transcript is related to development of chilling injury in the peel and may serve as a molecular marker.

The manipulation of ethylene synthesis and sensitivity may have an application in extending storage/shelf life of citrus fruits. Ethylene synthesis is generally limited by the supply of precursor (ACC, 1-aminocyclopropapane-1-carboxylic acid). The control of ethylene-forming enzyme or ACC synthase could control the induction of softening and ripening. In tomatoes, delayed ripening in some varieties is based on strategies to block the ethylene-biosynthetic pathway (ACC deaminase and antisense/cosuppressed ACC synthase) that is essential for ripening (Clark et al., 2004). Another approach could be identifying spontaneous mutants with relation to ripening in citrus fruits; they are already known in tomatoes. The DNA sequences of these genes can be transferred to cultivated varieties for slower ripening and better qualities.

The possibility of new genetically engineered citrus plants with the following attributes might be possible in long run.

1. Increased soluble solids and sugars in fruits: This is possible with modified sugar metabolism and ripening.

2. Increased resistance to insect-pests: Resistance – particularly to fruit flies, thrips, and fruit-sucking insect-pests – could go a long way in providing a respite to citrus growers incurring huge economic losses because of these pests.

The peels of citrus fruits are not eaten. Several insect-pests affect the market value of fruits, and insecticide sprays are needed to control these pests. It needs to be seen if a genetically modified fruit variety can be developed using *Basillus thuringiensis* (a soil bacteria), that causes the peel to produce toxic proteins to ward off insect-pests so that use of insecticides is reduced and peel blemishes are minimized.

3. Increased resistance to postharvest fungal pathogens: It may be possible to transform fruits with genes that encode enzymes to degrade fungal cells/mycelium.

4. Increased carotenoid content of the fruit rind and pulp: A higher intake of carotenoids is reported to reduce the risk of cancer (Colditz, 1987), and this nutritional advantage could be attractive from a marketing standpoint.

5. Better flavor: Secondary metabolites such as volatiles and aromatic compounds are quite responsive to cell-culture manipulation. Somaclonal variants could be generated with higher levels of desired compounds (Whitaker and Hashimoto, 1986).

6. Increased resistance to canker: The canker is destroying millions of citrus trees all over the world, rendering fruit unmarketable and unfit for export. This is a major quarantine problem. A transgenic acid lime resistant to canker would be a dream come true for the citrus industry worldwide.

7. Seedlessness: Genes that induce seedlessness in fruit are being studied (Koltunow et al., 2000).

Citrus fruit crops are commercially important with annual world production valued at more than 6.5 billion US$. The development of viable GM fresh citrus varieties could be a slow, expensive, and time-consuming process, but its advantages are many. Because citrus is a vegetatively propagated tree crop, the chances of changes in the genetic makeup of a new genetically transformed variety would be less. Although there are technical, economic, regulatory, and market hurdles in the use of genetic engineering in citrus culture, the potential is tremendous, particularly for disease- and insect-pest-resistant GM/transgenic citrus varieties. Considering demand, a GM citrus variety targeting a particular market for fresh fruit might become a success. A higher-quality fruit bringing tangible value to the consumer could improve the market acceptance of biotech citrus crops. The regulatory hurdles will ease as pressure for increased production and nutritive value to cater to growing population increase in the future. As of today, GM foods have not been shown to cause any allergic reactions (Royal Horticultural society, 2002). Studies have indicated that GM foods are not harmful, and the use of specific viral DNA sequences in GM plants have been shown to cause negligible or no risk to human health. It has also been reported that well-characterized transgenes in food intake and their possible transfer to mammalian cells is not likely to have deleterious effects on the biological system of humans. The European Union, North America, Japan, Australia, and other industrially developed nations would be the major markets for GM fruits, although there is a labeling movement for such fruits/products. The United States Environment Protection Agency (USEPA) has recently deregulated the two transgenic papaya varieties Sun Up (homozygous for the coat protein) and Rainbow (F1 hybrid of Sun Up and non-transgenic Kapoho) to be sold in the U.S. (USEPA, 2004). These varieties, which were developed for resistance against papaya ring-spot virus (Farreira et al., 2002), are promising against the disease, which almost wiped out the Hawaiian papaya industry. The cost/benefit ratio will be the determining factor as to whether transgenic or GM citrus is going to be a commercial reality.

Transformation technology has been developed and trait evaluation is underway for citrus tristeza virus (Moore et al., 2000). Similarly, resistance to fruit fly is possible; it has already been developed for codling moth in walnuts and apples (Clark et al., 2004).

REFERENCES

Clark, D., Klee, H., and Dandekar, A. (2004). Despite benefits, commercialization of transgenic horticultural crops lags. *Calif. Agric.* 58(2), 89–98.

Colditz, G.A. (1987). *Horticulture and human health.* (B. Quebedeaux, and F. Bliss, eds.). Prentice Hall Publisher, Eaglewood Cliffs, NJ, pp. 150–157.

Farreira, S.A., Pitz, K.Y., and Marishardt, R. (2002). Virus-coat-protein transgenic papaya provides practical control of papaya ring spot virus in Hawaii. *Pl. Disease* 86, 101–105.

Hasegawa, S. (1989). Biochemistry and biological removal of limonoid bitterness in citrus juices. In '*Quality factors of fruits and vegetables*' (J.J. Jen, ed.). ACS Symp. Series 405. *National meeting of ACS*, Los Angeles, California,1988, pp. 84–97.

Hasegawa, S., and Miyake, M. (1996). Biochemistry and biological functions of citrus limonoids. *Fd. Rev. Int.* (USA) 12, 413–435.

Ikoma, Y., Komatsu, A., Kita, M., Ogawa, K., Omura, M., Yano, M., and Moriguchi, T. (2001). Expression of a phytoene synthase gene and characteristic carotenoid accumulation during citrus fruit development. *Physiol. Plant.* 111, 232–238.

Koltunow, A.M., Smith, A., and Sykes, S.R. (2000). Molecular and conventional breeding strategy for seedless citrus. *Acta Hort.* No. 535.

Moore, G.A., Febres, V.J., Niblett, C.J., Luth, D., Macffrey, M., and Garnsey, S.M. (2000). Agrobacterium mediated transformation of grapefruit with gene from citrus tristeza virus. *Acta Hort.* 535.

Romig, W.R., and Orton, T.J. (1989). Application of biotechnology to the improvement of quality of fruit and vegetables. In '*Quality factors of fruits and vegetables*' (J.J. Jen, ed.), ACS Symp. Series 405. *National Meeting of ACS*, Los Angeles, California, 1988, pp. 381–393.

Roose, M.L., Feng, D. Cheng, R.I., Tayyar, R.I., Federici, R.S., and Kupper, R.S. (2000). Mapping the citrus genome. *Acta Hort.* No. 535.

Royal Horticultural Society, 2002. Genetically modified plants for food use and human health – an up-date, Royal Society, London. www.royalsoc.ac.uk/gmplants.

Sanchez-Ballesta, M.T., Lafuente, M.T., Granell, A., and Zacarias, I. (2001). Isolation and expression of a citrus cDNA related to peel damage caused by postharvest stress conditions. *Acta Hort.* 553, 293–295.

USEPA (2004). United States Environment Protection Agency. Biotech Regulations. http://usbiotechreg.nbil.gov; http://vm.cfsan.fda.gov.

Whitaker, R., and Hashimoto, T. (1986). *Handbook of plant cell culture* (D.A. Evans, W.R. Sharp, and P.V. Ammirato, eds.). MacMillan Publishing Co., New York, pp. 264–286.

22

WORLD FRESH CITRUS TRADE AND QUARANTINE ISSUES

International trade and marketing of citrus has expanded over the years as the transport infrastructure and postharvest techniques have improved and production technologies advanced. Oranges and tangerines/mandarins are the most important citrus fruits around the world with an output of 80 percent of total citrus production. The acid (lime/lemon) group constitutes 11–12 percent and pummelo/grapefruit group about 6–7 percent. This proportion is reflected in fresh fruit trade as well, since mandarins and oranges contribute most to trade volume. Fruits are available almost year-round either from Northern or Southern Hemisphere. International brands such as Sunkist (California), Capespan (South Africa, EU), Jaffa (Israel), Morocco (Morocco) and Florida orange (Florida) obviously come to mind when one talks about fresh citrus. Capespan markets brands such as Cape, Outspan, Bellanova, and Goldland, mainly in Europe. Some products are trademarked, such as ClemenGold, an easy-peeler hybrid of Murcott and Clementine. Brands have improved and stabilized the relations between producers/marketing people and consumers. These brands differentiate among particular fruit on the basis of quality and origin; this also guarantees the fruit quality.

North America (especially the U.S. and Mexico), Latin America (Brazil, Argentina, Uruguay), the Mediterranean countries, China, India, and South Africa are the major producing centers of citrus in the world. Non-citrus producing countries/regions such as Western Europe, Russia and many former Soviet republics, and Canada are the major consuming markets willing to pay a premium for quality fruit. Imports to China are likely to increase as well in coming years with its economic growth.

In world trade, the value of citrus (fresh and processed) reached over 6.5 billion US$ in 1990, with fresh citrus accounting for more than 50 percent of the amount (Chang, 1992). There has been a substantial increase in citrus production in China and Brazil since the beginning of the 1990s. This has been attributed to new plantations in China and bumper crops in Brazil. An increase in citrus production in India during 1990 and 2003–2004 was around 1 million tons. Spain, South Africa, and the U.S. are the major producers of sweet oranges for fresh fruit market. Spain, Morocco, China, India, Turkey, and Japan are the important producers of tangerine/mandarin-type fruits. The most of acid limes are produced in Mexico, followed by India. Argentina, the U.S., South Africa, Italy, and Spain are the leading countries in lemon production. The U.S., South Africa, Argentina, and Israel are the major producers of grapefruits in the world.

Of the world total citrus production (94.79 million tons), 26.63 million tons were processed, amounting to 28 percent during 2004–2005 (FAO, 2006). Of the 26.63 million tons processed, about 21 million tons were oranges, 1.83 million tons were tangerines, and 2.11 million tons were limes and lemons. This trend indicates that most of the mandarin/tangerines and more than half of the oranges are consumed fresh worldwide.

I. EXPORTS, IMPORTS, AND WORLD TRADE

A. Fresh Citrus Consumption Trend

Like all other agricultural crops, citrus is affected by the cycle of production and supply. Prices decline with increased supply, which slows down growth in terms of new plantings and improvements in productivity. However, keeping in mind changes resulting from liberalization and globalization and the new WTO regime in trade, it is likely that citrus production touches new heights in spite of unintentional contractions in supply because of biotic and abiotic stresses. With increased health awareness and changing lifestyles, healthful citrus fruits are likely to make a dent in people's food habits. Technological advances in production, transportation, and logistics; improvements in infrastructure facilities; and communication will allow importers to provide steady supplies of fresh citrus to consumers.

Consumption of fresh citrus is relatively high in developed countries. During the early 1990s in Western Europe, fresh citrus fruit consumption was about 40kg/person/year, while in Eastern Europe it was 5kg/person/year. The lowest consumption is in Asian and African countries. In southeast Asia, per capita citrus consumption during the late 1980s was about 4.2kg/person/year (Aubert, 1991), much lower than the threshold of 10kg/person/year. In China, fresh citrus consumption is 6kg/person/year. Thailand is the only country in Southeast Asia with consumption of 11.2 kg/capita/year after deducting exported mandarins and pummelos.

With increased production over the years, the consumption of fresh citrus increased in absolute terms, but it decreased on per capita basis. The factors responsible for this situation are mainly population growth and lack of buying power in developing and underdeveloped countries. In addition, price competitiveness with respect to other citrus fruit, a lack of convenience, fruit image, and difficulties in market access are important factors in the developed world (Chang, 1992). The access to markets is restricted by tariff and non-tariff barriers between different countries. Market-oriented strategies are needed to replace production-led ones. Many citrus-producing countries now have started developing varieties that take market preferences into consideration.

The citrus industry globally looks forward to supplying healthy, good-looking, high-value, high-quality fruit to demanding consumers. Europe determines the survival and growth of the industry because it is the biggest paying market for fresh as well as processed citrus. A continuous search for better varieties and better handling practices are necessary to maintain sustained growth. Success in retail trade depends on having the right type of fruit in the right place at the right time at the right price (Taylor, 1997). New mandarin types are attracting more consumers. These include Fortunes, Novas, Clemendores, and seedless Murcotts. Bright, reddish-orange, and seedless 'easy peelers' (tangerines) are preferred. The Spanish varieties Clausellina, Marisole, Fortuna, and Nova; and Moroccan varieties Nour, Nule, and Fina are the clementines that fetch better returns.

B. Major Exporting Countries

Roughly 11 million tons of fresh citrus fruits were exported in 2004–05 from different citrus-producing countries. Spain leads the world in fresh orange and tangerine/mandarin exports. About 1.53 million tons of tangerine/mandarins were exported by Spain during 2004–2005 (Table 22.1). Turkey, China, and Morocco are the other major mandarin exporters, each exporting about 0.25–0.30 million tons of fruit. Pakistan exported 94 000 tons of mandarins, mainly Kinnows, in 2002–2003. Spain, Argentina, and Mexico were the major exporters of limes and lemons in 2002–2003. Spain also exported the majority of oranges (1.17 million tons) in the world in 2004–2005.

Countries in the Mediterranean region, such as Italy, Spain, Morocco, Algeria, Cyprus, Greece, Turkey, Egypt, and Israel, contribute to more than half of the world trade in fresh citrus (Table 22.1). Fresh orange fruit exports are relatively higher from Mediterranean regions, especially Spain, compared to countries in the Southern Hemisphere such as Australia, South Africa, Brazil, and Argentina. Mexico, Argentina, South Africa, and Spain are major exporters of limes and lemons. Spain tops the world in mandarin exports: its exports are more than those of Morocco, Japan, and China combined. Spain is expected to expand its production of mandarins/tangerines. Japan exports mostly Satsumas to the

TABLE 22.1 Fresh Citrus Exports by Important Countries during 2004–2005 (Thousand Tons)

Countries	Total citrus	Oranges	Tangerines/ mandarins	Lime and lemons	Grapefruit/ pummelo
Spain	3117.0	1179.2	1533.3	371.5	32.8
South Africa	1080.0	765.2	–	120.0	200.0
Egypt	604.7	576.0	12.1	18.1	–
Turkey	877.0	168.6	282.2	341.0	87.1
Greece	235.4	210.0	18.1	5.2	–
Italy	158.3	90.2	39.2	26.4	–
Israel	172.9	37.2	44.8	–	87.9
Morocco	489.9	236.6	252.7	–	–
Argentina	642.0	150.0	75.0	380.2	37.0
Australia	125.0	125.0	29.6	1.1	–
China	386.0	35.0	351.0	–	15.0
United States	919.0	575.0	19.0	98.0	227.0
Mexico	413.0	20.0	–	382.0	11.0
Brazil	96.3	41.0	11.0	37.3	–
Uruguay	130.0	77.0	31.0	13.2	2.1

FAO (2006), N.A. – not Available – represent very small quantities.

U.S., Canada, and some – Southeast Asian countries. China exports Ponkan and Satsuma mandarins.

The U.S. is the leading exporter of the fresh citrus, mainly oranges (Navels) and grapefruits. The leading exporting state is California because of its excellent quality of oranges, lemons, and grapefruits. Florida exports white and red grapefruits to Pacific Rim countries and Europe. Mexico, Argentina, Brazil, and Uruguay are the other important exporters in the region. Mexico exports limes, lemons, and oranges to Japan. The U.S. leads the world in fresh grapefruit exports, with its major supplies going to Japan and South Korea. The U.S. is also an important exporter of oranges and lemons to Canada, Japan, and South Korea. The Inter-American Citrus Network (IACNET) has been set up with the active, self-supporting participation of the countries of Latin America and North America. These countries are trying to resolve problems of demand and supply within the group and in the world (Albrigo and Menini, 1992).

The Australian citrus industry is also growing, with 0.71 million tons of production and exports of about 6 percent of the citrus fruit in early 1990s. About 10 percent or more of Australian fresh citrus fruit was projected to be exported during 2000 and following years. Seedless easy peelers from Australia are in demand in Singapore, Malaysia, China, and New Zealand. Navel oranges are also in demand for their quality (Gallasch, 1992).

Israel is a prominent supplier of fresh citrus to the European Union. The Citrus Marketing Board of Israel (CMBI), with the established Jaffa brand, is a strong and credible supplier. Israeli fruit is sold through the largest supermarket chains in Europe. Shamouti, Valencia, Navels, red grapefruits, and easy peelers, along with exotics such as kumquats, sweeties, pummelos, and limes are the products marketed by CMBI (Weinberg, 1988). The Israeli Suntina mandarin is late-maturing and therefore strategically marketed in April, when most mandarin types from other parts of the world are slowly declining in supply. Israel grows a wide range of easy-peelers such as Orah (January–May), Topaz (March May), Temple (February–March), Morcott (February–March), and Winnola (February–April), and fruits are marketed with extended storage life using advanced storage technologies.

The Citrus Growers Association (CGA) of South Africa is an organization that is a cooperative of 400 growers, and has almost completely controlled that country's citrus production. Outspan has amalgamated with Unifruco and markets citrus as Capespan. The citrus industry of South Africa exists to supply the export market. About 60 percent of the produce is exported and exports earn more than 90 percent of farming income (Stanbury, 1997). The important fresh citrus fruits exported by South Africa are Star Ruby and Flame grapefruits, Navel and Valencia oranges, and specialty easy-peelers.

China's citrus production is growing steadily with a 1.3 million-ha area under citrus in 2001 (Xin, 2001). New plantations have yet to bear fruit. The quality and quantity of citrus fruits produced is influenced by government programs taking full advantage of modern technology. The total citrus production is expected to exceed 15 million tons annually in coming years. High-quality Navels and Ponkan production is going to increase with pummelo maintaining its proportion. China is going to increase storage facilities and its varietal structure (harvest season) to supply fruit after February–March from its own production, considering the demand for fresh citrus fruit. Mandarins constitute Satsumas (early) and Ponkan, which account for nearly 55 percent of the total production followed by oranges at 30 percent. Seedless pummelos and kumquats and other citrus comprise the final 10 percent. About 95 percent fruit is consumed fresh in the country.

From 1970 to 2000, China moved from producing 1 percent of the world's citrus to around 16 percent (in 2004–2005). Exports of tangerine/mandarin-type fruits from China have undergone rapid growth in 1990s, with Russia and countries of Southeast Asia being the major markets. Chinese citrus is mostly exported to Philippines, Malaysia, Vietnam, Indonesia, and Singapore in Southeast Asia. Problems in exports from China are poor quality (mostly appearance), cheap packaging, and variable eating quality.

In South Asia, Pakistan produces and exports large quantities of Kinnow mandarins to the countries of the Middle East, including the U.A.E., Saudi Arabia, and Kuwait. Fruits are also being sent to the U.K. Many other countries of South Asia (especially SAARC) have trade through land routes, such as the export of Bhutanese mandarins to India and Bangladesh. Khasi and Nagpur mandarins are exported from India to Nepal and Bangladesh in traditional wooden boxes with paddy straw cushioning over land routes.

The potential of exports, particularly mandarins and acid limes, has increased with the increase in production in India. Most of the citrus fruits produced in the country are consumed fresh domestically. Mandarins, sweet oranges, limes, and lemons have long been exported; 3912 tons of citrus worth 1.46 million rupees were exported way back in the early 1960s (DMI, 1965). During the 1970s, trial shipments of graded and waxed Coorg mandarins (from Coorg, Karnataka) were sent to Singapore in good condition in refrigerated cargo ships (Anandswamy and Venkatsubbaiah, 1976). Nearly 7000 metric tons of citrus valued at Rs. 42.6 million were exported during 1993–1994, and in 1995–1996 exports jumped to 18 367 tons valued at Rs. 143.0 million (APEDA, 1997). In 2002–2003, nearly 31 525 tons of fresh citrus were exported valued at 339.53 million rupees.

Mandarins constitute the bulk of citrus exports from India (Table 22.2). Efforts are continuing and several successful attempts have been made to export Kinnows and Nagpur mandarins in reefer containers. As the infrastructure is gradually becoming available, coupled with developments in scientific postharvest handling techniques, exports by reefer container are also rising. Nearly 300 metric tons of Nagpur mandarin fruits have been exported by refrigerated containers to Sri Lanka, the U.K., and other European countries in 1995–1996. About 374 tons of Kinnow fruit were also exported to these countries from the Ganganagar district of Rajasthan. In 1993–1994, Punjab Agro Industries Corporation exported 200 metric tons of Kinnow to the U.K., Dubai, Mauritius, and Sri Lanka.

The harvesting season can provide some advantage for the export of citrus from India, particularly in the case of the monsoon crop. Similarly, acid fruits (limes and lemons) can be exported during most part of the year. Considering the geographical location of India, most of the distant markets in Europe can be reached within 28 days by sea routes. Efficient services of reefer containers and containerships are required to minimize time. The markets of Southeast Asia in the East and Gulf countries in the West can be reached by sea route within 10 days of harvesting the fruit.

TABLE 22.2 Quantity (Metric Tons) and Value (Million Indian Rupees) of Fresh Citrus Exported from India

Citrus fruit	2000–2001 (Quantity)	2000–2001 (Value)	2001–2002 (Quantity)	2001–2002 (Value)	2002–2003 (Quantity)	2002–2003 (Value)
Mandarins	26 837.06	273.93	28 588.76	318.75	27 484.71	284.67
Limes/lemons	3526.96	57.48	4295.81	67.21	3156.08	44.71
Grapefruit	219.78	8.47	170.05	2.82	32.82	1.41
Other fresh citrus	78.85	0.78	116.35	2.752	852.13	8.74
Total	30 662.59	340.66	33 170.97	391.53	31 525.74	339.53

Source: APEDA (2003).

Some of the constraints in fresh citrus fruit exports from India are: (1) lack of availability of exportable quality fruit: Only 10–20 percent of citrus fruits, particularly Nagpur mandarins, produced in the country fulfill the quality requirement for export. In the case of Khasi and Kinnow mandarins this percentage is relatively high. The production of exportable fruit is likely to increase in the near future as the awareness of quality increases among growers, (2) limited varietal base: More varieties need to be developed as per market demand, (3) insufficient infrastructure: Packinghouses, precooling units, and transport vehicles are not available as needed, (4) lack of export promotion and market development campaigns: Marketing and advertising campaigns for citrus fruits are required, (5) very low productivity: Prices of citrus fruits are high in the domestic market because of low productivity; hence, export is not considered lucrative.

A package of improved agronomic practices for the production of exportable-quality Nagpur mandarins is being adopted by growers. Control measures for insects-pests and diseases are being followed and IPM programs have been introduced in the fields. The postharvest package of practices for handling fruit on the packing line were developed for Nagpur mandarin for export by reefer container (Ladaniya and Dass, 1993; Ladaniya and Singh, 1999). Quality-assurance programs based on ISO 9000 guidelines have also been developed for Kinnow and Nagpur mandarins (Chapter 19 part IV).

C. Major Importing Countries

The European Union, Canada, Russia, the Persian Gulf countries, Southeast Asian countries, China, and Japan are the major importers of fresh citrus in the world (Table 22.3). Countries of the former USSR also import sizable quantity of oranges, tangerines, and limes. There was an increase in imports of mandarin fruits in some European countries during 1990s. Major importers of mandarin/tangerine-type fruits are Germany, France, the U.K. The Netherlands, and the former U.S.S.R. Canada and many other European countries such as the Netherlands, Poland, and Belgium are also importing a sizable quantity of easy-peelers. France, the U.K. the Netherlands, Japan, and Germany are also major importers of fresh grapefruit from South Africa, the U.S., and Israel.

The European Union is now a very large block of countries with its own currency and its own rules and regulations to satisfy member countries. There is a kind of protectionism for agriculture to some extent for member countries; therefore produce from other countries has to face stiff import duties. The member countries have some privileges. They have no reference price and no import duty so their fruit is obviously cheaper. Available statistics indicate that the importation of mandarin fruits has increased in European countries. Varieties such as Clementine and Satsuma are predominant in trade because they are seedless, easy-peeler and have an attractive color.

Mediterranean countries that are members of the European Union import very small quantities of fresh citrus, since they are the major producers. This is

TABLE 22.3 Major Importing Countries of Fresh Citrus Fruit in the World (2004–2005) (Thousand Tons)

Country	Sweet orange	Tangerine/ mandarin	Lime and lemon	Grapefruit
Germany	613.2	404.4	143.1	65.8
France	397.5	320.6	117.1	94.4
Netherlands	348.9	158.9	95.3	89.8
U.K.	359.4	303.5	90.9	60.1
Japan	119.0	–	74.0	219.0
Belgium	189.8	89.2	59.7	57.0
Poland	–	159.0	–	–
Canada	232.4	106.2	35.8	97.7
Former U.S.S.R., mainly Russia	571.3	462.7	254.0	60.2
U.S.A.	–	–	326.0	–
Saudi Arabia	319.0	56.8	59.0	–
Malaysia	85.8	59.2	–	–
Singapore	41.0	19.0	–	–
China	164.1	14.3	25.0	–
South Korea	123.0	–	–	–

FAO (2006).

similar to some of the countries in the Southern Hemisphere: Australia, South Africa, Brazil, Argentina, and Uruguay.

The demands of the European market include (1) quality and freshness, (2) regularity of supplies (at the required size of the product) (3) correct net weight and correct count, (4) modern packaging, (5) advertising, (6) Sales promotion, (7) public relations, (8) bar coding (Pape, 1988). The Western European market is dynamic and competitive. It is competitive because new players from the Southern Hemisphere and new members of the EU are posing a challenge for established citrus fruit suppliers. The market is dynamic because needs are changing. New exotics or new types of citrus with attractive color, unique flavor, convenience in eating, a sweet taste, and easy-peeling characteristics are in demand. Attractive packaging and quality fruit are the main requirements of this market. The currency of this market is strong; most countries are interested in selling their citrus in this market. The wholesale prices of citrus fruit during 2005–2006 in some countries are given in Table 22.4. It has been the experience that rapid growth leads to increased production and a decline in prices. There has been considerable variation in prices over the years and seasons. The average prices for Florida fresh citrus per 4/5 bushel carton were 7.21, 7.31, 7.63, and 14.41 US$ for Navel, Valencia, red grapefruit, and honey tangerines respectively in 2002–2003 (Spreen, 2003).

TABLE 22.4 Wholesale Prices of Some Fresh Citrus Fruits per kg in Selected Countries during 2005–2006

Country	Orange	Lemon	Grapefruit	Clementine
Japan (yen)	180	200	156	–
Germany (Euro)	0.81	0.80	–	1.07
France (Euro)	0.65	–	–	0.66
U.S.A. (cents US) Navel orange	93.1	164.9	–	–
U.S.A. (cents US) Valencia orange	134.9	–	–	–

FAO (2006).

In Asia, the major importing countries are Japan, China, North and South Korea, Malaysia, Indonesia, Thailand, and Singapore, which all demand high-quality fruit. The Chinese market prefers brightly colored sweet fruit, as fruit is mainly utilized for gift-giving on special occasions such as the autumn and moon festivals. Navels and Honey Murcott are preferred. South Africa, the U.S., Egypt, and Australia are major suppliers of citrus to China. Most of the fruit enters through Hong Kong into China. The U.S. and South Africa are suppliers of red-fleshed seedless grapefruit. Australian Honey Murcotts are also preferred in this market. There are niche markets for grapefruit and sweet oranges in China around Shanghai and developing commercial areas. Imports in China during 2000–2001 were 180000 tons. During early 2001, the prices of citrus in China averaged Hong Kong $5.60–7.50/kg. Some fancy citrus achieved HK $9.0/kg ex agent. While Yuan Rmb 2–3/kg was the average price for domestic fruit, the imported fruit received Yuan Rmb 10/kg (Hanlon, 2001).

Japan is a major importer of lemons, orange, and grapefruit from the U.S. Nearly 103900 tons of lemons and 156700 tons of grapefruit were imported in Japan from the U.S. during 1990 (Kitagawa et al., 1992). The quantity of imported grapefruit increased to 219000 tons in 2004–2005. Lemons consumed in Japan are mostly exported from California. Japanese people like grapefruit, and the best-quality grapefruits are imported from the U.S. and Israel. The Japanese market is very quality-conscious and prices are reduced if fruit fails to meet standards (Kitagawa et al., 1992).

In the Asia-Pacific region, Singapore, Hong Kong (China), South Korea, Japan, Indonesia, Mauritius, and Malaysia are the potential markets. Demand in these markets has been increasing by 7–10 percent every year. Gulf countries such as Bahrain, Kuwait, and Saudi Arabia are the traditional importers of citrus fruit from India and Pakistan.

D. Fresh Citrus Trade Prospects

Fresh citrus has to compete with juice. Particularly in Germany, orange juice consumption increased in 1980s. Juice and concentrates are imported from Brazil.

Every liter of orange juice consumed by the consumer costs the sale of 5–6 oranges (Pape, 1988). Fresh citrus fruit exports can be increased through advertising and publicity about the healthful aspects of their consumption. The nutritional and medicinal value of fresh citrus fruit need to be highlighted.

Roughly 10 percent of the citrus produced in the world enters international trade as fresh fruit. The growth in exports is directly related to production on sustained basis. To meet the requirements of importing countries, it is essential for exporters to ensure that the quality of citrus fruits meets the stipulated standards.

Under the new WTO regime, countries have to initiate economic policy reforms for globalization that favor privatization in general. The agriculture sector in particular is bound to undergo changes in many countries with these reforms. Upon signing GATT (the General Agreement on Tariff and Trade) and becoming members of the WTO, countries will have to review, reorient, and reinforce economic policies. With respect to the external trade environment, negotiations with other countries under the auspices of the WTO will continue keeping in view the interests of fruit growers. The Uruguay Round Agreement on Agriculture brought agriculture into the discipline of GATT. It covers (1) market access and tariffs, (2) export competition, (3) domestic support. In this new WTO regime followed by bilateral and multilateral agreements between countries, domestic and export markets no longer are separate entities and export–import is bound to affect external trade and farmers' interests. Countries have to open up markets for external produce while keeping mutual benefits in mind. In this context, countries have to improve phytosanitary standards, marketing and export infrastructure, and fruit quality. State and federal/central government policies concerning agriculture with respect to subsidies on inputs; training staff and growers; and providing critical inputs such as water, electricity, and market intelligence will all determine how the countries benefit from the changed scenario.

Imports in China will increase considering its accession to WTO and the implementation of economic reforms and seasonality of Chinese citrus. China may also become an important market for fresh grapefruit and lemons in addition to oranges. According to Hanlon (2001), the constraints in expanding exports to China are: (1) lack of cool chains and bad roads in interior areas, (2) lack of legal access to position the product against the competitors.

In Africa, there is a trade of citrus between major producers such as Republic of South Africa and Mozambique and their neighboring countries in Southern Africa. In South Asia, India, and Pakistan have a strong potential to expand citrus exports. Mandarins, limes, and lemons can be exported from these countries (keeping in mind production and geographic location). As per the estimate of the author, at least 225 000 tons, or 5 percent of the India's citrus production of 4.5 million metric tons, can be exported initially with proper planning and promotion management.

E. Barriers in Citrus Trade

Tariff escalation, domestic support, stringent sanitary and phytosanitary measures, and seasonal tariffs are some issues that get in the way of free and fair trade.

To protect growers, seasonal tariffs are applied to discriminate against the arriving product. These are 5–10 time high as off-season tariffs on the same commodity. The EU protects its citrus-producing member nations by applying very high seasonal tariffs. At present (from January 1, 2007) there are 27 members in this block of countries, and Spain, Italy, Cyprus, Portugal, Greece, and Malta are the major citrus producers. The entry price system is also applied, in which higher import duties are levied on low-priced imported produce so that imported produce is not competitive. Even on processed products higher rates are applied. EU countries have producers' organizations that have postharvest infrastructure to withhold the product when required and market it when it is remunerative. Operational funds are also provided to finance the trade.

II. QUARANTINE ISSUES

The International Plant Protection Convention (IPPC) was established in 1951 under the aegis of FAO with the purpose of securing common and effective action to preclude introduction and further dissemination of pests/diseases of plants/plant materials. The convention envisaged that the inspection of exportable plants/plant materials would be carried out and a phytosanitary certificate (PSC) be issued by technically qualified people. A PSC may be issued in a prescribed format with additional declarations, if any, that conform to the regulations of importing country. The PSC is an official document stating that perishable material moving across international boundaries is free from pests/diseases. Since most countries are signatory to this convention, its regulations are binding on them.

Plant quarantine is a regulatory mechanism that exporting and importing countries use to ensure that diseases and insect-pests are not carried by the material being traded. This is a preventive mechanism so that diseases and pests are not introduced in new areas where they do not exist. Pathogens and pests on such material could be a threat to the crops grown in the importing country. Importing countries often place a quarantine barrier on produce originating from an area in which insects and diseases of concern are known to occur. In order to market produce, exporting nations with quarantined insect-pests and diseases must develop effective treatments that satisfy the importing country by eliminating insect-pests and diseases.

The SPS Agreement (Application of SPS measures in trade) of the WTO came into force on January 1, 1995. The WTO members' privileged forum to discuss SPS-related issues is the SPS committee (of the WTO). This committee discusses trade issues related to the SPS. Important issues in citrus SPS aspects are disease and insect-pests, particularly citrus canker and fruit flies, which can cross national borders via exported fruit. These issues come under the ambit of the SPS committee. In the past such discussions on the disputes were held under the auspices of this committee and amicably solved. One example is the issues between Argentina and the U.S. and also between some other Latin American countries and EC. The member countries have the right to request the establishment of the panel to resolve disputes related to the trade of citrus fruits among member nations. This comes under the WTO dispute settlement procedures. The parties in dispute also have a right to appeal the panel findings to the Appellate body of WTO.

The FAO/WHO Codex Alimentarious Commission and IPPC are identified in the agreement as the international standard-setting bodies to develop standards for food safety and plant health as well as guidelines for recommendations. Countries have the right to maintain SPS protection for their citrus industries, but such measures must be based on scientific facts that are themselves based on international standards. If a country has higher standards, or if international standards do not exist in certain cases, that country has to demonstrate that its concerns are based on risk assessment that is scientifically done on the basis of internationally approved methods.

There can be different measures of plant protection for certain insect-pests or diseases in importing countries. The importing country has to accept the measure as equivalent to its own if the exporting country provides evidence that quarantine measures taken achieve the level of protection of the importing country. There are several phytosanitary measures against fruit-flies – hot-water dip, vapor heat treatment, irradiation, cold treatment, and fumigation. The exporting and importing countries have to agree on a specific method of disinfestation or disinfection (in case of disease) of fruit for trade to take place. Organizations such as IACNET can help other citrus-producing countries with their expertise in SPS-related problems.

In some countries, areas may be free of the insect-pests that are of phytosanitary concern to the importing country. In the concept of regionalization, importing countries should take into account the situation of the exporting country, particularly the part of their territory that is pest- or disease-free. According to the SPS agreement, the exporting country has to provide necessary evidence of the claimed pest-free or disease-free areas by facilitating inspection and testing. The IPPC has defined the pest-free and disease-free areas in the 'principles of plant quarantine as related to international trade'. The citrus-exporting country has to declare its areas as insect-free and provide related evidences for smooth trade (Chapter 18, part I).

The increasing competition between countries to capture a greater share in the export market is likely to lead to more discrimination by importing countries. This is likely to occur through the use of more sophisticated analytical equipment to detect residues of non-approved chemicals, as well as approved chemicals that may exceed regulated limits. This is a part of quarantine as the countries are also likely to establish inspection protocols consistent with GATT.

A. Quarantine Regulations

In Japan, upon the arrival of refrigerated containers, plant quarantine officials inspect fruit; if fruit has more than 3 percent decay it has to be sorted. The importers sort out the fruit and the decayed fruits are disposed of (Kitagawa et al., 1988). Since fumigation is banned for citrus fruit entering in Japan, cold treatment is necessary to disinfest it from the fruit-flies of family Tephritidae. Cold treatment at temperature of 0.5–2°C for 12–20 days depending on type of citrus fruit is recommended. Importing countries also need a certificate from phytosanitory authorities of the exporting countries that citrus fruit consignment is free from the *Xanthomonas compestris* pathogen, which causes citrus canker. The demand for stringent sanitary and hygienic standards during production and in packinghouses have greatly increased under WTO regime. The main cause is the stipulation of Hazard Analysis Critical Control Point (HACCP) by many countries, including the US FDA. Packing units must follow Sanitary and Phytosanitary (SPS) measures, and get the unit registered. The unit should be accredited.

Phytosanitary certificates need to be produced for citrus fruit entering the EU. The European Economic Commission provides directives for pesticide residues (MRL) on citrus fruits. There is every possibility that SPS measures are unfairly used as non-tariff barriers against import of unwanted produce, citing examples of pests, diseases, and pesticide residues. Importing countries are likely to introduce regulations prescribing unreasonably low levels of pesticides and also object to the sensitivity of methods of estimation in exporting countries. The exporting countries, especially developing and underdeveloped ones, will have to bear the cost of the developing infrastructure without any expected commensurate return. In India, the plant quarantine department (functioning under the Ministry of Agriculture) is mandated to issue phytosanitary certificates. Various agricultural universities (plant pathology departments) and authorities of the State Agriculture Departments issue these certificates after inspection of the consignments. The Plant Quarantine Department has offices at all air- and sea-ports and all customs points at land borders.

REFERENCES

Albrigo, L.G., and Menini, U.G. (1992). Citrus expansion in the American continent: prospects and constraints in relation to the newly established Inter-American Citrus Network (IACNET). *Proc. Int. Soc. Citric.*, Acireale, Italy, Vol. 3, pp. 1201–1203.

Anandswamy, B., and Venkatsubbaiah, G. (1976). Wooden and corrugated shipping containers for the export of 'Coorg' oranges. *Indian Fd. Packer* 30(3), 44–49.

APEDA (1997a). *Export statistics for agro and food products: India 1995–96*. Agricultural and Processed Products Export Development Authority, GOI, New Delhi, pp. 247–249.

APEDA (2003). *Export statistics for agro and food products India 2002–03*. Agricultural and Processed Food Products Export Development Authority, GOI, New Delhi.

Aubert, B. (1991). What citriculture in South-East Asia for the year 2000. *Proc. Int. Citrus Symp.*, Guangzhou, China, pp. 37–58.

Chang, K. (1992). The evaluation of citrus demand and supply. *Proc. Int. Soc. Citric.*, Italy, Vol. 3, pp. 1153–1155.

DMI, (1965). *Marketing of citrus fruits in India*. Directorate of Marketing and Inspection, Government of India, Marketing Series No. 155.

FAO (2006). Citrus fruit fresh and processed. Annual Statistics, Commodities and Trade Division, FAO, Rome.

Gallasch, P.T. (1992). Recent trends in the Australian citrus industry. *Proc. Int. Soc. Citric.* Acireale, Italy, Vol. 3, pp. 1207–1211.

Hanlon, D. (2001). China's citrus and trade: observations and issues. *Proc. China/ FAO Symp.*, Beijing, China, May 2001, pp. 85–101.

Kitagawa, H., Matsui, T., and Kawada, K. (1988). Some problems in the marketing of citrus fruits in Japan. *Proc. 6th Int. Citrus Cong.*, Israel, Vol. 4, pp. 1581–1587.

Kitagawa, H., Kawada, K., and Esguerra, E.B. (1992). Some recent trends in production and importation of citrus fruits in Japan. *Proc. Int. Soc. Citric.*, Italy, Vol. 3, pp. 1212–1215.

Ladaniya, M.S., and Dass, H.C. (1993). 'Nagpur' mandarin export market scenario, technology status and strategies. Export oriented Horticulture Production Research and Strategies. *Proc. Nat. Seminar,* 5–6 December 1993, Nagpur organized by PDKV, Akola and V. Naik Pratishthan, Nagpur, pp. 1–9.

Ladaniya, M.S., and Singh, S. (1999). Export of 'Nagpur' mandarin. *Extension Bull.* No. 21, NRCC, Nagpur, p. 20.

Pape, K. (1988). Changes and trends in the citrus market of affluent countries. *Proc. 6th Int. Citrus Cong.*, Israel, pp. 1663–1677.

Spreen, T.H. (2001). Projection of world production and consumption of citrus up to 2010, China. FAO Citrus Symp., May 2001.

Spreen, T.H. (2003). Economic considerations of fresh citrus markets. IFAS, University of Florida. http://postharvest.ifas.ufl.edu/events/packinghouseDay/2003.

Stanbury, J.S. (1997). The nature and scope of South African citrus industry. *Proc. Int. Soc. Citric., 8th Int. Citrus Congress.* Vol. 1, pp. 7–11.

Taylor, M.S. (1997). The future of citrus fruits in the fresh produce world. A discerning customers view. *Proc. Int. Soc. Citric. The 8th Int. Citrus Congress* Vol. 1, pp. 15–18.

Weinberg, Y. (1988). Marketing of fresh citrus fruit – the market and citrus marketing Board of Israel in the next decade. *Proc. 6th Int. Citrus Congress,* Vol. 4, pp. 1567–1570.

Xin Lu, L. (2001). China's citrus production: Retrospect, present situation and future prospects. Proc. *China /FAO citrus Symp.* May 2001.

ANNEXURE I

Some important conversion formulae:

1. Celsius (ºC) = 5/9 × (ºF − 32)
2. Fahrenheit ºF = 9/5 × (ºC) + 32
3. Kelvin (K) = Celsius + 273.15
4. Calories = BTU × 252
5. Joules = calories × 4.187
6. Joules = BTU (British Thermal Units) × 1055
7. 1 BTU = 1.05 KJ (Kilo Joules)
8. 1 Watt.m.K (Watt per meter Kelvin) = 0.23 Cal.sec.m.°C
   = 0.577 BTU.hr.Ft. °F
   = 6.93 BTU.hr.inch. °F

Thermal Properties of Air and Water (Common Cooling Media)

Material	Temperature (°C)	Density (Kg/cum)	Sp. Heat (KJ/kg-K)	Thermal conductivity (W/m-K)
Air	0	1.203	1.00	0.0244
	20	1.205	1.005	0.0256
	40	1.127	1.005	0.071

(*continued*)

Thermal Properties of Air and Water (Common Cooling Media) *(continued)*

Material	Temperature (°C)	Density (Kg/cum)	Sp. Heat (KJ/kg-K)	Thermal conductivity (W/m-K)
Water	0.01	999.9	4.21	0.561
	5	1000	4.20	–
	20	998.3	4.18	0.611
	40	992.2	4.17	–

The quantity of heat that pass through water per unit time (when all other parameters are similar) is much higher than air, hence, water is much better heat transfer medium than air. In air-cooling, rate of heat transfer can be increased by increasing air velocity to some extent.

ANNEXURE II

Length

1 cm = 10 mm
1 m = 100 cm
1 ft = 30.48 cm
1 inch = 25.4 mm
1 mm = 1000 micron or micro metre
1 m = 1 000 000 micron
1 m = 39.37 inches
1 m = 3.281 feet

Area

1 m^2 = 10 000 cm^2
1 Ft^2 = 929 cm^2
1 Ft^2 = 144.59 $inch^2$
1 m^2 = 1550 $inch^2$
1 cm^2 = 0.155 $inch^2$
1 cm^2 = 0.09290 m^2
1 $inch^2$ = 645.25 mm^2
1 m^2 = 10.76 ft^2.

Volume

1 Bushel = 0.03524 cum (35.24 lit)
1 cum = 1000 lit. (35.31 cuft.)
1 lit = 0.2642 US gallons
1 cum = 1000 000 cc or cu cm
1 lit = 1000 cc
1 cu.ft = 28 320 cm^3
1 cu.ft = 28.32 lit
1 Gallon (British) = 4.54 lit.
4/5 bushel = 28.19 lit.
1 cum = 61023 cm inches

Velocity

1 m/sec = 3.281 ft/sec
1 ft/min. = 0.3048 m/min
1 ft/sec = 30.48 cm/sec

Mass

1 kg = 2.2 lbs
1 metric tonne = 1000 kg

1 ton (short) = 907.2 kg
1 ton (short) = 2000 lbs
1 Quintal = 100 kg
1 gram = 15.43 grains
1 gram = 0.03215 ounces(ozs)
1 kg = 1000 g
1 g = 1000 mg
1 mg = 1000 micro gram
1 micro gram = 1000 nanogram
1 nanogram = 1000 picogram

Density
1 kg/cum = 1000 g/cum
1 lb/cu inch = 27.68 g/cu cm
1 kg/cum = 0.062 lb/cuft

Force
1 kg force = 9.80 Newton (N)
1 lbf = 4.44 N
1 kgf = 2.20 lbf
1 gram f = 980.7 Dynes

Pressure
1 Atmosphere = 1.01 Bar
1 Atm = 101320 Pascal
1 Atm = 760 mmHg
1 Atm = 406.76 inch of water

Energy
1000 calories = 1 k calories
1 kCal = 3.968 BTU
1 kWatt.h = 3415 BTU
1 Watt.h = 0.860 kcal
1 kW = 1.341 Horse power
1 kW.h = 860 kcal
1 kcal.m^2.h. °C = 4.187 kJ. m^2.h. °C
1 HP = 746 Watt
1 J/sec = 1 Watt
1 BTU/min = 0.02356 HP
1 BTU = 252 calories
1 BTU = 0.2520 kcal
1 Kcal = 4.18 KJoule

Heat Transfer coefficient for convective heat
1 kcal.m^2.h. °C = 0.2048 BTU.ft^2.h. °F
1 BTU.ft^2.h. °F = 4.882 kcal.m^2.h. °C
1 BTU.ft^2.h. °F = 5.682 W/sqm/C
1 kcal.m^2.h. °C = 1.163 W/sqm/C

Thermal conductivity
1BTU.ft.h. °F = 1.488 kcal.m.h. °C
1 kcal.m.h. °C = 0.6720 BTU.ft.h. °F
1 W.m.K = 0.57 BTU.hr.ft. °F

Miscellaneous
1 Mole = 1 g mol. Wt of the substance
1 millimole = 1 mg mol. wt. of the substance
1 molar solution = 1 g mol.wt. in 1 lit
1 milli mole solution = 1 mg of mol wt in 1 lit.
1 mg /lit = 1 ppm
1 micro lit/ lit = 1 ppm
1 ppb = 1 part in 1000 000 000 parts
1 ppb = 1 mg in 1000 lit

Nitrogen – molecular weight = 28; volumetric composition in air 78.03%
Oxygen – molecular weight = 32; volumetric composition in air 20.99%
Carbon dioxide molecular weight = 44; volumetric composition in air 0.03%
Argon (A_2) and other gases = 0.89

GLOSSARY

Abscission: A natural physiological process in which an abscission layer, or separation layer, of thin-walled cells is formed between two plant parts that separate, as in a fruit or leaf separating from the branch. This process depends on balance of hormones and can be modified with external application of growth regulators. It is a genetically controlled process as the mature leaves and fruits drop naturally after a specific time. Various biotic and abiotic stresses also play roles in abscission.

Activated alumina: Aluminum oxide that is moisture-free ($Al_2 O_3$). It absorbs moisture when it comes in contact with moist air. It is used to remove moisture (as a desiccant). It is an adsorbent with no change in its composition.

Activated charcoal: An adsorptive material for gases, odors, and colors/pigments. It is a coal obtained from heating of wood to about 400°C with partial burning. It is used in cool storage chambers where fruit is stored to remove off-odors.

Adsorbent: Substance that is able to hold other substances on its surface through the force of adhesion. Examples: Silica gel, activated carbon.

Antagonist: An organism having the capacity to inhibit the growth or to interfere with the growth of other organisms. Antagonists are used in biological control methods for diseases.

Atomized spray: A fine spray of the wax emulsion or water used for waxing of fruits and humidifying storage chambers respectively.

Air changes: One air change is generally the amount of air volume equal to the storage area volume that is replaced with fresh air (in cold storage).

Astringent: Chemical compound, either natural or synthetic, that leads to contraction of tissues, controls bleeding, and relieves inflammation.

Atmospheric pressure: Air pressure equal to one atmosphere.

Anemometer: Instrument used to measure gas or air velocity. It is a rotating-vane type or resistance type with hot wires.

Barometer: Instrument with glass column/tube filled with water or mercury to measure air pressure.

BTU: British Thermal Unit of Heat; equivalent to heating one pound of water to raise the temperature by one degree Fahrenheit.

Chromatography: A method of separation of chemicals or gaseous mixtures into their constituents based on different adsorption or partition of compounds present in liquid or gaseous mixtures. The separated pure compounds appear as separate bands on an adsorption column.

Climacteric: A sudden rise in the respiration rate of the fruit coinciding with its maturity and ripening as a natural process. It mostly accompanies biochemical changes in composition of fruit, leading to fruit-eating quality such as the development of characteristic flavor, color, and texture. Climacteric can be artificially induced with application of ethylene gas, but the characteristic flavor and color will not develop if physiological growth and development of the fruit is not complete.

Coefficient of heat transfer: Indicates rate of heat transfer per unit area per degree temperature difference at a given surface area. (Units are KJ/sqm/hr/°C or Watt per sqm. hr/°C.)

Cool chain or cold chain: Cool temperature management around the perishable commodity, such as from the packinghouse or field to the consumer through various means using precooling, refrigerated transport vehicles, cold storage, and display units at retail market places without a break in the chain of low temperatures.

Cooling coil: Evaporator coil or expansion coil; a tubing through which refrigerant flows and changes phase from liquid to vapor with contact of air.

Compressor: A device in a mechanical refrigeration system to raise the pressure of gas.

Condenser: A type of heat exchanger in which gas under pressure changes its phase to liquid because of heat transfer.

Degreening: Process of breaking down green chlorophyll and synthesis of carotenoids in citrus-fruit peel by application of ethylene or ethephon, specific temperature, and relative humidity for better yellow or orange color.

Defrosting: Process of removing ice from cooling coils by melting ice.

Evaporative cooling: A reduction or depression in air temperature from evaporation of water in the air stream. Drop in temperature is generally equal to wet-bulb thermometer temperature. The greater the difference between dry- and wet-bulb thermometer, the greater the cooling. It is useful as it also increases relative humidity of the air in storage area.

Evaporator: A type of tubular heat exchanger in which refrigerant boils or evaporates to give refrigeration effect.

Expansion valve: A valve (throttle device) to reduce pressure between condenser and evaporator.

Gas manifold: A tube with multiple outlets on one side for distribution of gas mixture and single inlet on opposite side.

Glycosides: The derivatives of cyclic forms of sugars in which the hydroxyl group at C-1 condenses with the alcohol or phenol group or other non-sugar (aglycone) molecule.

Hydrometer: A glass apparatus to measure density of water or liquid; it floats in water/liquid depending on the solutes dissolved in it and thus provides ^{0}Brix measurement.

Hygrometer: An instrument that measures moisture content or RH in humid air.

Insulating material: Any material with thermal conductivity around 0.2 KJ/sqm/hr/°C or lower.

Manometer: A device to measure air pressure between two points; has a U-shaped glass tube filled with water.

Needle valve: A valve with a spindle that is fine-threaded and tapered for accurate control of air or fluid passing through it. Ordinary valves have a disk to control flow.

Parenchyma: The tissues in plants that are basic/fundamental ground tissues in which more specialized tissues differentiate. The parenchyma cells have variety of shapes, sizes, and wall characteristics and have intercellular spaces.

Parthenocarpy: Process in which fruit development takes place without fertilization of ovule.

Partial pressure: A portion/part of the total pressure and is exerted by one of the components of the gas mixture or air. Partial pressure of oxygen in air is about 21.

Precooling: Cooling of perishable commodities utilizing any of the several methods of cooling before it is stored or shipped.

Saturated air: In saturated air, partial pressure of water vapor equals the saturation pressure of water vapor at a given temperature.

Psychrometry: Deals with various properties of atmospheric air with respect to moisture content and temperature.

Respiration: A physiological process in which oxidative breakdown of organic substrates takes place with release of energy for metabolic activity. It can be an aerobic or anaerobic respiration. Anaerobic respiration takes place in absence of oxygen resulting in formation of ethanol, the release of energy being less in anaerobic respiration. Aerobic respiration requires free or dissolved oxygen for enzymatic breakdown of a substrate to release energy, water, and carbon dioxide.

Refrigerant: A medium of heat transfer flowing in tubing of evaporator, condenser, and compressor.

Rotameter: A device used to measure a air or liquid flow; the upper limit of the float indicates flow rate per minute.

Saturated fat: Fat formed from reaction of glycerol with any of the saturated fatty acids (such as stearic or palmitic acids). These fatty acids have no double bonds.

Senescence: A genetically controlled normal physiological process in which plants or plant parts grow old and die in a sequential manner of maturity, wilting, yellowing, and dropping. It is a part of growth and development and modified by external factors to a great extent.

Silica gel: A form of silicon dioxide used as desiccant to absorb moisture very quickly.

Specific gravity: Indicates the relative density of the material as compared to standard density of the material under given conditions.

Specific heat: The heat-retaining capacity of a material. It is expressed as KJ/Kg/C.

Thermal conductivity: The property of any material to transfer heat from one point to another.

Thermister: Temperature-measuring device with semiconductor having electrical resistance varying with temperature.

Thrombosis: A disorder in which blood clots are formed obstructing normal blood flow.

Transpiration: Physiological process in which water vapor diffuses out of stomata.

Turgur pressure: Pressure exerted on the cell wall by cell contents or cell organells inside, especially the vacuole, which increases in volume from an uptake of water. Cell wall stretches in a turgid condition; a flaccid condition is the opposite of turgid.

Unsaturated fat: Fat formed from the reaction of glycerol with any or all of the unsaturated fatty acids (oleic, linoleic, linolenic acid). These fatty acids have one or more double bonds.

Vapor pressure: Pressure exerted by water vapor in the air.

INDEX